VOLUME ONE HUNDRED AND EIGHT

VITAMINS AND HORMONES

Dehydroepiandrosterone

VOLUME ONE HUNDRED AND EIGHT

VITAMINS AND HORMONES

Dehydroepiandrosterone

Series Editor

GERALD LITWACK, PhD
Toluca Lake, North Hollywood, California

ELSEVIER

ACADEMIC PRESS
An imprint of Elsevier

Cover photo credit:
Bruno D. Arbo, Felipe S. Ribeiro, and Maria F. Ribeiro
Astrocyte Neuroprotection and Dehydroepiandrosterone
Vitamins and Hormones (2018) **108**, pp. 175–204.

Academic Press is an imprint of Elsevier
50 Hampshire Street, 5th Floor, Cambridge, MA 02139, United States
525 B Street, Suite 1650, San Diego, CA 92101, United States
The Boulevard, Langford Lane, Kidlington, Oxford OX5 1GB, United Kingdom
125 London Wall, London, EC2Y 5AS, United Kingdom

First edition 2018

ISBN: 978-0-12-814361-2
ISSN: 0083-6729

For information on all Academic Press publications
visit our website at https://www.elsevier.com/books-and-journals

Publisher: Zoe Kruze
Acquisition Editor: Ashlie Jackman
Editorial Project Manager: Joanna Collett
Production Project Manager: Vignesh Tamil
Cover Designer: Greg Harris

Typeset by SPi Global, India

CONTENTS

CONTRIBUTORS

Hadas Ahdoot-Levi
The Mina and Everard Goodman Faculty of Life Sciences, Bar-Ilan University, Ramat Gan, Israel

Kazutaka Aoki
Internal Medicine, Kanagawa Dental University, Yokosuka; Yokohama City University Graduate School of Medicine, Yokohama, Japan

Bruno D. Arbo
Instituto de Ciências Biológicas, Universidade Federal do Rio Grande (FURG), Rio Grande; Programa de Pós-Graduação em Ciências Biológicas: Farmacologia e Terapêutica, ICBS, Universidade Federal do Rio Grande do Sul (UFRGS), Porto Alegre, Brazil

Emilio Badoer
School of Health & Biomedical Sciences, RMIT University—Bundoora Campus, Melbourne, VIC, Australia

Jia-peng Bao
The Second Hospital of Medical College, Zhejiang University, Hangzhou, PR China

Maria Sonia Baquedano
Endocrine Department, Hospital de Pediatría Garrahan; National Scientific and Technical Research Council, Buenos Aires, Argentina

Tzofnat Bareli
The Mina and Everard Goodman Faculty of Life Sciences, Bar-Ilan University, Ramat Gan, Israel

Alicia Belgorosky
Endocrine Department, Hospital de Pediatría Garrahan; National Scientific and Technical Research Council, Buenos Aires, Argentina

M. Caretto
Department of Clinical and Experimental Medicine, University of Pisa, Pisa, Italy

Bindu N. Chimote
Vaunshdhara Clinic and Assisted Conception Centre, Nagpur, Maharashtra, India

Natchandra M. Chimote
Vaunshdhara Clinic and Assisted Conception Centre, Nagpur, Maharashtra, India

Maria R. Ciriolo
University of Rome 'Tor Vergata'; IRCCS San Raffaele 'La Pisana', Rome, Italy

Barbara J. Clark
Department of Biochemistry and Molecular Genetics, Center for Genetics and Molecular Medicine, University of Louisville School of Medicine, Louisville, KY, United States

Mariana Costanzo
Endocrine Department, Hospital de Pediatría Garrahan, Buenos Aires, Argentina

Sónia do Vale
Hospital de Santa Maria, Centro Hospitalar Lisboa Norte, EPE; Faculdade de Medicina, Universidade de Lisboa, Lisboa, Portugal

Carles Escera
Institute of Neurosciences; Brainlab—Cognitive Neuroscience Research Group, University of Barcelona; Institut de Recerca Sant Joan de Déu, Barcelona, Catalonia, Spain

A.R. Genazzani
Department of Clinical and Experimental Medicine, University of Pisa, Pisa, Italy

Julie Glowacki
Brigham and Women's Hospital, Harvard Medical School; Harvard School of Dental Medicine, Boston, MA, United States

Ronda Greaves
School of Health & Biomedical Sciences, RMIT University—Bundoora Campus; Murdoch Children's Research Institute, Parkville, Melbourne, VIC, Australia

Gabriela Guercio
Endocrine Department, Hospital de Pediatría Garrahan; National Scientific and Technical Research Council, Buenos Aires, Argentina

Kai Huang
Tongde Hospital of Zhejiang Province, Health Bureau of Zhejiang Province, Hangzhou, PR China

Motoyuki Iemitsu
Faculty of Sport and Health Science, Ritsumeikan University, Kusatsu, Japan

Carolyn M. Klinge
Department of Biochemistry and Molecular Genetics, Center for Genetics and Molecular Medicine, University of Louisville School of Medicine, Louisville, KY, United States

Rachel Maayan
Laboratory of Biological Psychiatry, Felsenstein Medical Research Center, Research Unit, Petah Tikva, Israel

P. Mannella
Department of Clinical and Experimental Medicine, University of Pisa, Pisa, Italy

Roxana Marino
Endocrine Department, Hospital de Pediatría Garrahan, Buenos Aires, Argentina

Michael P. Muehlenbein
Baylor University, Waco, TX, United States

Trevor M. Penning
Center of Excellence in Environmental Toxicology, Perelman School of Medicine, University of Pennsylvania, Philadelphia, PA, United States

Sean P. Prall
University of California, Los Angeles, CA, United States

Russell A. Prough
Department of Biochemistry and Molecular Genetics, Center for Genetics and Molecular Medicine, University of Louisville School of Medicine, Louisville, KY, United States

Marco A. Rivarola
Endocrine Department, Hospital de Pediatría Garrahan; National Scientific and Technical Research Council, Buenos Aires, Argentina

Tracey Quinn
The Ritchie Centre, Hudson Institute of Medical Research, Monash Medical Centre, Clayton, VIC, Australia

Felipe S. Ribeiro
Laboratório de Interação Neuro-Humoral, ICBS, Universidade Federal do Rio Grande do Sul (UFRGS), Porto Alegre, Brazil

Maria F. Ribeiro
Laboratório de Interação Neuro-Humoral, ICBS, Universidade Federal do Rio Grande do Sul (UFRGS), Porto Alegre, Brazil

Koji Sato
Graduate School of Human Development and Environment, Kobe University, Kobe, Japan

T. Simoncini
Department of Clinical and Experimental Medicine, University of Pisa, Pisa, Italy

Yasuo Terauchi
Yokohama City University Graduate School of Medicine, Yokohama, Japan

Rolando Vegliante
MN3T, The Microenvironmental Niche in Tumorigenesis and Targeted Therapy, Hopital Civil—Institut d'Hématologie et Immunologie, Strasbourg, France

David Walker
School of Health & Biomedical Sciences, RMIT University—Bundoora Campus, Melbourne, VIC, Australia

Abraham Weizman
Laboratory of Biological Psychiatry, Felsenstein Medical Research Center, Research Unit, Petah Tikva; Geha Mental Health Center, Tel-Aviv University, Tel Aviv, Israel

Li-dong Wu
The Second Hospital of Medical College, Zhejiang University, Hangzhou, PR China

Gal Yadid
The Mina and Everard Goodman Faculty of Life Sciences; The Leslie and Susan Gonda (Goldschmied) Multidisciplinary Brain Research Center, Bar-Ilan University, Ramat Gan, Israel

Shuanhu Zhou
Brigham and Women's Hospital, Harvard Medical School, Boston; Harvard Stem Cell Institute, Harvard University, Cambridge, MA, United States

PREFACE

Dehydroepiandrosterone (DHEA) and its sulfate derivative (DHEA-S) are unique hormones. The sulfate derivative represents the most concentrated hormone in blood. DHEA does not seem to have a classical receptor but has unique receptors on specific cells, information that has been recently uncovered. Because DHEA and its sulfate derivative decline in concentration with age, DHEA has been postulated to be an antiaging hormone. Consequently, it has been promoted as an oral supplement.

Of the 16 chapters in this volume, in the first two chapters we present an overview of research on this subject and a review of the mechanisms of action of this hormone. These chapters are followed by reports of basic research, by the actions of DHEA on specific tissues, and finally by a group of chapters relating DHEA and DHEA-S to disease or to clinical conditions.

Specifically, the first chapter is entitled: "Dehydroepiandrosterone research: past, current, and future" by C.M. Klinge, B.J. Clark, and R.A. Prough. In Chapter 2, the same group presents "Mechanisms of action of dehydroepiandrosterone" by B.J. Clark, R.A. Prough, and C.M. Klinge. Next, M.S. Baquedano, G. Guercio, M. Constanzo, R. Marino, M.A. Rivarola, and A. Belgorosky introduce "Mutation of *HSD3B2* gene and fate of dehydroepiandrosterone." Then, S.P. Prall and M.P. Muehlenbein provide a chapter entitled: "DHEA modulates immune function: a review of evidence." "DHEA in prenatal and postnatal life: implications for brain and behavior" is the topic of T. Quinn, R. Greaves, E. Badoer, and D. Walker. B.D. Arbo, F.S. Ribeiro, and M.F. Ribeiro report on "Astrocyte neuroprotection and dehydroepiandroisterone." Following this, K. Sato and M. Iemitsu discuss "The role of dehydroepiandrosterone (DHEA) in skeletal muscle." "Dehydroepiandrosterone (DHEA) and its sulfate (DHEA-S) in mammalian reproduction: known roles and novel paradigms" is authored by B.N. Chimote and N.M. Chimote. S. Zhou and J. Glowacki report on "Dehydroepiandrosterone and bone." This section is completed by a contribution by R. Vegliante and M.R. Ciriolo entitled: "Autophagy and autophagic cell death: uncovering new mechanisms whereby dehydroepiandrosterone promotes beneficial effects on human health."

Introducing the final section, T.M. Penning offers "Dehydroepiandrosterone (DHEA)-SO_4 depot and castration-resistant prostate cancer." "Dehydroepiandrosterone and cardiovascular disease" is reviewed by

P. Mannella, T. Simoncini, M. Caretto, and A.R. Genazzani. K. Aoki and T. Terauchi discuss "Effect of dehydroepiandrosterone (DHEA) on diabetes mellitus and obesity." "Dehydroepiandrosterone and experimental osteoarthritis" is reported by K. Huang, L.-d. Wu, and J.-p. Bao. Y. Gal, A.-L. Hadas, B. Zofnat, R.M. Maayan, and A. Weizman review "Dehydroepiandrosterone and addiction." Finally, the volume is completed by S. do Vale and C. Escera who introduce "Dehydroepiandrosterone and dehydroepiandrosterone-sulfate and emotional processing."

The illustration on the cover of this book is a reproduction of Fig. 1 of Chapter 6: "Astrocyte neuroprotection and dehydroepiandrosterone" by B.D. Arbo, F.S. Ribeiro, and M.F. Ribeiro.

I am grateful for the contributions of Joanna Collett of Elsevier, Oxford, UK and Vignesh Tamilselvvan of Elsevier, Chennai, India who were instrumental in the production of this book.

Gerald Litwack
December 30, 2017

CHAPTER ONE

Dehydroepiandrosterone Research: Past, Current, and Future

Carolyn M. Klinge, Barbara J. Clark, Russell A. Prough[1]
Department of Biochemistry and Molecular Genetics, Center for Genetics and Molecular Medicine, University of Louisville School of Medicine, Louisville, KY, United States
[1]Corresponding author: e-mail address: russ.prough@louisville.edu

Contents

Abstract

The discovery of "oestrus-producing" hormones was a major research breakthrough in biochemistry and pharmacology during the early part of the 20th century. The elucidation of the molecular weight and chemical structure of major oxidative metabolites of

Vitamins and Hormones, Volume 108
ISSN 0083-6729
https://doi.org/10.1016/bs.vh.2018.02.002

dehydroepiandrosterone (DHEA) led to the award of the Nobel Prize in 1939 to Adolf Frederick Johann Butenandt and Leopold Ruzicka. Considered a bulk androgen in the circulation, DHEA and its sulfated metabolite DHEA-S can be taken up by most tissues where the sterols are metabolized to active androgenic and estrogenic compounds needed for growth and development. Butenandt's interactions with the German pharmaceutical company Schering led to production of gram quantities of these steroids and other chemically modified compounds of this class. Sharing chemical expertise allowed Butenandt's laboratory at the Kaiser Wilhelm Institute to isolate and synthesize many steroid compounds in the elucidation of the pathway leading from cholesterol to testosterone and estrogen derivatives. As a major pharmaceutical company worldwide, Schering AG sought these new biological sterols as pharmacological agents for endocrine-related diseases, and the European medical community tested these compounds in women for conditions such as postmenopausal depression, and in men for increasing muscle mass. Since it was noted that circulating DHEA-S levels decline as a function of age, experimental pathology experiments in animals were performed to determine how DHEA may protect against cancer, diabetes, aging, obesity, immune function, bone density, depression, adrenal insufficiency, inflammatory bowel disease, diminished sexual function/libido, AIDS/HIV, chronic obstructive pulmonary disease, coronary artery disease, chronic fatigue syndrome, and metabolic syndrome. While the mechanisms by which DHEA ameliorates these conditions in animal models have been elusive to define, even less is known about its role in human disease, other than as a precursor to other sterols, e.g., testosterone and estradiol. Our groups have shown that DHEA and many of its oxidative metabolites serve as a low-affinity ligands for hepatic nuclear receptors, such as the pregnane X receptor, the constitutive androstane receptor, and estrogen receptors α/β (ERα/ERβ) as well as G protein-coupled ER (GPER1). This chapter highlights the founding research on DHEA from a historical perspective, provides an overview of DHEA biosynthesis and metabolism, briefly summarizes the early work on the beneficial effects attributed to DHEA in animals, and summarizes the human trials addressing the action of DHEA as a therapeutic agent. In general, most human studies involve weak correlations of circulating levels of DHEA and disease outcomes. Some support for DHEA as a therapeutic compound has been demonstrated for postmenopausal women, in vitro fertilization, and several autoimmune disorders, and adverse health effects, such as, acne, embryo virilization during pregnancy, and possible endocrine-dependent cancers.

1. DISCOVERY OF DEHYDROEPIANDROSTERONE AND ITS HORMONE METABOLITES

1.1 Nobel Prize for "Oestrus-Producing" Hormones

The discovery of the chemical composition of "oestrus-producing" hormones was a major research effort in biochemistry and pharmacology during the early part of the 20th century. The elucidation of the chemical structure, design of chemical synthesis, and the concept of a metabolic pathway leading from cholesterol to biologically active metabolites of dehydroepiandrosterone (DHEA) led to the award of the Nobel Prize in 1939 to Adolf Frederick Johann Butenandt for discovery and characterization of the sterol character of "Oestrus-producing" hormones and to Leopold Ruzicka for developing the unique chemical synthesis for androsterone. Butenandt and his research group discovered androsterone (Butenandt, 1931), testosterone (Butenandt & Hanisch, 1935), progesterone (Butenandt & Westphal, 1934), and also codiscovered estrone (Butenandt, 1929) with Edward Adelbert Doisy. The competition between the Butenandt and Doisy laboratories is discussed in the timely review by Simpson and Santen (2015). Although not included as an awardee in the 1939 Prize, Doisy was later recognized for his work on chemistry of vitamin K, and he was awarded the Nobel Prize in Medicine in 1943 with Henrik Carl Peter Dam who first discovered vitamin K. Due to a policy of the German Government at that time, Butenandt was forced by the "Greater German Reich" to decline the award in 1939 in Stockholm, but in 1949, he received the Nobel diploma and award from the Nobel Foundation. Butenandt and Dannenberg (1940) reported the presence of DHEA in urine and named the compound Prasterone. In 1944, DHEA-sulfate (DHEA-S) was first isolated in blood (Munson, Gallagher, & Koch, 1944) and later, Etienne-Emile Baulieu found DHEA and DHEA-S are produced by adrenal secretion (Baulieu et al., 1965). The sulfated species can be taken up by most tissues and subsequently desulfated to yield DHEA that can be metabolized to active androgenic and estrogenic compounds within the cell (Yen, 2001).

1.2 Relationship Between the Kaiser Wilhelm Institute Für Biochemie and the Schering AG Company

An advantage that Butenandt had was his extensive interactions with the Schering AG pharmaceutical company in Berlin, the site for both Butenandt's laboratory and Schering's research and production facility. This is well documented in the historical study of this relationship described by Gaudilliere (2005). First, contracts with Schering AG provided additional funds beyond that provided by the Foundation that supported the Kaiser Wilhelm Institute für Biochemie, Berlin-Dahlem, Germany (KWIB). This allowed Butenandt to fill his laboratory with excellent young German chemists, supplies, and equipment for experiments. Second, Schering AG allowed his group access to their pilot plant for purification of sex hormones from isolates of starting materials, thereby facilitating the effort of crystallization of steroids from human urine, blood, or tissues. This allowed Butenandt to have a large amount of natural starting materials for the crystallization process to obtain testosterone, progesterone, and dehydroandrosterone. In addition, as the Butenandt Laboratory crystallized new urinary or serum sterols and characterized them chemically, this information was shared with Schering AG that tested these lead compounds in biological assays (see Section 1.3). In addition, Schering AG chemists could begin devising chemical synthesis methods to obtain these compounds and their derivatives in bulk from cholesterol or bile acids. The availability of these chemical standards was in turn valued by Butenandt as he began to dissect the steroid biosynthetic pathway. This result can best be seen from his efforts to form dehydroandrosterone (what is now known as DHEA) from cholesterol. From this work, he proposed that males and females use the same biosynthetic pathway to form compounds with male or female gonad-specific properties predicted from his detailed biochemical pathway for androgen and estrogen production published in 1936 (Butenandt, 1936).

Since both Butenandt and Schering AG believed that these natural steroid hormones would be important pharmaceutical agents, they copatented many of these compounds in the 1930–1940s. This interesting Academe–Industry relationship was highly beneficial to the advancement of biochemical understanding of steroid hormones and provided important lead compounds to Schering AG that were promising new pharmaceuticals. Most of the early compounds isolated by the Butenandt laboratory were produced by collecting urine from humans or horses or from sow ovary and the isolation of natural products required concentrating the sterols many 1000-fold prior to crystallizing them. The large scale-up procedure required for isolating natural sterols often did not provide enough of the isolated steroid for industrial production. Therefore, Schering AG turned to chemical synthesis utilizing the processes published by Leopold Riczika, the coawardee for the 1939 Nobel Prize. Riczika, a Professor of Organic Chemistry at University of Utrecht, and later from the Eidgenössische Technische Hochschule in Zürich, developed a unique synthesis of these sterols from cholesterol.

1.3 The Early Studies in the Physiological Role of "Oestrus-Producing" Hormones to Identify Active Steroid Hormones

The isolation of these biologically active sterols required a biological endpoint to demonstrate whether they had any physiological function. A number of animal-based assays were developed, some performed at the KWIB and others at Schering AG or other contract companies associated with Schering AG. At that time, there were three major pharmaceutical companies interested in exploiting this emerging understanding of active sterols, namely, Schering at Berlin-Dahlem, Organon in Amsterdam, and Chemische Industrie Basel (CIBA) in Basel. All decided that there was a significant market for the sex steroids and set about to have their own facilities built to extract hormone from urine or organs for the market. Hoffman-La Roche and CIBA both devoted attention to unique synthesis of these steroids, leading to their support of Leopold Ruzicka's chemical synthesis efforts (Gaudilliere, 2005).

A number of assays were rapidly developed to test biological action of these newly identified sterols. The Allen–Doisy mouse test (Allen & Doisy, 1983) allowed researchers to inject various estrogens or other gonadal steroids into ovariectomized rodents and test whether they noted increased cornification of epithelial cells in vaginal smears or disappearance of leukocytes as a proof of estrogenic activity. Budenandt's group measured for androgen action by measuring the growth of the crest in castrated chickens after injection of androsterone and other steroids they isolated or synthesized. Similar animal-based assays that measured a known estrogen- or androgen-dependent phenotype were used in all of the pharmaceutical companies and laboratories. The Animal Facilities developed at Schering AG and other companies became independent research facilities as they improved techniques in these bioassays. This initially allowed sharing of reagents and expertise between the KWIB and Schering AG that diminished as Schering AG moved toward independence from its association with KWIB scientists. Eventually at the end of the World War II, Butenandt moved his research group to Tübingen in anticipation of the Russian invasion of eastern Germany. As described by Gaudilliere (2005), the close reliance between research institutes and industry was lost at the end of the war and Butenandt changed fields looking at insect pheromones and other natural products. The pheromone Bombykol (ironically, a steroid) was isolated from female silkworms and characterized by Butenandt in 1961 (Butenandt, Beckmann, & Stamm, 1961).

Although Butenandt had views of how these steroids could be used in medicine, Schering AG had its own collection of affiliated physicians who were applying these compounds in clinics in Germany (Gaudilliere, 2005). One example of medical experimentation was seen in the collaboration between Schering AG with two clinical research groups, namely, C. Kaufman's Frauenklinik der Charité in Berlin and Carl Clauberg at the gynecological clinic at Königsberg University. Kaufman published the use of Progynon (estriol glucuronide and estriol sulfate from women's urine) and Proluton (hydroxyprogesterone caproate) for the treatment of infertile women (Kaufmann, 1934). These compounds induced menses in women. Kaufmann apparently advocated these compounds as a way to manage many of the symptoms of postmenopausal women (Gaudilliere, 2005). Clauberg utilized these compounds in laboratory-based research to develop better biological assays for progesterone. Some of his work addressed control of fertility, heralding future work for the use of steroids in preventing conception (Clauberg, 1935). This work, described by Gaudilliere (2005), was the beginning of an expansion of medical use of these steroid hormones in medicine. In fact, Gaudilliere describes that Schering AG in 1945 constructed a new plant at the end of World War II to increase production of *Progynon*, *Proluton*, and *Testoviron* (Gaudilliere, 2005).

In preparing this chapter, we posited whether these early clinical studies demonstrating some beneficial effects upon administration of such hormonal preparations lead to the continued study of DHEA for the next 40 years. However, what was not known at that time was the normal functioning of the hormonal system in men and women and the ability of a single biosynthetic pathway to provide differential steroid levels to regulate expression of gender. With the initial successes in the clinic to benefit conditions, e.g., regulation of the menstrual cycle, infertility, and pregnancy, physicians throughout the world continued their "clinical" investigations with industry-produced steroids for medicine as part of their regular medical practice.

2. DHEA SYNTHESIS AND METABOLISM

2.1 DHEA Synthesis in the Adrenal *Zona Reticularis*

With the discovery of high concentrations of circulating DHEA-S in human blood, considerable interest arose to understand where and how DHEA and its sulfated form are produced. In an elegant study, Baulieu et al. (1965) discovered DHEA and DHEA-S are synthesized in the adrenal and thus began our understanding of the biochemical pathway. In the same year, MacDonald and Siteri demonstrated that estrogen production in pregnant women is derived from circulating DHEA (MacDonald & Siteri, 1965). In addition, MacDonald and coworkers developed isotope dilution methods to study the conversion of DHEA to various hormones in humans which subsequently allowed definition of how estrogen, testosterone, and dihydrotestosterone (DHT) are synthesized (Deshpande & Bulbrook, 1964; Gurpide,

MacDonald, Chapdelaine, Vande Wiele, & Lieberman, 1965). Many studies on the enzymatic processes for the production of estrogen, testosterone, and DHT from androgens, such as DHEA/DHEA-S, pursued. These metabolic pathways are briefly summarized later.

It is well established that DHEA is synthesized de novo from cholesterol mainly in the adrenal, ovary, and testis. In addition, there is limited production of DHEA in cells of the cerebral cortex and other cell types in the brain (Rossetti, Cambiasso, Holschbach, & Cabrera, 2016; Zwain & Yen, 1999). However, it is recognized that adrenal production accounts for the majority of the circulating levels of DHEA and DHEA-S. The adult adrenal cortex has three distinct zones characterized by specific gene expression patterns that lead to specific steroid output (reviewed in Turcu, Smith, Auchus, & Rainey, 2014). The *zona glomerulosa* (ZG), *fasciculata* (ZF), and *reticularis* (ZR) are the sites for mineralocorticoid (aldosterone), glucocorticoid (cortisol), and androgen (DHEA/DHEA-S) synthesis, respectively. The gonads are the major site for sex hormone production, and, in the testis and ovary, DHEA serves as an intermediate steroid within the testosterone and estradiol biosynthetic pathways, respectively. Peripheral tissues that uptake DHEA/DHEA-S from circulation typically generate more potent androgens and estrogens that act locally. In healthy adult males, less than 5% of the total circulating testosterone (T) levels is from peripheral metabolism of DHEA. However, under conditions of loss of testicular T production, DHT levels in the prostate are maintained by DHEA metabolism (Labrie et al., 2005). In healthy women, peripheral metabolism of DHEA is a major contributor to circulating sex hormone levels with 40%–75% of T in premenopausal women and over 90% of the estrogens in postmenopausal women derived from adrenal androgens.

The DHEA biosynthetic pathway in the adrenal *zona reticularis* is shown in Fig. 1. The first regulated step is the transfer of cholesterol across the mitochondrial membranes to the cytochrome P450 (CYP) side chain cleavage enzyme (CYP11A1 encoded by the *CYP11A1* gene). CYP11A1 is anchored in the inner membrane by an amino-terminal α-helix with the catalytic site oriented toward the matrix. The movement of cholesterol from the mitochondrial outer membrane to the CYP11A1 catalytic site is dependent on the action of the steroidogenic acute regulatory protein (STAR, *STARD1*) (Stocco, Zhao, Tu, Morohaku, & Selvaraj, 2017). CYP11A1 cleaves the side chain of cholesterol-producing pregnenolone using electrons from NADPH that are shuttled to the enzyme via FAD-containing adrenodoxin reductase and an iron–sulfur protein, adrenodoxin

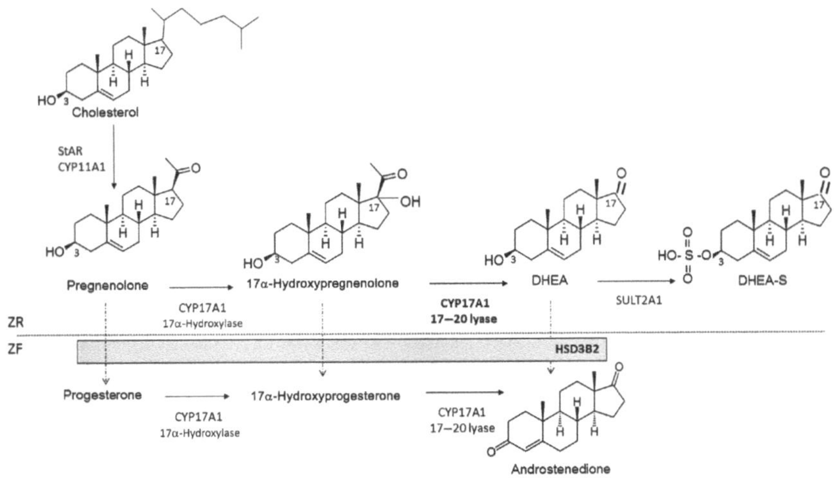

Fig. 1 DHEA biosynthetic pathway. DHEA/DHEA-S in circulation mainly originates from the adrenal. Shown are the key enzymes and steroid intermediates in the adrenal *zona reticularis* (ZR) and *zona fasciculata* (ZF). The pathway and enzyme activities are described in the text. *StAR*, steroidogenic acute regulatory protein; *CYP11A1*, cytochrome P450 side-chain cleavage; *CYP17A1*, cytochrome P450 17a-hydroxylase, 17,20 lyase (one enzyme, two activities); *HSD3B*, 3b-hydroxysteroid dehydrogenase; *SULT2A1*, sulfotransferase family 2A member 1.

(Mast et al., 2011). Pregnenolone produced in the mitochondria moves to the endoplasmic reticulum and to the catalytic site of cytochrome P450 17α-hydroxylase/17,20 lyase (CYP17A1 encoded by the *CYP17A1* gene) (reviewed in Auchus & Rainey, 2004). CYP17A1 is a type 1 integral membrane protein that has two distinct catalytic actions hydroxylation at C17 of the steroid nucleus forming 17α-hydroxypregnenolone, and hydrolysis of the C17–C20 bond of 17α-hydroxypregnenolone to generate DHEA. Again, this P450-dependent reaction requires electrons from NADPH, and for endoplasmic reticulum-resident P450s the cytochrome P450 oxidoreductase (encoded by the *POR* gene) that contains both FMN and FAD centers to shuttle electrons from NADPH to CYP17A1 (reviewed in Miller, 2005). These reactions generate a keto group at C17, which, along with the C3-β-hydroxyl group and a double bond between C5 and C6 of the B ring in the steroid nucleus, provides the unique sterol structure for DHEA: 5-androsten-3β-ol-17-one (Fig. 1).

As stated previously, the major circulating form of DHEA is DHEA-S. DHEA sulfonation in the adrenal is catalyzed by sulfotransferase family 2A member 1 (SULT2A1) that conjugates sulfuryl (SO3) to the C3-β-hydroxyl group of DHEA-producing DHEA-S (Wang, Cook, Falany, & Leyh, 2014). Three biochemical consequences of DHEA sulfonation are (1) DHEA-S has higher binding affinity for albumin and a 50-fold longer half-life in circulation compared to DHEA (Longcope, 1996); (2) DHEA-S uptake by a cell requires active transport by a member of the solute carrier organic anion transporter family (Mueller, Gilligan, Idkowiak, Arlt, & Foster, 2015); and (3) DHEA-S is classified as biologically inactive, e.g., unable to bind receptor(s), although this classification is being challenged by new studies (see chapter by Clark, Prough, & Klinge, 2017, in this volume).

2.2 Peripheral DHEA Metabolism

Generally, DHEA-S action is dependent upon cellular uptake and conversion back to DHEA. Members of the solute carrier organic anion transporter family that are capable of transporting sulfated steroids include solute organic anion transport (SOAT encoded by SLC10A6) and organic anion transporter polypeptide-1A2 (OATP-1A2 encoded by SLCO1A2). SOAT is predominant in adipose and placenta, while OATP-1A2 is high in brain, liver, and kidney. Steroid sulfatase (encoded by the *STS* gene) is a ubiquitously expressed enzyme that hydrolyzes the sulfuryl group from DHEA-S to produce DHEA. Thus, a cell that expresses an STS is capable of using DHEA-S to generate DHEA which may be further metabolized to sex hormones in a cell type-dependent manner (reviewed in Mueller et al., 2015) (Fig. 2). This intracellular control of androgen and estrogen levels and the subsequent biological response within the cell is defined as an intracrine mechanism of action (reviewed in Labrie, Martel, Belanger, & Pelletier, 2017). The major tissues associated with DHEA-S metabolism are placenta, ovary, testes, prostate, adipose, liver, and brain—tissues associated with sex hormone-nuclear receptor-driven processes.

DHEA serves as the precursor for the weak androgens, androstenedione (4-androstene-3β,17β-dione, Adione) and androstenediol (5-androstene-3β,17β-diol, Adiol). To generate Adione the 3β-hydroxy group of DHEA is oxidized to a keto group along with isomerization of the double bond from C6–C5 in the B ring to C5–C4 in the A ring of the steroid nucleus. DHEA, DHEA-S, and Adione are commonly referred to as C17 keto steroids and represent the major adrenal androgens in circulation (Figs. 1 and 2). Adiol is generated by reduction of the C17 keto group to a 17β-hydroxyl group and is a relatively minor adrenal androgen (Fig. 1). Adione and Adiol both serve as precursors for testosterone and estradiol synthesis, thereby

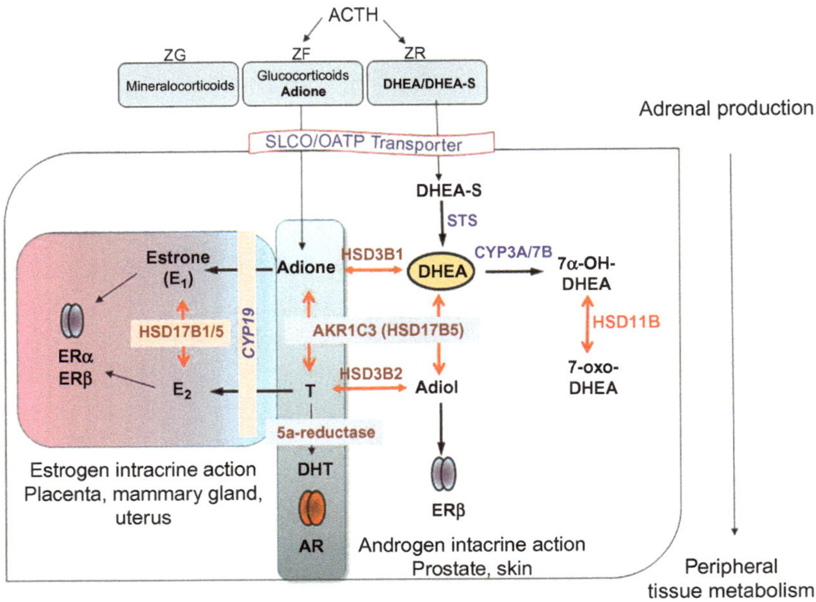

Fig. 2 Peripheral metabolism of adrenal androgens. This schematic depicts the metabolism of the major circulating adrenal-derived weak androgens, DHEA-S, DHEA, and androstenedione (Adione). The pathways and enzyme activities are described in the text. Cellular uptake of DHEA-S and Adione is mediated by members of the solute carrier organic anion transporter family (SLCO) and the organic anion transporter polypeptide family (OATP). DHEA-S is desulfated by steroid sulfatase (STS) prior to further metabolism. The conversion of DHEA and Adione to estrogens (estradiol and estrone) and/or testosterone is dependent upon the relative expression in a particular cell/tissue of the enzymes shown. DHEA metabolism to estrogens in estrogen responsive tissues and the sex hormone acts locally by binding estrogen receptor-alpha and/or estrogen receptor-beta (ERα and ERβ). DHEA metabolism to potent androgens testosterone (T) and dihydrotestosterone (DHT) occurs in androgen-responsive tissues and the sex hormones act locally by binding the androgen receptor (AR). ERs and AR are members of a family of nuclear receptors that bind as a homodimer to DNA and either activate or inhibit gene transcription. DHEA metabolism to 7α-hydroxy-DHEA and 7-oxo-DHEA occurs in the liver via the actions of cytochrome P450 3A (CYP3A) or 7B (CYP7B) and 11β-hydroxysteroid dehydrogenase (HSD11B), respectively. *Adiol*, 5-androsten-3β,17β-diol; *Adione*, 4-androstene-3β,17β-dione; *HSD3B*, 3β-hydroxysteroid dehydrogenase; *HSD17B*, 17β-hydroxysteroid dehydrogenase. HSD17B1/3/5 are different isoforms with tissue, substrate, and activity specificity.

providing the route for adrenal-derived DHEA/DHEA-S contribution to sex hormone action in peripheral tissues (Fig. 2). The adrenal is capable of metabolizing DHEA to androstenedione (Adione) and androstenediol (Adiol) that serve as precursors for the sex hormones testosterone (T) and estrogens (E1 and E2) (Fig. 1), yet this represents a very minor pathway for contribution to circulating steroid levels.

Members of the Short Chain Dehydrogenase family are responsible for DHEA, Adione, and Adiol metabolism. 3β-Hydroxysteroid dehydrogenase/Δ^5–Δ^4 isomerase (HSD3B) catalyzes the irreversible isomerization reaction that changes Δ^5 steroids (defined as containing a C6–C5 double bond in the steroid B ring) to Δ^4 steroids (defined as steroids containing a C5–C4 double bond in the steroid A ring) that include DHEA to Adione, and Adiol to T (Figs. 1 and 2). In the adrenal and gonads, HSD3B type 2 (HSD3B2) is the major isoform expressed. However, the adrenal *zona reticularis* has high CYP17A1-17,20 lyase activity, high SULT2A1 expression, and low HSD3B2 activity that drives the pathway toward DHEA/DHEA-S (Fig. 1). Higher HSD3B2 activity in the other adrenal zones favors mineralocorticoid and glucocorticoid synthesis from pregnenolone and 17α-hydroxypregnenolone, respectively. HSD3B type 1 (HSD3B1) is the major isoform expressed in steroid responsive tissues including placenta, prostate, and skin and is a key for intracrine action of DHEA by converting DHEA to Adione, and Adiol to T (Fig. 2). Importantly, 17β-hydroxysteroid dehydrogenase (HSD17B) family members catalyze reversible oxidation–reduction at C17 to interconvert between steroids containing either C17 keto or C17 hydroxyl groups. HSD17B isoforms type 1 and 3 preferentially catalyze the reduction of a C17 keto group to a β-hydroxyl group for generation of the potent sex hormones T and estradiol (E2) that are characterized by a C17 β-hydroxyl group. HSD17B3 is highly expressed in the testis for T production, while HSD17B1 is the isoform expressed placenta, ovary, and mammary gland with substrate specificity for estrogens over androgens. In the adrenal, C17 keto reduction is catalyzed by a member of the aldo-keto reductase family, AKR1C3 (alias HSD17B5) (Fig. 1). AKR1C3/HSD17B5 also is the major isoform that catalyzes the peripheral metabolism of DHEA to Adiol, Adione to T, and estrone to E2 (Fig. 2). Adione and T are substrates for the enzyme 5α-reductase (encoded by the *SRD5A2* gene) that reduces the double bond in the A ring to form 5α-Adione and DHT, respectively (Fig. 2). DHT is reported to have higher affinity binding to the nuclear steroid receptor, androgen receptor (AR), and increases AR stability relative to T binding, thereby making DHT the most potent androgen for AR activation (Zhou, Lane, Kemppainen, French, & Wilson, 1995). 5α-Reductase is highly expressed in prostate, seminal vesicles, hair follicles, and skin where intracrine actions of DHT promote AR-dependent signaling for growth and development.

Further, androgens are converted to estrogens by the action of aromatase (CYP19A1) that catalyzes a complex series of reactions that result in aromatization of the A ring and demethylation at C19. Aromatase converts Adione and T to estrone and estradiol, respectively (Fig. 2). Lastly, cytochrome P450-dependent metabolism (CYP) of DHEA generates monohydroxylated DHEA metabolites with 7α-hydroxy-DHEA shown to be further metabolized to 7-oxo-DHEA by 11β-hydroxysteroid dehydrogenase (HSD11B) (Fitzpatrick, Ripp, Smith, Pierce Jr, & Prough, 2001; Hennebert et al., 2009; Robinzon, Michael, Ripp, Winters, & Prough, 2003) (Fig. 2).

The importance of peripheral DHEA metabolism increases in disorders in gonadal steroid production and with age and declining gonadal function (Lois, Kassi, Prokopiou, & Chrousos, 2000). Conversely, polycystic ovarian disease (PCOS) and nonclassical 21-hydroxylase-deficient congenital adrenal hyperplasia (Goodarzi, Carmina, & Azziz, 2015) are disorders associated with elevated DHEA-S and T of adrenal origin. In these patients, the elevated DHEA may contribute to symptoms of PCOS disease by direct receptor activation and not simply supplying precursor steroid for T production (Daan et al., 2015; Moran, Arriaga, Arechavaleta-Velasco, & Moran, 2015; Pinola et al., 2015). To date, hormone-dependent cancers provide the best examples for possible detrimental intracrine actions of sex hormone synthesis from DHEA/DHEA-S (Chang, Ercole, & Sharifi, 2014; McNamara & Sasano, 2015). In addition, adipose tissue is being increasingly recognized as an important player in peripheral metabolism of adrenal-derived DHEA/DHEA-S. The intracrine effects of adipose androgen production from DHEA/DHEA-S may regulate adipocyte differentiation and energy metabolism (Fujioka et al., 2012; O'Reilly, House, & Tomlinson, 2014). The contribution of adipose tissue to hormone imbalance disorders is an area that requires further investigation, especially given the rise in obesity.

2.3 Regulation of DHEA Synthesis and Metabolism

The developmental pattern for DHEA/DHEA-S synthesis is cyclical with the fetal adrenal producing high levels (providing the steroid precursor for placental estrogen synthesis) that fall to nadir levels after birth and remain low for the first 6 years of life. Elevations in serum DHEA/DHEA-S are coincident with the expansion and differentiation of the adrenal *zona reticularis*, a change that begins typically around 6–8 years of age and defines adrenarche (Auchus & Rainey, 2004; Rege et al., 2016). DHEA/DHEA-S levels continue to rise and peak between the second and third decade of life. In healthy young adults, the circulating DHEA-S:DHEA ratio is 1000:1 with standard values ranging from 1 to 10 μM DHEA-S compared to 2 to 30 nM for DHEA (Andus et al., 2003; Mueller et al., 2015) with women typically having lower levels than men. With age these levels decline and are 10-fold lower by 60–80 years of age (Labrie, 2010a; Labrie, Belanger, Cusan, Gomez, & Candas, 1997; Labrie et al., 2005). The association between age-related health disorders and the decline in DHEA sparked interest in the potential for restoring DHEA/DHEA-S levels as a "fountain of youth" treatment for aging-associated disorders (Baulieu, 1996). In 1994, Congress passed the Dietary Supplement Health and Education Act that opened the door for DHEA to be sold as an over-the-counter nutritional supplement (outside of the FDA's jurisdiction), and DHEA was/is marketed as a therapeutic agent to promote many health benefits. However, the use of DHEA as a nutritional supplement is hotly debated since a causal relationship between DHEA levels and health benefits has not been clearly established despite four decades of research, and the potential for DHEA to increase sex hormone-driven cancer risk via local metabolism to sex hormones within the tumor elicits a cautionary note. The chapter by Clark et al. (2017, in this volume) focuses on the mechanism(s) for DHEA (and DHEA metabolites) action that are beginning to provide rationales for both the potential health benefits of this steroid and potential detrimental tumor-promoting actions.

Adrenocorticotropin hormone (ACTH) is capable of stimulating DHEA synthesis, as observed by increased circulating DHEA levels following ACTH stimulus in rodent and human studies. Further, DHEA levels follow a circadian pattern that parallels ACTH with levels of both hormones peaking in the morning (Hammer et al., 2005). ACTH binds to melanocortin-2 receptors (MC2R) expressed in the *zona fasciculata* and *reticularis* and activates cAMP-dependent protein kinase A (PKA)-dependent signaling. PKA signaling results in phosphorylation of key transcription factors such as SF-1, CREB/CREM, GATA4/6, and AP1 family members that activate *STAR*, *CYP11A1*, *CYP17A1*, and *SULT2A1* gene expression to maintain an increase in steroid output (Jimenez, Saner, Mayhew, & Rainey, 2003; Ruggiero & Lalli, 2016). Thus, the molecular mechanisms that control StAR, CYP11A1, CYP17A1, and SULT2A1 expression in the adrenal ZR would be predicted to control DHEA/DHEA-S synthesis (Udhane & Fluck, 2016).

3. BENEFICIAL EFFECTS OF DHEA SUPPLEMENTATION

3.1 Early Studies That Sparked Research on DHEA Supplementation

Early studies in humans documented the levels of DHEA and DHEA-S in many disease states (Barrett-Connor, Khaw, & Yen, 1986; Bird, Masters, & Clark, 1984; Bulbrook & Hayward, 1965). For example, Bulbrook and Hayward (1965) and Bulbrook, Hayward, and Spicer (1971) reported that patients with breast cancer who excreted lower quantities of urinary 11-deoxy 17-oxosteroids relative to the 17-hydroxycorticosteroids have higher rates of breast cancer recurrence after mastectomy. Further, women who excrete lower levels of these androgens, now known to largely be DHEA-S, are at greater risk for breast cancer than women with higher levels of excretion (Bulbrook et al., 1971). These findings suggested lower DHEA excretion in urine (i.e., lower levels of circulating 17-hydroxycorticosteroids/urine) could be a predictive tool for breast cancer occurrence.

3.2 Summary of Effects of DHEA in Animal Models

A possible correlation between circulating DHEA-S levels and cancer risk encouraged further studies on DHEAs role in additional health issues and stimulated animal studies to address the mechanism of action of DHEA in these models. Kalimi and Regelson (1990, 2000) have coauthored two books on DHEA and wrote a review (Gursoy, Hu, Cardounel, Regelson, & Kalimi, 2011) that extensively covers DHEA beneficial effects in animal models. For the sake of brevity, we provide a summary of the animal work in table form (Table 1). We apologize to the many authors we have not cited, but the voluminous literature on DHEA effects on animal models prevent full review.

As an example of the breath of these studies, the Schwartz research group at Temple University has examined on the role of DHEA administration on the management of cancer, obesity (Schwartz & Cleary, 1987), cancer chemoprotection (Pashko & Schwartz, 1983), and brain injury (Malik et al., 2003). This chapter will focus mainly on the human clinical trials.

3.3 Inhibition of Glucose 6-Phosphate Dehydrogenase by DHEA and Other Sterols

DHEA and other steroids have long been known to be potent inhibitors of glucose 6-phosphate dehydrogenase (G6PD), leading to decrease in both

Table 1 Beneficial Effects of DHEA Supplementation in Animal Models of Disease

Disease State	Animal Model	Conclusion on DHEA Action	References
Obesity	Rat (diet vs diet plus DHEA)	Weight gain decreased relative to control	Prough, Webb, Wu, Lapenson, and Waxman (1994), Schwartz and Cleary (1987), and Wu et al. (1989)
Chemoprevention of carcinogenesis	MNU breast cancer	Inhibition of tumor formation	Lubet et al. (1998)
	TPA-promoted skin cancer	DMBA DNA binding decreased	Pashko and Schwartz (1983)
Immune function	IL2 regulation in mouse	Enhanced immunity in mice	Araneo, Woods, and Daynes (1993) and Daynes, Dudley, and Araneo (1990)
Diabetes	Diabetes in rats	Reversed insulin insufficiency	Coleman, Leiter, and Schwizer (1982)
Atherosclerosis	Cholesterol ester accumulation	Reduced cholesterol in murine macrophages	Taniguchi, Yanase, Kobayashi, Takayanagi, and Nawata (1996)
		Reduced atherosclerosis in rabbits	Gordon, Bush, and Weisman (1988)

NADPH levels for reduction reactions and ribose phosphate levels for nucleotide synthesis (Menzel, Gobbert, & Oertel, 1970; Shantz, Talalay, & Gordon, 1989). G6PD inhibition may account for the cytostatic properties of DHEA. This theory was supported by studies showing that administration of ribose phosphate derivatives to female Wistar rats reduced the ability of dietary DHEA to inhibit liver tumor growth in vivo (Garcea et al., 1988). The G6PD inhibitory effect of DHEA, or other sterols with higher affinity to G6PD, involved reduced production of NADPH and ribose phosphates and remains an interesting chemoprotective mechanism that might be exploited to hinder tumor cell growth (Hamilton et al., 2012; Shantz et al., 1989).

4. BENEFICIAL EFFECTS OF DHEA SUPPLEMENTATION ON HUMAN DISEASE

After many reports on the positive effects of DHEA and DHEA-S supplementation on various animal models of disease (Table 1), a logical next question to address is whether the positive effects observed in animals could be recapitulated in humans and were there any negative side effects of DHEA supplementation? Table 2 lists the DHEA supplementation studies in aging humans demonstrating beneficial effects on symptoms of systemic lupus erythematosus (SLE) (Van Vollenhoven & McGuire, 1996), depression (Wolkowitz et al., 1997), AIDS (Centurelli & Abate, 1997), ulcerative colitis (Andus et al., 2003), vaginal atrophy (Labrie, 2010a, 2010b), diminished ovarian reserve (Fouany & Sharara, 2013; Nagels et al., 2015), and mood (do Vale et al., 2015), as well as its use as an adjuvant during immunization against tetanus and influenza (Evans et al., 1996). DHEA supplementation has also been associated with small decreases in adipose stores (Corona et al., 2013), a site for intracrine action of DHEA. This section will first summarize studies conducted by Baulieu and coworkers on DHEA supplementation effects on bone density (Baulieu et al., 2000), pulmonary hypertension and COPD (Debonneuil, Quillard, & Baulieu, 2006; Dumas de La Roque et al., 2012), and mental and physical aging (Baulieu, 1996; Mazat et al., 2001). Other selected areas covered include DHEA effects in female fertility and cardiovascular health.

4.1 Protocol DHEAge

"Protocol DHEAge" was a large (280 subjects) randomized, double-blind, placebo-controlled clinical trial. DHEA was administered either at a dosage of 50 mg daily or a placebo (per os) for a year (Baulieu et al., 2000). No harmful side effects due to DHEA administration were observed at 1 year, and, in addition to the expected increase in DHEA/DHEA-S levels in blood, modest increases in estrogens and testosterone were observed. DEXA scans revealed that bone density improved in women, and decreased osteoclastic activity was observed in subjects >70 years of age. Increases in libido in women >70 years of age were observed as well as improvement of skin texture. However, given the known role of estrogens and androgens in human breast and prostate cancer, it was deemed "medically justified to keep aging subjects who take DHEA ≤50 mg/day under appropriate clinical and biological control at reasonable time intervals." We note that this paper has

Table 2 Beneficial Effects of DHEA Supplementation in Human Disease Confirmed

Disease State	Parameter Measured	Conclusion on DHEA Action	References
Pulmonary hypertension	6-Min walk test	Improved	Dessouroux, Akwa, and Baulieu (2008) and Dumas de La Roque et al. (2012)
	Carbon monoxide diffusing capacity of the lung (DLCO % predicted) increased significantly	Improved	
	Pulmonary hemodynamics	Improved	
Systemic lupus erythematosus	Lupus flares	Improvement	Van Vollenhoven and McGuire (1996)
Crohn's disease	Crohn's active index	Decreased	Andus et al. (2003)
	Ulcerative colitis	Decreased	
Vaginal atrophy	Hormone deficiency of menopause	Improved by local application	Labrie (2010a, 2010b)
	Sexual dysfunction	Improvement	
Infertility	Ovarian reserve	Improvement for fertility	Fouany and Sharara (2013) and Nagels, Rishworth, Siristatidis, and Kroon (2015)
Obesity	Body weight decline	Slight improvement in males only	Corona et al. (2013) and Morales, Haubrich, Hwang, Asakura, and Yen (1998)
Immune function	Response to immunization	Improvement	Evans et al. (1996)
	AIDs suppression	Improvement	Centurelli and Abate (1997)
Mood	Depression	Improvement	do Vale et al. (2015) and Wolkowitz et al. (1997)

been cited in 45 other papers in PubMed, including a meta-analysis that refuted the findings of the Baulieu study noting "low confidence" in reports that DHEA administration results in significant improvement in libido or sexual function, serum lipids, serum glucose, weight, body mass index, or bone mineral density in postmenopausal women with normal adrenal function due to imprecision, risk of bias, and inconsistencies across randomized controlled trials (Elraiyah et al., 2014).

4.2 The PAQUID Research Program

The PAQUID Research Program was a prospective cohort study of mental and physical aging in a representative sample of elderly subjects residing in south-west France (the Gironde "Département") (Berr, Lafont, Debuire, Dartigues, & Baulieu, 1996; Mazat et al., 2001). The subjects were followed for 4 and 10 years after their baseline-circulating DHEA levels were measured, looking at a variety of functional, psychological, and mental status measures. Men displayed more rapid declines in circulating DHEA after age 65 than women, confirming the notion of age-related declines in circulating DHEA levels in both sexes. A number of measures of normal activities of daily living, such as drug medication utilization, confinement, dyspnea, depression, poor subjective perception of health, and life satisfaction were assessed over a 4-year period. Women who had lower levels of DHEA/DHEA-S demonstrated on average lower levels of these functional markers, while there was less correlation in men. For men, lower DHEA-S levels were significantly associated with increased short-term mortality at 2 and 4 years after baseline measurement. At 10 years, there were reliable associations of lower circulating DHEA-S levels with increased death in male smokers; death was associated with various types of cardiovascular disease endpoints (Mazat et al., 2001).

4.3 DHEA and Pulmonary Hypertension

Another area of interest to Baulieu and coworkers (Debonneuil et al., 2006; Dumas de La Roque et al., 2012) was the impact of DHEA supplementation on pulmonary hypertension in patients with chronic obstructive pulmonary disease (PH-COPD). In a murine survival study in which mice were made hypoxic with 50% oxygen for prolonged periods, DHEA supplementation protected against pulmonary hypertension (Debonneuil et al., 2006). This was followed by a human subjects pilot experiment where eight subjects with PH-COPD received 200 mg DHEA daily for 3 months

(Dumas de La Roque et al., 2012), displayed improved performance on a 6-min walk test, increased lung carbon monoxide diffusing capacity (DLCO % predicted), and improved pulmonary hemodynamics, thereby, reversing chronic hypoxia-induced pulmonary hypertension. This group also studied the effects of DHEA on human pulmonary arterial smooth muscle cells to identify its mechanism of action under hypoxia in vitro. Interestingly, DHEA, nonmetabolizable 3β-methyl-Δ^5-androsten-17-one, T, and E2 were all active in stimulating the increase in HIF-1α (Dessouroux et al., 2008). These sterols appear to increase accumulation of HIF-1α protein through a posttranslational process, as HIF-1α mRNA levels were unchanged in the presence of DHEA and the other compounds. Surprisingly, there are, to our knowledge, no reports, other than ours (Teng et al., 2014), of DHEA regulation of miRNA expression in any tissue. This is a clear gap in knowledge, and future studies will need to assess direct DHEA effects vs effects of DHEA metabolites.

4.4 DHEA and Female Fertility

The number and quality of oocytes retrieved during in vitro fertilization (IVF) is an important parameter for successful fertilization. There have been many studies over the past decade on DHEA supplementation for IVF procedures to improve oocyte retrieval in women who do not respond well to gonadotropin-induced oocyte maturation. The results appear promising yet are not consistent across trials. Two separate trials support DHEA supplementation can improve oocyte retrieval (Kara, Aydin, Aran, Turktekin, & Ozdemir, 2014; Tsui et al., 2015), yet improved pregnancy rates were reported for only one trial (Tsui et al., 2015). A meta-analysis of eight trials reported between 2006 and 2014 concluded that DHEA supplementation improves pregnancy rates without significant improvement in oocyte retrieval or quality (Li et al., 2015). A newly reported prospective study showed that DHEA supplementation at 25 mg three times per day for 8 weeks resulted in increased serum T levels, and AR and FSHR mRNA levels in granulosa cells, that were correlated with an increase in the number and quality of oocytes compared to the control group without DHEA supplementation; however, there was no improvement for the in vitro fertility outcomes (Hu et al., 2017).

4.5 DHEA and Cardiovascular Function and Well-Being

Yen and Coworkers (Yen, 2001) have performed several human studies evaluating the role of DHEA supplementation on overall well-being and cardiovascular benefits in human subjects. The strongest set of data deal with the correlation in men over age 50 of circulating DHEA-S levels being an independent and inversely related to death from any cause and death from cardiovascular disease (Barrett-Connor et al., 1986), a result similar to that of Baulieu and coworkers (Mazat et al., 2001). The studies from Yen's group also identified changes in serum insulin-like growth factor 1 (IGF-1) and IGF-binding protein-1 (IGF1-BP) caused by DHEA supplementation in aged humans (Morales, Nolan, Nelson, & Yen, 1994). Increased bioavailability of IGF-1 was proposed to enhance the function of the immune system accounting for the benefits of DHEA in HIV and other immune therapies (Khorram, Vu, & Yen, 1997). However, Baulieu and coworkers (Raynaud-Simon et al., 2001) performed several other clinical trials that showed a lack of correlation between plasma DHEA/DHEA-S and IGF-1 levels. Similarly, only modest effects on body weight maintenance were observed in aging males, but little effect for aging females (Morales et al., 1998). In addition, the link between these biomarkers and the mechanism of DHEA action was not conclusively demonstrated and the results were modest in magnitude.

Most critics decry the lack of large numbers of subjects in the clinical trials cited, the modest changes seen, and express concern about hormone-sensitive cancers in men and women (Labrie, 2010a). However, the studies by Yen (Yen, 2001) and by Baulieu and coworkers (Baulieu et al., 2000; Mazat et al., 2001) showed no adverse effects from taking DHEA-S at a dosage of 50 mg daily. If DHEA were to have medical utility, much more sophisticated randomized, placebo-controlled trials in large cohorts of aging humans would be required to prove efficacy and safety. This approach will be weighed against the expense to continue such studies, unless it were supported by a pharmaceutical company documenting efficacy for a regulatory agency.

5. THE FUTURE

From a casual evaluation of the human studies on DHEA, there are a few areas where one might suggest the appropriate application of DHEA and DHEA-S as a therapeutic regimen. Clearly, the declines in endocrine function seen during menopause in women (Labrie, 2010b), such as vaginal atrophy, are an area where DHEA-S therapy may have significant impact. Considerable clinical application has been provided in gynecological medicine since the 1940s due to the direct clinical application of these steroids to ameliorating symptoms of menopause (Labrie, 2010a). Indeed, Vagifem (E2) is routinely prescribed to alleviate symptoms of vaginal epithelial atrophy in postmenopausal women (Chollet, 2011). Several other areas, such as pulmonary hypertension, SLE, Crohn's disease and ulcerative colitis, IVF, and immune modulation may be worthy of consideration of further research (see Table 2), but many more clinical trials in aged individuals will be required and compared to other therapies before the full utility of DHEA/DHEA-S treatment will be proven. For many of the other conditions, other therapeutic or practice applications will need to come forward instead of DHEA therapy. In addition, there are known adverse effects of DHEA supplementation, including acne, embryo virilization in pregnancy, and endocrine-sensitive cancers. This does not mean that understanding the biology and normal biochemistry and physiology of DHEA and DHEA-S is not important. Understanding the various receptors and binding proteins that DHEA interacts with is important for understanding how it functions and possibly can be a tool in understanding new "receptors" for DHEA as described in a chapter of this volume (Clark et al., 2017). However, in the face of the changing health care system in the United States, it does not seem likely that this steroid as an "orphan drug" will ever be used extensively in human medicine except for those applications where efficacy has strong promise as reviewed earlier. World War II and the effect it had on Schering AG prevented it from being tested rigorously in the clinic like other pharmaceutical agents.

ACKNOWLEDGMENT

The authors thank Boaz Robinzon, Ph.D., for his critical evaluation of this chapter.

REFERENCES

Allen, E., & Doisy, E. A. (1983). Landmark article Sept 8, 1923. An ovarian hormone. Preliminary report on its localization, extraction and partial purification, and action in test animals. By Edgar Allen and Edward A. Doisy. *JAMA*, *250*(19), 2681–2683.

Andus, T., Klebl, F., Rogler, G., Bregenzer, N., Scholmerich, J., & Straub, R. H. (2003). Patients with refractory Crohn's disease or ulcerative colitis respond to dehydroepiandrosterone: A pilot study. *Alimentary Pharmacology & Therapeutics*, *17*(3), 409–414.

Araneo, B. A., Woods, M. L., 2nd, & Daynes, R. A. (1993). Reversal of the immunosenescent phenotype by dehydroepiandrosterone: Hormone treatment provides an adjuvant effect on the immunization of aged mice with recombinant hepatitis B surface antigen. *The Journal of Infectious Diseases*, *167*(4), 830–840.

Auchus, R. J., & Rainey, W. E. (2004). Adrenarche—Physiology, biochemistry and human disease. *Clinical Endocrinology*, *60*(3), 288–296.

Barrett-Connor, E., Khaw, K. T., & Yen, S. S. (1986). A prospective study of dehydroepiandrosterone sulfate, mortality, and cardiovascular disease. *The New England Journal of Medicine, 315*(24), 1519–1524. https://doi.org/10.1056/NEJM198612113152405.

Baulieu, E. E. (1996). Dehydroepiandrosterone (DHEA): A fountain of youth? *The Journal of Clinical Endocrinology and Metabolism, 81*(9), 3147–3151. https://doi.org/10.1210/jcem.81.9.8784058.

Baulieu, E. E., Corpechot, C., Dray, F., Emiliozzi, R., Lebeau, M. C., Mauvais Jarvis, P., et al. (1965). An adrenal-secreted "Androgen": Dehydroisoandrosterone sulfate. Its metabolism and a tentative generalization on the metabolism of other steroid conjugates in man. *Recent Progress in Hormone Research, 21*, 411–500.

Baulieu, E. E., Thomas, G., Legrain, S., Lahlou, N., Roger, M., Debuire, B., et al. (2000). Dehydroepiandrosterone (DHEA), DHEA sulfate, and aging: Contribution of the DHEAge study to a sociobiomedical issue. *Proceedings of the National Academy of Sciences of the United States of America, 97*(8), 4279–4284.

Berr, C., Lafont, S., Debuire, B., Dartigues, J. F., & Baulieu, E. E. (1996). Relationships of dehydroepiandrosterone sulfate in the elderly with functional, psychological, and mental status, and short-term mortality: A french community-based study. *Proceedings of the National Academy of Sciences of the United States of America, 93*(23), 13410–13415.

Bird, C. E., Masters, V., & Clark, A. F. (1984). Dehydroepiandrosterone sulfate: Kinetics of metabolism in normal young men and women. *Clinical and Investigative Medicine*, 7(2), 119–122.

Bulbrook, R. D., & Hayward, J. L. (1965). The possibility of predicting the response of patients with early breast cancer to subsequent endocrine ablation. *Cancer Research, 25*(7), 1135–1139.

Bulbrook, R. D., Hayward, J. L., & Spicer, C. C. (1971). Relation between urinary androgen and corticoid excretion and subsequent breast cancer. *Lancet, 2*(7721), 395–398.

Butenandt, A. (1929). Über "Progynon" ein krystallisiertes weibliches sexualhormon. *Die Naturwissenschaften, 17*(45), 879.

Butenandt, A. (1931). Uber die chemische untersuchung der sexualhormone. *Angewandte Chemie, 44*, 905–916.

Butenandt, A. (1936). Ergebnisse und Probleme in der biochemischen Erforschung der Keimdrusenhormone. *Naturwissenschafen, 24*, 529–536.

Butenandt, A., Beckmann, R., & Stamm, D. (1961). On the sexattractant of silk-months. II. Constitution and configuration of bombvkol. *Hoppe-Seyler's Zeitschrift für Physiologische Chemie, 324*, 84–87.

Butenandt, A., & Dannenberg, H. (1940). Über das wahre Δ^1-Androsten-ol-(17)-on-(3), ein isomeres des testosterons. *Berichte Der Deutschen Chemischen Gesellschaft (A and B Series), 73*(3), 206–208.

Butenandt, A., & Hanisch, G. (1935). About testosterone. Conversion of dehydroandrosterons into androstendiol and testosterone; a way for the structure assignment of testosterone from cholestrol. *Hoppe-Seyler's Zeitschrift für Physiologische Chemie, 237*(2), 89–97.

Butenandt, A., & Westphal, U. (1934). Zur isolierung und charakterisierung des corpusluteum-hormons. *Chemische Berichte, 67*, 1440–1445.

Centurelli, M. A., & Abate, M. A. (1997). The role of dehydroepiandrosterone in AIDS. *Annals of Pharmacotherapy, 31*(5), 639–642.

Chang, K. H., Ercole, C. E., & Sharifi, N. (2014). Androgen metabolism in prostate cancer: From molecular mechanisms to clinical consequences. *British Journal of Cancer, 111*(7), 1249–1254. https://doi.org/10.1038/bjc.2014.268.

Chollet, J. A. (2011). Efficacy and safety of ultra-low-dose vagifem (10 mcg). *Patient Preference and Adherence, 5*, 571–574. https://doi.org/10.2147/PPA.S22940.

Clark, B. J., Prough, R. A., & Klinge, C. M. (2017). *Mechanisms of action of dehydroepiandrosterone* (In Press). Oxford, UK: Elsevier.

Clauberg, C. (1935). Experimentelle Untersuchungen zur hormonalen temporaren Sterilislerung und zur Behebung hormonal-bedingter Sterilitat. *Zeitschrift fur Gerburtshilfe und Gyndkologie, 112*, 4–33.

Coleman, D. L., Leiter, E. H., & Schwizer, R. W. (1982). Therapeutic effects of dehydroepiandrosterone (DHEA) in diabetic mice. *Diabetes, 31*(9), 830–833.

Corona, G., Rastrelli, G., Giagulli, V. A., Sila, A., Sforza, A., Forti, G., et al. (2013). Dehydroepiandrosterone supplementation in elderly men: A meta-analysis study of placebo-controlled trials. *The Journal of Clinical Endocrinology and Metabolism, 98*(9), 3615–3626. https://doi.org/10.1210/jc.2013-1358.

Daan, N. M., Jaspers, L., Koster, M. P., Broekmans, F. J., de Rijke, Y. B., Franco, O. H., et al. (2015). Androgen levels in women with various forms of ovarian dysfunction: Associations with cardiometabolic features. *Human Reproduction, 30*(10), 2376–2386. https://doi.org/10.1093/humrep/dev195.

Daynes, R. A., Dudley, D. J., & Araneo, B. A. (1990). Regulation of murine lymphokine production in vivo. II. Dehydroepiandrosterone is a natural enhancer of interleukin 2 synthesis by helper T cells. *European Journal of Immunology, 20*(4), 793–802. https://doi.org/10.1002/eji.1830200413.

Debonneuil, E. H., Quillard, J., & Baulieu, E. E. (2006). Hypoxia and dehydroepiandrosterone in old age: A mouse survival study. *Respiratory Research*, 7, 144. https://doi.org/10.1186/1465-9921-7-144.

Deshpande, N., & Bulbrook, R. D. (1964). A method for the simultaneous determination of 17-oxosteroids and 17-hydroxycorticosteroids in human plasma. *The Journal of Endocrinology, 28*, 289–296.

Dessouroux, A., Akwa, Y., & Baulieu, E. E. (2008). DHEA decreases HIF-1alpha accumulation under hypoxia in human pulmonary artery cells: Potential role in the treatment of pulmonary arterial hypertension. *The Journal of Steroid Biochemistry and Molecular Biology, 109*(1–2), 81–89. https://doi.org/10.1016/j.jsbmb.2007.12.001.

do Vale, S., Selinger, L., Martins, J. M., Bicho, M., do Carmo, I., & Escera, C. (2015). Dehydroepiandrosterone (DHEA) and dehydroepiandrosterone-sulfate (DHEAS) and emotional processing—A behavioral and electrophysiological approach. *Hormones and Behavior, 73*, 94–103. https://doi.org/10.1016/j.yhbeh.2015.06.005.

Dumas de La Roque, E., Savineau, J. P., Metivier, A. C., Billes, M. A., Kraemer, J. P., Doutreleau, S., et al. (2012). Dehydroepiandrosterone (DHEA) improves pulmonary hypertension in chronic obstructive pulmonary disease (COPD): A pilot study. *Annales d'endocrinologie, 73*(1), 20–25. https://doi.org/10.1016/j.ando.2011.12.005.

Elraiyah, T., Sonbol, M. B., Wang, Z., Khairalseed, T., Asi, N., Undavalli, C., et al. (2014). Clinical review: The benefits and harms of systemic dehydroepiandrosterone (DHEA) in postmenopausal women with normal adrenal function: A systematic review and meta-analysis. *The Journal of Clinical Endocrinology and Metabolism, 99*(10), 3536–3542. https://doi.org/10.1210/jc.2014-2261.

Evans, T. G., Judd, M. E., Dowell, T., Poe, S., Daynes, R. A., & Araneo, B. A. (1996). The use of oral dehydroepiandrosterone sulfate as an adjuvant in tetanus and influenza vaccination of the elderly. *Vaccine, 14*(16), 1531–1537.

Fitzpatrick, J. L., Ripp, S. L., Smith, N. B., Pierce, W. M., Jr., & Prough, R. A. (2001). Metabolism of DHEA by cytochromes P450 in rat and human liver microsomal fractions. *Archives of Biochemistry and Biophysics, 389*(2), 278–287. https://doi.org/10.1006/abbi.2001.2341.

Fouany, M. R., & Sharara, F. I. (2013). Is there a role for DHEA supplementation in women with diminished ovarian reserve? *Journal of Assisted Reproduction and Genetics, 30*(9), 1239–1244. https://doi.org/10.1007/s10815-013-0018-x.

Fujioka, K., Kajita, K., Wu, Z., Hanamoto, T., Ikeda, T., Mori, I., et al. (2012). Dehydroepiandrosterone reduces preadipocyte proliferation via androgen receptor. *American Journal of Physiology. Endocrinology and Metabolism*, *302*(6), E694–704. https://doi.org/10.1152/ajpendo.00112.2011.

Garcea, R., Daino, L., Frassetto, S., Cozzolino, P., Ruggiu, M. E., Vannini, M. G., et al. (1988). Reversal by ribo- and deoxyribonucleosides of dehydroepiandrosterone-induced inhibition of enzyme altered foci in the liver of rats subjected to the initiation—Selection process of experimental carcinogenesis. *Carcinogenesis*, *9*(6), 931–938.

Gaudilliere, J. P. (2005). Better prepared than synthesized: Adolf butenandt, schering Ag and the transformation of sex steroids into drugs (1930-1946). *Studies in History and Philosophy of Biological and Biomedical Sciences*, *36*(4), 612–644. https://doi.org/10.1016/j.shpsc.2005.09.006.

Goodarzi, M. O., Carmina, E., & Azziz, R. (2015). DHEA, DHEAS and PCOS. *The Journal of Steroid Biochemistry and Molecular Biology*, *145*, 213–225. https://doi.org/10.1016/j.jsbmb.2014.06.003.

Gordon, G. B., Bush, D. E., & Weisman, H. F. (1988). Reduction of atherosclerosis by administration of dehydroepiandrosterone. A study in the hypercholesterolemic New Zealand white rabbit with aortic intimal injury. *The Journal of Clinical Investigation*, *82*(2), 712–720. https://doi.org/10.1172/JCI113652.

Gurpide, E., MacDonald, P. C., Chapdelaine, A., Vande Wiele, R. L., & Lieberman, S. (1965). Studies on the secretion and interconversion of the androgens. II. Methods for estimation of rates of secretion and of metabolism from specific activities of urinary metabolites. *The Journal of Clinical Endocrinology and Metabolism*, *25*(12), 1537–1556. https://doi.org/10.1210/jcem-25-12-1537.

Gursoy, E., Hu, Y., Cardounel, A., Regelson, W., & Kalimi, M. (2011). Biological effects of DHEA: A review. In M. Kalimi & W. Regelson (Eds.), *Dehydroepiandrosterone (DHEA) biochemical, physiological and clinical aspects* (pp. 1–29). Berlin, New York: Walter de Gruyter.

Hamilton, N. M., Dawson, M., Fairweather, E. E., Hamilton, N. S., Hitchin, J. R., James, D. I., et al. (2012). Novel steroid inhibitors of glucose 6-phosphate dehydrogenase. *Journal of Medicinal Chemistry*, *55*(9), 4431–4445. https://doi.org/10.1021/jm300317k.

Hammer, F., Subtil, S., Lux, P., Maser-Gluth, C., Stewart, P. M., Allolio, B., et al. (2005). No evidence for hepatic conversion of dehydroepiandrosterone (DHEA) sulfate to DHEA: In vivo and in vitro studies. *The Journal of Clinical Endocrinology and Metabolism*, *90*(6), 3600–3605. https://doi.org/10.1210/jc.2004-2386.

Hennebert, O., Montes, M., Favre-Reguillon, A., Chermette, H., Ferroud, C., & Morfin, R. (2009). Epimerase activity of the human 11beta-hydroxysteroid dehydrogenase type 1 on 7-hydroxylated C19-steroids. *The Journal of Steroid Biochemistry and Molecular Biology*, *114*(1–2), 57–63. https://doi.org/10.1016/j.jsbmb.2008.12.015.

Hu, Q., Hong, L., Nie, M., Wang, Q., Fang, Y., Dai, Y., et al. (2017). The effect of dehydroepiandrosterone supplementation on ovarian response is associated with androgen receptor in diminished ovarian reserve women. *Journal of Ovarian Research*, *10*(1), 32. https://doi.org/10.1186/s13048-017-0326-3.

Jimenez, P., Saner, K., Mayhew, B., & Rainey, W. E. (2003). GATA-6 is expressed in the human adrenal and regulates transcription of genes required for adrenal androgen biosynthesis. *Endocrinology*, *144*(10), 4285–4288. https://doi.org/10.1210/en.2003-0472.

Kalimi, M. Y., & Regelson, W. (1990). *The biologic role of dehydroepiandrosterone (DHEA)*. Berlin, New York: Walter de Gruyter.

Kalimi, M., & Regelson, W. (2000). *Dehydroepiandrosterone (DHEA). Biochemical, physiological and clinical aspects*. Berlin, New York: Walter de Gruyter.

Kara, M., Aydin, T., Aran, T., Turktekin, N., & Ozdemir, B. (2014). Does dehydroepiandrosterone supplementation really affect IVF-ICSI outcome in women with poor ovarian reserve? *European Journal of Obstetrics, Gynecology, and Reproductive Biology, 173*, 63–65. https://doi.org/10.1016/j.ejogrb.2013.11.008.

Kaufmann, C. (1934). Therapeutics with hormones of the ovary: (Section of therapeutics and pharmacology). *Proceedings of the Royal Society of Medicine, 27*(7), 849–863.

Khorram, O., Vu, L., & Yen, S. S. (1997). Activation of immune function by dehydroepiandrosterone (DHEA) in age-advanced men. *The Journals of Gerontology. Series A, Biological Sciences and Medical Sciences, 52((*1), M1–7.

Labrie, F. (2010a). DHEA, important source of sex steroids in men and even more in women. *Progress in Brain Research, 182*, 97–148. https://doi.org/10.1016/S0079-6123(10)82004-7.

Labrie, F. (2010b). Intravaginal DHEA, by a strictly local action, exerts beneficial effects on both vaginal atrophy symptoms and sexual dysfunction. *Hormone Molecular Biology and Clinical Investigation, 4*(1), 499–507. https://doi.org/10.1515/HMBCI.2010.064.

Labrie, F., Belanger, A., Cusan, L., Gomez, J. L., & Candas, B. (1997). Marked decline in serum concentrations of adrenal C19 sex steroid precursors and conjugated androgen metabolites during aging. *The Journal of Clinical Endocrinology and Metabolism, 82*(8), 2396–2402. https://doi.org/10.1210/jcem.82.8.4160.

Labrie, F., Luu-The, V., Belanger, A., Lin, S. X., Simard, J., Pelletier, G., et al. (2005). Is dehydroepiandrosterone a hormone? *The Journal of Endocrinology, 187*(2), 169–196. https://doi.org/10.1677/joe.1.06264.

Labrie, F., Martel, C., Belanger, A., & Pelletier, G. (2017). Androgens in women are essentially made from DHEA in each peripheral tissue according to intracrinology. *The Journal of Steroid Biochemistry and Molecular Biology, 168*, 9–18. https://doi.org/10.1016/j.jsbmb.2016.12.007.

Li, J., Yuan, H., Chen, Y., Wu, H., Wu, H., & Li, L. (2015). A meta-analysis of dehydroepiandrosterone supplementation among women with diminished ovarian reserve undergoing in vitro fertilization or intracytoplasmic sperm injection. *International Journal of Gynaecology and Obstetrics, 131*(3), 240–245. https://doi.org/10.1016/j.ijgo.2015.06.028.

Lois, K., Kassi, E., Prokopiou, M., & Chrousos, G. P. (2000). Adrenal androgens and aging. In L. J. De Groot, G. Chrousos, K. Dungan, K. R. Feingold, A. Grossman, J. M. Hershman, C. Koch, M. Korbonits, R. McLachlan, M. New, J. Purnell, R. Rebar, F. Singer, & A. Vinik (Eds.), *Endotext.* South Dartmouth, MA: MDText.com, Inc.

Longcope, C. (1996). Dehydroepiandrosterone metabolism. *Journal of Endocrinology, 150*(Suppl 3), S125–127.

Lubet, R. A., Gordon, G. B., Prough, R. A., Lei, X. D., You, M., Wang, Y., et al. (1998). Modulation of methylnitrosourea-induced breast cancer in Sprague Dawley rats by dehydroepiandrosterone: Dose-dependent inhibition, effects of limited exposure, effects on peroxisomal enzymes, and lack of effects on levels of Ha-Ras mutations. *Cancer Research, 58*(5), 921–926.

MacDonald, P. C., & Siteri, P. K. (1965). Origin of estrogen in women pregnant with an anencephalic fetus. *The Journal of Clinical Investigation, 44*, 465–474. https://doi.org/10.1172/JCI105160.

Malik, A. S., Narayan, R. K., Wendling, W. W., Cole, R. W., Pashko, L. L., Schwartz, A. G., et al. (2003). A novel dehydroepiandrosterone analog improves functional recovery in a rat traumatic brain injury model. *Journal of Neurotrauma, 20*(5), 463–476. https://doi.org/10.1089/089771503765355531.

Mast, N., Annalora, A. J., Lodowski, D. T., Palczewski, K., Stout, C. D., & Pikuleva, I. A. (2011). Structural basis for three-step sequential catalysis by the cholesterol side chain cleavage enzyme CYP11A1. *The Journal of Biological Chemistry, 286*(7), 5607–5613. https://doi.org/10.1074/jbc.M110.188433.

Mazat, L., Lafont, S., Berr, C., Debuire, B., Tessier, J. F., Dartigues, J. F., et al. (2001). Prospective measurements of dehydroepiandrosterone sulfate in a cohort of elderly subjects: Relationship to gender, subjective health, smoking habits, and 10-year mortality. *Proceedings of the National Academy of Sciences of the United States of America, 98*(14), 8145–8150. https://doi.org/10.1073/pnas.121177998.

McNamara, K. M., & Sasano, H. (2015). The intracrinology of breast cancer. *The Journal of Steroid Biochemistry and Molecular Biology, 145*, 172–178. https://doi.org/10.1016/j.jsbmb.2014.04.004.

Menzel, P., Gobbert, M., & Oertel, G. W. (1970). Effects of dehydroepiandrosterone and its conjugates upon the activity of glucose-6-phosphate dehydrogenase. V. Inhibition of glucose-6-phosphate dehydrogenase in human placenta. *Hormone and Metabolic Research, 2*(4), 225–227. https://doi.org/10.1055/s-0028-1095076.

Miller, W. L. (2005). Minireview: Regulation of steroidogenesis by electron transfer. *Endocrinology, 146*(6), 2544–2550. https://doi.org/10.1210/en.2005-0096.

Morales, A. J., Haubrich, R. H., Hwang, J. Y., Asakura, H., & Yen, S. S. (1998). The effect of six months treatment with a 100 mg daily dose of dehydroepiandrosterone (DHEA) on circulating sex steroids, body composition and muscle strength in age-advanced men and women. *Clinical Endocrinology, 49*(4), 421–432.

Morales, A. J., Nolan, J. J., Nelson, J. C., & Yen, S. S. (1994). Effects of replacement dose of dehydroepiandrosterone in men and women of advancing age. *The Journal of Clinical Endocrinology and Metabolism, 78*(6), 1360–1367. https://doi.org/10.1210/jcem.78.6.7515387.

Moran, C., Arriaga, M., Arechavaleta-Velasco, F., & Moran, S. (2015). Adrenal androgen excess and body mass index in polycystic ovary syndrome. *The Journal of Clinical Endocrinology and Metabolism, 100*(3), 942–950. https://doi.org/10.1210/jc.2014-2569.

Mueller, J. W., Gilligan, L. C., Idkowiak, J., Arlt, W., & Foster, P. A. (2015). The regulation of steroid action by sulfation and desulfation. *Endocrine Reviews, 36*(5), 526–563. https://doi.org/10.1210/er.2015-1036.

Munson, P. L., Gallagher, T. F., & Koch, F. C. (1944). Isolation of dehydroisoandrosterone sulfate from normal male urine. *The Journal of Biological Chemistry, 152*, 67–77.

Nagels, H. E., Rishworth, J. R., Siristatidis, C. S., & Kroon, B. (2015). Androgens (dehydroepiandrosterone or testosterone) for women undergoing assisted reproduction. *Cochrane Database of Systematic Reviews, 26*(11), CD009749. https://doi.org/10.1002/14651858.CD009749.pub2.

O'Reilly, M. W., House, P. J., & Tomlinson, J. W. (2014). Understanding androgen action in adipose tissue. *The Journal of Steroid Biochemistry and Molecular Biology, 143*, 277–284. https://doi.org/10.1016/j.jsbmb.2014.04.008.

Pashko, L. L., & Schwartz, A. G. (1983). Effect of food restriction, dehydroepiandrosterone, or obesity on the binding of 3H-7,12-dimethylbenz(a)anthracene to mouse skin DNA. *Journal of Gerontology, 38*(1), 8–12.

Pinola, P., Piltonen, T. T., Puurunen, J., Vanky, E., Sundstrom-Poromaa, I., Stener-Victorin, E., et al. (2015). Androgen profile through life in women with polycystic ovary syndrome: A nordic multicenter collaboration study. *The Journal of Clinical Endocrinology and Metabolism, 100*(9), 3400–3407. https://doi.org/10.1210/jc.2015-2123.

Prough, R. A., Webb, S. J., Wu, H. Q., Lapenson, D. P., & Waxman, D. J. (1994). Induction of microsomal and peroxisomal enzymes by dehydroepiandrosterone and its reduced metabolite in rats. *Cancer Research, 54*(11), 2878–2886.

Raynaud-Simon, A., Lafont, S., Berr, C., Dartigues, J. F., Baulieu, E. E., & Le Bouc, Y. (2001). Plasma insulin-like growth factor I levels in the elderly: Relation to plasma dehydroepiandrosterone sulfate levels, nutritional status, health and mortality. *Gerontology, 47*(4), 198–206. https://doi.org/10.1159/000052799.

Rege, J., Karashima, S., Lerario, A. M., Smith, J. M., Auchus, R. J., Kasa-Vubu, J. Z., et al. (2016). Age-dependent increases in adrenal cytochrome b5 and serum 5-androstenediol-3-sulfate. *The Journal of Clinical Endocrinology and Metabolism, 101*(12), 4585–4593. https://doi.org/10.1210/jc.2016-2864.

Robinzon, B., Michael, K. K., Ripp, S. L., Winters, S. J., & Prough, R. A. (2003). Glucocorticoids inhibit interconversion of 7-hydroxy and 7-oxo metabolites of dehydroepiandrosterone: A role for 11beta-hydroxysteroid dehydrogenases? *Archives of Biochemistry and Biophysics, 412*(2), 251–258.

Rossetti, M. F., Cambiasso, M. J., Holschbach, M. A., & Cabrera, R. (2016). Oestrogens and progestagens: Synthesis and action in the brain. *Journal of Neuroendocrinology, 28*(7). https://doi.org/10.1111/jne.12402.

Ruggiero, C., & Lalli, E. (2016). Impact of ACTH signaling on transcriptional regulation of steroidogenic genes. *Frontiers in Endocrinology (Lausanne)*, 7, 24. https://doi.org/10.3389/fendo.2016.00024.

Schwartz, A. G., & Cleary, M. P. (1987). The effect of dehydroepiandrosterone on body weight gain. *The Journal of Nutrition, 117*(2), 406–407.

Shantz, L. M., Talalay, P., & Gordon, G. B. (1989). Mechanism of inhibition of growth of 3T3-L1 fibroblasts and their differentiation to adipocytes by dehydroepiandrosterone and related steroids: Role of glucose-6-phosphate dehydrogenase. *Proceedings of the National Academy of Sciences of the United States of America, 86*(10), 3852–3856.

Simpson, E., & Santen, R. J. (2015). Celebrating 75 years of oestradiol. *Journal of Molecular Endocrinology, 55*(3), T1–20. https://doi.org/10.1530/JME-15-0128.

Stocco, D. M., Zhao, A. H., Tu, L. N., Morohaku, K., & Selvaraj, V. (2017). A brief history of the search for the protein(s) involved in the acute regulation of steroidogenesis. *Molecular and Cellular Endocrinology, 441*, 7–16. https://doi.org/10.1016/j.mce.2016.07.036.

Taniguchi, S., Yanase, T., Kobayashi, K., Takayanagi, R., & Nawata, H. (1996). Dehydroepiandrosterone markedly inhibits the accumulation of cholesteryl ester in mouse macrophage J774-1 cells. *Atherosclerosis, 126*(1), 143–154.

Teng, Y., Litchfield, L. M., Ivanova, M. M., Prough, R. A., Clark, B. J., & Klinge, C. M. (2014). Dehydroepiandrosterone-induces miR-21 transcription in HepG2 cells through estrogen receptor beta and androgen receptor. *Molecular and Cellular Endocrinology, 392*(1–2), 23–36. https://doi.org/10.1016/j.mce.2014.05.007.

Tsui, K. H., Lin, L. T., Chang, R., Huang, B. S., Cheng, J. T., & Wang, P. H. (2015). Effects of dehydroepiandrosterone supplementation on women with poor ovarian response: A preliminary report and review. *Taiwanese Journal of Obstetrics & Gynecology, 54*(2), 131–136. https://doi.org/10.1016/j.tjog.2014.07.007.

Turcu, A., Smith, J. M., Auchus, R., & Rainey, W. E. (2014). Adrenal androgens and androgen precursors-definition, synthesis, regulation and physiologic actions. *Comprehensive Physiology, 4*(4), 1369–1381. https://doi.org/10.1002/cphy.c140006.

Udhane, S. S., & Fluck, C. E. (2016). Regulation of human (adrenal) androgen biosynthesis-new insights from novel throughput technology studies. *Biochemical Pharmacology, 102*, 20–33. https://doi.org/10.1016/j.bcp.2015.10.010.

Van Vollenhoven, R. F., & McGuire, J. L. (1996). Studies of dehydroepiandrosterone (DHEA) as a therapeutic agent in systemic lupus erythematosus. *Annales de Medecine Interne (Paris), 147*(4), 290–296.

Wang, T., Cook, I., Falany, C. N., & Leyh, T. S. (2014). Paradigms of sulfotransferase catalysis: The mechanism of SULT2A1. *The Journal of Biological Chemistry, 289*(38), 26474–26480. https://doi.org/10.1074/jbc.M114.573501.

Wolkowitz, O. M., Reus, V. I., Roberts, E., Manfredi, F., Chan, T., Raum, W. J., et al. (1997). Dehydroepiandrosterone (DHEA) treatment of depression. *Biological Psychiatry, 41*(3), 311–318.

Wu, H. Q., Masset-Brown, J., Tweedie, D. J., Milewich, L., Frenkel, R. A., Martin-Wixtrom, C., et al. (1989). Induction of microsomal NADPH-cytochrome P-450 reductase and cytochrome P-450IVA1 (P-450LA omega) by dehydroepiandrosterone in rats: A possible peroxisomal proliferator. *Cancer Research*, *49*(9), 2337–2343.

Yen, S. S. (2001). Dehydroepiandrosterone sulfate and longevity: New clues for an old friend. *Proceedings of the National Academy of Sciences of the United States of America*, *98*(15), 8167–8169. https://doi.org/10.1073/pnas.161278698.

Zhou, Z. X., Lane, M. V., Kemppainen, J. A., French, F. S., & Wilson, E. M. (1995). Specificity of ligand-dependent androgen receptor stabilization: Receptor domain interactions influence ligand dissociation and receptor stability. *Molecular Endocrinology*, *9*(2), 208–218. https://doi.org/10.1210/mend.9.2.7776971.

Zwain, I. H., & Yen, S. S. (1999). Neurosteroidogenesis in astrocytes, oligodendrocytes, and neurons of cerebral cortex of rat brain. *Endocrinology*, *140*(8), 3843–3852. https://doi.org/10.1210/endo.140.8.6907.

CHAPTER TWO

Mechanisms of Action of Dehydroepiandrosterone

Barbara J. Clark, Russell A. Prough, Carolyn M. Klinge[1]
Department of Biochemistry and Molecular Genetics, Center for Genetics and Molecular Medicine, University of Louisville School of Medicine, Louisville, KY, United States
[1]Corresponding author: e-mail address: carolyn.klinge@louisville.edu

Contents

Abstract

Dehydroepiandrosterone (3β-hydroxy-5-androsten-17-one, DHEA) and its sulfated metabolite DHEA-S are the most abundant steroids in circulation and decline with age. Rodent studies have shown that DHEA has a wide variety of effects on liver,

Vitamins and Hormones, Volume 108
ISSN 0083-6729
https://doi.org/10.1016/bs.vh.2018.02.003

kidney, adipose, reproductive tissues, and central nervous system/neuronal function. The mechanisms by which DHEA and DHEA-S impart their physiological effects may be direct actions on plasma membrane receptors, including a DHEA-specific, G-protein-coupled receptor in endothelial cells; various neuroreceptors, e.g., aminobutyric-acid-type A, *N*-methyl-D-aspartate (NMDA), and sigma-1 (S1R) receptors; by binding steroid receptors: androgen and estrogen receptors (ARs, ERα, or ERβ); or by their metabolism to more potent sex steroid hormones, e.g., testosterone, dihydrotestosterone, and estradiol, which bind with higher affinity to ARs and ERs. DHEA inhibits voltage-gated T-type calcium channels. DHEA activates peroxisome proliferator-activated receptor (PPARα) and CAR by a mechanism apparently involving PP2A, a protein phosphatase dephosphorylating PPARα and CAR to activate their transcriptional activity. We review our recent study showing DHEA activated GPER1 (G-protein-coupled estrogen receptor 1) in HepG2 cells to stimulate miR-21 transcription. This chapter reviews some of the physiological, biochemical, and molecular mechanisms of DHEA and DHEA-S activity.

1. INTRODUCTION

Dehydroepiandrosterone (3β-hydroxy-5-androsten-17-one, DHEA) is synthesized in the *zona reticularis* of the adrenal cortex, in ovarian theca cells, in Leydig cells in the testes, and in brain. DHEA and its 3β-sulfated metabolite DHEA-S are abundant in circulation and decline with age (Labrie, 2010). DHEA is an established precursor for the sex steroids, estrogens, and androgens (Fig. 1). DHEA and DHEA-S account for ~75% of estrogens formed in premenopausal women and ~50% of androgens in men (Maggio et al., 2015). After menopause, extragonadal estrogens and androgens are synthesized from adrenal DHEA, DHEA-S, and androstenedione (Maggio et al., 2015). DHEA synthesis and metabolism are reviewed in chapter "Dehydroepiandrosterone Research: Past, current and future" by Klinge et al. In addition to serving as precursors for other steroids, DHEA and DHEA-S act as ligands for members of the nuclear receptor family of proteins and G-protein-coupled receptors (GPCRs) in a variety of tissues throughout the body. Surprisingly, a search in NURSA identified no transcriptome studies of DHEA (https://www.nursa.org/nursa/molecules/ligand.jsf?doi=10.1621/PNXJ6CHX7S) included in this website. This chapter summarizes the current state of knowledge of the molecular mechanisms by which DHEA acts in cells and the tissue-selective activities of DHEA observed in animal models.

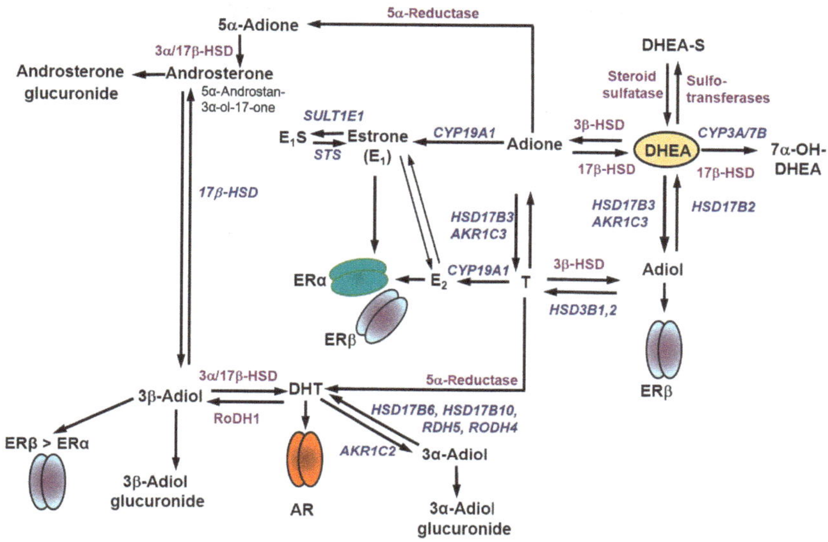

Fig. 1 DHEA metabolism to androgens and estrogens. The pathways show some of the enzymes involved in DHEA metabolism to estrogens and androgens. This pathway information is based on Labrie et al. (2005), Mostaghel (2014), and Prough, Clark, and Klinge (2016). Abbreviations: *3α-Adiol*, 5α-androstane-3α,17β-diol; *3β-Adiol*, 5α-androstane-3β,17β-diol; *Adiol*, 5-androsten-3β,17β-diol; *Adione*, 4-androstene-3β,17β-dione; *Androsterone*, 5α-androstan-3α-ol-17-one.

2. OVERVIEW OF STEROID AND NUCLEAR RECEPTORS AND THE TRANSCRIPTIONAL ACTIVITY

Steroid hormones, i.e., androgens, estrogens, glucocorticoids, mineralocorticoids, and progestins, bind with high affinity to their cognate receptors, the steroid hormone receptors (SRs). SRs are members of a family of nuclear receptors (NRs) that have a common gene structure, bind to DNA, and activate or inhibit gene transcription. SRs are considered Class I NRs, whereas NRs, including thyroid hormone receptor (TR), vitamin D3 receptor (VDR), retinoic acid receptor (RAR), lipid X receptor (LXR), and the peroxisome proliferator-activated receptors (PPARs), that heterodimerize with retinoid X receptor (RXR) to bind DNA are called Class II NRs (Olefsky, 2001). The evolutionary relationship between the SR and NR has been deduced from the conservation in their DNA-binding domains (DBDs) and in their less-conserved ligand-binding domains (LBDs) and it indicates that this large group of proteins arose from a common ancestral molecule (Thornton, 2001; Thornton, Need, & Crews, 2003). Classically, we think of ligand-activated SRs dissociating from cytoplasmic HSP90 chaperone complexes and entering the nucleus where they bind with high affinity to specific DNA sequences: hormone response elements (HREs). However, current models based on chromatin immunoprecipitation and RNA sequencing (ChIP-seq) have demonstrated that some proportion of unliganded SRs and almost all class II NR reside in the nucleus, bound to DNA (Caizzi et al., 2014; Miano et al., 2016; Vinckevicius & Chakravarti, 2012). Glucocorticoid receptor (GR), mineralocorticoid receptor (MR), progesterone receptor (PR), and androgen receptor (AR) bind the same DNA response element having a palindromic (symmetrical) sequence 5′-GGTACAnnnTGTTCT-3′, where n = any nucleotide (Beato, Chalepakis, Schauer, & Slater, 1989). AR, PR, GR, and MR show subtle differences in DNA base contact points to this HRE (Beato et al., 1989). Estrogen receptors (ERs, ERα and ERβ) bind a different DNA sequence called an estrogen response element (ERE): 5′-GGTCAnnnTGACC-3′ (Klein-Hitpass, Ryffel, Heitlinger, & Cato, 1988), and differ in only 2 bp from the GRE (Klock, Strahle, & Schutz, 1987). Extension of the length of the ERE palindrome, e.g., 5′-CAGGTCAnnnTGACCTG-3′, and the sequences immediately flanking the ERE helped determine ERα-binding affinity (reviewed in Klinge, 2001). Class II NRs pregnane X receptor (PXR) and constitutive androstane receptor (CAR) are termed "xenobiotic receptors" because of their promiscuous ligand-binding properties which include structurally related steroids and their metabolites (Banerjee, Robbins, & Chen, 2013). In addition to ligand binding, SRs and NRs are regulated by covalent modification including phosphorylation (Anbalagan & Rowan, 2015; Lannigan, 2003; Rajbhandari et al., 2012), methylation (Zhang et al., 2013), and acetylation (Abdel-Hafiz & Horwitz, 2014; Kim, Woo, Chong, Homenko, & Kraus, 2006). Ligand-activated SRs and NRs interact with coregulator proteins (coactivators and corepressors) as well as chromatin-remodeling complexes to regulate target gene expression (Dasgupta, Lonard, & O'Malley, 2014; Klinge, 2000; O'Malley, 2009; York et al., 2013) In addition, a small percentage of total SRs are also associated with the plasma membrane (PM) and PM proteins leading to activation of intracellular phosphorylation cascades mediated by MAPK and PI3K as well as other rapid signaling processes (reviewed in Banerjee, Chambliss, Mineo, & Shaul, 2014; Lang, Alevizopoulos, & Stournaras, 2013; Mueck, Ruan, Seeger, Fehm, & Neubauer, 2014; Tasker, Di, & Malcher-Lopes, 2006; Watson, Jeng, & Guptarak, 2011). This is called the "nongenomic" activity of these SRs, but this is somewhat of a misnomer since activation of these PM-associated SRs leads ultimately to changes in gene expression.

3. DHEA ACTION VIA NUCLEAR RECEPTOR BINDING AND ACTIVATION

DHEA binds Class I nuclear receptors NRs, i.e., the steroid hormone receptors, and some of the Class II NRs, i.e., those that dimerize with retinoid X receptor (RXR) (Evans & Mangelsdorf, 2014; Table 1; reviewed in Prough et al., 2016; Traish, Kang, Saad, & Guay, 2011; Webb et al., 2006). DHEA binds AR, estrogen receptors α and β (ERα and ERβ, *ESR1* and *ESR2*), pregnane X receptor/steroid, and xenobiotic receptor (PXR/SXR, *NR1I2*). The affinities for the DHEA–NR binding are much lower than those of cognate ligands for the receptors. Thus, it is important to consider the metabolism of DHEA to the androgens, e.g., testosterone (T) and dihydrotestosterone (DHT), with higher binding affinity for AR and estrogens, most notably estradiol (E_2), with higher affinity for ERα and ERβ as a confounder in addressing the molecular mechanisms of DHEA action on NRs.

Table 1 DHEA-Binding Affinities for Steroid (AR, ERα, and ERβ) and Class II (PXR/SXR and PPARα) Nuclear Receptors

AR	ERα	ERβ	PXR/SXR	PPARα
2 μM (Lu, Mo, Hu, Garippa, & Simon, 2003) 1 μM (Chen et al., 2005)	1.2 μM (Chen et al., 2005) 245 nM (Kuiper et al., 1997) 5 μM (Adams, Garcia, & Rochefort, 1981; Maggiolini, Donze, Jeannin, Ando, & Picard, 1999) >500 μM (Michael Miller et al., 2013)	500 nM (Chen et al., 2005) 163 nM (Kuiper et al., 1997) 500 μM (Michael Miller et al., 2013)	50–100 μM (Webb, Geoghegan, Prough, & Michael Miller, 2006)	No binding affinity determined (Altman, Motton, Kota, & Rutledge, 2008)

The K_d values in this table were included in our previous review (Prough et al., 2016).

4. REGULATION OF STEROID AND NUCLEAR RECEPTOR LEVELS BY DHEA

There is a significant clinical as well as research interest in DHEA as a potential therapy to prevent osteoporosis in postmenopausal women (Christianson & Shen, 2013). However, clinical trials show androgenic effects of DHEA administration (Scheffers, Armstrong, Cantineau, Farquhar, & Jordan, 2015). Long-term treatment (3 months) of ovariectomized (ovex) BALB/c mice with DHEA was reported to increase the mRNA transcripts of ERα and ERβ in the uterus, and ERα, ERβ, and AR transcripts in mouse osteoblasts (Wang, Wang, Wang, & Li, 2009). DHEA increased E_2 concentrations in bone but not in uterus, due to increased aromatase, but no mechanism was provided for the increase in ERα, ERβ, and AR transcript levels.

DHEA (1 mg/kg body weight for 6 weeks) was reported to increase the protein levels PPARα and PPARδ in the gastrocnemius muscle of obese male Wistar rats (Horii, Sato, Mesaki, & Iemitsu, 2016). In contrast, no change in PPARγ protein was detected, but transcript levels for 3β-hydroxyacyl-CoA dehydrogenase (*HADH*) and cytochrome *c* oxidase IV (*COXIV*), target genes of PPARα and PPARδ, respectively, were increased with DHEA. DHEA also increased the levels of the coactivator PGC-1α (*PPARGC1A*) in muscle of the obese rats and downstream specific gene targets. The authors suggested that DHEA may be converted to T in the muscle, but the underlying mechanism for the increase in PPARα, PPARδ, and PGC-1α was not examined (Horii et al., 2016).

5. DHEA ACTIVATION OF NUCLEAR ESTROGEN RECEPTORS

A classic way by which to demonstrate that a chemical is an estrogen is to transfect an ER-null cell with an expression vector for *ESR1* or *ESR2* (ERα or ERβ), a luciferase reporter gene containing one or more tandem EREs 5′ to the firefly luciferase gene, and a *Renilla* luciferase reporter as a transfection control. Then the transfected cells are treated with increasing concentrations of the chemical vs E_2 as a positive control and the vehicle, e.g., ethanol, as a negative control, and measure luciferase reporter activity. Transfection of GT1–7 mouse hypothalamic neuronal cells with ERα and an ERE-luciferase reporter demonstrated that DHEA (1 μM) and

10 nM E_2 resulted in equivalent luciferase activity, suggesting ~1000-fold lower affinity of DHEA for ERα (Bruder, Sobek, & Oettel, 1997). The authors did not detect metabolism of DHEA to E_2 in the GT1–7 cells. An alternative reporter assay involves using cells with endogenous ER. The term *ER* indicates that whether the cells expressed solely ERα or ERβ or both ER subtypes was not stated. DHEA (1 nM to 1 μM) activated ERE-driven luciferase activity in an ER-dependent manner in MCF-7 breast cancer cells and Ishikawa endometrial cancer cells (Maggiolini et al., 1999). The involvement of ER was demonstrated by inhibition of the DHEA response by cotreatment of the cells with the antiestrogen 4-hydroxytamoxifen (4-OHT), which competes with "estrogens" for the ligand-binding pocket of ERs and thus inhibits ER-transcriptional activity (Maggiolini et al., 1999). A caveat of this study is that the authors did not examine whether DHEA was metabolized to estrogens. This is a concern because MCF-7 cells express all enzymes required to metabolize DHEA to E_2 (Fig. 1; Wiebe & Lewis, 2003; Zhang, Wang, Chen, & Lin, 2015), including aromatase protein (Phuong, Lim, Kim, & Kang, 2014). More recently, DHEA-S and estrone sulfate (E1S) were reported to activate endogenous ERα activity in MVLN cells stably expressing ERE-luciferase reporter; however, this activity was inhibited, but not completely blocked by the steroid sulfatase (STS) inhibitor STX64, implicating DHEA and estrone as the active metabolites (Bjerregaard-Olesen, Ghisari, Kjeldsen, Wielsøe, & Bonefeld-Jørgensen, 2016; Fig. 1).

DHEA also activated ERα in a yeast reporter assay in which yeast were transfected with ERα and an ERE reporter (Bruder et al., 1997). This result suggests that mammalian coregulators are not needed for DHEA-induced ERα transcriptional activity. In a mammalian two-hybrid luciferase reporter assay in transiently transfected COS-1 cells, DHEA (5 μM) activated the LBD of ERβ, but not ERα (Chen et al., 2005). This result suggests that DHEA may activate ERα by a mechanism other than direct LBD binding and/or that cell-specific factors, whether enzymes to "activate" DHEA or the transcriptional response, e.g., coactivators, regulate whether DHEA activates ER action. For example, as indicated previously, DHEA increases PGC-1α in muscle (Horii et al., 2016) and PGC-1α is also an ER coactivator (Klinge, 2000). Another possibility is that DHEA may activate MAPK which phosphorylates and activates ERα in a ligand-independent manner, e.g., by phosphorylation of Ser-118 in the activation function 1 (AF-1) domain of ERα (Anbalagan & Rowan, 2015). This suggestion is supported by reports that DHEA activates a DHEA-specific PM receptor and activates intracellular MAPK signaling in endothelial cells (Liu & Dillon, 2004; Liu et al., 2008), as will be described in more detail later in this chapter.

6. DHEA ACTIVATION OF NUCLEAR ANDROGEN RECEPTOR

DHEA binds AR with an apparent K_d of 1–2 μM (Table 1). Early studies showed that AR activation by DHEA had antiestrogenic activity in ERα+ MCF-7 breast cancer cells since AR antagonist flutamide blocked DHEA's inhibition of E_2-stimulated cell growth (Boccuzzi et al., 1993). Interestingly, DHEA (0.5 μM) and the ER antagonist tamoxifen (10 μM) had an additive inhibitory effect on MCF-7 cell proliferation, but no mechanism was identified (Boccuzzi et al., 1993).

Prostate cancer is an androgen-dependent disease that is initially treated by androgen deprivation with good response, followed by the development of castration-resistant, AR-dependent disease progression (reviewed in Shanle et al., 2013). Castration-resistant tumors have increased expression of the enzymes involved in androgen synthesis and metabolism, e.g., AKR1C3, CYP17A1, CYP19A1, HSD3B1, HSD17B3, SRD5A1 (skin), SRD5A3 (bile acids), StAR (*STARD1*), and UGT2B17 and reduced levels of CYP3A4, CYP3A5, CYP3A7, and SRD5A2 (prostate) (reviewed in Shanle et al., 2013). Adrenal DHEA is converted to active androgens in the prostate, circumventing "castration" therapies (Labrie, 2011). It is possible that LBD mutations of AR, such as those detected in prostate tumors (Wadosky & Koochekpour, 2016), result in greater activation of mutant AR by DHEA. One study demonstrates that mutant AR H874Y showed higher transcriptional activation by DHEA than wild-type AR in CWR-R1 and 22Rv1 human prostate cancer cells (Tan et al., 1997). Other studies showed that AR T878A was also activated by DHEA (reviewed in Wadosky & Koochekpour, 2016). DHEA activation of ERβ predominates over AR activation in prostate and other tissues in which both NRs are expressed (reviewed in Arnold & Blackman, 2005). As indicated previously in this chapter, since the prostate has the enzyme machinery to be capable of DHEA metabolism to androgens that have high affinity for AR and activating it (Mostaghel, 2014), the precise mechanism of activation of AR is difficult to interrogate without specific metabolic inhibitors (reviewed in Arnold, 2009). DHEA-S and E1S were reported to activate AR activity in CHO-K1 cells cotransfected with an MMTV-luciferase reporter in which AR binds multiple response elements in the MMTV promoter (Bjerregaard-Olesen, Ghisari, Kjeldsen, Wielsøe, & Bonefeld-Jørgensen, 2016). The DHEA-S activity was blocked by the STS inhibitor STX64, implicating DHEA as the active AR ligand; however, E1S activity was not inhibited by STX64, suggesting that E1S may directly activate AR (Bjerregaard-Olesen et al., 2016).

7. DHEA ACTIVATION OF MINERALOCORTICOID RECEPTOR

Aldosterone binding to MR in the kidney regulates blood pressure and extracellular volume by controlling renal potassium excretion and sodium reabsorption (Fuller & Young, 2005; Thomas & Harvey, 2011). MR is also expressed in endothelial cells, atrial cardiomyocytes, subcutaneous adipocytes, vascular smooth muscle cells, epidermis, and hair follicle from skin (reviewed in Jaisser & Farman, 2016); in the brain, e.g., hippocampus (reviewed in Talabér, Jondal, & Okret, 2013); and in macrophages (reviewed in Fuller, 2015). Aberrant MR activity has been implicated in a variety of diseases (reviewed in Jaisser & Farman, 2016).

One report showed that DHEA-mediated signaling in human umbilical vein endothelial cells (HUVECs), human skin fibroblasts, and either human or rat aortic vascular smooth muscle cells (VSMCs) was mediated by MR activation (Lindschau et al., 2011). DHEA (100 nM) rapidly increased activation of ERK1/2 and phosphorylation and nuclear localization of FOXO1 in rat VSMCs. Transfection of the cells with MR siRNA or treatment with MR antagonist eplerenone blocked DHEA-induced ERK1/2 and FOXO1 phosphorylation and activation. In contrast, DHEA did not activate MR-luciferase reporter activity in transiently transfected CHO-K1 cells, indicating that DHEA is not a direct ligand of MR (Lindschau et al., 2011). While there appear to be no further studies on DHEA activation of MR, DHEA-MR cross talk may be mediated by other membrane receptors, e.g., EGFR (Meinel, Gekle, & Grossmann, 2014).

DHEA (100 μM) was reported to inhibit the chronotropic and the hypertrophic responses of isolated rat cardiomyocytes induced by stimulation with 10 nM aldosterone (to activate MR specifically) to 1 μM dexamethasone (to activate GR specifically) in vitro (Mannic et al., 2013).

DHEA rapidly inhibited the chronotropic response to aldosterone by a decrease in T-type calcium channel ($Ca_V3.2/\alpha 1H$) expression and activity, as assessed by quantitative PCR and the patch-clamp technique. In addition, DHEA had an antioxidant effect as examined using a fluorescent cell redox probe, dichlorofluorescein, although the mechanisms for these DHEA effects are unknown (Mannic et al., 2013; Mannic, Viguie, & Rossier, 2015).

8. DHEA INHIBITION OF GLUCOCORTICOID RECEPTOR ACTIVITY

Glucocorticoids are obesogenic through effects in tissues including adipocytes, liver, and the central nervous system (CNS) (Dallman, Warne, Foster, & Pecoraro, 2007; Lee, Pramyothin, Karastergiou, & Fried, 2014; Peckett, Wright, & Riddell, 2011; Razzoli, Pearson, Crow, & Bartolomucci, 2017). DHEA inhibits glucocorticoid-induced increases in tyrosine aminotransferase (*TAT*), ornithine decarboxylase 1 (*ODC1*), and interleukin 4 (*IL4*) gene expression and glucocorticoid-induced inhibition of *IL2* in vivo (reviewed in Kalimi, Shafagoj, Loria, Padgett, & Regelson, 1994), but no specific mechanisms accounting for this activity have been reported (reviewed in Arlt, 2004). DHEA (5 μM) was reported to protect mouse hippocampal cell line HT-22 from the toxicity induced by glutamate (5 mM) by decreasing nuclear GR levels (Cardounel, Regelson, & Kalimi, 1999). DHEA, but not Adione, Adiol, or T, inhibited proliferation of human PAZ6 preadipocytes, a cell line derived from human brown adipose tissue (BAT), and adipogenesis in omental but not subcutaneous preadipocytes (Rice et al., 2010). More recent studies using telomerase-immortalized Chub-S7 human subcutaneous adipocytes showed that OATP-D, an influx transporter of DHEA-S, and STS were expressed in Chub-S7 in which DHEA was converted to androstenediol, but not to DHEA-S (McNelis et al., 2013). DHEA was shown to have antiglucocorticoid activity in the Chub-D7 cell model of differentiation from preadipocytes to mature adipocytes, whereas DHEA-S had no effect and DHEA's effects were not blocked by AR antagonist flutamide or by ER antagonist Faslodex. DHEA inhibited the expression and activity of 11β-HSD1, thus blocking preadipocyte differentiation and GR antagonist RU-486 had an additive inhibitory effect (McNelis et al., 2013). No mechanism for DHEA's inhibition of 11β-HSD1 gene expression was uncovered.

9. DHEA ACTIVATION OF PPARs

DHEA and other sterols regulate the activity of various hepatic NRs including constitutive androstane receptor (CAR), pregnane X receptor (PXR), and PPARα that regulate the transcription of foreign compound metabolizing enzymes and the CYP genes which are regulated by sterols like DHEA, as reviewed first by Webb et al. (2006) and later by us (Prough et al., 2016).

PPARs, i.e., PPARα (*NR1C1*), PPARβ/δ (*NR1C3*), and PPARγ (*NR1C2*), form heterodimers with RXR and are bound to their response elements (5′-AGGCAnAGGCTA-3′, a direct repeat separated by one nucleotide "n," any nucleotide, termed DR1) in the absence of ligand complexed with corepressors (Berger & Moller, 2002). Binding of PPAR ligands including fatty acids and eicosanoids leads to dismissal of the compressor complex and association of coactivator complexes leading to chromatin remodeling and activation of target gene transcription (Feige, Gelman, Michalik, Desvergne, & Wahli, 2006). PPARα is expressed in heart, skeletal muscle, liver, pancreas, BAT, and endothelial cells where it regulates genes involved in fatty acid β-oxidation, gluconeogenesis, and inflammatory responses (Lefebvre, Chinetti, Fruchart, & Staels, 2006). PPARα is considered a "nutritional sensor" to modulate FA catabolism, lipogenesis, and ketone body synthesis (Pawlak, Lefebvre, & Staels, 2015). PPARβ/δ, referred to as PPARδ, is expressed in skeletal muscle, adipose tissue, skin, and liver, but plays a major role in muscle by regulating mitochondrial metabolism and in hepatocytes, hepatic stellate cells, and Kupffer cells to modulate nonalcoholic steatohepatitis (NASH) (Shan et al., 2008) and nonalcoholic fatty liver disease (NAFLD) (Liss & Finck, 2017). PPARγ is highly expressed in adipose tissue and macrophages and has low expression in liver where it plays a role in NASH and NAFLD (Tanaka, Aoyama, Kimura, & Gonzalez, 2017). Selective PPARα modulators (SPPARMs) in combination with PPARγ agonists may be beneficial in humans with liver disease including NASH and NAFLD (Pawlak et al., 2015). PPARγ activation improves insulin sensitivity and glucose homeostasis and attenuates inflammation (reviewed in Marion-Letellier, Savoye, & Ghosh, 2016).

High doses of DHEA-S increase the number and size of peroxisomes in the liver and kidney of rodents (reviewed in Peters et al., 1996). Notably, DHEA is inactive and must be sulfated to DHEA-S. The transcription of

CYP4A was increased by DHEA-S and this response is lost in PPARα knockout mice, implicating DHEA-S activation of PPARα (Lee et al., 1995). However, studies in the Prough lab found that neither DHEA, DHEA-S, nor any of the DHEA metabolites tested increased PPARα-dependent luciferase reporter activity in human HepG2 hepatoma cells, while nafenopin, a peroxisome proliferator, increased activity more than 40-fold (Webb et al., 2006). DHEA and its metabolites (Adiol, DHEA-S, 7α-hydroxy-DHEA, 7-oxo-DHEA, and 7-oxo-Adiol) induced *Cyp4a1* mRNA expression in primary rat hepatocytes. Thus, the authors concluded induction of *Cyp4a1* by DHEA and DHEA-S, and its metabolite requires a transcription factor or modulator in addition to PPARα that is present in the primary rat hepatocytes, but not HepG2 cells (Webb et al., 2006).

10. DHEA STIMULATES PROTEIN PHOSPHATASE 2 CATALYTIC SUBUNIT ALPHA TRANSCRIPTION

Subsequent studies in primary rat hepatocytes showed that DHEA and the PPARα agonist WY14643 increased PPARα mRNA and protein (two- to fourfold) while decreasing PPARα protein phosphorylation of serine 6, 12, and 21, suggesting that these peroxisome-proliferating agents decrease PPARα phosphorylation which increases transcriptional activity (Tamasi et al., 2008). Further investigation employing site-directed mutagenesis to convert those serine residues to alanine resulted in a >60% increase in PPARα transcriptional activity, supporting the idea that dephosphorylation of PPARα stimulates its transcriptional activity (Tamasi et al., 2008). Similarly, other PPARα agonists nafenopin and WY-14643, and DHEA decreased the phosphorylation of PPARα on serine 6 and 21 in primary rat hepatocytes, while treatment with Okadaic acid, a potent inhibitor of PP2A, a multisubunit protein phosphatase, caused PPARα to become fully phosphorylated at those residues and this inhibited DHEA-stimulated PPARα-luciferase activity. DHEA caused an eightfold increase in protein phosphatase 2 catalytic subunit alpha (*PPP2CA*) (the catalytic subunit of PP2A) mRNA in primary rat hepatocytes (Tamasi et al., 2008). These studies demonstrated that DHEA transcriptionally upregulates *PPP2CA* and implicates protein kinases in regulating PPARα transcriptional through direct protein phosphorylation in primary rat hepatocytes and HepG2 cells (Fig. 2).

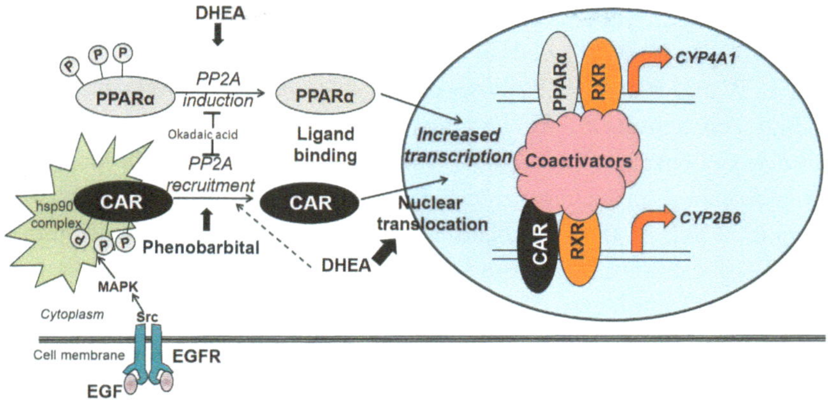

Fig. 2 DHEA activates PPARα by dephosphorylation. Phenobarbital was reported to stimulate PP2A recruitment to the cytoplasmic CAR:HSP90 complex where it dephosphorylated CAR to induce nuclear translocation (Kobayashi, Hashimoto, Honkakoski, & Negishi, 2015; Mutoh et al., 2013; Yoshinari, Kobayashi, Moore, Kawamoto, & Negishi, 2003). Likewise, DHEA induces PP2A that dephosphorylates PPARα which appears to enhance its ligand-activated transcription (Tamasi et al., 2008). Since DHEA stimulates CAR's nuclear translocation (Kohalmy et al., 2007), we suggested a role for PP2A in dephosphorylating HSP90-associated, cytoplasmic PPARα and CAR, leading to their nuclear localization and transactivation (Prough et al., 2016). *Solid arrows* = reported facts; *dashed arrow* = speculation based on experimental evidence. *Modified from fig. 2 in Prough, R. A., Clark, B. J., & Klinge, C. M. (2016). Novel mechanisms for DHEA action.* Journal of Molecular Endocrinology, *56(3), R139–R155. https://doi.org/10.1530/jme-16-0013.*

11. DHEA ACTIVATES CAR

Human CAR (*NR1I3*) is highly expressed in the liver and intestine, whereas it has lower expression in the heart, muscle, kidneys, and lung (Tanaka et al., 2017). CAR uses RXR as its heterodimer partner and its activity is inhibited by binding of two androstane metabolites, i.e., 5α-androst-16-en-3α-ol (androstenol) and 5α-androstan-3α-ol (androstanol), acting as inverse agonists (Forman et al., 1998). Endogenous CAR ligands include bilirubin and bile acids. CAR is a xenobiotic sensor that regulates the transcription of genes encoding xenobiotics-metabolizing enzymes and drug transporters, e.g., sulfotransferase, glucuronosyltransferase, and ATP-binding cassette subfamily B member 1 (*ABCB1*) (reviewed in Tanaka et al., 2017). CAR binds two DNA response elements: (1) the phenobarbital-responsive enhancer module (PBREM) and (2) xenobiotic-responsive enhancer module (XREM) that are located in the promoter regions of

CAR target genes, e.g., *CYP2B6*, and have DR4 elements +/− half-sites (5′-AGGTCA-3′) (reviewed in Chai, Cherian, Wang, & Chen, 2016; Yang & Wang, 2014). CAR shows remarkable species selectivity in its ligand binding and activation profiles, e.g., E_2 and 3,3′,5,5′-tetrachloro-1,4-bis (pyridyloxy)benzene (TCPOBOP)-activated mouse, but not human CAR, and androstanol, progesterone, and T repress mouse, but not human CAR (Yang & Wang, 2014). Phenobarbital activates human CAR by stimulating its nuclear localization in response to dephosphorylation of CAR at threonine 38 (Thr-38) (Mutoh et al., 2009, 2013; Fig. 2).

Early studies reported that mice consuming a diet containing DHEA or DHEA-S (0.4% (w/v) for 15 days) had hepatomegaly, peroxisome proliferation, increased *Cyp2b10* transcript expression (Wu et al., 1989), and activation of liver CARβ as assessed by DNA-binding activity in vitro in an electrophoretic mobility shift assay (Fujita et al., 2002). However, only nuclear localization of RXR and not CARβ was detected after DHEA feeding. Studies from the Prough lab in collaboration with Urs Meyer and Katalin Monostory documented that DHEA, but neither 7α-hydroxy-DHEA nor 7-oxo-DHEA, activates CAR's transcriptional activity in primary human hepatocytes (Kohalmy et al., 2007). They reported that the CAR antagonist androstanol ablated the DHEA stimulation of *CYP2B6* transcription. They reported that the mechanism by which DHEA activated CAR was, like phenobarbital or (TCPOBOP in mice), dephosphorylation of CAR at Thr-38 by protein phosphatase 2A (PP2A) (Kohalmy et al., 2007). Phosphorylated CAR is localized in a cytoplasmic HSP90 complex that is disrupted by PP2A-dependent dephosphorylation of Thr-38 resulting in protein conformational changes exposing the CAR's nuclear translocation signal (Mutoh et al., 2009; Sueyoshi, Moore, Sugatani, Matsumura, & Negishi, 2008). Further studies from Negishi's lab demonstrated that phenobarbital blocks epidermal growth factor (EGF) binding to the EGF receptor (EGFR) (Mutoh et al., 2013). Activated EGFR in turn activates its downstream signaling cascades Src kinase and ERK that keep CAR Thr-38 phosphorylated and sequestered in the HSP90 cytoplasmic protein complex (Kobayashi et al., 2015). In addition, EGF stimulates CAR homodimerization, constraining CAR in its inactive form (Shizu et al., 2017). De novo synthesis increases PP2A, thus dephosphorylating CAR Thr-38 for nuclear localization and target gene regulation (Fig. 2). We suggested that PP2A plays a common role for both PPARα (Tamasi et al., 2008) and CAR activation (Mutoh et al., 2013), which we speculated may also stimulated by DHEA to activate CAR (Prough et al., 2016; Fig. 2).

12. DHEA ACTIVATION OF PXR

PXR (*NR1I2*) is a Class I NR that forms a heterodimer with RXR and is present in the intestine, colon, kidney, heart, spinal cord, peripheral mononuclear blood cells, brain, stomach, prostrate, breast, ovary, hepatocytes, stellate cells, and Kupffer cells (Tanaka et al., 2017). Like CAR, PXR is a xenobiotic receptor. CAR and PXR share common ligands and gene targets, e.g., phenobarbital and genes including *ABCB1*, glutathione *S*-transferases (*GSTs*), UDP-glucuronosyltransferases (UGTs), and sulfotransferases (SULTs) (Banerjee, Robbins, & Chen, 2015). PXR binds response elements with DR3 and DR4 sequences and everted repeats separated by 6 or 7 bp, suggesting flexibility in the DBD, and activates a number of genes including CYP3A4, stearoyl-CoA desaturase 1 (*SCD*), FA elongase (*ELOV6*), the FA transporter *CD36*, and transcription factors PPARα and SREBP1 (Tanaka et al., 2017). Human PXR binds pregnane-related steroids, rifampicin, and SR12813 and is antagonized by sulforaphane, coumestrol, ketoconazole, metformin, and camptothecin (Chai et al., 2016).

Dietary DHEA induced *Cyp3a23* in rat liver and, since DHEA is a pregnane derivative, we postulated that DHEA and its metabolites may act as PXR ligands (Ripp, Fitzpatrick, Peters, & Prough, 2002). DHEA feeding of PPARα-null mice did not induce *Cyp4a1*, but *Cyp3a11* was increased at both the mRNA and protein activity levels in both wild-type and PPARα-null mice (Ripp et al., 2002). DHEA did not induce *Cyp3a11* in PXR-null mice, suggesting that DHEA and/or its metabolites increase *Cyp3a11* via PXR activation (S. J. Webb and R. A. Prough, unpublished results). DHEA, Adiol, and Adione activated PXR in HepG2 cells as indicated in a luciferase reporter assay in which two copies of the PXR-responsive element (PXRE) from the rat *Cyp3a23* gene were cloned into the promoter (Ripp et al., 2002). However, low μM concentrations of oxidized DHEA metabolites did not stimulate PXRE-driven luciferase activity (Ripp et al., 2002). Like CAR, human and murine PXRs show ligand selectivity (Chai et al., 2016). Pregnenolone 16α-carbonitrile activates mouse CAR, but not human CAR (Kliewer et al., 1998). The Prough lab reported DHEA metabolite-specific differences between mouse and human PXR. For example, DHEA and Adione were good activators of human PXR (Ripp et al., 2002). In contrast, mouse PXR showed higher selectivity for Adiol than for DHEA and Adione. These results reflect differences in the active sites of the human and mouse PXR (Chai et al., 2016; Jones et al., 2000).

13. DHEA INCREASES IGF-I-RECEPTOR

DHEA is considered to have antiosteoporosis effects and increases bone mineral density in women and men (Christianson & Shen, 2013; Traish et al., 2011; Weiss et al., 2009). Insulin-like growth factor 1 (IGF-I) is essential for growth of the skeleton and maintenance of the bone mass (Crane & Cao, 2014). Treatment of primary human mesenchymal stem cells (hMSCs) with 10 nM DHEA stimulated osteoblast differentiation and inhibited IL-6 secretion, an osteolytic cytokine, by increasing IGF-I gene transcription and protein (Liang et al., 2016). Use of experimental inhibitors suggested that DHEA's mechanism in the hMSCs involved IGF-I receptor, P38 MAPK, ERK1/2, and PI3K, but not CREB or PKA (Liang et al., 2016). However, the investigators did not examine whether DHEA binds IGF-I-R. No follow-up studies appear in PubMed.

14. GPCRs ARE ACTIVATED BY DHEA AND DHEA-S

A number of PM receptors have been reported to bind and be activated by DHEA in a cell-specific manner (Table 2 and Fig. 3). Dillon's laboratory first reported DHEA's activation of a GPCR in 2002 (Liu & Dillon, 2002). This group has demonstrated that DHEA binds a DHEA-specific GPCR in caveolae in the PM of vascular endothelial cells (ECs) with a $K_d \sim 49$ pM and this interaction results in the activation of MAPK and eNOS (Liu & Dillon, 2002, 2004; Liu et al., 2008, 2010; Olivo et al., 2010; Simoncini et al., 2003). In addition, they demonstrated that a membrane-impermeable DHEA-BSA conjugate (1 nM) activated MAPK and induced EC proliferation, suggesting that DHEA activation of the DHEA-GPCR does not require DHEA internalization. Further, both DHEA and the DHEA-BSA conjugate (1 nM, each) induced H_2O_2 production in a Gi/o protein-dependent manner (Iruthayanathan, O'Leary, Paul, & Dillon, 2011). Using a biotinylated DHEA conjugate, the Dillon group identified DHEA-binding proteins of 55, 80, and 150 kDa in the PM of bovine aortic ECs (BAECs) (Liu et al., 2010; Olivo et al., 2010). Other investigators demonstrated that neither PM-associated ERα nor ERβ or AR in BAECs is involved in DHEA (10 nM) activation of MAPK and downstream signaling (Williams et al., 2004).

DHEA-S (1 μM) rapidly (5–15 min) induced the release of granule-associated β-HEX from RBL-2H3 cells (Mizota et al., 2005). E_2 and

Table 2 DHEA and DHEA-S Activation of Plasma Membrane (PM) Receptors

PM Receptor	Cell/Tissue	Affinity/Concentration	Activity/Comments	References
DHEA-S-specific PM "membrane components"	Human cervical fibroblasts	$K_d = 12\,nM$ $B_{max} = 1.25\,fmol/mg$ protein	[^{3}H]DHEA-S, but not DHEA or E1-S; maximal uptake at 22°C within 2 min. No conversion to DHEA or estrogens. No binding activity with skin fibroblasts	Imai, Ohno, and Tamaya (1992)
G-protein-coupled Galpha i2,3-coupled	Bovine aortic endothelial cells (BAEC); human umbilical vein endothelial cells (HUVEC)	$K_d = 48.7\,pM$ $B_{max} = 500\,fmol/mg$ protein; [^{3}H]DHEA-binding sites were not competed with T, E_2, or androstenedione	10 nM DHEA activated eNOS in 15 min – No effect on intracellular Ca^{2+} fluxes – Activated MAPK – Lower binding in heart and liver, very low in kidney	Liu and Dillon (2002, 2004)
G-protein-coupled $G\alpha_{q/11}$ receptor	RBL-2H3 mast cells	Activity blocked by a PLC inhibitor, an antagonist to InsP3 receptor	DHEA-S (1 μM)-induced release of granule-associated β-HEX from RBL-2H3 cells	Mizota, Yoshida, Uchida, Fujita, and Ueda (2005)
G-protein-coupled $G\alpha_{q/11}$ receptor	Mouse spermatogenic GC-2 cells		DHEA-S (1 nM)-activated phosphorylation of ERK1/2, c-Src, ATF-1, and CREB	Shihan, Kirch, and Scheiner-Bobis (2013)
G-protein coupled	Human carotid vascular smooth muscle cell (hCASMC)		DHEA inhibited activation of the AKT/GSK-3β axis	Bonnet et al. (2009)
GPER/GPR30		7β-Hydroxy-epiandrosterone	DHEA-stimulated SKBR3 and MCF-7 cell proliferation	Niro, Pereira, Pélissier, Morfin, and Hennebert (2012)

Continued

Table 2 DHEA and DHEA-S Activation of Plasma Membrane (PM) Receptors—cont'd

PM Receptor	Cell/Tissue	Affinity/Concentration	Activity/Comments	References
GPER (*GPER1*)	HepG2 cells	10 nM DHEA	Stimulated EGFR, Src, ER1/2, PI3K signaling; miR-21 transcription	Teng et al. (2015)
NMDA receptor	Rat hippocampal slices Mouse embryonic neocortical neurons Primary human neural stem cells	50 nM to 1 μM DHEA 1 pM	Increased intracellular Ca^{2+}	Monnet, Mahe, Robel, and Baulieu (1995), Compagnone and Mellon (1998), and Suzuki, Wright, Marwah, Lardy, and Svendsen (2004)
Voltage-gated T-type Ca^{2+} channels: $Ca_V3.1$, $Ca_V3.2$, or $Ca_V3.3$	Transfected NG108-15 cells—somatic mouse neuroblastoma/rat gliomablastoma cell line	10 μM DHEA	Inhibited peak current	Chevalier et al. (2012)
Sigma 1 (σ1) receptor (S1R)	C6 rat glioma cells PC12 rat adrenal pheochromocytoma cells	DHEA $K_d > 50\,\mu M$	PI3K/AKT, Src, PKA, PKC, and MAPK pathways	Moriguchi et al. (2013) and Rybczynska et al. (2009)
TrkA and $p75^{NTR}$ membrane receptors of neurotrophin NGF family	PC12 cells	DHEA K_d: 7.4 ± 1.75 nM and 5.6 ± 0.55 nM for TrkA and $p75^{NTR}$ receptors, respectively, in HEK-293 cells	Induced TrkA phosphorylation and NGF receptor-mediated signaling; Shc, Akt, and ERK1/2 kinases downstream to TrkA receptors and TRAF6, RIP2, and RhoGDI interactors of $p75^{NTR}$ receptors	Anagnostopoulou et al. (2013), Gravanis et al. (2012), Lazaridis et al. (2011), and Pediaditakis et al. (2015)

$GABA_A$ receptors	HEK-293 cells transfected with $GABA_A$ R		DHEA blocked activity without affecting $GABA_A$ agonist binding	Svob Strac, Jazvinscak, Erhardt, Mirkovic Kos, and Pericic, (2012)
IGF-I-R	Human mesenchymal stem cells	10 nM DHEA	IGF-I-receptor, PI3K, P38 MAPK, and ERK1/2 MAPK, but not PKA or CREB	Liang et al. (2016)
Dendritic brain microtubule-associated protein MAP2C	In vitro experiments with MAP2C expressed in *E. coli* and purified; methods used: circular dichroism and isothermal titration calorimetry	DHEA $K_d = 27\ \mu M$	Promotes tubulin polymerization to stabilize microtubules	Laurine et al. (2003)

Most of the information in this table was included in our previous review (Prough et al., 2016); however, new information has been added.

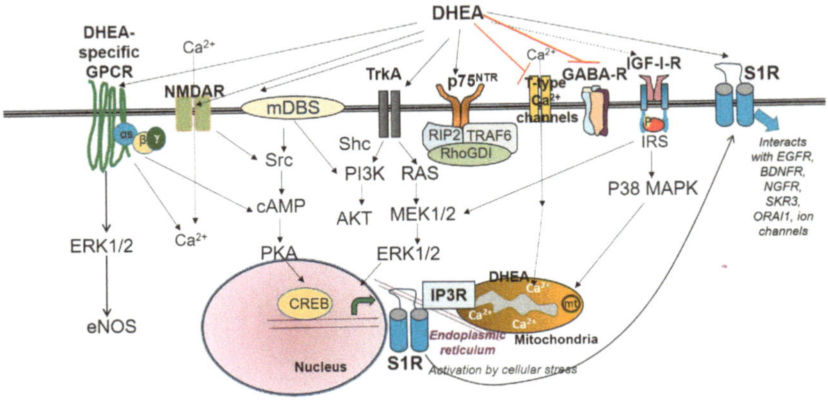

Fig. 3 Summary of DHEA activation or inhibition of plasma membrane receptors. This model summarizes DHEA interaction with plasma membrane receptors (Table 2) as described in the text. S1R is an interface between the endoplasmic reticulum and mitochondria (referred to as mitochondrial-associated endoplasmic reticulum membrane) (Bernard-Marissal, Medard, Azzedine, & Chrast, 2015; Cheng et al., 2008; Lewis, Tsai, & Su, 2016). Inositol-1,4,5-trisphosphate (IP3) receptor (IP3R) modulates Ca^{2+} homeostasis (Ruscher & Wieloch, 2015). As described in the text, others have demonstrated that activation of endoplasmic reticulum stress activates S1R which is inserted in the endoplasmic reticulum membranes, thus triggering its movement to the PM (Chu & Ruoho, 2015). This model is not specific to any one cell type; rather, it depicts many DHEA receptors described in different cells. *Modified from fig. 3 in Prough, R. A., Clark, B. J., & Klinge, C. M. (2016). Novel mechanisms for DHEA action.* Journal of Molecular Endocrinology, *56(3), R139–R155. https://doi.org/10.1530/jme-16-0013.*

membrane-impermeable progesterone-BSA (each at 10 μM) blocked DHEA-S activity, but neither fulvestrant (an ER antagonist) nor RU486 (an antiprogestin) blocked this effect, suggesting that neither PM-associated ER nor PR was involved in the DHEA-S effect. Likewise, Wortmannin (a PI3K inhibitor) and pertussis toxin (PTX, a G-protein inhibitor) failed to block DHEA-stimulated granule-associated β-HEX release. Rather, DHEA-S activity was blocked by inhibitors of phospholipase C and inositol trisphosphate (InsP3) receptor and by pretreatment with an antisense oligomer for Gq/11. The authors concluded that DHEA-S interacts with a Gq/11 protein-coupled membrane receptor (Mizota et al., 2005; Fig. 4).

PM-associated G-protein Gnα11 (the equivalent of Gαq/11) was implicated in rapid (30–180 min) DHEA-S (1 nM) activation of an Src/RasRaf/ERK1/2 signaling cascade in mouse spermatogenic GC-2 cells which lack STS, suggesting a direct rapid effect of DHEA-S at the PM rather than DHEA or a DHEA metabolite (Shihan et al., 2013; Fig. 4). AR knockdown

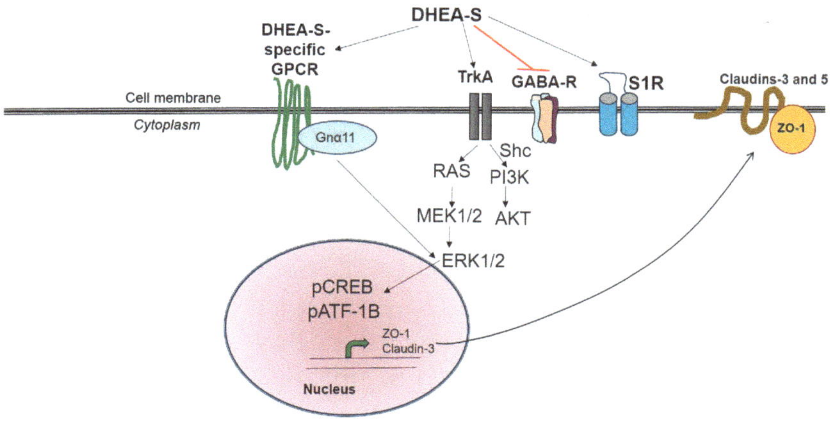

Fig. 4 Summary of DHEA-S activation or inhibition of plasma membrane receptors. This model summarizes DHEA-S interaction with plasma membrane receptors as described in the text. DHEA-S activates a GPCR–Gnα11 signaling pathway in GC-2 (Shihan et al., 2013; Shihan, Chan, Konrad, & Scheiner-Bobis, 2015) and TM4 (Papadopoulos, Dietze, Shihan, Kirch, & Scheiner-Bobis, 2016) cells resulting in activation of ERK1/2 signaling and phosphorylation of CREB and ATF-1B. Work from Scheiner-Bobis' group has also demonstrated that treatment of bEnd3 cells with 1 μM DHEA-S for 48 h activated a GPCR–Gnα11 signaling pathway resulting in increased transcription of claudin-3, claudin-5, and zonula occludens-1 (ZO-1) proteins and increased transendothelial resistance across cell monolayers, as depicted here based on the graphical abstract in Papadopoulos and Scheiner-Bobis (2017). As in Fig. 3, this model is not specific to any one cell type; rather, it depicts many DHEA receptors described in different cells. *Modified from fig. 3 in Prough, R. A., Clark, B. J., & Klinge, C. M. (2016). Novel mechanisms for DHEA action.* Journal of Molecular Endocrinology, *56(3), R139–R155. https://doi.org/10.1530/jme-16-0013.*

did not block DHEA-S-stimulated ERK1/2 activation, but silencing of Gnα11 (the equivalent of Gαq/11) inhibited DHEA-S-induced phosphorylation of ERK1/2, c-Src, ATF-1, and CREB (Shihan et al., 2013). This group of investigators reported that T binds ZIP9, a Zn^{2+} transporter, which acts as a membrane-bound AR that directly interacts with Gnα11 in GC-2 cells leading to activation of ERK1/2, CREB, and ATF-1 (Shihan et al., 2015). Whether DHEA-S or DHEA activates ZIP9 in GC-2 or other cells has not been tested.

New data show that DHEA-S (1 μM for 120 min), but not DHEA, activated ERK1/2, CREB, and ATF-1 phosphorylation in the mouse Sertoli cell line TM4 (Papadopoulos et al., 2016; Fig. 4). These DHEA-S-stimulated phosphorylation events were not inhibited by the STS inhibitor STX64, suggesting that DHEA-S does not need to be desulfated to activate this signaling pathway. Abrogation of AR expression by siAR transfection did not block DHEA-S-stimulated ERK1/2 activation nor did siAR block longer term DHEA-S (48 h) stimulation of claudin-3 and claudin-5 expression. Transfection of TM4 cells with siRNA for Gnα11 reduced Gnα11 but not Gnαq and abrogated the DHEA-S-stimulated phosphorylation of ERK1/2, CREB, and ATF-1, implicating an unidentified GPCR linked to Gnα11 in the DHEA-S response in TM4 cells (Papadopoulos et al., 2016), similar to the earlier report in GC-2 cells (Shihan et al., 2015).

DHEA easily crosses the brain–blood barrier and is also produced locally in the brain where both DHEA and DHEA-S have neuromodulatory and neuroprotective effects (reviewed in Maggio et al., 2015; Stárka, Dušková, & Hill, 2015). DHEA-S (1 μM for 48 h) was recently reported to stimulate the expression of zonula occludens-1 (ZO-1) and claudin-3 and claudin-5 proteins that are critical components of tight junctions (TJ) of the blood–brain barrier (BBB), in mouse brain endothelial cell line bEnd.3 (Papadopoulos & Scheiner-Bobis, 2017; Fig. 4). These increases in TJ proteins were not blocked when the bEnd.3 cells were transfected with siAR, but were inhibited by silencing Gnα11 expression. Incubation of the bEnd.3 cells with T had no effect. The authors concluded that a membrane DHEA-S-specific GPCR coupled with Gnα11 mediates the increase in TJ proteins in bEnd.3 (Papadopoulos & Scheiner-Bobis, 2017), similar to their earlier reports of DHEA-S activation of Gnα11 in GC-2 (Shihan et al., 2015) and TM4 cells (Papadopoulos et al., 2016).

These reports demonstrate that DHEA and DHEA-S activate Gαq/11 signaling in EC, BAEC, GC-2, TM4, and bEnd.3 cells, suggesting a common GPCR-activated signaling pathway, although the GPCR that binds DHEA and/or DHEA-S may not be identical between different cell types. This remains to be determined.

DHEA has also been reported to inhibit PM-initiated cell signaling. DHEA and BSA-conjugated DHEA inhibited the proliferation of human carotid vascular smooth muscle cell (hCASMC) by blocking activation of the AKT/GSK-3β axis through a plasma membrane GPCR that did not involve ERα or AR (Bonnet et al., 2009). Similarly, DHEA (100 μM for 48 h) activated Src/STAT3 and decreased STAT3 target genes, i.e., *NFATC2, BMPR2, PIM1*, and Survivin in pulmonary artery smooth muscle cells isolated from patients with idiopathic pulmonary arterial hypertension (Paulin et al., 2011). No mechanism for DHEA-STAT3 activation was identified. Likewise, although DHEA (100 μM) suppressed an angiotensin II (Ang II, 1 μM)-stimulated inflammatory response in rat aortic vascular smooth muscle cells (VSMCs) by inhibiting ROS production, P38 MAPK and ERK1/2 activation, preventing degradation of cytoplasmic IκB, and subsequently inhibiting NFκB activation (Chen, Xu, & Huang, 2014). However, the mechanism by which DHEA achieved these STAT3-suppressive or antiinflammatory effects was not identified and, to our knowledge, no follow-up studies have been reported.

15. DHEA ACTIVATION OF G-PROTEIN ESTROGEN RECEPTOR GPER AND miR-21 TRANSCRIPTION

GPER (*GPER1*), first named GPR30 (Carmeci, Thompson, Ring, Francke, & Weigel, 1997), is an integral PM protein coupled to Gαs in its inactive state and, when activated, stimulates adenylate cyclase, Src, and EGFR signaling (Barton et al., 2018; Gaudet, Cheng, Christensen, & Filardo, 2015). The DHEA metabolite, 7β-hydroxy-epiandrosterone, was suggested to bind GPER because it inhibited GPER agonist G-1-induced MCF-7 and SKBR3 breast cancer cell proliferation (Niro et al., 2012). We reported that DHEA (10 nM) increased the transcription of the oncogenic microRNA (oncomiR) miR-21 in HepG2 cells in part via GPER signaling (Teng et al., 2015). We note that 10 nM DHEA is a physiologically relevant level found in human serum (Labrie, 2010). MicroRNAs (miRNAs) are initially transcribed as primary microRNAs (pri-miRs) and are subsequently processed to precursor microRNAs (pre-miRs) and ultimately to mature miRNAs. In HepG2 cells, DHEA had a biphasic stimulatory effect on pri-miR-21 transcription with an initial peak at 1 h followed by a decrease and then an increase again after 3 h that was sustained to 12 h (Teng et al., 2014, 2015). The DHEA-induced increase in pri-miR-21 and mature miR-21 was inhibited by actinomycin D (ActD, a transcriptional inhibitor) but not by the protein synthesis inhibitor cycloheximide, suggesting a primary transcriptional effect of DHEA. We demonstrated that the initial rapid DHEA-induced increase in miR-21 transcription detected after 1 h of DHEA treatment involved GPER, estrogen receptor α-36 (ERα36), EGFR signaling, and G protein activation of c-Src, ERK1/2, and PI3K (Fig. 5). The DHEA- and BSA-conjugated DHEA, which is PM impermeable, simulation of pri-miR-21 transcription at 1 h was abrogated by the GPER-specific antagonist G-15 (Teng et al., 2015). G-1 and DHEA increased GPER and ERα36 mRNA and protein levels in HepG2 cells. DHEA also increased Src and ERK1/2 phosphorylation in a GPER-responsive manner. ERα Ser-118 phosphorylation was increased by DHEA (Teng et al., 2015).

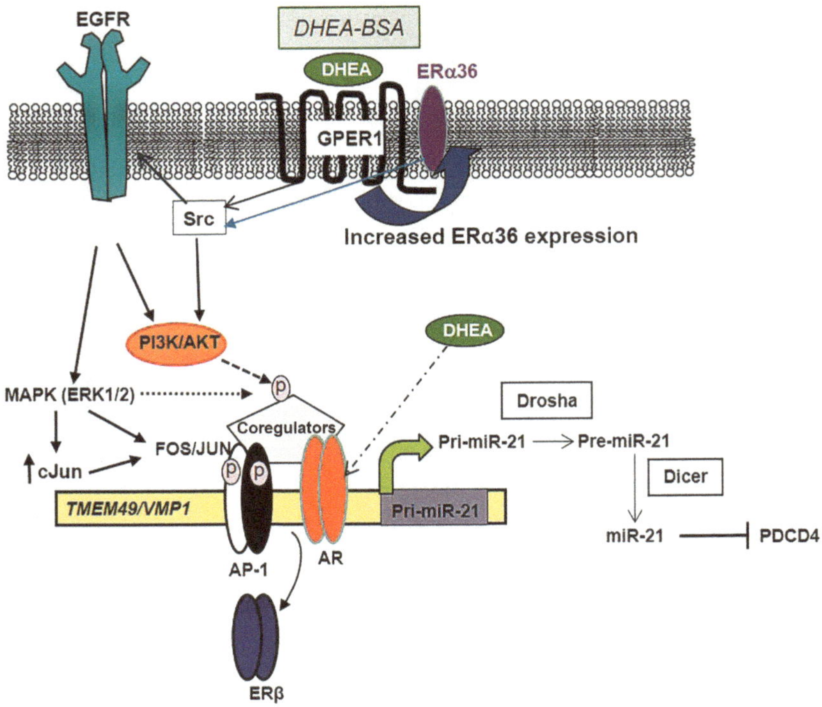

Fig. 5 DHEA activation of GPER (*GPER1*) stimulates miR-21 transcription. *Solid lines* represent conclusions based on studies and those of other investigators, while *dashed lines* indicate possible mechanisms based on other studies. We suggested that DHEA activates GPER and increases ERα36 expression (mechanism not diagrammed, but indicated by a *curved dark blue arrow*) leading to activation of Src, EGFR, MAPK, and PI3K/AKT. These events increase transcription of pri-miR-21. This pathway reflects the established pathway GPER1 activity (Barton et al., 2018; Gaudet et al., 2015). Since DHEA-BSA, like DHEA, activated pri-miR-21 transcription and antibodies to GPER and ERα36 ablated DHEA-induced pri-miR-21 transcription, we suggested a plasma membrane location for GPER. GPER activates EGFR and stimulates intracellular kinase pathway activation (Prossnitz & Barton, 2011; Prossnitz & Maggiolini, 2009). We observed increased AR, c-Fos, and c-Jun recruitment to the miR-21 promoter in cells treated with DHEA, whereas ERβ recruitment was reduced (Teng et al., 2014, 2015). We suggested that MAPK and PI3K/AKT may phosphorylate (*dashed lines*, *pink circle*) and activate AP-1 or coregulators (coactivators such as SRC-1) interacting with AP-1 and AR to modify chromatin structure and increase pri-miR-21 transcription. Although the AR antagonist bicalutamide did not block DHEA-activated pri-miR-21 transcription and we used physiologically relevant 10 nM DHEA, well below $K_d \sim 1$–$2\,\mu$M for DHEA–AR binding (Table 1), a caveat of our study is that in addition to membrane GPER activation, DHEA may also bind and activate AR (*dashed/dotted line*). *Modified from fig. 8 in Teng, Y., Radde, B. N., Litchfield, L. M., Ivanova, M. M., Prough, R. A., Clark, B. J., et al. (2015). Dehydroepiandrosterone activation of G-protein-coupled estrogen receptor rapidly stimulates microRNA-21 transcription in human hepatocellular carcinoma cells.* Journal of Biological Chemistry, *290(25), 15799–15811. https://doi.org/10.1074/jbc.M115.641167.*

ERα Ser-118 is an established target of ERK1/2 (Anbalagan & Rowan, 2015). GPER activation of ERK1/2 increases the expression of downstream transcription factors (Prossnitz & Maggiolini, 2009). For example, GPER agonist G-1 increased c-Fos and c-Jun expression in an ERK1/2-dependent manner in PC-3 prostate cancer cells (Chan et al., 2010). We reported that DHEA increased c-Jun, but we observed no effect of DHEA on c-Fos protein levels in HepG2 cells (Teng et al., 2015). Thus, we invoked cell-specific differences in downstream signaling that may account for this difference in a downstream target of GPER activation.

ERα36 is a PM-associated, N-terminal truncated, splice variant of ERα that has been proposed to mediate the membrane-initiated effects of E_2 (Wang & Yin, 2015). Overexpression of GPER stimulated ERα36 transcription in COS-7 and HEK-293 cells (Chaudhri et al., 2012). GPER activation of the Src/MEK1/2/AP-1 pathway increased ERα36 (Kang et al., 2010; Wang & Yin, 2015). As stated earlier, we observed that DHEA or G-1 (each at 10 nM for 1 h) increased ERα36 expression and transfection of the HepG2 cells with siGPER abrogated the increase in ERα36 (Teng et al., 2015). However, the DHEA induction of ERα36 was transient as levels returned to baseline control after 6 h of DHEA or G-1 treatment. This result suggested that different cellular mechanisms are involved in the observed rapid GPER activation with $\leq$1 h of DHEA vs the longer term DHEA effects that appeared to involve its cellular metabolites after 6 h of treatment (Teng et al., 2014).

Since efficient signal transduction requires compartmentalization of signaling molecules in microdomains at the PM (Martinez-Outschoorn, Sotgia, & Lisanti, 2015), we examined if methyl-β-cyclodextrin (MβCD), an agent that removes cholesterol from the PM (Rodal et al., 1999), would affect DHEA signaling in HepG2 cells. Using pri-miR-21 and mature miR-21 transcript expression as readout, we observed that MβCD inhibited DHEA signaling in HepG2 cells (Teng et al., 2015). We suggested that rapid DHEA signaling in HepG2 cells requires an intact lipid raft structure to organize signaling molecules involved in this process.

In conclusion, we suggested that a physiologically relevant concentration of DHEA, i.e., 10 nM, rapidly increases pri-miR-21 transcription in HepG2 cells by activating GPER and ERα36 and subsequent Src, EGFR, and ERK1/2 signaling. We do not know whether the increase in Ser-118-ERα phosphorylation of c-Jun expression results in their recruitment to the miR-21 promoter, although the promoter contains binding sites for these transcription factors (Chen & Bourguignon, 2014; Wickramasinghe et al., 2009). The second phase of DHEA-induced pri-miR-21 transcription, after 6h, was mediated by metabolites of DHEA that activate AR and ERβ recruitment to the miR-21 promoter (Teng et al., 2014).

16. DHEA ACTIVITY IN CNS AND NEURONAL CELLS INVOLVES NEUROTRANSMITTER RECEPTORS

As stated earlier in this chapter, DHEA and DHEA-S are well-established neurosteroids that cross the BBB and are produced locally within the brain (Stárka et al., 2015). DHEA and DHEA-S regulate normal development (Nguyen, 2018) and may play roles in neuropsychiatric diseases, e.g., depression (Hough et al., 2017), bipolar disorder, and schizophrenia (Buoli et al., 2016), and protect against neurodegenerative diseases (reviewed in Porcu et al., 2016). The age-related decline in DHEA and DHEA-S has been suggested to play a role in age-related physiological decline, including decreased cognitive function; however, these suggestions are widely disputed (de Menezes et al., 2016; Maninger, Wolkowitz, Reus, Epel, & Mellon, 2009).

DHEA and DHEA-S are allosteric modulators of the neurotransmitter receptors γ-aminobutyric-acid-type A ($GABA_A$), *N*-methyl-D-aspartate (NMDA) receptor, and sigma-1 receptors (S1R) (Table 2). The activation of these receptors supports a role for rapid, nongenomic activities of DHEA/DHEA-S and may offer mechanism to explain the observed beneficial effects of DHEA in neurological disorders.

The $GABA_A$ receptors ($GABA_A$ R) are heteropentameric, ligand-gated chloride and bicarbonate channels encoded by 19 $GABA_A$ R genes, which exhibit complex pharmacology and are activated by a range of clinically used therapeutics (reviewed in Calvo & González, 2016; Chua & Chebib, 2017). DHEA-S acted as an allosteric antagonist of a recombinant $GABA_A$ R by binding at or near the picrotoxin/*t*-butylbicyclophosphorothionate site on the $GABA_A$ receptor complex and abrogated currents induced by flunitrazepam and *t*-butylbicycloorthobenzoate binding $GABA_A$ R in transfected HEK-293 cells (Sachidanandan & Bera, 2015; Svob Strac et al., 2012).

The *N*-methyl-D-aspartate receptors (NMDARs) are a large heterotetrameric family of ligand-gated ionotropic glutamate receptors that bind neurotransmitters and allosteric effectors to regulate transmembrane ion channels involved in neuronal development, learning, memory, and degenerative diseases (Chakraborty, Murphy, & Coleman, 2017; Regan, Romero-Hernandez, & Furukawa, 2015). Mutation and altered activity of NMDA receptors have been associated with autism and epilepsy (Burnashev & Szepetowski, 2015) and with major depressive and bipolar disorders (Ghasemi et al., 2014). Abnormal localization or decreased NMDAR activity is seen in congenital mental retardation, schizophrenia, Huntington's disease, and amyotrophic lateral sclerosis (reviewed in Chakraborty et al., 2017). DHEA-S (50 nM to 1 μM) activated NMDAR in rat hippocampal slices (Monnet et al., 1995) and mouse embryonic neocortical neurons (Compagnone & Mellon, 1998). NMDAR showed specificity for DHEA (1 pM), and not DHEA-S (up to 1 μM), for rapid NMDAR-induced increases in intracellular calcium (Ca^{2+}) in mouse embryonic neocortical neurons (Compagnone & Mellon, 1998; Fig. 3).

The sigma 1 receptor (S1R) is a 25 kDa protein, ligand-regulated, non-ATP-binding, membrane-bound chaperone protein that has been associated with endoplasmic reticulum, nuclear envelope, and mitochondrial membranes in neurons, astrocytes, oligodendrocytes, and microglia (Chu & Ruoho, 2015). S1R is activated by cell stress and acts as a chaperone to preserve neuron survival in neurodegenerative diseases (Prause et al., 2013). The S1R binds dimethyltryptamine as an endogenous ligand, neurosteroids (progesterone, DHEA, and cholesterol), as well as sphingosine cholesterol (Chu & Ruoho, 2015), and many different ligands including dextrobenzomorphans, neuroleptics, and psychostimulants such as cocaine and methamphetamine (Rousseaux & Greene, 2015). In the endoplasmic reticulum, S1R has a dual role: it binds to the inositol trisphosphate receptor (IP3R) and modulates cellular calcium homeostasis and functions as a chaperone protein in response to endoplasmic reticulum stress (Hashimoto, 2013; Maurice, 2004; Ruscher & Wieloch, 2015). S1R also interacts with PM-bound receptors, GPCRs, NMDARs, and ion channels to regulate their functions (Chu & Ruoho, 2015). Recent studies show that S1R is overexpressed in color rectal and breast cancers and correlates with tumor grade and lower overall survival (Gueguinou et al., 2017). In breast and colorectal cancer cells, S1R interacts with SK2, a Ca^{2+}-activated K^{+} channel (*KCNN3*), and ORAI1, a voltage-independent Ca^{2+} channel, in the PM for Ca^{2+} influx that stimulates cell migration (Gueguinou et al., 2017). This observation suggests that DHEA may enhance cell migration in breast and colorectal cancer cells, but this has not been examined.

S1R activation by DHEA reversed memory deficits in mice induced by prenatal exposure to cocaine (Meunier & Maurice, 2004) or by stroke induced by transient bilateral common carotid artery ligation (Yabuki et al., 2015). DHEA-S interacts with S1R in C6 glioma cells, with low affinity, the apparent K_d was >50 μM (Rybczynska et al., 2009). DHEA activated both NMDAR and S1R to stimulate the proliferation of primary human neural stem cells derived from the fetal cortex (ltNSC^{ctx} cells) (Suzuki et al., 2004).

Other potential DHEA and DHEA-S receptors have been reported in PC12 rat sympathoadrenal cells (an established model for the study of neuronal cell apoptosis and survival) since the protective effects of these neurosteroids were seen with EC_{50} of 1.8 and 1.1 nM, respectively (Charalampopoulos et al., 2006).

DHEA activated a PM DHEA-binding site (mDBS) (Fig. 3) and sequentially activated pSrc, MAPK, PKC, AKT, cAMP, and PKA, resulting in increased NFκB and CREB, and Bcl-2/Bcl-XL gene expression in PC12 rat sympathoadrenal cells derived from a pheochromocytoma of the rat adrenal medulla (Charalampopoulos et al., 2006; Suzuki et al., 2004). In serum-deprived PC12 cells, 100 nM DHEA or DHEA-S rapidly (2 min) increased cAMP (5–10 min) phosphorylation of Src, PI3K, AKT (30 min), but not MAPK phosphorylation (Charalampopoulos, Margioris, & Gravanis, 2008). These DHEA and DHEA-S effects in PC12 cells were blocked by pertussis toxin, suggesting a G-protein-coupled signaling mechanism. Later studies demonstrated that a photoreactive DHEA analog was bound specifically to the PM of PC12 cells and activated adenylyl cyclase leading to increased cAMP in PC12 and HepG2 cells (Waschatko, Kojro, Zahnow, & Gehrig-Burger, 2011). Subsequent investigation identified PM TrkA and p75NTR receptors as the binding targets for DHEA and DHEA-S in PC12 cells (Lazaridis et al., 2011). Overexpression of TrkA or p75NTR in HEK-293 cells revealed that [^{3}H]DHEA bound these receptors with K_d values of 7.4 ± 1.7 nM and 5.6 ± 0.5 nM, respectively (Gravanis et al., 2012).

TrkA and p75NTR are neurotrophin nerve growth factor (NGF)-binding transmembrane tyrosine kinase receptors (Lazaridis et al., 2011). DHEA (100 nM for 20 min) rapidly induced TrkA and p75NTR signaling, seen as increased phospho-ERK and AKT, in transfected HEK-293 and PC12 cells (Lazaridis et al., 2011). [^{3}H]DHEA binds Ltrk, ApTrk, or AmphiTrk invertebrate neurotrophin receptors expressed in HEK-293 cells with K_d: 3.6 ± 0.4 nM, 1.14 ± 0.11 nM, and 0.47 ± 0.18 nM, respectively (Pediaditakis et al., 2015). DHEA (100 nM for 24 h) inhibited apoptosis stimulated by serum deprivation in DU145 prostate and Caco2 colorectal carcinoma cells in a TrkA- and p75NTR-dependent manner (Anagnostopoulou et al., 2013). DHEA increased TrkA receptor expression and phosphorylation in PC12 cells and DU145 prostate cancer cells and stimulated the interaction of RhoGDI and RIP2 effectors with p75NTR (Anagnostopoulou et al., 2013).

16.1 DHEA Binds MAP2c

The microtubule-associated proteins (MAPs) promote tubulin polymerization for microtubule stabilization and MAP2 is highly expressed in brain (Sánchez, Díaz-Nido, & Avila, 2000). MAP2c is a dendritic marker that is critical for dendritic plasticity (Konar & Thakur, 2015). MAP2c interacts with Brb2, Src, and PKA (Lim et al., 2013). DHEA was reported to interact with a hydrophobic pocket of MAP2c in vitro with a $K_d = 27\,\mu M$ (Laurine et al., 2003).

17. DHEA INHIBITS VOLTAGE-GATED T-TYPE Ca^{2+} CHANNELS: $Ca_V3.1$, $Ca_V3.2$, OR $Ca_V3.3$

T-type Ca^{2+} channels (Ca_V3 family), one of three voltage-dependent Ca^{2+} channels, are modulated independently of GPCRs and protein kinases by several endogenous compounds including fatty acids, endocannabinoids, lipoamino acids, and arachidonic acid metabolites (Chemin, Cazade, & Lory, 2014). Since DHEA and DHEA-S decline with age, there is a great interest in knowing whether DHEA is beneficial to prevent or treat cardiovascular disease (CVD); however, the putative benefit of DHEA in CVD is highly controversial (reviewed in Mannic et al., 2015; Savineau, Marthan, & Dumas de la Roque, 2013). A recent meta-analysis found that patients with CVD with lower DHEA-S levels may have poorer prognosis than those with higher DHEA-S levels (Wu et al., 2017). DHEA ($10\,\mu M$) inhibited $Ca_V3.1$, $Ca_V3.2$, or $Ca_V3.3$ in transfected NG108-15 cells, and a somatic mouse neuroblastoma/rat gliomablastoma cell line (Chevalier et al., 2012). The authors of this study concluded that inhibition of these Ca^{2+} channels is responsible for the ability of DHEA to inhibit pulmonary artery contraction and the physiological effects of DHEA in cardiovascular and neuronal diseases. However, as indicated earlier, DHEA activity also involves S1R, PPARα, its metabolism to estrogens, and unknown PM-mediated mechanisms in cardiovascular tissues (reviewed in Mannic et al., 2015). Other activities associated with the beneficial effects of DHEA may be related to reduction in fat mass and a trend toward an increase in lean mass, its ability to inhibit platelet aggregation, and facilitate fibrinolysis (Wu et al., 2017).

18. DHEA ACTIVITY IN KIDNEY INVOLVES S1R AND OTHER MECHANISMS

Acute kidney injury, characterized by loss of the ability of the kidneys to concentrate urine, conserve electrolytes, and excrete wastes, is induced by renal ischemia–reperfusion injury (IRI), is often seen in patients in the ICU, and is the main cause of delayed graft function or loss after kidney transplantation (Schrier, Wang, Poole, & Mitra, 2004). DHEA (4 mg/kg/day for 25 h) treatment of male Wistar rats was reported to reduce renal IRI (Vannay et al., 2009). Patients on dialysis for chronic kidney failure showed lower DHEA-S and the decrease in DHEA-S was associated with all-cause mortality in males (Kakiya et al., 2012). S1R protein is expressed in the renal cortex, medulla, and papilla of rat kidney and DHEA was recently reported to act as a low-affinity SR1 agonist to attenuate IRI inhibition of AKT-NOS signaling and improve NO-mediated kidney perfusion (Hosszu et al., 2017).

DHEA can ameliorate some of the physiological effects caused by hyperglycemia in diabetes and increases insulin sensitivity in liver and muscle in rodent models of diabetes (Aoki et al., 2003). Conversely, one study reported that DHEA increased plasma glucose and creatinine and reduced glomerular filtration rates in streptozotocin-induced diabetic male Wistar rats, suggesting that DHEA treatment was not beneficial in this model (Jahn et al., 2011). However, DHEA was shown to decrease dexamethasone-induced renal glucose synthesis by inhibiting glucose-6-phosphatase activity in rabbits and in primary cultured rabbit kidney-cortex tubules (Kiersztan et al., 2016). How DHEA inhibited glucose-6-phosphatase activity was not determined. Recent studies on the effect of DHEA on insulin signaling were evaluated in the liver and muscle tissue of male IRS1$^{-/-}$ and IRS2$^{-/-}$ vs control C57BL6 mice fed an HFD $\pm$ 0.2% DHEA or 0.2% androstenedione for 4 weeks (Aoki, Tajima, Taguri, & Terauchi, 2016). Since DHEA administration increases pAKT/AKT (Jacob et al., 2011) and phosphorylation of IRS1, IRS2, and PKCζ (Campbell et al., 2004), which are upstream of AKT, the authors suspected that the IRS1$^{-/-}$ and IRS2$^{-/-}$ mice would show a loss of dietary DHEA protective effects. However, dietary DHEA decreased the body weights and increased the liver weight/body weight ratio of the C57BL6, IRS1$^{-/-}$, and IRS2$^{-/-}$ mice. DHEA had no effect on GLUT4 protein expression in muscle. The main findings were a tissue-specific impact of IRS1$^{-/-}$ and

$IRS2^{-/-}$ knockout on the DHEA-activated AKT–PKCζ pathway. $IRS1^{-/-}$ and $IRS2^{-/-}$ knockout blocked DHEA-stimulated AKT and PKCζ phosphorylation in muscle, but not liver of the mice on the HFD. Androstenedione administration increased Akt and PKCζ phosphorylation in the muscle tissue, but not liver, of the HFD-fed mice. The authors concluded that the DHEA-induced increase in pAKT/AKT in the liver of C57BL6 mice requires IRS1 and IRS2 (Aoki et al., 2016).

19. DHEA AFFECTS MITOCHONDRIAL FUNCTION

Age-related mitochondrial dysfunction contributes to diseases such as neurodegeneration, cardiovascular disease, obesity, and diabetes (Mc et al., 2017; Yin, Sancheti, Liu, & Cadenas, 2016). DHEA has a variety of anti-aging properties (Traish et al., 2011). DHEA (1 μM) (but not DHEA-S) protected mitochondrial function in isolated rat brain mitochondria from stresses including anoxia–reoxygenation, possibly by a direct effect on increasing mitochondrial membrane fluidity and suppressing cytochrome *c* release (Morin, Zini, Simon, & Tillement, 2002). DHEA (166 μM) acted like rotenone to inhibit respiratory Complex I in isolated rat kidney mitochondria, thus reducing mitochondrial membrane potential, ATP production, and increasing reactive oxygen species (ROS) and Ca^{2+} release (Correa, García, García, & Chávez, 2003). On the other hand, DHEA (10 and 100 nM for 24 h) increased mitochondrial membrane potential, oxygen consumption rate (OCR), ATP synthesis, and ROS production in human SH-SY5Y neuroblastoma cells and primary mouse cortical neurons (Grimm, Schmitt, Lang, Mensah-Nyagan, & Eckert, 2014), although no mechanism was elucidated. Treatment of SK-Hep-1 human hepatoma cells with DHEA (200 μM for 48 h) reduced OCR, ATP production, and mitochondrial membrane potential with a compensatory increase in mitochondrial mass (Cheng, Chi, Wu, & Ho, 2016). The authors used a stable isotope labeling by amino acids in cell culture (SILAC)-based approach on isolated mitochondria and identified decreased protein expression of Complex I subunits, e.g., MT-ND1, NDUFA2, NDUFA11, NDUFB8, and NDUFS4; SDHB (Complex II), COX6B1 and MT-CO2 (Complex IV), UQCRC2, and FASTKD2. They suggested that the decrease in repression of FASTKD2 expression may partly mediate the cellular effects of DHEA since transient overexpression of FASTKD2 in transfected SK-Hep-1 cells partially restored DHEA-induced inhibition of OCR and suppressed ROS generation (Cheng et al., 2016). Fas-activated serine/threonine kinases

(FASTKD1–5) have an N-terminal mitochondrial targeting domain and FASTKD2 is localized in the mitochondrial inner compartment where it is a mitochondrial (mt) RNA-binding protein required for mtRNA translation and cellular respiration (Popow et al., 2015). How DHEA down-regulates FASTKD2 expression was not examined.

20. CONCLUSIONS AND FUTURE DIRECTIONS

Based on this review of the literature, we conclude that DHEA has a structure that is similar to pregnane structures that are important in endocrinology and neuroendocrinology. Therefore, many receptors are activated by DHEA, but at much higher levels than the cognate ligands. These lower affinity-binding events likely account for the wide range of activities ascribed to DHEA. It is evident from this review of the observed effects of DHEA and DHEA-S in rodents and humans that mechanisms, specifically related to plasma membrane receptors, remain poorly defined. Examples of gaps in knowledge identified in this review chapter include (1) how DHEA increases PGC-1α, PPARδ, and PPARα in muscle; (2) why DHEA and tamoxifen are synergistically antiproliferative in MCF-7 cells; (3) the nature of the membrane MR activated by DHEA in HUVECs and VSMCs; (4) how DHEA inhibits T-type calcium channel ($Ca_V3.2/\alpha1H$) in cardiomyocytes; (5) how DHEA inhibits 11β-HSD1 gene expression in adipocytes; (6) the nature of specific DHEA-S vs DHEA-specific GPCR and their downstream pathways in various cell types; (7) the identity and cellular signaling pathway that appears to be common for DHEA and DHEA-S activation of Gαq/11 signaling in EC, BAEC, GC-2, TM4, and bEnd.3 cells; and (8) does DHEA directly affect mitochondria and by what mechanism?

REFERENCES

Abdel-Hafiz, H. A., & Horwitz, K. B. (2014). Post-translational modifications of the progesterone receptors. *The Journal of Steroid Biochemistry and Molecular Biology*, *140*, 80–89. https://doi.org/10.1016/j.jsbmb.2013.12.008.

Adams, J., Garcia, M., & Rochefort, H. (1981). Estrogenic effects of physiological concentrations of 5-androstene-3β,17β-diol and its metabolism in MCF7 human breast cancer cells. *Cancer Research*, 41(11 Pt 1), 4720–4726. https://doi.org/10.1016/S0039-128X(02)00064-8.

Altman, R., Motton, D. D., Kota, R. S., & Rutledge, J. C. (2008). Inhibition of vascular inflammation by dehydroepiandrosterone sulfate in human aortic endothelial cells: Roles of PPARα and NF-κB. *Vascular Pharmacology*, *48*(2–3), 76–84. https://doi.org/10.1016/j.vph.2007.12.002.

Anagnostopoulou, V., Pediaditakis, I., Alkahtani, S., Alarifi, S. A., Schmidt, E.-M., Lang, F., et al. (2013). Differential effects of dehydroepiandrosterone and testosterone in prostate and colon cancer cell apoptosis: The role of nerve growth factor (NGF) receptors. *Endocrinology, 154*(7), 2446–2456. https://doi.org/10.1210/en.2012-2249.

Anbalagan, M., & Rowan, B. G. (2015). Estrogen receptor alpha phosphorylation and its functional impact in human breast cancer. *Molecular and Cellular Endocrinology, 418*(Pt 3), 264–272. https://doi.org/10.1016/j.mce.2015.01.016.

Aoki, K., Nakajima, A., Mukasa, K., Osawa, E., Mori, Y., & Sekihara, H. (2003). Prevention of diabetes, hepatic injury, and colon cancer with dehydroepiandrosterone. *Journal of Steroid Biochemistry and Molecular Biology, 85*(2–5), 469–472.

Aoki, K., Tajima, K., Taguri, M., & Terauchi, Y. (2016). Effect of dehydroepiandrosterone (DHEA) on Akt and protein kinase C zeta (PKCζ) phosphorylation in different tissues of C57BL6, insulin receptor substrate (IRS)1 −/−, and IRS2 −/− male mice fed a high-fat diet. *The Journal of Steroid Biochemistry and Molecular Biology, 159*, 110–120. https://doi.org/10.1016/j.jsbmb.2016.03.011.

Arlt, W. (2004). Dehydroepiandrosterone replacement therapy. *Seminars in Reproductive Medicine, 22*(4), 379–388. https://doi.org/10.1055/s-2004-861554.

Arnold, J. T. (2009). DHEA metabolism in prostate: For better or worse? *Molecular and Cellular Endocrinology, 301*(1–2), 83–88. https://doi.org/10.1016/j.mce.2008.10.019.

Arnold, J. T., & Blackman, M. R. (2005). Does DHEA exert direct effects on androgen and estrogen receptors, and does it promote or prevent prostate cancer? *Endocrinology, 146*(11), 4565–4567.

Banerjee, S., Chambliss, K. L., Mineo, C., & Shaul, P. W. (2014). Recent insights into non-nuclear actions of estrogen receptor alpha. *Steroids, 81*, 64–69. https://doi.org/10.1016/j.steroids.2013.11.002.

Banerjee, M., Robbins, D., & Chen, T. (2013). Modulation of xenobiotic receptors by steroids. *Molecules, 18*(7), 7389–7406. https://doi.org/10.3390/molecules18077389.

Banerjee, M., Robbins, D., & Chen, T. (2015). Targeting xenobiotic receptors PXR and CAR in human diseases. *Drug Discovery Today, 20*(5), 618–628. https://doi.org/10.1016/j.drudis.2014.11.011.

Barton, M., Filardo, E. J., Lolait, S. J., Thomas, P., Maggiolini, M., & Prossnitz, E. R. (2018). Twenty years of the G protein-coupled estrogen receptor GPER: Historical and personal perspectives. *The Journal of Steroid Biochemistry and Molecular Biology, 176*, 4–15. https://doi.org/10.1016/j.jsbmb.2017.03.021.

Beato, M., Chalepakis, G., Schauer, M., & Slater, E. P. (1989). DNA regulatory elements for steroid hormones. *Journal of Steroid Biochemistry, 32*, 737–747.

Berger, J., & Moller, D. E. (2002). The mechanisms of action of PPARs. *Annual Review of Medicine, 53*, 409–435. https://doi.org/10.1146/annurev.med.53.082901.104018.

Bernard-Marissal, N., Medard, J. J., Azzedine, H., & Chrast, R. (2015). Dysfunction in endoplasmic reticulum-mitochondria crosstalk underlies SIGMAR1 loss of function mediated motor neuron degeneration. *Brain, 138*(Pt 4), 875–890. https://doi.org/10.1093/brain/awv008.

Bjerregaard-Olesen, C., Ghisari, M., Kjeldsen, L. S., Wielsøe, M., & Bonefeld-Jørgensen, E. C. (2016). Estrone sulfate and dehydroepiandrosterone sulfate: Transactivation of the estrogen and androgen receptor. *Steroids, 105*, 50–58. https://doi.org/10.1016/j.steroids.2015.11.009.

Boccuzzi, G., Di Monaco, M., Brignardello, E., Leonardi, L., Gatto, V., Pizzini, A., et al. (1993). Dehydroepiandrosterone antiestrogenic action through androgen receptor in MCF-7 human breast cancer cell line. *Anticancer Research, 13*(6A), 2267–2272.

Bonnet, S., Paulin, R., Sutendra, G., Dromparis, P., Roy, M., Watson, K. O., et al. (2009). Dehydroepiandrosterone reverses systemic vascular remodeling through the inhibition of the Akt/GSK3-β/NFAT axis. *Circulation, 120*(13), 1231–1240. https://doi.org/10.1161/circulationaha.109.848911.

Bruder, J. M., Sobek, L., & Oettel, M. (1997). Dehydroepiandrosterone stimulates the estrogen response element. *The Journal of Steroid Biochemistry and Molecular Biology*, *62*(5–6), 461–466. https://doi.org/10.1016/s0960-0760(97)00056-3.

Buoli, M., Caldiroli, A., Cumerlato Melter, C., Serati, M., de Nijs, J., & Altamura, A. C. (2016). Biological aspects and candidate biomarkers for psychotic bipolar disorder: A systematic review. *Psychiatry and Clinical Neurosciences*, *70*(6), 227–244. https://doi.org/10.1111/pcn.12386.

Burnashev, N., & Szepetowski, P. (2015). NMDA receptor subunit mutations in neurodevelopmental disorders. *Current Opinion in Pharmacology*, *20*, 73–82. https://doi.org/10.1016/j.coph.2014.11.008.

Caizzi, L., Ferrero, G., Cutrupi, S., Cordero, F., Ballaré, C., Miano, V., et al. (2014). Genome-wide activity of unliganded estrogen receptor-α in breast cancer cells. *Proceedings of the National Academy of Sciences of the United States of America*, *111*(13), 4892–4897. https://doi.org/10.1073/pnas.1315445111.

Calvo, D. J., & González, A. N. B. (2016). Dynamic regulation of the $GABA_A$ receptor function by redox mechanisms. *Molecular Pharmacology*, *90*(3), 326.

Campbell, C. S. G., Caperuto, L. C., Hirata, A. E., Araujo, E. P., Velloso, L. A., Saad, M. J., et al. (2004). The phosphatidylinositol/AKT/atypical PKC pathway is involved in the improved insulin sensitivity by DHEA in muscle and liver of rats in vivo. *Life Sciences*, *76*(1), 57–70. https://doi.org/10.1016/j.lfs.2004.06.017.

Cardounel, A., Regelson, W., & Kalimi, M. (1999). Dehydroepiandrosterone protects hippocampal neurons against neurotoxin-induced cell death: Mechanism of action. *Proceedings of the Society for Experimental Biology and Medicine*, *222*(2), 145–149.

Carmeci, C., Thompson, D. A., Ring, H. Z., Francke, U., & Weigel, R. J. (1997). Identification of a gene (GPR30) with homology to the G-protein-coupled receptor superfamily associated with estrogen receptor expression in breast cancer. *Genomics*, *45*(3), 607–617.

Chai, S. C., Cherian, M. T., Wang, Y.-M., & Chen, T. (2016). Small-molecule modulators of PXR and CAR. *Biochimica et Biophysica Acta*, *1859*(9), 1141–1154. https://doi.org/10.1016/j.bbagrm.2016.02.013.

Chakraborty, A., Murphy, S., & Coleman, N. (2017). The role of NMDA receptors in neural stem cell proliferation and differentiation. *Stem Cells and Development*, *26*(11), 798–807. https://doi.org/10.1089/scd.2016.0325.

Chan, Q. K. Y., Lam, H. M., Ng, C. F., Lee, A. Y. Y., Chan, E. S. Y., Ng, H. K., et al. (2010). Activation of GPR30 inhibits the growth of prostate cancer cells through sustained activation of Erk1/2, c-jun/c-fos-dependent upregulation of p21, and induction of G2 cell-cycle arrest. *Cell Death and Differentiation*, *17*(9), 1511–1523. https://doi.org/10.1038/cdd.2010.20.

Charalampopoulos, I., Alexaki, V. I., Tsatsanis, C., Minas, V., Dermitzaki, E., Lasaridis, I., et al. (2006). Neurosteroids as endogenous inhibitors of neuronal cell apoptosis in aging. *Annals of the New York Academy of Sciences*, *1088*, 139–152.

Charalampopoulos, I., Margioris, A. N., & Gravanis, A. (2008). Neurosteroid dehydroepiandrosterone exerts anti-apoptotic effects by membrane-mediated, integrated genomic and non-genomic pro-survival signaling pathways. *Journal of Neurochemistry*, *107*(5), 1457–1469.

Chaudhri, R. A., Olivares-Navarrete, R., Cuenca, N., Hadadi, A., Boyan, B. D., & Schwartz, Z. (2012). Membrane estrogen signaling enhances tumorigenesis and metastatic potential of breast cancer cells via estrogen receptor-α36 (ERα36). *Journal of Biological Chemistry*, *287*(10), 7169–7181. https://doi.org/10.1074/jbc.M111.292946.

Chemin, J., Cazade, M., & Lory, P. (2014). Modulation of T-type calcium channels by bioactive lipids. *Pflügers Archiv—European Journal of Physiology*, *466*(4), 689–700. https://doi.org/10.1007/s00424-014-1467-5.

Chen, L., & Bourguignon, L. Y. (2014). Hyaluronan-CD44 interaction promotes c-Jun signaling and miRNA21 expression leading to Bcl-2 expression and chemoresistance in breast cancer cells. *Molecular Cancer, 13*, 52. https://doi.org/10.1186/1476-4598-13-52.

Chen, F., Knecht, K., Birzin, E., Fisher, J., Wilkinson, H., Mojena, M., et al. (2005). Direct agonist/antagonist functions of dehydroepiandrosterone. *Endocrinology, 146*(11), 4568–4576.

Chen, J., Xu, L., & Huang, C. (2014). DHEA inhibits vascular remodeling following arterial injury: A possible role in suppression of inflammation and oxidative stress derived from vascular smooth muscle cells. *Molecular and Cellular Biochemistry, 388*(1–2), 75–84. https://doi.org/10.1007/s11010-013-1900-7.

Cheng, M.-L., Chi, L.-M., Wu, P.-R., & Ho, H.-Y. (2016). Dehydroepiandrosterone-induced changes in mitochondrial proteins contribute to phenotypic alterations in hepatoma cells. *Biochemical Pharmacology, 117*, 20–34. https://doi.org/10.1016/j.bcp.2016.08.005.

Cheng, Z.-X., Lan, D.-M., Wu, P.-Y., Zhu, Y.-H., Dong, Y., Ma, L., et al. (2008). Neurosteroid dehydroepiandrosterone sulphate inhibits persistent sodium currents in rat medial prefrontal cortex via activation of sigma-1 receptors. *Experimental Neurology, 210*(1), 128–136. https://doi.org/10.1016/j.expneurol.2007.10.004.

Chevalier, M., Gilbert, G., Lory, P., Marthan, R., Quignard, J. F., & Savineau, J. P. (2012). Dehydroepiandrosterone (DHEA) inhibits voltage-gated T-type calcium channels. *Biochemical Pharmacology, 83*(11), 1530–1539. https://doi.org/10.1016/j.bcp.2012.02.025.

Christianson, M. S., & Shen, W. (2013). Osteoporosis prevention and management: Nonpharmacologic and lifestyle options. *Clinical Obstetrics and Gynecology, 56*(4), 703–710. https://doi.org/10.1097/GRF.0b013e3182a9d15a.

Chu, U. B., & Ruoho, A. E. (2015). Biochemical pharmacology of the sigma-1 receptor. *Molecular Pharmacology, 89*(1), 142.

Chua, H. C., & Chebib, M. (2017). Chapter one—GABAA receptors and the diversity in their structure and pharmacology. In P. G. Dominic & D. R. Lachlan (Eds.), Advances in pharmacology (Vol. 79, pp. 1–34): Academic Press.

Compagnone, N. A., & Mellon, S. H. (1998). Dehydroepiandrosterone: A potential signalling molecule for neocortical organization during development. *Proceedings of the National Academy of Sciences of the United States of America, 95*(8), 4678–4683.

Correa, F., García, N., García, G., & Chávez, E. (2003). Dehydroepiandrosterone as an inducer of mitochondrial permeability transition. *The Journal of Steroid Biochemistry and Molecular Biology, 87*(4–5), 279–284. https://doi.org/10.1016/j.jsbmb.2003.09.002.

Crane, J. L., & Cao, X. (2014). Function of matrix IGF-1 in coupling bone resorption and formation. *The Journal of Molecular Medicine (Berlin, Germany), 92*(2), 107–115. https://doi.org/10.1007/s00109-013-1084-3.

Dallman, M. F., Warne, J. P., Foster, M. T., & Pecoraro, N. C. (2007). Glucocorticoids and insulin both modulate caloric intake through actions on the brain. *The Journal of Physiology, 583*(Pt 2), 431–436. https://doi.org/10.1113/jphysiol.2007.136051.

Dasgupta, S., Lonard, D. M., & O'Malley, B. W. (2014). Nuclear receptor coactivators: Master regulators of human health and disease. *Annual Review of Medicine, 65*(1), 279–292. https://doi.org/10.1146/annurev-med-051812-145316.

de Menezes, K. J., Peixoto, C., Nardi, A. E., Carta, M. G., Machado, S., & Veras, A. B. (2016). Dehydroepiandrosterone, its sulfate and cognitive functions. *Clinical Practice & Epidemiology in Mental Health, 12*, 24–37. https://doi.org/10.2174/1745017901612010024.

Evans, R. M., & Mangelsdorf, D. J. (2014). Nuclear receptors, RXR, and the big bang. *Cell, 157*(1), 255–266. https://doi.org/10.1016/j.cell.2014.03.012.

Feige, J. N., Gelman, L., Michalik, L., Desvergne, B., & Wahli, W. (2006). From molecular action to physiological outputs: Peroxisome proliferator-activated receptors are nuclear receptors at the crossroads of key cellular functions. *Progress in Lipid Research, 45*(2), 120–159. https://doi.org/10.1016/j.plipres.2005.12.002.

Forman, B. M., Tzameli, I., Choi, H., Chen, J., Simha, D., Seol, W., et al. (1998). Androstane metabolites bind to and deactivate the nuclear receptor CAR-beta. *Nature, 395*, 612–615.
Fujita, A., Furutama, D., Tanaka, T., Sakai, R., Koyama, A., Hanafusa, T., et al. (2002). In vivo activation of the constitutive androstane receptor beta (CARbeta) by treatment with dehydroepiandrosterone (DHEA) or DHEA sulfate (DHEA-S). *FEBS Letters, 532*(3), 373–378.
Fuller, P. J. (2015). Novel interactions of the mineralocorticoid receptor. *Molecular and Cellular Endocrinology, 408*, 33–37. https://doi.org/10.1016/j.mce.2015.01.027.
Fuller, P. J., & Young, M. J. (2005). Mechanisms of mineralocorticoid action. *Hypertension, 46*(6), 1227–1235. https://doi.org/10.1161/01.hyp.0000193502.77417.17.
Gaudet, H. M., Cheng, S. B., Christensen, E. M., & Filardo, E. J. (2015). The G-protein-coupled estrogen receptor, GPER: The inside and inside-out story. *Molecular and Cellular Endocrinology*, 418(3), 207–219. https://doi.org/10.1016/j.mce.2015.07.016.
Ghasemi, M., Phillips, C., Trillo, L., De Miguel, Z., Das, D., & Salehi, A. (2014). The role of NMDA receptors in the pathophysiology and treatment of mood disorders. *Neuroscience & Biobehavioral Reviews, 47*, 336–358. https://doi.org/10.1016/j.neubiorev.2014.08.017.
Gravanis, A., Calogeropoulou, T., Panoutsakopoulou, V., Thermos, K., Neophytou, C., & Charalampopoulos, I. (2012). Neurosteroids and microneurotrophins signal through NGF receptors to induce prosurvival signaling in neuronal cells. *Science Signaling, 5*(246), pt8. https://doi.org/10.1126/scisignal.2003387.
Grimm, A., Schmitt, K., Lang, U. E., Mensah-Nyagan, A. G., & Eckert, A. (2014). Improvement of neuronal bioenergetics by neurosteroids: Implications for age-related neurodegenerative disorders. *Biochimica et Biophysica Acta, 1842*(12 Pt A), 2427–2438. https://doi.org/10.1016/j.bbadis.2014.09.013.
Gueguinou, M., Crottes, D., Chantome, A., Rapetti-Mauss, R., Potier-Cartereau, M., Clarysse, L., et al. (2017). The SigmaR1 chaperone drives breast and colorectal cancer cell migration by tuning SK3-dependent Ca2+ homeostasis. *Oncogene, 36*(25), 3640–3647. https://doi.org/10.1038/onc.2016.501.
Hashimoto, K. (2013). Sigma-1 receptor chaperone and brain-derived neurotrophic factor: Emerging links between cardiovascular disease and depression. *Progress in Neurobiology, 100*, 15–29. https://doi.org/10.1016/j.pneurobio.2012.09.001.
Horii, N., Sato, K., Mesaki, N., & Iemitsu, M. (2016). DHEA administration activates transcription of muscular lipid metabolic enzymes via PPARalpha and PPARdelta in obese rats. *Hormone and Metabolic Research, 43*(3), 207–212. https://doi.org/10.1055/s-0035-1564132.
Hosszu, A., Antal, Z., Lenart, L., Hodrea, J., Koszegi, S., Balogh, D. B., et al. (2017). Sigma 1-receptor agonism protects against renal ischemia-reperfusion injury. *Journal of the American Society of Nephrology, 28*(1), 152–165. https://doi.org/10.1681/asn.2015070772.
Hough, C. M., Lindqvist, D., Epel, E. S., Denis, M. S., Reus, V. I., Bersani, F. S., et al. (2017). Higher serum DHEA concentrations before and after SSRI treatment are associated with remission of major depression. *Psychoneuroendocrinology*, 77, 122–130. https://doi.org/10.1016/j.psyneuen.2016.11.035.
Imai, A., Ohno, T., & Tamaya, T. (1992). Dehydroepiandrosterone sulfate-binding sites in plasma membrane from human uterine cervical fibroblasts. *Experientia, 48*(10), 999–1002.
Iruthayanathan, M., O'Leary, B., Paul, G., & Dillon, J. S. (2011). Hydrogen peroxide signaling mediates DHEA-induced vascular endothelial cell proliferation. *Steroids, 76*(13), 1483–1490. https://doi.org/10.1016/j.steroids.2011.08.002.
Jacob, M. H. V. M., Janner, D. D. R., Araújo, A. S. D. R., Jahn, M. P., Kucharski, L. C. R., Moraes, T. B., et al. (2011). Dehydroepiandrosterone improves hepatic antioxidant reserve and stimulates Akt signaling in young and old rats. *The Journal of Steroid Biochemistry and Molecular Biology, 127*(3–5), 331–336. https://doi.org/10.1016/j.jsbmb.2011.07.007.

Jahn, M. P., Gomes, L. F., Jacob, M. H. V. M., Janner, D. d. R., Araújo, A. S. d. R., Belló-Klein, A., et al. (2011). The effect of dehydroepiandrosterone (DHEA) on renal function and metabolism in diabetic rats. *Steroids*, *76*(6), 564–570. https://doi.org/10.1016/j.steroids.2011.02.006.

Jaisser, F., & Farman, N. (2016). Emerging roles of the mineralocorticoid receptor in pathology: Toward new paradigms in clinical pharmacology. *Pharmacological Reviews*, *68*(1), 49–75. https://doi.org/10.1124/pr.115.011106.

Jones, S. A., Moore, L. B., Shenk, J. L., Wisely, G. B., Hamilton, G. A., McKee, D. D., et al. (2000). The pregnane X receptor: A promiscuous xenobiotic receptor that has diverged during evolution. *Molecular Endocrinology*, *14*(1), 27–39.

Kakiya, R., Shoji, T., Hayashi, T., Tatsumi-Shimomura, N., Tsujimoto, Y., Tabata, T., et al. (2012). Decreased serum adrenal androgen dehydroepiandrosterone sulfate and mortality in hemodialysis patients. *Nephrology, Dialysis, Transplantation*, *27*(10), 3915–3922. https://doi.org/10.1093/ndt/gfs162.

Kalimi, M., Shafagoj, Y., Loria, R., Padgett, D., & Regelson, W. (1994). Anti-glucocorticoid effects of dehydroepiandrosterone (DHEA). *Molecular and Cellular Biochemistry*, *131*(2), 99–104.

Kang, L., Zhang, X., Xie, Y., Tu, Y., Wang, D., Liu, Z., et al. (2010). Involvement of estrogen receptor variant ER-{alpha}36, not GPR30, in nongenomic estrogen signaling. *Molecular Endocrinology*, *24*(4), 709–721. https://doi.org/10.1210/me.2009-0317.

Kiersztan, A., Nagalski, A., Nalepa, P., Tempes, A., Trojan, N., Usarek, M., et al. (2016). DHEA-induced modulation of renal gluconeogenesis, insulin sensitivity and plasma lipid profile in the control- and dexamethasone-treated rabbits. Metabolic studies. *Biochimie*, *121*, 87–101. https://doi.org/10.1016/j.biochi.2015.11.019.

Kim, M. Y., Woo, E. M., Chong, Y. T. E., Homenko, D. R., & Kraus, W. L. (2006). Acetylation of estrogen receptor alpha by p300 at lysines 266 and 268 enhances the deoxyribonucleic acid binding and transactivation activities of the receptor. *Molecular Endocrinology*, *20*(7), 1479–1493. https://doi.org/10.1210/me.2005-0531.

Klein-Hitpass, L., Ryffel, G. U., Heitlinger, E., & Cato, A. C. (1988). A 13 bp palindrome is a functional estrogen responsive element and interacts specifically with estrogen receptor. *Nucleic Acids Research*, *16*(2), 647–663.

Kliewer, S. A., Moore, J. T., Wade, L., Staudinger, J. L., Watson, M. A., Jones, S. A., et al. (1998). An orphan nuclear receptor activated by pregnanes defines a novel steroid signaling pathway. *Cell*, *92*, 73–82.

Klinge, C. M. (2000). Estrogen receptor interaction with co-activators and co-repressors. *Steroids*, *65*, 227–251. https://doi.org/10.1016/S0039-128X(99)00107-5.

Klinge, C. M. (2001). Estrogen receptor interaction with estrogen response elements. *Nucleic Acids Research*, *29*(14), 2905–2919.

Klock, G., Strahle, U., & Schutz, G. (1987). Oestrogen and glucocorticoid responsive elements are closely related but distinct. *Nature*, *329*, 734–736.

Kobayashi, K., Hashimoto, M., Honkakoski, P., & Negishi, M. (2015). Regulation of gene expression by CAR: An update. *Archives of Toxicology*, *89*(7), 1045–1055. https://doi.org/10.1007/s00204-015-1522-9.

Kohalmy, K., Tamasi, V., Kobori, L., Sarvary, E., Pascussi, J.-M., Porrogi, P., et al. (2007). Dehydroepiandrosterone induces human CYP2B6 through the constitutive androstane receptor. *Drug Metabolism and Disposition: The Biological Fate of Chemicals*, *35*(9), 1495–1501. https://doi.org/10.1124/dmd.107.016303.

Konar, A., & Thakur, M. K. (2015). Neuropsin expression correlates with dendritic marker MAP2c level in different brain regions of aging mice. *Molecular Neurobiology*, *51*(3), 1130–1138. https://doi.org/10.1007/s12035-014-8780-z.

Kuiper, G. G., Carlsson, B., Grandien, J., Enmark, E., Haggblad, J., Nilsson, S., et al. (1997). Comparison of the ligand binding specificity and transcript tissue distribution of estrogen receptors α and β. *Endocrinology*, *138*, 863–870.

Labrie, F. (2010). DHEA, important source of sex steroids in men and even more in women. In M. Luciano (Ed.), *Progress in brain research: Vol. 82* (pp. 97–148). Elsevier.

Labrie, F. (2011). Blockade of testicular and adrenal androgens in prostate cancer treatment. *Nature Reviews. Urology*, *8*(2), 73–85.

Labrie, F., Luu-The, V., Bélanger, A., Lin, S.-X., Simard, J., Pelletier, G., et al. (2005). Is dehydroepiandrosterone a hormone? *Journal of Endocrinology*, *187*(2), 169–196. https://doi.org/10.1677/joe.1.06264.

Lang, F., Alevizopoulos, K., & Stournaras, C. (2013). Targeting membrane androgen receptors in tumors. *Expert Opinion on Therapeutic Targets*, *17*(8), 951–963. https://doi.org/10.1517/14728222.2013.806491.

Lannigan, D. A. (2003). Estrogen receptor phosphorylation. *Steroids*, *68*(1), 1–9.

Laurine, E., Lafitte, D., Gregoire, C., Seree, E., Loret, E., Douillard, S., et al. (2003). Specific binding of dehydroepiandrosterone to the N terminus of the microtubule-associated protein MAP2. *Journal of Biological Chemistry*, *278*(32), 29979–29986. https://doi.org/10.1074/jbc.M303242200.

Lazaridis, I., Charalampopoulos, I., Alexaki, V.-I., Avlonitis, N., Pediaditakis, I., Efstathopoulos, P., et al. (2011). Neurosteroid dehydroepiandrosterone interacts with nerve growth factor (NGF) receptors, preventing neuronal apoptosis. *PLoS Biology*, *9*(4), e1001051.

Lee, S. S., Pineau, T., Drago, J., Lee, E. J., Owens, J. W., Kroetz, D. L., et al. (1995). Targeted disruption of the alpha isoform of the peroxisome proliferator-activated receptor gene in mice results in abolishment of the pleiotropic effects of peroxisome proliferators. *Molecular and Cellular Biology*, *15*, 3012–3022.

Lee, M.-J., Pramyothin, P., Karastergiou, K., & Fried, S. K. (2014). Deconstructing the roles of glucocorticoids in adipose tissue biology and the development of central obesity. *Biochimica et Biophysica Acta*, *1842*(3), 473–481. https://doi.org/10.1016/j.bbadis.2013.05.029.

Lefebvre, P., Chinetti, G., Fruchart, J. C., & Staels, B. (2006). Sorting out the roles of PPAR alpha in energy metabolism and vascular homeostasis. *Journal of Clinical Investigation*, *116*(3), 571–580. https://doi.org/10.1172/jci27989.

Lewis, A., Tsai, S. Y., & Su, T. P. (2016). Detection of isolated mitochondria-associated ER membranes using the sigma-1 receptor. *Methods in Molecular Biology*, *1376*, 133–140. https://doi.org/10.1007/978-1-4939-3170-5_11.

Liang, X., Glowacki, J., Hahne, J., Xie, L., LeBoff, M. S., & Zhou, S. (2016). Dehydroepiandrosterone stimulation of osteoblastogenesis in human MSCs requires IGF-I signaling. *Journal of Cellular Biochemistry*, *117*(8), 1769–1774. https://doi.org/10.1002/jcb.25475.

Lim, W., Jeong, W., Kim, J., Yoshimura, Y., Bazer, F. W., Han, J. Y., et al. (2013). Expression and regulation of beta-defensin 11 in the oviduct in response to estrogen and in ovarian tumors of chickens. *Molecular and Cellular Endocrinology*, *366*(1), 1–8. https://doi.org/10.1016/j.mce.2012.10.031.

Lindschau, C., Kirsch, T., Klinge, U., Kolkhof, P., Peters, I., & Fiebeler, A. (2011). Dehydroepiandrosterone-induced phosphorylation and translocation of FoxO1 depend on the mineralocorticoid receptor. *Hypertension*, *58*(3), 471–478. https://doi.org/10.1161/hypertensionaha.111.171280.

Liss, K. H. H., & Finck, B. N. (2017). PPARs and nonalcoholic fatty liver disease. *Biochimie*, *136*, 65–74. https://doi.org/10.1016/j.biochi.2016.11.009.

Liu, D., & Dillon, J. S. (2002). Dehydroepiandrosterone activates endothelial cell nitric-oxide synthase by a specific plasma membrane receptor coupled to Galpha i2,3. *Journal of Biological Chemistry*, *277*(24), 21379–21388.

Liu, D., & Dillon, J. S. (2004). Dehydroepiandrosterone stimulates nitric oxide release in vascular endothelial cells: Evidence for a cell surface receptor. *Steroids*, *69*(4), 279–289.

Liu, D., Iruthayanathan, M., Homan, L. L., Wang, Y., Yang, L., Wang, Y., et al. (2008). Dehydroepiandrosterone stimulates endothelial proliferation and angiogenesis through extracellular signal-regulated kinase 1/2-mediated mechanisms. *Endocrinology*, *149*(3), 889–898. https://doi.org/10.1210/en.2007-1125.

Liu, D., O'Leary, B., Iruthayanathan, M., Love-Homan, L., Perez-Hernandez, N., Olivo, H. F., et al. (2010). Evaluation of a novel photoactive and biotinylated dehydroepiandrosterone analog. *Molecular and Cellular Endocrinology*, 328(1–2), 56–62. https://doi.org/10.1016/j.mce.2010.07.002.

Lu, S.-F., Mo, Q., Hu, S., Garippa, C., & Simon, N. G. (2003). Dehydroepiandrosterone upregulates neural androgen receptor level and transcriptional activity. *Journal of Neurobiology*, *57*(2), 163–171. https://doi.org/10.1002/neu.10260.

Maggio, M., De Vita, F., Fisichella, A., Colizzi, E., Provenzano, S., Lauretani, F., et al. (2015). DHEA and cognitive function in the elderly. *The Journal of Steroid Biochemistry and Molecular Biology*, *145*(0), 281–292. https://doi.org/10.1016/j.jsbmb.2014.03.014.

Maggiolini, M., Donze, O., Jeannin, E., Ando, S., & Picard, D. (1999). Adrenal androgens stimulate the proliferation of breast cancer cells as direct activators of estrogen receptor alpha. *Cancer Research*, *59*, 4864–4869.

Maninger, N., Wolkowitz, O. M., Reus, V. I., Epel, E. S., & Mellon, S. H. (2009). Neurobiological and neuropsychiatric effects of dehydroepiandrosterone (DHEA) and DHEA sulfate (DHEAS). *Frontiers in Neuroendocrinology*, *30*(1), 65–91. https://doi.org/10.1016/j.yfrne.2008.11.002.

Mannic, T., Mouffok, M., Python, M., Yoshida, T., Maturana, A. D., Vuilleumier, N., et al. (2013). DHEA prevents mineralo- and glucocorticoid receptor-induced chronotropic and hypertrophic actions in isolated rat cardiomyocytes. *Endocrinology*, *154*(3), 1271–1281. https://doi.org/10.1210/en.2012-1784.

Mannic, T., Viguie, J., & Rossier, M. F. (2015). In vivo and in vitro evidences of dehydroepiandrosterone protective role on the cardiovascular system. *International Journal of Endocrinology and Metabolism*, *13*(2), e24660. https://doi.org/10.5812/ijem.24660.

Marion-Letellier, R., Savoye, G., & Ghosh, S. (2016). Fatty acids, eicosanoids and PPAR gamma. *European Journal of Pharmacology*, *785*, 44–49. https://doi.org/10.1016/j.ejphar.2015.11.004.

Martinez-Outschoorn, U. E., Sotgia, F., & Lisanti, M. P. (2015). Caveolae and signalling in cancer. *Nature Reviews. Cancer*, *15*(4), 225–237. https://doi.org/10.1038/nrc3915.

Maurice, T. (2004). Neurosteroids and sigma1 receptors, biochemical and behavioral relevance. *Pharmacopsychiatry*, *37*(Suppl 3), S171–182. https://doi.org/10.1055/s-2004-832675.

Mc, A., Mark, T., Guimera, A. M., Hodgson, D., McDonald, N., Mooney, K. M., et al. (2017). Modelling the molecular mechanisms of aging. *Bioscience Reports*, *37*(1).

McNelis, J. C., Manolopoulos, K. N., Gathercole, L. L., Bujalska, I. J., Stewart, P. M., Tomlinson, J. W., et al. (2013). Dehydroepiandrosterone exerts antiglucocorticoid action on human preadipocyte proliferation, differentiation, and glucose uptake. *American Journal of Physiology. Endocrinology and Metabolism*, *305*(9), E1134–1144. https://doi.org/10.1152/ajpendo.00314.2012.

Meinel, S., Gekle, M., & Grossmann, C. (2014). Mineralocorticoid receptor signaling: Crosstalk with membrane receptors and other modulators. *Steroids*, *91*, 3–10. https://doi.org/10.1016/j.steroids.2014.05.017.

Meunier, J., & Maurice, T. (2004). Beneficial effects of the sigma1 receptor agonists igmesine and dehydroepiandrosterone against learning impairments in rats prenatally exposed to cocaine. *Neurotoxicology and Teratology*, *26*(6), 783–797. https://doi.org/10.1016/j.ntt.2004.07.003.

Miano, V., Ferrero, G., Reineri, S., Caizzi, L., Annaratone, L., Ricci, L., et al. (2016). Luminal long non-coding RNAs regulated by estrogen receptor alpha in a ligand-independent manner show functional roles in breast cancer. *Oncotarget*, 7(3), 3201–3216. https://doi.org/10.18632/oncotarget.6420.

Michael Miller, K. K., Al-Rayyan, N., Ivanova, M. M., Mattingly, K. A., Ripp, S. L., Klinge, C. M., et al. (2013). DHEA metabolites activate estrogen receptors alpha and beta. *Steroids*, *78*(1), 15–25. https://doi.org/10.1016/j.steroids.2012.10.002.

Mizota, K., Yoshida, A., Uchida, H., Fujita, R., & Ueda, H. (2005). Novel type of Gq/11 protein-coupled neurosteroid receptor sensitive to endocrine disrupting chemicals in mast cell line (RBL-2H3). *British Journal of Pharmacology*, *145*(4), 545–550. https://doi.org/10.1038/sj.bjp.0706213.

Monnet, F. P., Mahe, V., Robel, P., & Baulieu, E. E. (1995). Neurosteroids, via sigma receptors, modulate the [3H]norepinephrine release evoked by N-methyl-D-aspartate in the rat hippocampus. *Proceedings of the National Academy of Sciences of the United States of America*, *92*(9), 3774–3778.

Moriguchi, S., Shinoda, Y., Yamamoto, Y., Sasaki, Y., Miyajima, K., Tagashira, H., et al. (2013). Stimulation of the sigma-1 receptor by DHEA enhances synaptic efficacy and neurogenesis in the hippocampal dentate gyrus of olfactory bulbectomized mice. *PLoS One*, *8*(4), e60863, https://doi.org/10.1371/journal.pone.0060863.

Morin, C., Zini, R., Simon, N., & Tillement, J. P. (2002). Dehydroepiandrosterone and [alpha]-estradiol limit the functional alterations of rat brain mitochondria submitted to different experimental stresses. *Neuroscience*, *115*(2), 415–424.

Mostaghel, E. A. (2014). Beyond T and DHT—Novel steroid derivatives capable of wild type androgen receptor activation. *International Journal of Biological Sciences*, *10*(6), 602–613. https://doi.org/10.7150/ijbs.8844.

Mueck, A. O., Ruan, X., Seeger, H., Fehm, T., & Neubauer, H. (2014). Genomic and non-genomic actions of progestogens in the breast. *The Journal of Steroid Biochemistry and Molecular Biology*, *142*, 62–67. https://doi.org/10.1016/j.jsbmb.2013.08.011.

Mutoh, S., Osabe, M., Inoue, K., Moore, R., Pedersen, L., Perera, L., et al. (2009). Dephosphorylation of threonine 38 is required for nuclear translocation and activation of human xenobiotic receptor CAR (NR1I3). *Journal of Biological Chemistry*, *284*(50), 34785–34792. https://doi.org/10.1074/jbc.M109.048108.

Mutoh, S., Sobhany, M., Moore, R., Perera, L., Pedersen, L., Sueyoshi, T., et al. (2013). Phenobarbital indirectly activates the constitutive active androstane receptor (CAR) by inhibition of epidermal growth factor receptor signaling. *Science Signaling*, *6*(274), ra31. https://doi.org/10.1126/scisignal.2003705.

Nguyen, T. V. (2018). Developmental effects of androgens in the human brain. *Journal of Neuroendocrinology*, *30*(2), e12486, https://doi.org/10.1111/jne.12486.

Niro, S., Pereira, E., Pélissier, M.-A., Morfin, R., & Hennebert, O. (2012). The DHEA metabolite 7β-hydroxy-epiandrosterone exerts anti-estrogenic effects on breast cancer cell lines. *Steroids*, 77(5), 542–551. https://doi.org/10.1016/j.steroids.2012.01.019.

Olefsky, J. M. (2001). Nuclear receptor minireview series. *Journal of Biological Chemistry*, *276*(40), 36863–36864.

Olivo, H. F., Perez-Hernandez, N., Liu, D., Iruthayanathan, M., O'Leary, B., Homan, L. L., et al. (2010). Synthesis and application of a photoaffinity analog of dehydroepiandrosterone (DHEA). *Bioorganic & Medicinal Chemistry Letters*, 20(3), 1153–1155. https://doi.org/10.1016/j.bmcl.2009.12.019.

O'Malley, B. W. (2009). The "fourth dimension" of gene transcription. *Molecular Endocrinology*, *23*(5), 587–589. https://doi.org/10.1210/me.2009-0015.

Papadopoulos, D., Dietze, R., Shihan, M., Kirch, U., & Scheiner-Bobis, G. (2016). Dehydroepiandrosterone sulfate stimulates expression of blood-testis-barrier proteins claudin-3 and -5 and tight junction formation via a Gnalpha11-coupled receptor in Sertoli cells. *PLoS One*, *11*(3), e0150143, https://doi.org/10.1371/journal.pone.0150143.

Papadopoulos, D., & Scheiner-Bobis, G. (2017). Dehydroepiandrosterone sulfate augments blood-brain barrier and tight junction protein expression in brain endothelial cells. *Biochimica et Biophysica Acta*, *1864*(8), 1382–1392. https://doi.org/10.1016/j.bbamcr.2017.05.006.

Paulin, R., Meloche, J., Jacob, M. H., Bisserier, M., Courboulin, A., & Bonnet, S. (2011). Dehydroepiandrosterone inhibits the Src/STAT3 constitutive activation in pulmonary arterial hypertension. *American Journal of Physiology—Heart and Circulatory Physiology*, *301*(5), H1798–H1809. https://doi.org/10.1152/ajpheart.00654.2011.

Pawlak, M., Lefebvre, P., & Staels, B. (2015). Molecular mechanism of PPARalpha action and its impact on lipid metabolism, inflammation and fibrosis in non-alcoholic fatty liver disease. *Journal of Hepatology*, *62*(3), 720–733. https://doi.org/10.1016/j.jhep.2014.10.039.

Peckett, A. J., Wright, D. C., & Riddell, M. C. (2011). The effects of glucocorticoids on adipose tissue lipid metabolism. *Metabolism*, *60*(11), 1500–1510. https://doi.org/10.1016/j.metabol.2011.06.012.

Pediaditakis, I., Iliopoulos, I., Theologidis, I., Delivanoglou, N., Margioris, A. N., Charalampopoulos, I., et al. (2015). Dehydroepiandrosterone: An ancestral ligand of neurotrophin receptors. *Endocrinology*, *156*(1), 16–23. https://doi.org/10.1210/en.2014-1596.

Peters, J. M., Zhou, Y. C., Ram, P. A., Lee, S. S., Gonzalez, F. J., & Waxman, D. J. (1996). Peroxisome proliferator-activated receptor alpha required for gene induction by dehydroepiandrosterone-3 beta-sulfate. *Molecular Pharmacology*, *50*(1), 67–74.

Phuong, N. T. T., Lim, S. C., Kim, Y. M., & Kang, K. W. (2014). Aromatase induction in tamoxifen-resistant breast cancer: Role of phosphoinositide 3-kinase-dependent CREB activation. *Cancer Letters*, *351*(1), 91–99. https://doi.org/10.1016/j.canlet.2014.05.003.

Popow, J., Alleaume, A.-M., Curk, T., Schwarzl, T., Sauer, S., & Hentze, M. W. (2015). FASTKD2 is an RNA-binding protein required for mitochondrial RNA processing and translation. *RNA*, *21*(11), 1873–1884. https://doi.org/10.1261/rna.052365.115.

Porcu, P., Barron, A. M., Frye, C. A., Walf, A. A., Yang, S. Y., He, X. Y., et al. (2016). Neurosteroidogenesis today: Novel targets for neuroactive steroid synthesis and action and their relevance for translational research. *Journal of Neuroendocrinology*, *28*(2), 12351. https://doi.org/10.1111/jne.12351.

Prause, J., Goswami, A., Katona, I., Roos, A., Schnizler, M., Bushuven, E., et al. (2013). Altered localization, abnormal modification and loss of function of sigma receptor-1 in amyotrophic lateral sclerosis. *Human Molecular Genetics*, *22*(8), 1581–1600. https://doi.org/10.1093/hmg/ddt008.

Prossnitz, E. R., & Barton, M. (2011). The G-protein-coupled estrogen receptor GPER in health and disease. *Nature Reviews. Endocrinology*, 7(12), 715–726.

Prossnitz, E. R., & Maggiolini, M. (2009). Mechanisms of estrogen signaling and gene expression via GPR30. *Molecular and Cellular Endocrinology*, *308*(1-2), 32–38.

Prough, R. A., Clark, B. J., & Klinge, C. M. (2016). Novel mechanisms for DHEA action. *Journal of Molecular Endocrinology*, *56*(3), R139–R155. https://doi.org/10.1530/jme-16-0013.

Rajbhandari, P., Finn, G., Solodin, N. M., Singarapu, K. K., Sahu, S. C., Markley, J. L., et al. (2012). Regulation of estrogen receptor α N-terminus conformation and function by peptidyl prolyl isomerase Pin1. *Molecular and Cellular Biology*, *32*(2), 445–457. https://doi.org/10.1128/mcb.06073-11.

Razzoli, M., Pearson, C., Crow, S., & Bartolomucci, A. (2017). Stress, overeating, and obesity: Insights from human studies and preclinical models. *Neuroscience & Biobehavioral Reviews*, *76*, 154–162, Pt A. https://doi.org/10.1016/j.neubiorev.2017.01.026.

Regan, M. C., Romero-Hernandez, A., & Furukawa, H. (2015). A structural biology perspective on NMDA receptor pharmacology and function. *Current Opinion in Structural Biology*, *33*, 68–75. https://doi.org/10.1016/j.sbi.2015.07.012.

Rice, S. P. L., Zhang, L., Grennan-Jones, F., Agarwal, N., Lewis, M. D., Rees, D. A., et al. (2010). Dehydroepiandrosterone (DHEA) treatment in vitro inhibits adipogenesis in human omental but not subcutaneous adipose tissue. *Molecular and Cellular Endocrinology*, *320*(1–2), 51–57. https://doi.org/10.1016/j.mce.2010.02.017.

Ripp, S. L., Fitzpatrick, J. L., Peters, J. M., & Prough, R. A. (2002). Induction of CYP3A expression by dehydroepiandrosterone: Involvement of the pregnane X receptor. *Drug Metabolism and Disposition*, *30*(5), 570–575. https://doi.org/10.1124/dmd.30.5.570.

Rodal, S. K., Skretting, G., Garred, O., Vilhardt, F., van Deurs, B., & Sandvig, K. (1999). Extraction of cholesterol with methyl-beta-cyclodextrin perturbs formation of clathrin-coated endocytic vesicles. *Molecular Biology of the Cell*, *10*(4), 961–974.

Rousseaux, C. G., & Greene, S. F. (2015). Sigma receptors [sigmaRs]: Biology in normal and diseased states. *Journal of Receptor and Signal Transduction Research*, *36*(4), 1–62. https://doi.org/10.3109/10799893.2015.1015737.

Ruscher, K., & Wieloch, T. (2015). The involvement of the sigma-1 receptor in neurodegeneration and neurorestoration. *Journal of Pharmacological Sciences*, *127*(1), 30–35. https://doi.org/10.1016/j.jphs.2014.11.011.

Rybczynska, A. A., Elsinga, P. H., Sijbesma, J. W., Ishiwata, K., de Jong, J. R., de Vries, E. F., et al. (2009). Steroid hormones affect binding of the sigma ligand 11C-SA4503 in tumour cells and tumour-bearing rats. *European Journal of Nuclear Medicine and Molecular Imaging*, *36*(7), 1167–1175. https://doi.org/10.1007/s00259-009-1076-2.

Sachidanandan, D., & Bera, A. K. (2015). Inhibition of the GABAA receptor by sulfated neurosteroids: A mechanistic comparison study between pregnenolone sulfate and dehydroepiandrosterone sulfate. *Journal of Molecular Neuroscience*, *56*(4), 868–877. https://doi.org/10.1007/s12031-015-0527-4.

Sánchez, C., Díaz-Nido, J., & Avila, J. (2000). Phosphorylation of microtubule-associated protein 2 (MAP2) and its relevance for the regulation of the neuronal cytoskeleton function. *Progress in Neurobiology*, *61*(2), 133–168. https://doi.org/10.1016/S0301-0082(99)00046-5.

Savineau, J. P., Marthan, R., & Dumas de la Roque, E. (2013). Role of DHEA in cardiovascular diseases. *Biochemical Pharmacology*, *85*(6), 718–726. https://doi.org/10.1016/j.bcp.2012.12.004.

Scheffers, C. S., Armstrong, S., Cantineau, A. E., Farquhar, C., & Jordan, V. (2015). Dehydroepiandrosterone for women in the peri- or postmenopausal phase. *Cochrane Database of Systematic Reviews*, 1, Cd011066. https://doi.org/10.1002/14651858.CD011066.pub2.

Schrier, R. W., Wang, W., Poole, B., & Mitra, A. (2004). Acute renal failure: definitions, diagnosis, pathogenesis, and therapy. *The Journal of Clinical Investigation*, *114*(1), 5–14. https://doi.org/10.1172/JCI22353.

Shan, W., Palkar, P. S., Murray, I. A., McDevitt, E. I., Kennett, M. J., Kang, B. H., et al. (2008). Ligand activation of peroxisome proliferator-activated receptor beta/delta (PPARbeta/delta) attenuates carbon tetrachloride hepatotoxicity by downregulating proinflammatory gene expression. *Toxicological Sciences*, *105*(2), 418–428. https://doi.org/10.1093/toxsci/kfn142.

Shanle, E. K., Zhao, Z., Hawse, J., Wisinski, K., Keles, S., Yuan, M., et al. (2013). Research resource: Global identification of estrogen receptor β target genes in triple negative breast cancer cells. *Molecular Endocrinology*, *27*(10), 1762–1775. https://doi.org/10.1210/me.2013-1164.

Shihan, M., Chan, K.-H., Konrad, L., & Scheiner-Bobis, G. (2015). Non-classical testosterone signaling in spermatogenic GC-2 cells is mediated through ZIP9 interacting with Gnα11. *Cellular Signalling*, *27*(10), 2077–2086. https://doi.org/10.1016/j.cellsig.2015.07.013.

Shihan, M., Kirch, U., & Scheiner-Bobis, G. (2013). Dehydroepiandrosterone sulfate mediates activation of transcription factors CREB and ATF-1 via a Gα11-coupled receptor in the spermatogenic cell line GC-2. *Biochimica et Biophysica Acta*, *1833*(12), 3064–3075. https://doi.org/10.1016/j.bbamcr.2013.08.015.

Shizu, R., Osabe, M., Perera, L., Moore, R., Sueyoshi, T., & Negishi, M. (2017). Phosphorylated nuclear receptor CAR forms a homodimer to repress its constitutive activity for ligand activation. *Molecular and Cellular Biology*, *37*(10). https://doi.org/10.1128/mcb.00649-16.

Simoncini, T., Mannella, P., Fornari, L., Varone, G., Caruso, A., & Genazzani, A. R. (2003). Dehydroepiandrosterone modulates endothelial nitric oxide synthesis via direct genomic and nongenomic mechanisms. *Endocrinology*, *144*(8), 3449–3455.

Stárka, L., Dušková, M., & Hill, M. (2015). Dehydroepiandrosterone: A neuroactive steroid. *The Journal of Steroid Biochemistry and Molecular Biology*, *145*(0), 254–260. https://doi.org/10.1016/j.jsbmb.2014.03.008.

Sueyoshi, T., Moore, R., Sugatani, J., Matsumura, Y., & Negishi, M. (2008). PPP1R16A, the membrane subunit of protein phosphatase 1beta, signals nuclear translocation of the nuclear receptor constitutive active/androstane receptor. *Molecular Pharmacology*, *73*(4), 1113–1121. https://doi.org/10.1124/mol.107.042960.

Suzuki, M., Wright, L. S., Marwah, P., Lardy, H. A., & Svendsen, C. N. (2004). Mitotic and neurogenic effects of dehydroepiandrosterone (DHEA) on human neural stem cell cultures derived from the fetal cortex. *Proceedings of the National Academy of Sciences of the United States of America*, *101*(9), 3202–3207. https://doi.org/10.1073/pnas.0307325101.

Svob Strac, D., Jazvinscak, J. M., Erhardt, J., Mirkovic Kos, K., & Pericic, D. (2012). Modulation of recombinant GABA(A) receptors by neurosteroid dehydroepiandrosterone sulfate. *Pharmacology*, *89*(3–4), 163–171. https://doi.org/10.1159/000336058.

Talabér, G., Jondal, M., & Okret, S. (2013). Extra-adrenal glucocorticoid synthesis: Immune regulation and aspects on local organ homeostasis. *Molecular and Cellular Endocrinology*, *380*(1–2), 89–98. https://doi.org/10.1016/j.mce.2013.05.007.

Tamasi, V., Miller, K. K. M., Ripp, S. L., Vila, E., Geoghagen, T. E., & Prough, R. A. (2008). Modulation of receptor phosphorylation contributes to activation of peroxisome proliferator activated receptor α by dehydroepiandrosterone and other peroxisome proliferators. *Molecular Pharmacology*, *73*(3), 968–976. https://doi.org/10.1124/mol.107.036780.

Tan, J.-A., Sharief, Y., Hamil, K. G., Gregory, C. W., Zang, D.-Y., Sar, M., et al. (1997). Dehydroepiandrosterone activates mutant androgen receptors expressed in the androgen-dependent human prostate cancer xenograft CWR22 and LNCaP cells. *Molecular Endocrinology*, *11*(4), 450–459. https://doi.org/10.1210/me.11.4.450.

Tanaka, N., Aoyama, T., Kimura, S., & Gonzalez, F. J. (2017). Targeting nuclear receptors for the treatment of fatty liver disease. *Pharmacology & Therapeutics*, *179*, 142–157. https://doi.org/10.1016/j.pharmthera.2017.05.011.

Tasker, J. G., Di, S., & Malcher-Lopes, R. (2006). Rapid glucocorticoid signaling via membrane-associated receptors. *Endocrinology*, *147*(12), 5549–5556. https://doi.org/10.1210/en.2006-0981.

Teng, Y., Litchfield, L. M., Ivanova, M. M., Prough, R. A., Clark, B. J., & Klinge, C. M. (2014). Dehydroepiandrosterone-induces miR-21 transcription in HepG2 cells through estrogen receptor β and androgen receptor. *Molecular and Cellular Endocrinology*, *392*(1–2), 23–36.

Teng, Y., Radde, B. N., Litchfield, L. M., Ivanova, M. M., Prough, R. A., Clark, B. J., et al. (2015). Dehydroepiandrosterone activation of G-protein-coupled estrogen receptor rapidly stimulates microRNA-21 transcription in human hepatocellular carcinoma cells. *Journal of Biological Chemistry*, *290*(25), 15799–15811. https://doi.org/10.1074/jbc.M115.641167.

Thomas, W., & Harvey, B. J. (2011). Mechanisms underlying rapid aldosterone effects in the kidney. *Annual Review of Physiology, 73*, 335–357. https://doi.org/10.1146/annurev-physiol-012110-142222.

Thornton, J. W. (2001). Evolution of vertebrate steroid receptors from an ancestral estrogen receptor by ligand exploitation and serial genome expansions. *Proceedings of the National Academy of Sciences of the United States of America*, 98, 5671–5676.

Thornton, J. W., Need, E., & Crews, D. (2003). Resurrecting the ancestral steroid receptor: Ancient origin of estrogen signaling. *Science, 301*(5640), 1714–1717.

Traish, A. M., Kang, H. P., Saad, F., & Guay, A. T. (2011). Dehydroepiandrosterone (DHEA)—A precursor steroid or an active hormone in human physiology (CME). *The Journal of Sexual Medicine, 8*(11), 2960–2982. https://doi.org/10.1111/j.1743-6109.2011.02523.x.

Vannay, A., Fekete, A., Langer, R., Toth, T., Sziksz, E., Vasarhelyi, B., et al. (2009). Dehydroepiandrosterone pretreatment alters the ischaemia/reperfusion-induced VEGF, IL-1 and IL-6 gene expression in acute renal failure. *Kidney and Blood Pressure Research, 32*(3), 175–184. https://doi.org/10.1159/000224629.

Vinckevicius, A., & Chakravarti, D. (2012). Chromatin immunoprecipitation: Advancing analysis of nuclear hormone signaling. *Journal of Molecular Endocrinology, 49*(2), R113–123. https://doi.org/10.1530/jme-12-0016.

Wadosky, K. M., & Koochekpour, S. (2016). Molecular mechanisms underlying resistance to androgen deprivation therapy in prostate cancer. *Oncotarget*, 7(39), 64447–64470. https://doi.org/10.18632/oncotarget.10901.

Wang, L., Wang, Y. D., Wang, W. J., & Li, D. J. (2009). Differential regulation of dehydroepiandrosterone and estrogen on bone and uterus in ovariectomized mice. *Osteoporosis International, 20*(1), 79–92. https://doi.org/10.1007/s00198-008-0631-1.

Wang, Z.-Y., & Yin, L. (2015). Estrogen receptor alpha-36 (ER-α36): A new player in human breast cancer. *Molecular and Cellular Endocrinology, 418*, 193–206. https://doi.org/10.1016/j.mce.2015.04.017.

Waschatko, G., Kojro, E., Zahnow, M., & Gehrig-Burger, K. (2011). Photo-DHEA—A functional photoreactive dehydroepiandrosterone (DHEA) analog. *Steroids, 76*(5), 502–507. https://doi.org/10.1016/j.steroids.2011.01.009.

Watson, C. S., Jeng, Y.-J., & Guptarak, J. (2011). Endocrine disruption via estrogen receptors that participate in nongenomic signaling pathways. *The Journal of Steroid Biochemistry and Molecular Biology, 127*(1–2), 44–50. https://doi.org/10.1016/j.jsbmb.2011.01.015.

Webb, S. J., Geoghegan, T. E., Prough, R. A., & Michael Miller, K. K. (2006). The biological actions of dehydroepiandrosterone involves multiple receptors. *Drug Metabolism Reviews, 38*(1–2), 89–116.

Weiss, E. P., Shah, K., Fontana, L., Lambert, C. P., Holloszy, J. O., & Villareal, D. T. (2009). Dehydroepiandrosterone replacement therapy in older adults: 1- and 2-y effects on bone. *The American Journal of Clinical Nutrition, 89*(5), 1459–1467. https://doi.org/10.3945/ajcn.2008.27265.

Wickramasinghe, N., Manavalan, T., Dougherty, S., Riggs, K., Li, Y., & Klinge, C. (2009). Estradiol downregulates miR-21 expression and increases miR-21 target gene expression in MCF-7 breast cancer cells. *Nucleic Acids Research, 37*(8), 2584–2595.

Wiebe, J. P., & Lewis, M. J. (2003). Activity and expression of progesterone metabolizing 5alpha-reductase, 20alpha-hydroxysteroid oxidoreductase and 3alpha(beta)-hydroxysteroid oxidoreductases in tumorigenic (MCF-7, MDA-MB-231, T-47D) and nontumorigenic (MCF-10A) human breast cancer cells. *BMC Cancer, 3*, 9.

Williams, M. R., Dawood, T., Ling, S., Dai, A., Lew, R., Myles, K., et al. (2004). Dehydroepiandrosterone increases endothelial cell proliferation in vitro and improves endothelial function in vivo by mechanisms independent of androgen and estrogen receptors. *Journal of Clinical Endocrinology and Metabolism, 89*(9), 4708–4715.

Wu, T. T., Chen, Y., Zhou, Y., Adi, D., Zheng, Y. Y., Liu, F., et al. (2017). Prognostic value of dehydroepiandrosterone sulfate for patients with cardiovascular disease: A systematic review and meta-analysis. *Journal of the American Heart Association*, *6*(5). https://doi.org/10.1161/jaha.116.004896.

Wu, H. Q., Masset-Brown, J., Tweedie, D. J., Milewich, L., Frenkel, R. A., Martin-Wixtrom, C., et al. (1989). Induction of microsomal NADPH-cytochrome P-450 reductase and cytochrome P-450IVA1 (P-450LA omega) by dehydroepiandrosterone in rats: A possible peroxisomal proliferator. *Cancer Research*, *49*(9), 2337–2343.

Yabuki, Y., Shinoda, Y., Izumi, H., Ikuno, T., Shioda, N., & Fukunaga, K. (2015). Dehydroepiandrosterone administration improves memory deficits following transient brain ischemia through sigma-1 receptor stimulation. *Brain Research*, *1622*, 102–113. https://doi.org/10.1016/j.brainres.2015.05.006.

Yang, H., & Wang, H. (2014). Signaling control of the constitutive androstane receptor (CAR). *Protein & Cell*, *5*(2), 113–123. https://doi.org/10.1007/s13238-013-0013-0.

Yin, F., Sancheti, H., Liu, Z., & Cadenas, E. (2016). Mitochondrial function in ageing: coordination with signalling and transcriptional pathways. *The Journal of Physiology*, *594*(8), 2025–2042. https://doi.org/10.1113/jp270541.

York, B., Sagen, J. V., Tsimelzon, A., Louet, J.-F., Chopra, A. R., Reineke, E. L., et al. (2013). Research resource: Tissue- and pathway-specific metabolomic profiles of the steroid receptor coactivator (SRC) family. *Molecular Endocrinology*, *27*(2), 366–380. https://doi.org/10.1210/me.2012-1324.

Yoshinari, K., Kobayashi, K., Moore, R., Kawamoto, T., & Negishi, M. (2003). Identification of the nuclear receptor CAR:HSP90 complex in mouse liver and recruitment of protein phosphatase 2A in response to phenobarbital. *FEBS Letters*, *548*(1-3), 17–20.

Zhang, X., Tanaka, K., Yan, J., Li, J., Peng, D., Jiang, Y., et al. (2013). Regulation of estrogen receptor α by histone methyltransferase SMYD2-mediated protein methylation. *Proceedings of the National Academy of Sciences of the United States of America*, *110*(43), 17284–17289. https://doi.org/10.1073/pnas.1307959110.

Zhang, C.-Y., Wang, W.-Q., Chen, J., & Lin, S.-X. (2015). Reductive 17beta-hydroxysteroid dehydrogenases which synthesize estradiol and inactivate dihydrotestosterone constitute major and concerted players in ER+ breast cancer cells. *The Journal of Steroid Biochemistry and Molecular Biology*, *150*, 24–34. https://doi.org/10.1016/j.jsbmb.2014.09.017.

CHAPTER THREE

Mutation of *HSD3B2* Gene and Fate of Dehydroepiandrosterone

Maria Sonia Baquedano*,†, Gabriela Guercio*,†, Mariana Costanzo*, Roxana Marino*, Marco A. Rivarola*,†, Alicia Belgorosky*,†,1
*Endocrine Department, Hospital de Pediatría Garrahan, Buenos Aires, Argentina
†National Scientific and Technical Research Council, Buenos Aires, Argentina
1Corresponding author: e-mail address: abelgo12345@gmail.com

Contents

Abstract

3βHSD2 enzyme is crucial for adrenal and gonad steroid biosynthesis. In enzyme deficiency states, due to recessive loss-of-function *HSD3B2* mutations, steroid flux is altered and clinical manifestations result. Deficiency of 3βHSD2 activity in the adrenals precludes normal aldosterone and cortisol synthesis and the alternative backdoor and 11-oxygenated C19 steroid pathways and the flooding of cortisol precursors along the Δ5 pathway with a marked rise in DHEA and DHEAS production. In gonads, it precludes normal T and estrogen synthesis. Here, we review androgen-dependent male differentiation of the external genitalia in humans and link this to female development and steroidogenesis in the developing adrenal cortex. The molecular mechanisms governing postnatal adrenal cortex zonation and ZR development were also revised. This chapter will review relevant clinical, hormonal, and genetic aspects of 3βHSD2 deficiency with emphasis on the significance of alternate fates encountered by steroid hormone precursors in the adrenal gland and gonads. Our current knowledge of the

Vitamins and Hormones, Volume 108
ISSN 0083-6729
https://doi.org/10.1016/bs.vh.2018.05.002

process of steroidogenesis and steroid action is derived from pathological conditions. In humans the 3βHSD2 deficiency represents a model of nature that reinforces our knowledge about the role of the steroidogenic alternative pathway in sex differentiation in both sexes. However, the physiological role of the high serum DHEAS levels in fetal life as well as after adrenarche remains to be elucidated.

1. ROLE OF 3β-HYDROXYSTEROID DEHYDROGENASE IN STEROID FORMATION

Type 2 3β-hydroxysteroid dehydrogenase Δ^4–Δ^5 isomerase (3βHSD2) is a bifunctional microsomal and mitochondrial nicotinamide adenine dinucleotide (NAD)+-dependent membrane-bound enzyme that catalyzes the conversion of the three principal endogenous Δ^5-steroid precursors, pregnenolone, 17α-hydroxypregnenolone, and dehydroepiandrosterone (DHEA) into their respective Δ^4-ketosteroids, namely progesterone, 17α-hydroxyprogesterone (17OHP), and androstenedione (A4), all with similar efficiency (apparent K_m and apparent V_{max}) (Lee, Miller, & Auchus, 1999; Rhéaume et al., 1991). The first reaction is the oxidation of the 3β-hydroxyl group to the ketone by dehydrogenase activity; during this process, NAD+ is reduced to NADH. The intermediate Δ^5,3-ketosteroid remains tightly bound to the enzyme with nascent NADH, and the presence of NADH in the cofactor-binding site activates the Δ^4–Δ^5-isomerase activity, residing in the same enzyme that adapts different conformation (Thomas, Boswell, Scaccia, Pletnev, & Umland, 2005; Thomas, Frieden, Nash, & Strickler, 1995). Therefore, 3βHSD2 is essential for the biosynthesis of all classes of active steroid hormones, including aldosterone, cortisol in the adrenal cortex, and sex steroids in the adrenals and gonads.

1.1 Human Steroidogenesis

Steroidogenesis entails processes by which cholesterol is converted to biologically active steroid hormones. These processes are repeated in each steroidogenic tissue with cell-type-specific patterns that are dictated by the cell-specific expression of specific steroidogenic enzymes.

The cleavage of cholesterol to pregnenolone is the first, rate-limiting, and hormonally regulated step (John, John, Boggaram, Simpson, & Waterman, 1986) in the biosynthesis of steroid hormones common to all steroidogenic cells. The cholesterol side-chain cleavage enzyme, P450scc (*CYP11A1*), supported by its electron transport system consisting of reduced nicotinamide adenine dinucleotide phosphate, adrenodoxin reductase (*FDXR*), and adrenodoxin (*FDX1*) (Miller, 2005), catalyzes three sequential reactions: 20α-hydroxylation, 22-hydroxylation, and cleavage of the 20–22 carbon bond of cholesterol to yield pregnenolone. Because P450scc resides on the inner mitochondrial membrane (Black, Harikrishna, Szklarz, & Miller, 1994), steroid acute regulatory (StAR) protein, as part of a multiprotein complex, termed transduceosome (Rone, Fan, & Papadopoulos, 2009), facilitates the movement of cholesterol from the outer to the inner mitochondrial membrane, thus providing the substrate for steroid hormone biosynthesis (Lin et al., 1995). The specific repertoire of enzymes distal to CYP11A1 in a cell determines the fate of pregnenolone metabolism and defines the function of that cell.

A schematic overview of steroidogenesis, including the classic androgen synthesis pathway and the two alternative androgen synthesis pathways (11-oxygenated 19 carbon steroid and alternative pathway to DHT), is shown in Fig. 1.

1.2 Adrenal Steroidogenesis

The coexpression of 3β-hydroxysteroid dehydrogenase/Δ^4–Δ^5 isomerase, 3βHSD (*HSD3B1* (Doi et al., 2014) and *HSD3B2*), and aldosterone synthase, P450c11AS (*CYP11B2*) in the zona glomerulosa (ZG) leads to aldosterone production under regulation by the renin/angiotensin system. The adrenal zona fasciculata (ZF) does not express angiotensin II receptors or P450c11AS, but instead expresses the ACTH receptor, MC2R, and 11β-hydroxylase, P450c11 (*CYP11B1*). The coexpression of 3βHSD type 2 (*HSD3B2*) and cytochrome P450 17α-hydroxylase/17,20-lyase, P450c17 (*CYP17A1*) along with P450c11 in the ZF, results in the production of cortisol under the influence of ACTH. By contrast, the zona reticularis (ZR) expresses relatively very little 3βHSD2 and large amounts of P450c17, cytochrome *b*5, which selectively activates the 17,20-lyase activity of P450c17 (Auchus, Lee, & Miller, 1998), and DHEA-sulfotransferase (*SULT2A1*) (Suzuki et al., 2000). Consequently, the C19 steroid DHEA is produced, much of which is sulfated to DHEAS (Miller & Auchus, 2011; Turcu, Smith, Auchus, & Rainey, 2014). This Δ^5 pathway is the preferred route to C19 steroid production in humans as the 17,20-lyase activity of human *CYP17A1* does not efficiently convert 17α-hydroxyprogesterone (17OHP) to androstenedione (A4) (Auchus, Lee, & Miller, 1998; Fig. 1). Androgen production in human adrenal cortex

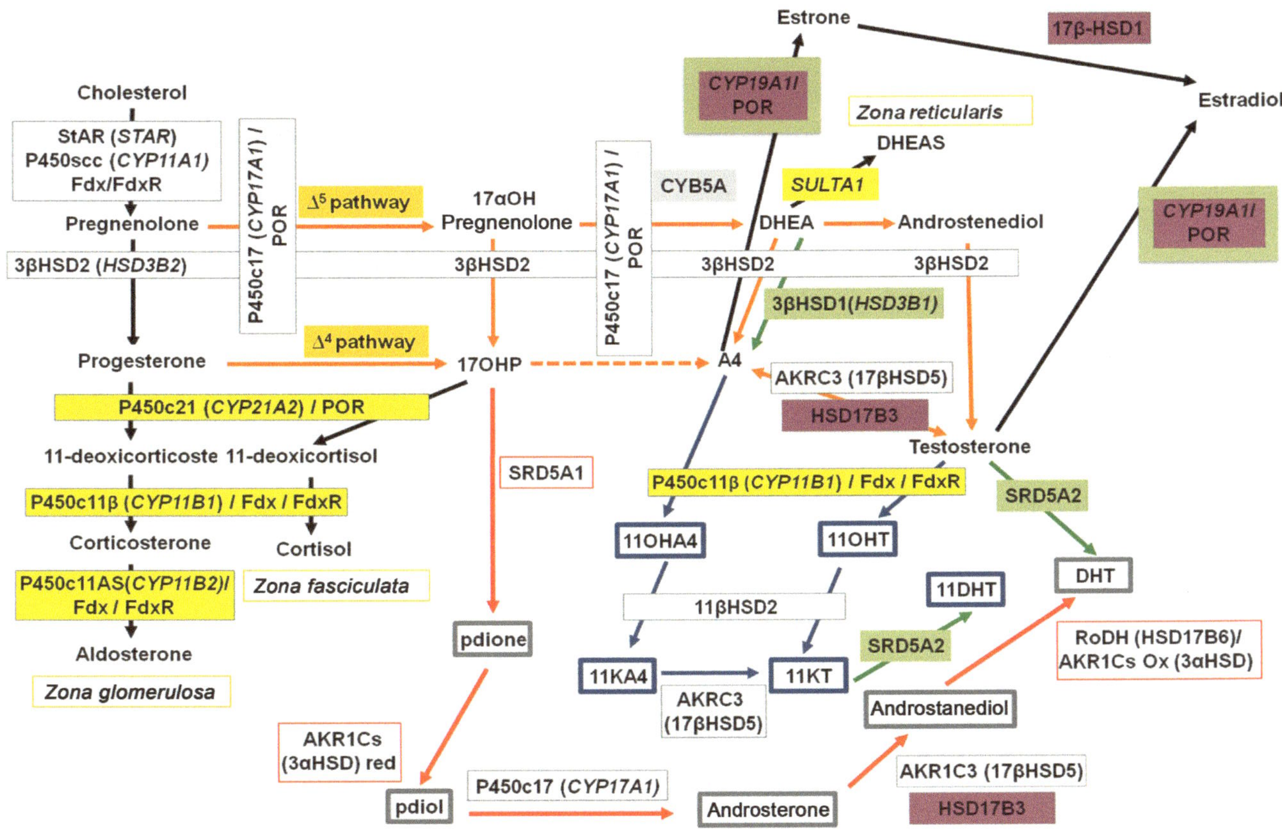

Fig. 1 See legend on opposite page.

is zonally and developmentally regulated. The 3βHSD2-deficient ZR is indistinct during infancy, but a continuous layer of reticularis cells starts to develop and thicken around 4–5 years (Remer, Boye, Hartmann, & Wudy, 2005). This process, known as adrenarche, is followed by a rise in circulating concentrations of DHEAS, with clinical signs physiologically observed between the ages of 6 and 10 (Auchus, 2011; Belgorosky, Baquedano, Guercio, & Rivarola, 2008). The human adrenal gland also makes little A4 and testosterone (T). It has been proposed that A4 is produced from DHEA in a layer of cells in the interface of the ZF and ZR which expresses CYP17A1, cytochrome *b*5, and 3βHSD2 (Nakamura et al., 2015, 2011). The expression of aldo-ketoreductase 1C3 (AKR1C3), also known as 17β-hydroxysteroid dehydrogenase type 5 (17βHSD5), in the ZR would be responsible for the conversion of DHEA and A4 to androstenediol and T, respectively (Nakamura, Hornsby, et al., 2009; Fig. 1).

Furthermore, the advent of comprehensive analyses of steroid profiles by sensitive and specific liquid chromatography (LC) and gas chromatography/tandem mass spectrometry (GC-MS/MS) revealed adrenal alternative androgen synthesis pathways (Homma et al., 2006; Jones et al., 2017; Kamrath, Hochberg, Hartmann, Remer, & Wudy, 2012; O'Reilly et al., 2017; Rege et al., 2013; Turcu et al., 2016). 17OHP is also a substrate for an alternative, so-called backdoor pathway to androgen synthesis, which generates DHT without the intermediacy of DHEA, A4, or T (Auchus, 2004). This pathway depends on the 5α- and 3α-reduction of 17OHP to

Fig. 1 Pathways of human steroid hormone synthesis. The biosynthetic pathways shown are relevant to the gonads and the adrenal cortex. In *white boxes* are shown enzymes that are common to adrenals and gonads; in *purple*, those related to gonadal steroids, and in *yellow boxes* those related to adrenal-specific pathways. The classical androgen synthesis pathway is showed in *orange*. Δ^5 pathway involving pregnenolone through DHEA, and Δ^4 pathway involving progesterone through testosterone are indicated. Minor routes indicated by *dashed arrows*. The alternative "backdoor" androgen pathway is in *red*, and the 11oxC19-steroid pathway is showed in *blue*. In *green* are depicted enzymes involved in routes to androgens and estrogens in the periphery. A small amount of DHEA is converted in the adrenal by HSD3B2, and larger amounts derive from peripheral metabolism, mediated by HSD3B1. *11KA4*, 11-ketoandrostenedione; *11KT*, 11-ketotestosterone; *11OHA4*, 11β-hydroxyandrostenedione; *11OHT*, 11β-hydroxytestosterone; *17OHP*, 17α-hydroxyprogesterone; *A4*, androstenedione; *CYB5A*, cytochrome *b*5; *DHEA*, dehydroepiandrosterone; *DHEAS*, dehydroepiandrosterone sulfate; *DHT*, 5α-dihydrotestosterone; *pdiol*, 17-hydroxyallopregnenolone (5α-pregnane-3α,17α-diol-20-one); *pdione*, 17-hydroxydihydroprogesterone (5α-pregnane-17α-ol-3,20-dione); *T*, testosterone.

produce 17-hydroxyallopregnenolone (pdiol), followed by 17,20-lyase activity of P450c17 and 17βHSD dehydrogenase activity, yielding androstanediol, and finally a 3α-oxidation step to DHT (Fig. 1) (Auchus, 2004). In contrast to the classical pathway, human CYP17A1 has a very high affinity for backdoor pathway intermediate, pdiol, which is an excellent substrate for its 17,20-lyase activity, not dependent from cytochrome *b*5, leading to androsterone production (Auchus, 2004; Gupta, Guryev, & Auchus, 2003). Recent evidence showed that the human adrenal cortex would express the enzymes to complete all the steps in the backdoor pathway to DHT (Baquedano et al., 2014; Flück et al., 2011; Marti et al., 2017) and that this pathway would contribute to the androgen production in pathological states in which 17OHP accumulates, including 21α-hydroxylase and POR deficiencies (Homma et al., 2006; Jones et al., 2017; Kamrath, Hochberg, Hartmann, Remer, & Wudy, 2012). Moreover, recent studies have demonstrated that human adrenal also produces a unique set of 11-oxygenated C19 steroid (11oxC19) (Fig. 1) (O'Reilly et al., 2017;Rege et al., 2013; Turcu et al., 2016). The first step of the 11-oxygenated androgen pathway is dependent on the adrenal CYP11B1-catalyzed 11β-hydroxylation of A4 to 11β-hydroxyandrostenedione (11OHA4), which is a major product of adrenal steroidogenesis (Rege et al., 2013; Turcu et al., 2016). Low quantities of the 11-oxygenated C19 steroid 11-ketoandrostenedione (11KA4), 11β-hydroxytestosterone (11OHT), and 11-ketotestosterone (11KT) are also produced by the human adrenal (Rege et al., 2013; Turcu et al., 2016; Fig. 1). 11KT and its peripheral 5α-reduced metabolite, 11-keto-5α-dihydrotestosterone (11KDHT), are potent agonists of the human androgen receptor (AR) with affinities and potencies similar to that of T and 5α-dihydrotestosterone (DHT), respectively (Pretorius et al., 2016; Rege et al., 2013; Storbeck et al., 2013).

1.3 Gonadal Steroidogenesis

Testicular synthesis of T follows a pathway that is similar to C19-steroid production in the adrenal ZR, but the stimulus for steroidogenesis is transduced by the LH receptor rather than MC2R and Leydig cells express abundant 3βHSD2 and 17βHSD3, but no SULT2A1 (Miller & Auchus, 2011). Thus, DHEA produced in the testis is not sulfated but is readily converted to A4 and/or androstenediol and then T via 3βHSD2 and 17βHSD3 (Fig. 1). As in the adrenal, the principal pathway to C19-steroids is via Δ^5 steroids to DHEA. Testicular T is converted to DHT in peripheral tissues, such as prostate and genital skin, by steroid 5α-reductase type 2 (*SRDA2*) (Auchus & Miller, 2012). Gonadal T is not an important precursor for adrenal 11β-hydroxylated 11OHT and 11KT (Turcu et al., 2016).

In the ovary, enzymatic steps of steroidogenesis are partitioned between the granulosa and theca cells, which surround the oocyte and form a follicle. Furthermore, the patterns of steroidogenesis vary during the cycle: estradiol is the principal product in the follicular phase, and progesterone is produced in the luteal phase. Granulosa cells do not express P450c17 (Voutilainen, Tapanainen, Chung, Matteson, & Miller, 1986). Thus, steroidogenesis is initiated in granulosa cells under the influence of LH, which stimulates the P450scc expression (Voutilainen, Tapanainen, Chung, Matteson, & Miller, 1986). Pregnenolone and progesterone from granulosa cells diffuse into adjacent theca cells, where P450c17 and 3βHSD2 catalyze A4 synthesis. Small amounts of this A4 are secreted or converted to T by AKR1C3/17βHSD5, but most A4 returns to the granulosa cells where P450 aromatase (*CYP19A1*) and 17βHSD1 convert it to estrone and then to estradiol, respectively, under the influence of FSH.

Recent gene and protein expression studies revealed that human gonads express all genes necessary for DHT production via the backdoor pathway (Flück et al., 2011; Marti et al., 2017). Furthermore, the expression of the backdoor pathway in the fetal testis appears to play an important role during male sexual differentiation (Flück et al., 2011; see below).

1.4 Gonadal and Adrenocortical Organogenesis. Human Sexual Differentiation

The adrenal cortex and gonads derive from a common embryological adrenogonadal precursor lineage. In human embryos, these adrenogonadal progenitors, referred to as the adrenogonadal primordium, can be identified from about 3 weeks postconception (wpc) as a thickening on the inner surface of the mesonephros, the equivalent to the primitive kidney system (Else & Hammer, 2005; Parker et al., 2002). As this primordium grows, these cells subsequently differentiate and the gonadal primordium and adrenal primordium then separate. This occurs at 33 dpc (Goto et al., 2006).

In the 7th wpc, the bipotential gonad begins to differentiate into a testis or an ovary, depending on the genetic sex of the individual and, during 8–12 wpc, the sexual dimorphism of the external genitalia is established. The male gonad initiates the expression of the sex-determining region of the Y chromosome (*SRY*) gene in the cells destined to become Sertoli cells, which secrete anti-Müllerian hormone leading to regression of the mesonephric ducts, which would otherwise give rise to the female internal genitalia (Wilhelm, Palmer, & Koopman, 2007). A few days later, a second wave of differentiation within the testis gives rise to a population of Leydig cells that secrete androgens, which virilize the male internal and external genitalia (Sobel, Zhu, & Imperato-McGinley, 2004). The evidence shows that both testicular T converted to DHT in genital skin by 5α-reductase type 2 and DHT from testicular backdoor pathway are needed for normal male sexual development (Flück et al., 2011). DHT signals to the bipotential external genitalia via the AR to elicit differentiation irrevocably along a male pathway: the urethral folds fuse in the midline to close the urogenital sinus, the genital swellings develop into the scrotum, and the phallus enlarges into a penis. Thus, during this narrow window of sexual differentiation (8–12 wpc), maintaining the appropriate intrauterine hormone environment is critical (Asby, Arlt, & Hanley, 2009; Hanley & Arlt, 2006).

Meanwhile, the cells of the adrenal primordium continue to proliferate, and at about 48 dpc, chromaffin cells originating from the neural crest migrate to the area where the adrenogonadal primordium is developing (Xing, Lerario, Rainey, & Hammer, 2015). At the same time, the adrenal cortex becomes encapsulated and by 50–52 dpc, two discrete zones are detectable in the adrenal cortex, the inner fetal zone (FZ), and the smaller outer zone, or definitive zone (DZ) (Goto et al., 2006; Xing, Lerario, Rainey, & Hammer, 2015). All key steroidogenic enzymes, including StAR, CYP11A, CYP17, and CYP21, are first detected robustly at 50–52 dpc within the nascent inner FZ and weaker within the outer DZ, except for CYP17 that is absent (Goto et al., 2006). Importantly, 3βHSD2 is detected in a wave of expression from 8 wpc until the end of the first trimester at the interface between the DZ and FZ. The presence of 3βHSD2 facilitates de novo cortisol biosynthesis, but it would also be relevant to androgen biosynthesis by either classical or alternative pathways. In the former, DHEA is converted to androstenedione by 3βHSD2; in the latter, 3βHSD2 would be a means of providing progesterone from pregnenolone (Auchus, 2004; Hanley & Arlt, 2006). Indeed, Goto and coworkers have demonstrated that the anterior pituitary–adrenal axis does appear mature as early as 50–52 dpc and that ACTH stimulates the secretion of A4 and T by the fetal adrenal cortex during the first trimester (Goto et al., 2006). This intimately ties the developing adrenal gland to the process of sexual differentiation and led to the hypothesis that the transient fetal cortisol production is capable of attenuating ACTH stimulation of the adrenal and subsequent upregulation of androgen production, thereby acting as a safeguard mechanism for normal female sexual development (Goto et al., 2006).

Regardless of the source of androgen production, the target tissue responds by male sexual differentiation of the external genitalia during a period spanning from approximately 8 weeks of fetal development until the end of the first trimester. From the second trimester onward, the potential for further virilization, even in the presence of androgen excess, is limited at least partly by diminishing expression of the AR in females (Shapiro, Huang, & Wu, 2000) and the onset of fetal P450 aromatase (*CYP19A1*) expression in both sexes (Grumbach & Auchus, 1999).

During the second and third trimesters, the adrenal glands continue to grow rapidly, largely due to an increase in the size of the FZ, which produce large amounts of DHEA and DHEAS (Ishimoto & Jaffe, 2011). A third cortical zone becomes evident by 14 weeks (Goto et al., 2006). This transitional zone (TZ) is located between the DZ and the FZ. After 23–24 weeks' gestation, 3βHSD2 expression is again detectable in the DZ and TZ (Ishimoto & Jaffe, 2011). Immediately before birth, there is a second peak in cortisol production that is required for fetal organ maturation.

After birth, a strong remodeling of the adrenal gland occurs; the fetal zone undergoes apoptosis, followed by encapsulation of the medulla and final zonation of the adult cortex. These morphologic changes are accompanied by a rapid drop in DHEA and DHEAS production due to the involution of the fetal zone. In preadrenarche children, the ZG and the ZF are clearly present, but only focal islands of ZR cells, insufficient to influence serum DHEAS levels, can be identified at the ages 3–5 years (Belgorosky, Baquedano, Guercio, & Rivarola, 2008; Dardis, Saraco, Rivarola, & Belgorosky, 1999; Gell et al., 1998; Nakamura, Gang, Suzuki, Sasano, & Rainey, 2009; Suzuki et al., 2000). After adrenarche, there is a development and thickening of a continuous ZR associated with detectable increases in circulating DHEA and DHEAS (de Peretti & Forest, 1976; Endoh, Kristiansen, Casson, Buster, & Hornsby, 1996; Havelock, Auchus, & Rainey, 2004; Hui et al., 2009).

2. HUMAN 3βHSD2 DEFICIENCY

In enzyme deficiency states, steroid flux is altered and clinical manifestations often result (Ghayee & Auchus, 2007). Deficiency of 3βHSD2 activity in the adrenals precludes normal aldosterone and cortisol synthesis and the flooding of cortisol precursors along the Δ5 pathway with a marked rise in pregnenolone, 17α-hydroxypregnenolone, DHEA, DHEAS, and androstenediol production (Fig. 2). Deficiency of 3βHSD2 activity in gonads precludes normal T and estrogen synthesis. The blockage caused

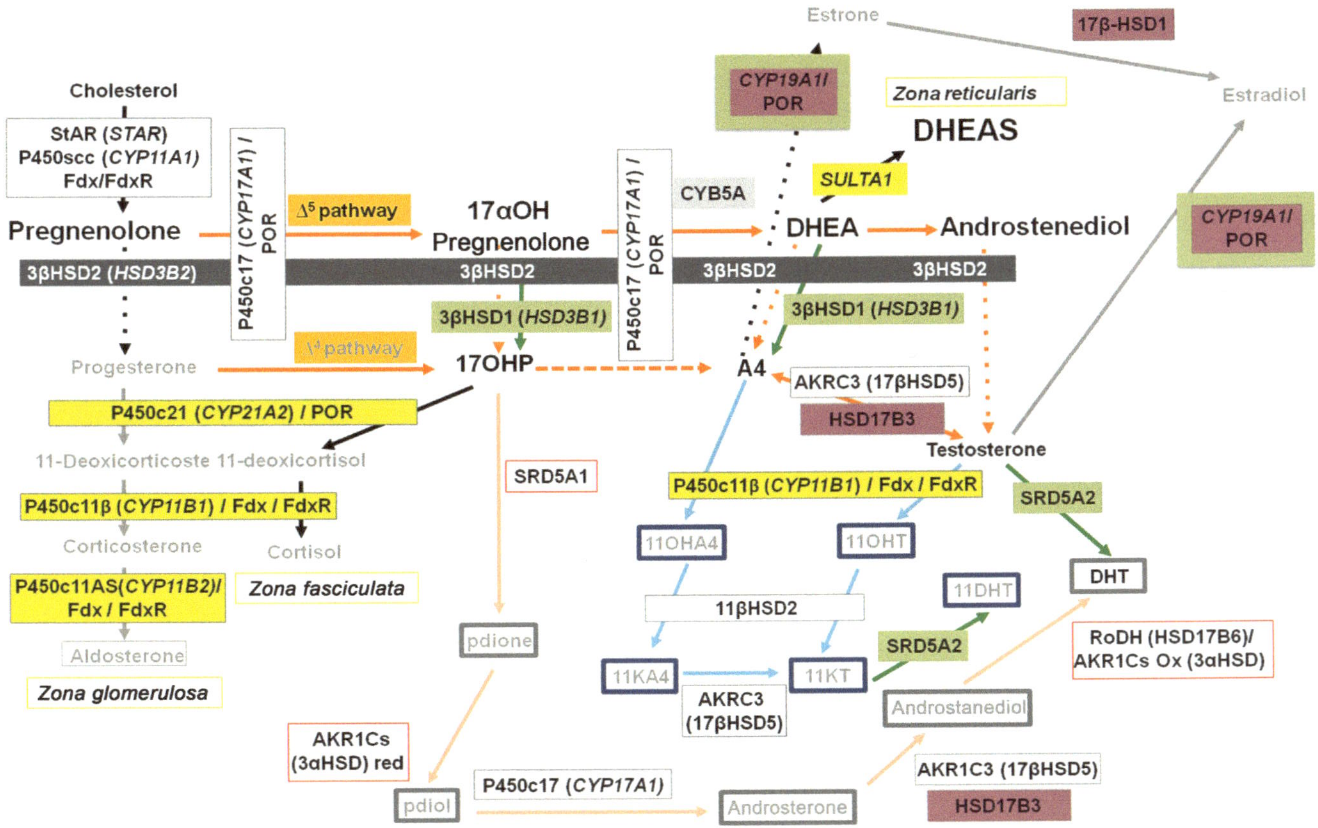

Fig. 2 See legend on opposite page.

by 3βHSD2 deficiency precludes flux to the backdoor pathway and 11-oxygenated C19 steroid synthesis due to very low intraadrenal 17OHP and A4 substrate levels, respectively (Fig. 2). However, peripheral 3βHSD type 1 (3βHSD1) conversion of Δ^5 steroids to Δ^4 steroids can result in high serum 17OHP and A4 (see below) (Fig. 2).

2.1 Molecular Genetics of Human 3βHSD2 Deficiency

In humans there are two closely linked 3βHSD genes and cognate enzymes. Type 1 (*HSD3B1*) and type 2 (*HSD3B2*) 3βHSD genes map to the short arm of chromosome 1 (1p13.1) with identical intron/exon organizations (Luu et al., 1989; Rhéaume et al., 1991). The structure of each of the *HSD3B1* and *HSD3B2* genes consists of four exons, which are included within a DNA fragment of 7.8 kb (Lorence, Corbin, Kamimura, Mahendroo, & Mason, 1990; Fig. 3A). *HSD3B1* gene encodes an enzyme of 372 amino acids predominantly expressed in placenta, breast, liver, brain, and some other peripheral tissues (Simard et al., 2005). *HSD3B2* gene, which encodes a protein of 371 amino acids (Fig. 3B), shares 93.5% identity with the type 1 and is almost exclusively expressed in the adrenals and gonads (Lorence et al., 1990; Rhéaume et al., 1991). The purified 3βHSD2 enzyme has a Michaelis constant (K_m) of 47 μM and a maximal velocity (V_{max}) of 82 nmol/min mg for 3βHSD substrate (DHEA) and a K_m of 88 μM and V_{max} of 970 nmol/min mg for the isomerase substrate (5-androstene-3,17-dione), while 3βHSD1 has a K_m of 3.7 μM and a V_{max} of 43 nmol/min mg for 3βHSD substrate (DHEA) and a K_m of 28 μM and V_{max} of 598 nmol/min mg for the isomerase substrate (5-androstene-3,17-dione) (Thomas, Mason, Blanco, & Veisaga, 2001). The higher affinity of 3βHSD1 could facilitate

Fig. 2 Altered steroid pathways in 3βHSD2 deficiency. The block of adrenal cortisol and aldosterone synthesis hunts precursors to 19-carbon steroids. Deficiency of 3βHSD2 activity in gonads precludes normal testosterone and estrogen synthesis. Peripheral 3βHSD1-catalyzed conversion of Δ^5 steroids to Δ^4 steroids results in high 17OHP and A4 in 46,XX patients. However, the alternative androgen synthesis pathways (11-oxygenated 19 carbon steroids and the "backdoor" pathway to DHT) are precluded due to very low intraadrenal Δ^4 steroids. Steroids that are produced in excess have larger font. Steroids that are deficient are graphically not emphasized, with minor pathways shown with *dotted arrows*. *11KA4*, 11-ketoandrostenedione; *11KT*, 11-ketotestosterone; *11OHA4*, 11β-hydroxyandrostenedione; *11OHT*, 11β-hydroxytestosterone; *17OHP*, 17α-hydroxyprogesterone; *A4*, androstenedione; *CYB5A*, cytochrome *b5*; *DHEA*, dehydroepiandrosterone; *DHEAS*, dehydroepiandrosterone sulfate; *DHT*, 5α-dihydrotestosterone; *pdiol*, 17-hydroxyallopregnenolone (5α-pregnane-3α,17α-diol-20-one); *pdione*,17-hydroxydihydroprogesterone (5α-pregnane-17α-ol-3,20-dione); *T*, testosterone.

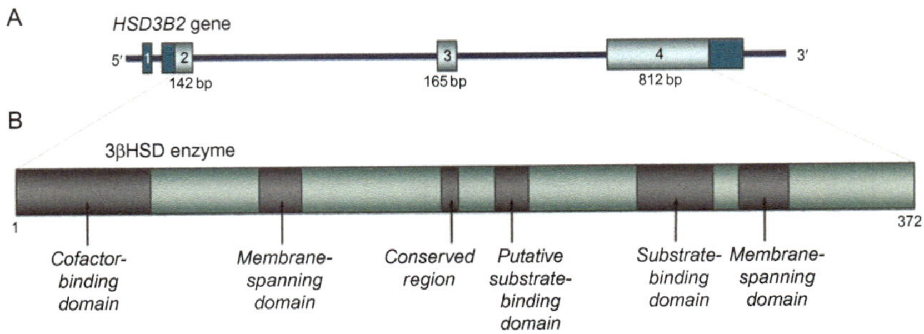

Fig. 3 (A) Genomic organization for *HSD3B2* gene. *Numbered boxes* represent exons. *Lines* represent introns. The HSD3B2 gene has four exons, whereof exon 1 and the 5′ part of exon 2 are not translated. (B) A cartoon of the different functional domains of the 3βHSD2 enzyme.

steroid formation from relatively low concentrations of substrates usually present in peripheral tissues.

The *HSD3B2* gene is mutated in 3βHSD deficiency, while *HSD3B1* gene mutations have never been described, possibly because placental 3βHSD1 deficiency would preclude progesterone production during pregnancy. Mutations throughout the *HSD3B2* gene have been reported. To date, a total of 50 mutations (including frameshift, nonsense, in-frame deletion, splicing, and missense mutations) have been identified in the *HSD3B2* gene in 82 individuals from 66 families suffering from classical 3βHSD deficiency as shown in Table 1. Gene conversions with the *HSD3B1* gene have not been described, and founder mutations account for only small cluster of cases in remote areas.

2.2 Clinical Features of Human 3βHSD2 Deficiency

Recessive loss-of-function *HSD3B2* mutations cause a rare form of congenital adrenal hyperplasia (CAH) that impairs both adrenal and gonadal steroidogenesis. It is one of the rarest forms of CAH with an incidence of only 1 in approximately 1,000,000 births and no ethnic pockets with higher prevalence have been described. Cortisol deficiency leads to an increase in corticotropin (ACTH) secretion through the feedback mechanism that results in increased production of Δ^5 steroid precursors, such as pregnenolone, 17α-hydroxypregnenolone, DHEA, and androstenediol from the adrenal cortex. Aldosterone deficiency causes the salt wasting, and combined cortisol and aldosterone deficiencies cause adrenal crisis if the affected subjects are not early identified and treated (Auchus & Chang, 2010; Auchus & Miller, 2012; Pang, 2001; Simard, Moisan, & Morel, 2002).

Table 1 *HSD3B2* Gene Mutations and Their Residual In Vitro Activity Reported in 3βHSD2-Deficient Patients

Mutation	Variant ID	Activity In Vitro (% of wt)	Domain of the protein	Evidence of Protein Instability	References
Missense					
p.Leu6Phe c.16C > T	CM000720	Δ5-P to P 47.8%, DHEA to Δ4-A 39.3%		Yes	Zhang et al. (2000)
p.Ala10Glu c.29C > A	CM001196 rs28934880	DHEA to Δ4-A 0%	Putative NAD-binding domain	Yes	Alos et al. (2000)
p.Ala10Val c.29C > T	CM993167	DHEA to Δ4-A 29.1%	Putative NAD-binding domain	No	Moisan et al. (1999)
p.Gly12Glu c.35G > A	rs756607591	ND		ND	Benkert et al. (2015)
p.Gly15Asp c.44G > A	CM952220	Δ5-P to P 0%	Putative NAD-binding domain	No	Rhéaume et al. (1995)
p.Ala82Thr c.244G > A	CM940954 rs757033996	DHEA to Δ4-A 7.6%		No	Mendonça et al. (1994)
p.Ala82Pro c.244G > C	CM124707	ND	Membrane-binding domain	Yes	Rabbani, Mahdieh, Haghi Ashtiani, Setoodeh, & Rabbani, 2012)
p.Ala82Asp c.245C > A		ND		ND	Nordenström, Forest, and Wedell (2007)
p.Asn100Ser c.299A > G	CM950653	Δ5-P to P 2.7%, DHEA to Δ4-A 11%		No	Mébarki et al. (1995)

Continued

Table 1 *HSD3B2* Gene Mutations and Their Residual In Vitro Activity Reported in 3βHSD2-Deficient Patients—cont'd

Mutation	Variant ID	Activity In Vitro (% of wt)	Domain of the protein	Evidence of Protein Instability	References
p.Leu108Trp c.323T > G	CM940955	Δ5-P to P 0%		Yes	Sanchez, Mébarki, et al. (1994), Sanchez, Rhéaume, et al. (1994)
p.Gly129Arg c.385G > A	CM940956	Δ5-P to P 2%, DHEA to Δ4-A 4.7%	Cofactor binding domain	No	Rhéaume et al. (1994)
p.Glu142Lys c.424G > A	CM930411 rs80358219	Δ5-P to P 0%		No	Simard et al. (1993)
p.Pro155Leu c.464C > T	CM993168 rs779418168	DHEA to Δ4-A 0%	YXXXK sequence located in the active site of short-chain alcohol dehydrogenases	No	Moisan et al. (1999)
p.Ala167Val c.500C > T	CM985528	ND		ND	Nayak, Lee, and Witchel (1998)
p.Leu173Arg c.518T > G	CM940957 rs762479018	DHEA to Δ4-A 52.8%		No	Moisan et al. (1999)
p.Thr181Ile c.542C > A		ND		ND	Johannsen et al. (2005)
p.Pro186Leu c.557C > T	CM940958	Δ5-P to P 0%		Yes	Sanchez, Mébarki, et al. (1994), Sanchez, Rhéaume, et al. (1994)

p.Tyr190Cys c.569A > G		Δ5-17P to 17-OHP <5%, Δ5-P to P <1%, DHEA to Δ4-A <1%	Between the two putative substrate-binding domains. 3D model: extremely close to the putative substrate-binding pocket	No	Takasawa et al. (2014)
p.Leu205Pro c.614T > C	CM950654	DHEA to Δ4-A 0%		No	Moisan et al. (1999)
p.Ser213Thr c.638G > C	CM040433 rs137867568	DHEA to Δ4-A 40%		ND	Codner et al. (2004)
p.Ser213Gly c.637A > G	CM995305 rs759422374	DHEA to Δ4-A 58.4 ± 0.6%		No	Moisan et al. (1999)
p.Lys216Glu c.646A > G	CM995306	DHEA to Δ4-A 58.9%		No	Moisan et al. (1999)
p.Ser218Pro c.652T > C		Δ5-17P to 17-OHP <10%, Δ5-P to P <1%, DHEA to Δ4-A <1%	Between the two putative substrate-binding domains. 3D model: extremely close to the putative substrate-binding pocket	No	Takasawa et al. (2014)
p.Pro222Gln c.665C > A	CM993169 rs765547422	DHEA to Δ4-A 0%		No	Moisan et al. (1999)
p.Pro222His		DHEA to Δ4-A 0%		No	Moisan et al. (1999)
p.Pro222Thr c.664C > A	CM023090 rs80358220	Δ5-P to P 0%, DHEA to Δ4-A 0%	Adjacent flanking substrate-binding domain	Yes	Pang et al. (2002)

Continued

Table 1 *HSD3B2* Gene Mutations and Their Residual In Vitro Activity Reported in 3βHSD2-Deficient Patients—cont'd

Mutation	Variant ID	Activity In Vitro (% of wt)	Domain of the protein	Evidence of Protein Instability	References
p.Leu236Ser c.707T > C	CM993170 rs35887327	DHEA to Δ4-A 100%		No	Moisan et al. (1999)
p.Ala245Pro c.733G > C	CM930412	Δ5-P to P 11.9%, DHEA to Δ4-A 13.1%		No	Simard et al. (1993)
p.Gly250Val c.749G > T	CM1210438	Δ5-P to P 20%, DHEA to Δ4-A 27%	Loop next to a beta-sheet structure in the cofactor domain	No	Baquedano et al. (2015)
p.Tyr253Asn c.757T > A	CM930413	Δ5-P to P 0%, DHEA to Δ4-A 0%		No	Simard et al. (1993)
p.Tyr254Asp c.760T > G	CM940959	Δ5-P to P 0%, DHEA to Δ4-A 0%	Substrate-binding domain	No	Sanchez, Mébarki, et al. (1994), Sanchez, Rhéaume, et al. (1994)
p.Thr259Arg c.776C > G	CM950656	DHEA to Δ4-A 0%	Substrate-binding domain	Yes	Moisan et al. (1999)
p.Thr259Met c.776C > T	CM993171 rs80358221	DHEA to Δ4-A 0%	Substrate-binding domain	Yes	Moisan et al. (1999)
p.Ser284Arg c.852C > G	CM040434	DHEA to Δ4-A 32%		ND	Codner et al. (2004)
p.Gly294Val c.881G > T	CM993172	DHEA to Δ4-A 20.5%	Membrane-spanning domain	No	Moisan et al. (1999)

p.Tyr339Cys c.1016A > G		ND		ND	Bahíllo-Curieses, Loidi Fernández de Trocóniz, del Cañizo López, and Martínez-Sopena (2016)
p.Pro341Leu c.1022C > T	CM081307	Δ5-P to P 2%, DHEA to Δ4-A 6%	C-Terminal part of the protein	Yes	Welzel et al. (2008)
p.Ter373Cys c.1119A > C	CM021276	Δ5-P to P 27%, DHEA to Δ4-A 27%	C-Terminal part of the protein	Yes	Pang et al. (2002)
Nonsense					
p.Glu25Ter c.73G > T		No predicted activity		ND	Huang et al. (2014)
p.Glu135Ter c.403G > T	CM970746	No predicted activity		ND	Marui et al. (1998)
p.Trp171Ter c.512G > A	CM920363 rs80358216	No predicted activity		ND	Rhéaume et al. (1992)
p.Trp230Ter c.690G > A		No predicted activity		ND	Nordenström, Forest, and Wedell (2007)
p.Arg249Ter c.745C > T	CM950655 rs80358217	No predicted activity		ND	Tajima et al. (1995)
p.Tyr308Ter c.924C > G	CM950657	No predicted activity		ND	Tajima et al. (1995)

Continued

Table 1 *HSD3B2* Gene Mutations and Their Residual In Vitro Activity Reported in 3βHSD2-Deficient Patients—cont'd

Mutation	Variant ID	Activity In Vitro (% of wt)	Domain of the protein	Evidence of Protein Instability	References
p.Gln334Ter c.1000C > T	CM125506	No predicted activity	C-Terminal part of the protein	ND	Jeandron and Sahakitrungruang (2012)
p.Arg335Ter c.1003C > T	CM081308 rs148200568	Δ5-P to P 2 %, DHEA to Δ4-A 2%	C-Terminal part of the protein	Yes	Welzel et al. (2008)
p.Trp355Ter c.1064G > A	CM081309	Δ5-P to P 1.5 %, DHEA to Δ4-A 2%	C-Terminal part of the protein	Yes	Welzel et al. (2008)
Deletions					
p.Ala168Valfs c.503delC		No predicted activity		ND	Probst-Scheidegger, Udhane, l'Allemand, Flück, and Camats (2016)
p.Trp230_Ala238del c.687_713del	CG994940	DHEA to Δ4-A 0%		Yes	Moisan et al. (1999)
p.Asn266ThrfsTer6 c.797delA	CD000265	No predicted activity		ND	McCartin et al. (2000)
p.Lys273ArgfsTer7 c.818_819delAA	CD941706 rs754609778	No predicted activity		ND	Simard et al. (1994)
p. Met290CysfsTer10 c.867delG	CD994557 rs767167623	No predicted activity		ND	Moisan et al. (1999)

p.Thr318LysfsTer50 c.953delC	CD962037	No predicted activity		ND	Zhang, Sakkal-Alkaddour, Chang, Yang, and Pang (1996)
p.Val319AlafsTer49 c.956delT		No predicted activity	3D modeling predicts a conformational change in the substrate-binding pocket	ND	Bizzarri et al. (2016)
p.Lys368SerfsTer129 c.1103delA		No predicted activity		ND	Johannsen et al. (2005)
Insertions					
p.Thr187HisfsTer17 c.558insC	CI920941 rs770815049	No predicted activity		ND	Rhéaume et al. (1992)
Indels					
p.Val248AsnfsTer2 c.742_746delinsaact	CX931241	No predicted activity		ND	Chang, Kappy, et al. (1993)
Splicing					
c.308-6G>A	CS942333 rs755048400	No predicted activity		ND	Rhéaume et al. (1994)

All mutations were designated following the recommendations of the Human Genome Variation Society and HUGO Gene Nomenclature Committee (den Dunnen et al., 2016). Nucleotides change considering reference sequences: NM_001166120.1, NP_001159592.1, and CCDS902.1. *DHEA*, dehydroepiandrosterone; *ND*, not determined; *P*, progesterone; *Δ4-A*, androstenedione; *Δ5-P*, pregnenolone; *Δ5-17P*, 17OH pregnenolone; *17-OHP*, 17 OH progesterone.

The clinical phenotype in the affected patients is significantly heterogeneous with different degrees of genital ambiguity and adrenal failure in both sexes (Table 2 and 3).

Clinical symptoms of adrenal failure could be present early after birth. Mineralocorticoid deficiency with salt wasting, hypotension, and hyperkalemia is observed in 60%–70% of the reported cases (Simard, Moisan, & Morel, 2002). No functional 3βHSD2 isoenzyme is expressed in the adrenals and gonads of these patients. The nonsalt-wasting form of 3βHSD2 deficiency results from variations in the *HSD3B2* gene causing an incomplete loss of enzymatic activity (Baquedano et al., 2015; Moisan et al., 1999).

Testicular steroidogenic defect leads to different degrees of under virilization in 46,XY affected subjects. The phenotypic spectrum includes patients with severe under virilization and female appearance. Most frequently affected individuals present with genital ambiguity, hypospadias, microphallus, and cryptorchidism (Auchus & Miller, 2012; Pang, 2001).

Mild genital virilization with clitoromegaly with or without partial labial fusion, and even normal female genitalia has been reported in the 46,XX affected subjects (Auchus & Chang, 2010; Benkert et al., 2015; Pang, 2001). This lesser genital virilization, compared to 21-hydroxylase and 11-hydroxylase deficiencies, is related to peripheral conversion of the DHEA excess through the homologous enzyme 3βHSD1 into more potent androgens. The absence of adrenal 3βHSD activity precludes elevation of intra-adrenal 17OHP and the production of backdoor-derived DHT and 11-oxygenated androgens, which would be the major source of virilizing androgens in other forms of CAH (Jones et al., 2017; Kamrath, Hochberg, Hartmann, Remer, & Wudy, 2012; Turcu et al., 2016). The underrepresentation of 46,XX patients in this autosomal recessive disorder may be explained by the milder genital phenotype that could delay diagnosis and/or lead to unrecognized cause of infant death (Alos et al., 2000; Moisan et al., 1999).

Peripheral conversion does not rescue cortisol and aldosterone deficiency, because the 21-carbon steroid precursors do not efficiently return to the adrenal for subsequent hydroxylations. Peripheral cytochromes P450 cannot substitute for the adrenal 21α- and 11α-hydroxylases to make significant amounts of cortisol, but progesterone 21α-hydroxylation through CYP3A4 and CYP2C9 in the liver can partially compensate for the mineralocorticoid deficiency by making 11-deoxycorticosterone in older children and adults (Casey & MacDonald, 1982).

As mentioned earlier, patients with less severe enzymatic defects can maintain a normal hydroelectrolyte balance. If not treated, they will develop symptoms of androgen excess later in life (body odor, pubarche, axilarche,

Table 2 Ethnic Origin, Genotype, and Clinical Findings of 46,XY Described Patients

					At Diagnosis				
Families/ Patients	**Population**	**Genotype**	**Dead Relatives**	**Newborn Screening (17OHP4)**	**Chronological Age**	**Clinical Phenotype**	**Salt Wasting**	**Long-Term Follow-up**	**References**
I.1	Japanese	p.[Arg249Ter]; [Arg249Ter]			Birth	Skin pigmentation. Hypospadias with micropenis, bifid scrotum, palpable gonads	Yes	Gynecomastia at 7.6 years. Spontaneous pubertal development	Yoshimoto et al. (1997)
II.1	American	p.[Trp171Ter]; [Thr187HisfsTer17]	Yes		1 month	Mild hypospadias	Yes	Spontaneous puberty. Gynecomastia. Fathered two children	Parks, Bermudez, Anast, Bongiovanni, and New (1971), Rhéaume et al. (1992)
III.1	Mexican Hispanic	p.[Val248AsnfsTer2]; [Val248AsnfsTer2]			Birth	Peineal hypospadias, micropenis, bifid scrotum, scrotal testes	Yes	Gynecomastia at 12 years	Chang, Kappy, et al. (1993)
IV.1	Dutch	p.[Tyr253Asn]; [Thr187HisfsTer17]			Birth	Urethral diverticula, hypospadias	Yes		Simard et al. (1993)
V.1	American	p.[Glu142Lys]; [Trp171Ter]			1 week	Perineal hypospadias, bifid scrotum	Yes		Simard et al. (1993)
VI.1	Turkish	p.[Ala245Pro]; [Ala245Pro]	Yes		4 years	Scrotal hypospadias, bifid scrotum	No		Simard et al. (1993)

Continued

Table 2 Ethnic Origin, Genotype, and Clinical Findings of 46,XY Described Patients—cont'd

			At Diagnosis						
Families/ Patients	**Population**	**Genotype**	**Dead Relatives**	**Newborn Screening (17OHP4)**	**Chronological Age**	**Clinical Phenotype**	**Salt Wasting**	**Long-Term Follow-up**	**References**
VII.1	Pakistani	p.[Lys273ArgfsTer7]; [Lys273ArgfsTer7]			Birth	Hypospadias	Yes		Simard et al. (1994)
VIII.1	Afghan	p.[Lys273ArgfsTer7]; [Lys273ArgfsTer7]			Birth	Sever hypospadias, bifid scrotum, palpable gonads	Yes		Simard et al. (1994)
IX.1	Afghan	p.[Lys273ArgfsTer7]; [Lys273ArgfsTer7]	Yes		Birth	Severe hypospadias, bifid scrotum	Yes		Simard et al. (1994)
X.1	Brazilian	p.[Ala82Thr]; [Ala82Thr]			Birth	Ambiguous genitalia	Compensated salt loss (elevated renin activity)	Assigned female. Gonadectomized	Mendonça et al. (1994), Moisan et al. (1999)
X.2	Brazilian	p.[Ala82Thr]; [Ala82Thr]			Birth	Perineal hypospadias, bifid scrotum	Compensated salt loss (elevated renin activity)	Assigned female. Spontaneous typically male puberty at 14 years. No gynecomastia. Reassigned male at 17 years	Mendonça et al. (1994), Moisan et al. (1999)

XI.1	Spanish/ Portuguese	p.[Leu108Trp]; [Pro186Leu]	Birth	Perineal hypospadias, bifid scrotum	Yes		De Peretti, Forest, Feit, and David (1980), Sanchez, Mébarki, et al. (1994), Sanchez, Rhéaume, et al. (1994), Moisan et al. (1999)
XII.1	American	c.[385G>A]; [308-6G>A] p.[Gly129Arg]	6 years	Premature pubarche. History of multiple surgeries for perineal hypospadias and bilateral cryptorchidism	No	Spontaneous puberty at 12 years	Pang et al. (1983), Chang, Kulin, et al. (1993), Rhéaume et al. (1994), Moisan et al. (1999)
XIII.1	Scottish	p.[Leu173Arg]; [Leu173Arg]	5 years	Ambiguous genitalia	No	Assigned female at birth. Gonadectomized	Russell et al. (1994), Moisan et al. (1999)
XIV.1	Algerian	p.[Gly15Asp]; [Gly15Asp]	Birth	Perineoscrotal hypospadias, bifid scrotum	Yes		Gendrel, Chaussain, Roger, and Job (1979), Rhéaume et al. (1995), Moisan et al. (1999)
XV.1	French, English	p.[Asn100Ser]; [Asn100Ser]	20 months	Perineal hypospadias, palpable testes	Compensated salt loss (elevated renin activity)	Assigned female. Gonadectomized	De Peretti et al. (1980), Mébarki et al. (1995), Moisan et al. (1999)

Continued

Table 2 Ethnic Origin, Genotype, and Clinical Findings of 46,XY Described Patients—cont'd

					At Diagnosis				
Families/ Patients	Population	Genotype	Dead Relatives	Newborn Screening (17OHP4)	Chronological Age	Clinical Phenotype	Salt Wasting	Long-Term Follow-up	References
XVI.1	Japanese	p.[Leu205Pro]; [Leu205Pro]			3 months	Hyperpigmentation, severe hypospadias, bifid scrotum	Yes		Katsumata et al. (1995), Moisan et al. (1999)
XVII.1	Japanese	p.[Thr259Arg]; [Thr259Arg]			N/A	Pigmentation, hypospadias, palpable testes in bifid scrotum	Yes		Tajima et al. (1995), Moisan et al. (1999)
XVIII.1	Japanese	p.[Arg249Ter]; [Arg249Ter]			N/A	Pigmentation, hypospadias, bifid scrotum	Yes		Tajima et al. (1995)
XIX.1	Japanese	p.[Arg249Ter]; [Arg249Ter]			N/A	Pigmentation, severe undervirilization	Yes		Tajima et al. (1995)
XX.1	Japanese	p.[Tyr308Ter]; [Tyr308Ter]	Yes		N/A	Pigmentation, perineal hypospadias, bifid scrotum	Yes		Tajima et al. (1995)
XXI.1	French	p.[Pro155Leu]; [Gly294Val]			Birth	Perineal hypospadias, palpable testes in scrotum	No		Moisan et al. (1999), Gendrel et al. (1979)
XXI.2	French	p.[Pro155Leu]; [Gly294Val]			Birth	Perineal hypospadias, scrotal testes	No		Moisan et al. (1999), Gendrel et al. (1979)

XXII.1	Egyptian	p.[Ala10Val];[Ala10Val]	6 months	Perineoscrotal hypospadias, scrotal testes	No		Moisan et al. (1999)
XXII.2	Egyptian	p.[Ala10Val];[Ala10Val]	4 months	Perineoscrotal hypospadias, micropenis, scrotal testes	No		Moisan et al. (1999)
XXIII.1	Algerian	p.[Pro222Gln];[Pro222Gln]	Birth	Perineal hypospadias, micropenis, scrotal testes	Yes		Moisan et al. (1999)
XXIV.1	French, American	p.[Leu236Ser];[Met290CysfsTer10]	Birth	Perineal hypospadias with micropenis, no palpable testis	No		Moisan et al. (1999)
XXV.1	French	p.[Tyr259Met];[Met290CysfsTer10]	Birth	Perineal hypospadias, scrotal testes	Yes		Moisan et al. (1999)
XXVI.1	Sri-Lankan	p.[Trp230_Ala238del];[Trp230_Ala238del]	Birth	Perineal hypospadias with micropenis, scrotal testes	Yes		Moisan et al. (1999)
XXVII.1	Sri-Lankan	p.[Trp230_Ala238del];[Trp230_Ala238del]	Birth	Perineal hypospadias	Yes		Moisan et al. (1999)
XXVIII.1	English	p.[Asn266ThrfsTer6];[Asn100Ser]	11 years	Perineal hypospadias, micropenis with normal testes	No	Premature pubarche (7 years)	McCartin et al. (2000)
XXVIII.2	English	p.[Asn266ThrfsTer6];[Asn100Ser]	9 years	Perineal hypospadias with normal testes	Compensated salt loss (elevated renin activity)		McCartin et al. (2000)

Continued

Table 2 Ethnic Origin, Genotype, and Clinical Findings of 46,XY Described Patients—cont'd

					At Diagnosis				
Families/ Patients	Population	Genotype	Dead Relatives	Newborn Screening (17OHP4)	Chronological Age	Clinical Phenotype	Salt Wasting	Long-Term Follow-up	References
XXIX.1	ND	p.[Ala82Thr];[Ala82Thr]	Yes		10 months	Perineal hypospadias, inguinal testes	Compensated salt loss (elevated renin activity)		McCartin et al. (2000)
XXX.1	Pakistani	p.[Leu6Phe];[Leu6Phe]			Birth	Hyperpigmented scrotum, severe hypospadias, scrotal gonads	No		Zhang et al. (2000)
XXXI.1	Taiwanese	p.[Thr259Met];[Thr259Met]			Birth	Perineal hypospadias, chordee	Yes	Assigned female. Gonadectomized	Zhang et al. (2000), Moisan et al. (1999)
XXXII.1	French-Canadian	p.[Ala10Glu];[Ala10Glu]			13 days	Perineal hypospadias with micropenis, palpable testes in bifid scrotum	Yes	Spontaneous puberty at 10.5 years. No gynecomastia. Testicular adrenal rest. Azoospermia at 18.5 years	Alos et al. (2000)
XXXIII.1	ND	p.[Ser284Arg];ND[a]			7 years	Midshaft hypospadias	No		Codner et al. (2004)
XXXIV.1	ND	p.[Ser213Thr];ND[a]			12.9 years	Scrotal hypospadias, bilateral cryptorchidism, chordee, precocious pubarche	No		Codner et al. (2004)

XXXV.1	Brazilian	p.[Pro222Gln]; [Pro222Gln]		Birth	Perineal hypospadias, palpable testes in bifid scrotum	Yes	Mermejo et al. (2005)
XXXVI.1	Lebanese	p.[Pro341Leu]; [Pro341Leu]		Birth	Micropenis, broad urogenital sinus, palpable gonads	Yes	Welzel et al. (2008)
XXXVII.1	Turkish	p.[Arg335Ter]; [Arg335Ter]	Elevated	Birth	Micropenis, scrotal hypospadias, a broad urogenital sinus, palpable testes	Yes	Welzel et al. (2008)
XXXVII.2	Turkish	p.[Arg335Ter]; [Arg335Ter]	Elevated	Birth	Micropenis, penoscrotal hypospadias, maldescended testes. Incomplete cleft lip	Yes	Welzel et al. (2008)
XXXVIII.1	Bangladeshi	p.[Trp355Ter]; [Trp355Ter]		Birth	Micropenis, perineal hypospadias, bifid scrotum, palpable testes	Yes	Welzel et al. (2008)
XXXIX.1	Caucasian	p.[Pro222Gln]; [Pro222Gln]		2 months	Normal male, unilateral cryptorchidism	Yes	Lusa et al. (2010)
XL.1	Iranian	p.[Ala82Pro]; [Ala82Pro]		20 days	Mild hyperpigmentation, micropenis with chordee, bifid scrotum, scrotal testis, perineal hypospadias	Yes	Rabbani, Mahdieh, Haghi Ashtiani, Setoodeh, and Rabbani (2012)

Continued

Table 2 Ethnic Origin, Genotype, and Clinical Findings of 46,XY Described Patients—cont'd

			At Diagnosis						
Families/ Patients	**Population**	**Genotype**	**Dead Relatives**	**Newborn Screening (17OHP4)**	**Chronological Age**	**Clinical Phenotype**	**Salt Wasting**	**Long-Term Follow-up**	**References**
XLI.1	Brazilian	p.[Pro222Gln]; [Pro222Gln]		Elevated	3 months	2.5 cm phallus, penoscrotal proximal hypospadias, incompletely fused labioscrotal folds, bilaterally palpable gonads	Yes		de Araújo, de Oliveira, Gameleira, Cruz, and Lofrano-Porto (2014)
XLII.1	Sri-Lankan	p.[Trp230_Ala238del]; [Trp230_Ala238del]			Birth	Severe hypospadias, cryptorchidism, undervirilization. Perinatal asphyxia	Yes	Severe cerebral palsy. pH 3 at 6.3 years. Spontaneous puberty. Gynecomastia. Testicular biopsy at 15.5 years. Few germ cells with spermatogenic arrest, no evidence of malignancy	Burckhardt et al. (2015)

XLIII.1-6	Old Order Amish of North America	p.[Gly12Glu]; [Gly12Glu]	Variable	Variable	Hypospadias	Yes	Postnatal virilization (premature pubarche). Gynecomastia. Testicular adrenal rests in 2	Benkert et al. (2015)
XLIV.1	Italian	p.[Val319AlafsTer49]; [Val319AlafsTer49]		Birth	Perineal hypospadias, palpable testes within the labial scrotal folds and a micropenis.	Yes		Bizzarri et al. (2016)
XLV.1	Spanish	p.[Ala82Thr]; [Tyr339Cys]	Elevated	15 days	Female assigned at birth. Mild clitoromegaly, urogenital sinus	No		Bahíllo-Curieses, Loidi Fernández de Trocóniz, del Cañizo López, and Martínez-Sopena (2016)
XLVI.1	Caucasian	UPD Crh1 p.[Glu142Lys]; [Glu142Lys]	Bordeline elevation	7 days	Perineal hypospadias, bifid scrotum, penile chordee, scrotal testes. Hypotonia, nonspecific dysmorphic facial findings	Yes		Panzer, Ekhaguere, Darbro, Cook, and Shchelochkov (2017)

[a]Heterozygous reported case, see text for details.
Families are designated with roman numbers and siblings are designated as II.1, II.2.

Table 3 Ethnic Origin, Genotype, and Clinical Findings of 46,XX Described Patients

Families/ Patients	Population	Genotype	Dead Relatives	Newborn Screening (17OHP4)	Chronological Age	Clinical Phenotype	Salt Wasting	Long-Term Follow-up	References
XLVII.1	Swiss	p.[Trp171Ter]; [Trp171Ter]	Yes		Birth	No signs of virilization. Skin pigmentation	Yes	No spontaneous breast development at 14.7 years	Zachmann, Vollmin, Murset, Curtius, and Prader (1970), Zachmann, Forest, and De Peretti (1979), Rhéaume et al. (1992)
XLVIII.1	Swiss	p.[Trp171Ter]; [Trp171Ter]			Unknown	No signs of virilization	Yes		Rhéaume et al. (1992)
X.3	Brazilian	p.[Ala82Thr]; [Ala82Thr]			31 years	Clinically normal	No	Spontaneous puberty and menarche, regular menses	Mendonça et al. (1994), Moisan et al. (1999)
XII.2	American	c[385G>A]; [308-6G>A] p.[Gly129Arg]			8 years	Normal genitalia at birth. Postnatal virilization: Premature pubarche (4 years), acne, mild clitoromegaly (8 years)	No	Spontaneous puberty and menarche, Irregular menses	Pang et al. (1983), Chang, Kulin, et al. (1993), Rhéaume et al. (1994)

XLIX.1	Brazilian	p.[Ala82Thr];[Ala82Thr]	5 years	Premature pubarche	No		Mendonça et al. (1994), Moisan et al. (1999)
L.1	Brazilian	p.[Thr259Met];[Thr259Met]	41 years	Incidentally discovered bilateral accessory adrenal tissue in the ovaries and in the paraaortic region. Clitoromegaly, severe virilization	No		Paula et al. (1994), Moisan et al. (1999)
XIII.2	Scottish	p.[Leu173Arg];[Leu173Arg]	2 years	Clinically normal	No		Russell et al. (1994), Moisan et al. (1999)
LI.1	American	p.[Tyr254Asp];ND[a]	17 years	Primary amenorrhea, slight clitoromegaly, moderate hirsutism, enlarged ovaries. Hyperpigmented skin	Compensated salt loss (elevated renin activity)	Spontaneous puberty at 11 years	Rosenfield et al. (1980), Sanchez, Mébarki, et al. (1994a), Sanchez, Rhéaume, et al. (1994b), Moisan et al. (1999)
XVI.2	Japanese	p.[Leu205Pro];[Leu205Pro]	11 months	Hyperpigmentation, mild clitoromegaly	Yes		Katsumata et al. (1995), Moisan et al. (1999)
XVII.2	Japanese	p.[Thr259Arg];[Thr259Arg]	2 weeks	Normal genitalia with severe pigmentation	Yes		Tajima et al. (1995), Moisan et al. (1999)

Continued

Table 3 Ethnic Origin, Genotype, and Clinical Findings of 46,XX Described Patients—cont'd

Families/ Patients	Population	Genotype	Dead Relatives	Newborn Screening (17OHP4)	Chronological Age	Clinical Phenotype	Salt Wasting	Long-Term Follow-up	References
LII.1	Pakistani	p.[Lys273ArgfsTer7]; [Thr318LysfsTer50]			Birth	Pigmentation, mildly enlarged clitoris	Yes	Spontaneous puberty, menarche (13 years), secondary amenorrhea, acne and hirsutism	Zhang, Sakkal-Alkaddour, Chang, Yang, and Pang (1996) Pang et al. (2002)
LIII.1	Chilean	p.[Glu135Ter]; [Glu135Ter]			20 months	Normal, hyper pigmented external genitalia	Yes		Marui et al. (1998)
XXV.2	French	p.[Tyr259Met]; [Met290CysfsTer10]			Birth	Normal genitalia	Yes		Moisan et al. (1999)
LIV.1	American	p.[Ser213Gly]; [Lys216Glu]			7 years	Premature pubarche at 4 years, growth acceleration	No		Moisan et al. (1999)
LV.1	Brazilian	p.[Pro222His]; [Gly129Arg]			7 years	Normal genitalia, premature pubarche	No		Marui et al. (2000), Moisan et al. (1999)
LV.2	Brazilian	p.[Pro222His]; [Gly129Arg]			6.7 years	Normal genitalia, premature pubarche	No		Marui et al. (2000), Moisan et al. (1999)
LVI.1	Brazilian	p.[Thr259Met]; [Thr259Met]			7.8 years	Clitoromegaly, precocious pubarche	No		Marui et al. (2000), Moisan et al. (1999)

XXIII.2	Algerian	p.[Pro222Gln]; [Pro222Gln]		1 month	Mild clitoromegaly	Yes		Moisan et al. (1999)
LVII.1	French Canadian	p.[Ala10Glu]; [Ala10Glu]		3 weeks	Normal external genitalia. Enlarged adrenals	Yes	Premature pubarche at 4 years. Spontaneous puberty (8 years) and menarche (10.5 years). Multicystic enlarged ovaries	Alos et al. (2000)
LVIII.1	Eastern European	p.[Pro222Thr]; [Pro222Thr]		4 weeks	Normal external genitalia	Yes		Pang et al. (2002)
LIX.1	Caucasian	p.[Glu142Lys]; [Ter373Cys]		7 years	Normal external genitalia	No	Premature pubarche (5.5 years). Increased somatic growth and acne (7.3 years). Telarche (9 years)	Pang et al. (2002)
LX.1	Caucasian	p.[Lys368SerfsTer129]; [Thr181Ile]	Elevated	7.6 years	Precocious pubarche	Yes	Spontaneous puberty (10.1 years)	Johannsen et al. (2005)
LX.2	Caucasian	p.[Lys368SerfsTer129]; [Thr181Ile]	Elevated	3.7 years	Sister	Yes		Johannsen et al. (2005)

Continued

Table 3 Ethnic Origin, Genotype, and Clinical Findings of 46,XX Described Patients—cont'd

Families/ Patients	Population	Genotype	Dead Relatives	Newborn Screening (17OHP4)	Chronological Age	Clinical Phenotype	Salt Wasting	Long-Term Follow-up	References
LXI.1	Russian	p.[Ala82Asp]; [Trp230Ter]		Elevated	Birth	Neonatal screening	Yes		Nordenström, Forest, and Wedell (2007)
LXII.1	American/ Salvadoran	p.[Gln334Ter]; [Gln334Ter]		Elevated	8 days	Normal external genitalia. Skin pigmentation	Yes		Jeandron and Sahakitrungruang (2012)
LXII.2	Salvadoran	p.[Gln334Ter]; [Gln334Ter]			28 days	Normal external genitalia	Yes		Jeandron and Sahakitrungruang (2012)
LXIII.1	Japanese	p.[Tyr190Cys]; [Ser218Pro]			Birth	Mild virilization with clitoromegaly. Skin pigmentation	Compensated salt loss (elevated renin activity)		Takasawa et al. (2014)
LXIV.1	Chinese	p.[Glu25Ter]; [Glu25Ter]			1 month	Skin pigmentation. Clitoromegaly	Yes	Spontaneous puberty (9 years) and menarche (12 years). Oligomenorrhea. Postmenarcheal recurrent ovarian cysts	Huang et al. (2014)

LXV.1	ND (adopted)	p.[Gly250Val]; [Gly250Val]		7 months	Precocious pubarche PH II–III and postnatal clitoromegaly. Two orifices, and nonpalpable gonads	Compensated salt loss (elevated renin activity)		Baquedano et al. (2015)
XLIII.7–16	Old Order Amish of North America	p.[Gly12Glu]; [Gly12Glu]	Variable	Variable	Normal female external genitalia. No ambiguous genitalia	Yes	Premature pubarche. Two adult females developed polycystic ovaries	Benkert et al. (2015)
LXVI.1	Switzerland	p.[Trp171Ter]; [Ala168Valfs]	Elevated	8 days	Normal female external genitalia without virilization or palpable gonads	Yes		Probst-Scheidegger, Udhane, l'Allemand, Flück, and Camats (2016)

[a]Heterozygous reported case, see text for details.
Families are designated with roman numbers and siblings are designated as II.1, II.2.

rapid somatic growth, and advance skeletal maturation) (Baquedano et al., 2015; Codner et al., 2004; Mendonça et al., 1994; Pang et al., 2002; Rhéaume et al., 1994). Signs of androgen excess worsen with age, probably due to an increase in DHEA production with the development of the ZR in the adrenal cortex.

Puberty in affected boys showed a broad clinical heterogeneity, but spontaneous pubertal development, with or without enlarged testes and gynecomastia, has been reported even in the presence of a severe salt-wasting form (Alos et al., 2000; Mendonça et al., 1994; Parks, Bermudez, Anast, Bongiovanni, & New, 1971; Rhéaume et al., 1992). Indeed, an affected male with proven severe 3βHSD2 deficiency has fathered children (Rhéaume et al., 1992). Testicular adrenal rest tumors (TART) were also reported in male affected individuals (Benkert et al., 2015).

In affected girls, ovarian estrogen synthesis is compromised and the development of secondary sexual characteristics is impaired (Auchus & Chang, 2010). However, spontaneous pubertal development and menarche have been reported in a few of the affected females with severe mutation in the *HSD3B2* gene without any residual enzyme activity (Alos et al., 2000; Huang et al., 2014; Moisan et al., 1999). Hirsutism and menstrual disorders were described in older affected females (Rosenfield et al., 1980).

Taking together the findings observed in affected boys and girls suggest that Δ^4 steroids could originate from gonadal 3βHSD1 activity, because even though only very low levels of 3βHSD1 mRNA could be detected in normal gonads, this enzyme possesses a 5- to 10-fold higher activity than the type 2 isoenzyme and could be stimulated by the increased gonadotropin secretion that results from low sex steroid levels.

A nonclassic or late-onset clinical form was described in girls presenting with precocious pubarche and a PCO-like syndrome (Marui et al., 2000; Pang et al., 1985). The genetic basis for this clinical form has not been properly described, and use of this terminology and diagnosis is discouraged, and should be reserved only for patients with markedly elevated ratios of Δ^5 steroid/Δ^4 steroids and confirmatory molecular studies.

The biochemical marker of this steroidogenic disorder is the increased serum level of Δ^5 steroids, especially 17-hydroxypregnenolone, DHEA, and DHEAS. Peripheral conversion of these Δ^5 precursors by 3βHSD1 isoenzyme leads to an increase in serum levels of 17OHP and A4. Even though androgen excess is less severe than in other forms of CAH, especially than 21α-hydroxylase deficiency, the increase in 17OHP might lead to misinterpretations of neonatal screening results. An increase in precursor/product ratio

(Δ^5/Δ^4) more than 10 SDS above the normal reference value is very useful in the diagnosis of this enzymatic disorder (17-hydroxypregnenolone/17-hydroxyprogesterone, DHEA/androstenedione, pregnenolone/progesterone) (Auchus & Chang, 2010; Auchus & Miller, 2012; Lutfallah et al., 2002). However, confirmatory molecular studies are needed.

Initial management and long-term follow-up of the affected patients, as in other DSD conditions, require the participation of a multidisciplinary team with experience in the field.

Initial treatment is directed to restore fluid and electrolyte balance. Affected infants should receive glucocorticoid, mineralocorticoid, and salt supplementation as in 21α-hydroxylase deficiency. Usually lower hydrocortisone doses are needed in this period, as androgen excess is easier to control. With the development of the ZR, androgen excess becomes more problematic and glucocorticoid and mineralocorticoid treatment should continue. T and DHEAS are used to monitor treatment.

In 3βHSD2 deficiency the variable impairment in gonadal steroidogenesis could lead to primary hypogonadism that would need sex steroid replacement therapies to develop and sustain secondary sexual characteristics.

Testicular ultrasound should be considered to detect the presence of TART (Benkert et al., 2015).

2.3 Genotype–Phenotype Relationships. Structure–Function Relationships

Mutations in the *HSD3B2* gene result in a wide spectrum of molecular repercussions, which are associated with the different phenotypic manifestations of classical 3βHSD2 deficiency. In almost all the reported cases, the severity of salt wasting usually showed good correlation with the functional consequences of *HSD3B2* mutations as assayed in in vitro cell systems (Baquedano et al., 2015; Moisan et al., 1999; Simard, Moisan, & Morel, 2002; Simard et al., 2005; Tables 1–3). Evidences showed that it is more appropriate to assess the enzymatic activity of transiently expressed mutant proteins using intact cells rather than homogenates from cells, because addition of exogenous cofactor can drive a reaction that may not occur in vivo (Moisan et al., 1999). However, the relationship between the genotype and the gonadal phenotype in severe 3βHSD2 deficiency is more complex and fertility is difficult to predict (Mendonca et al., 1987; Mendonça et al., 1994; Rhéaume et al., 1992; Zachmann, Forest, & De Peretti, 1979).

As previously mentioned, *HSD3B2* gene defects appeared to follow an autosomal recessive mode of inheritance, as with other steroidogenic enzyme defects. However, heterozygous *HSD3B2* gene mutations in a female patient with nonsalt-losing 3βHSD deficiency diagnose at puberty (Sanchez, Rhéaume, et al., 1994) and in two male patients with apparent idiopathic hypospadias without clear steroidogenic pattern of 3βHSD deficiency (Codner et al., 2004) have been reported. In vitro experiments argued for its functional impact (Codner et al., 2004; Sanchez, Rhéaume, et al., 1994) and dominant-negative properties (Codner et al., 2004). Nevertheless, the absence of symptoms in the heterozygous patients' relatives argues against such a dominant effect. Noteworthy, a putative second mutation located in the promoter or within intronic regions potentially affecting the normal expression of this gene or generating an alternative aberrant splicing site cannot be refuted. Associated genetic or environmental/hormonal influences that may contribute to the clinical phenotype should be ruled out in those patients.

The amino acid residues that are the sites of missense mutations are generally in highly conserved regions in members of the vertebrate 3βHSD isoenzymes, suggesting the crucial role of these residues for the activity and/stability of the enzyme. Because the crystal structure of 3βHSD has not been determined, several 3D model structures of the 3βHSD2 were generated by the comparative modeling technique to study the potential impact of mutations on the 3βHSD2 structure (Baquedano et al., 2015; Rabbani, Mahdieh, Haghi Ashtiani, Setoodeh, & Rabbani, 2012). Mutations that impair steroid binding, nicotinamide cofactor binding, and enzyme stability have all been described (Fig. 3; Table 1), and in many instances, the enzymatic defect is complex and multifactorial.

3. CONCLUDING REMARKS

3βHSD2 is essential for the biosynthesis of all classes of active steroid hormones, including aldosterone, cortisol in the adrenal cortex, and sex steroids in the adrenals and gonads. In the classical steroidogenic pathway, known as front door steroidogenic pathway, the Δ^5 steroids represent the preferred route for C19 steroid production in humans. In the last decade, an alternative, so-called backdoor pathway to androgen synthesis has been described. 17OHP is the substrate from which can be generated DHT without the intermediacy of DHEA, A4, or T. It has been proposed that this pathway would contribute to the androgen production in pathological states in which 17OHP accumulates, including 21α-hydroxylase and POR deficiencies. Furthermore, the expression of the backdoor pathway in the fetal testis appears to play an important role during male sexual differentiation.

Recent studies have demonstrated that human steroidogenic tissues also produce a unique set of 11-oxygenated C19 steroids. 11-Oxygenated C19 steroid and its peripheral 5α-reduced metabolites are potent agonists of the human AR with affinities and potencies similar to that of T and DHT. Androgen production in the adrenal cortex is zonally and developmentally regulated. The 3βHSD2-deficient ZR is indistinct during infancy, but a continuous layer of reticularis cells starts to develop and thicken around 4–5 years of age. This process, known as adrenarche, is followed by a rise in circulating concentrations of DHEAS, with clinical signs physiologically observed between the ages of 6 and 10. In the testes the synthesis of T by Leydig cells follows a pathway that is similar to C19-steroid production in the adrenal ZR. Leydig cells express abundant 3βHSD2 and 17βHSD3, but no SULT2A1 is required to produce sulfo conjugate compounds. Thus, DHEA produced in the testis is not sulfurylated but is readily converted to A4 and/or androstenediol and then T via 3βHSD2 and 17βHSD3. In the ovary, small amounts of A4 are secreted or converted to T in Teca cells, but most A4 is converted to estradiol by granulosa cells, under the influence of FSH.

Recessive loss-of-function HSD3B2 mutations cause a rare form of CAH that impairs both adrenal and gonadal steroidogenesis. Deficiency of 3βHSD2 activity in the adrenals precludes normal aldosterone and cortisol synthesis and the flooding of cortisol precursors along the Δ^5 pathway with a marked rise in pregnenolone, 17-hydroxypregnenolone, DHEA, DHEAS, and androstenediol production which represent serum biomarkers for suspecting the diagnosis of 3βHSD2 deficiency. In gonads, deficiency of 3βHSD2 activity precludes normal T and estrogen synthesis. The blockage caused by 3βHSD2 deficiency precludes flux to the backdoor pathway and 11-oxygenated C19 steroid synthesis due to very low intraadrenal 17OHP and A4 substrate levels, respectively. Peripheral 3βHSD1 conversion of Δ5 steroids to Δ4 steroids can result in high serum 17OHP and A4.

The clinical phenotype in the affected patients is significantly heterogeneous with different degrees of genital ambiguity and adrenal failure in both sexes. Clinical symptoms of adrenal failure could be present early after birth. Mineralocorticoid deficiency with salt wasting, hypotension, and hyperkalemia is observed in 60%–70% of the reported cases. No functional 3βHSD2 isoenzyme is expressed in the adrenals and gonads of these patients. The nonsalt-wasting form of 3βHSD2 deficiency results from variations in the HSD3B2 gene causing an incomplete loss of enzymatic activity. Testicular steroidogenic defect leads to different degrees of under virilization.

In 46,XX affected patients, mild genital virilization with clitoromegaly with or without partial labial fusion and even normal female genitalia has been reported. The absence of adrenal 3βHSD activity precludes elevation of intraadrenal 17OHP and the production of backdoor-derived DHT and 11-oxygenated androgens, which would be the major source of virilizing androgens in other forms of CAH.

Finally, in humans the 3βHSD2 deficiency represents a model of nature that reinforces our knowledge about the role of the steroidogenic alternative pathway in sex differentiation in both sexes. In addition, the physiological role of the high serum DHEAS levels in fetal life as well as after adrenarche remains to be elucidated.

ACKNOWLEDGMENTS

This chapter was supported by grants from Consejo Nacional de Investigaciones Científicas y Técnicas (CONICET) and Fondo para la Investigación Científica y Tecnológica (FONCYT).

DISCLOSURE

All authors have nothing to declare.

REFERENCES

Alos, N., Moisan, A. M., Ward, L., Desrochers, M., Legault, L., Leboeuf, G., et al. (2000). A novel A10E homozygous mutation in the HSD3B2 gene causing severe salt-wasting 3beta-hydroxysteroid dehydrogenase deficiency in 46,XX and 46,XY French-Canadians: Evaluation of gonadal function after puberty. *The Journal of Clinical Endocrinology and Metabolism*, *85*(5), 1968–1974. https://doi.org/10.1210/jcem.85.5.6581.

Asby, D. J., Arlt, W., & Hanley, N. A. (2009). The adrenal cortex and sexual differentiation during early human development. *Reviews in Endocrine & Metabolic Disorders*, *10*(1 spec. issue), 43–49. https://doi.org/10.1007/s11154-008-9098-9.

Auchus, R. J. (2004). The backdoor pathway to dihydrotestosterone. *Trends in Endocrinology and Metabolism: TEM*, *15*(9), 432–438. https://doi.org/10.1016/j.tem.2004.09.004.

Auchus, R. J. (2011). The physiology and biochemistry of adrenarche. *Endocrine Development*, *20*, 20–27. https://doi.org/10.1159/000321209.

Auchus, R. J., & Chang, A. Y. (2010). 46,XX DSD: The masculinised female. *Best Practice & Research Clinical Endocrinology & Metabolism*, *24*(2), 219–242. https://doi.org/10.1016/j.beem.2009.11.001.

Auchus, R. J., Lee, T. C., & Miller, W. L. (1998). Cytochrome b5 augments the 17,20-lyase activity of human P450c17 without direct electron transfer. *The Journal of Biological Chemistry*, *273*(6), 3158–3165. Retrieved from http://www.ncbi.nlm.nih.gov/pubmed/9452426.

Auchus, R., & Miller, W. (2012). Defects in androgen biosynthesis causing 46,XY disorders of sexual development. *Seminars in Reproductive Medicine*, *30*(5), 417–426. https://doi.org/10.1055/s-0032-1324726.

Bahíllo-Curieses, M. P., Loidi Fernández de Trocóniz, L., del Cañizo López, A., & Martínez-Sopena, M. J. (2016). Partial 3ß-hydroxysteroid dehydrogenase type 2 deficiency:

Diagnosis of a novel mutation after positive newborn screening for 21-hydroxylase deficiency. *Medicina Clínica, 146*(2), 92–93. https://doi.org/10.1016/j.medcli.2015.04.004.

Baquedano, M. S., Ciaccio, M., Marino, R., Perez Garrido, N., Ramirez, P., Maceiras, M., et al. (2015). A novel missense mutation in the *HSD3B2* gene, underlying nonsalt-wasting congenital adrenal hyperplasia. New insight into the structure-function relationships of 3β-hydroxysteroid dehydrogenase type II. *The Journal of Clinical Endocrinology & Metabolism, 100*(1), E191–E196. https://doi.org/10.1210/jc.2014-2676.

Baquedano, M. S., Madjinca, S., Aliberti, P., Saraco, N. I., Warman, M., Rivarola, M. A., et al. (2014). Expression of androgen receptor (AR) and the alternative "Backdoor" pathway to dihydrotestosterone (DHT) in normal and 21-hydroxylase deficient human adrenal cortex. Abstract SAT-0753. *Endocrine Reviews, 35*(3 Suppl), E320.

Belgorosky, A., Baquedano, M. S., Guercio, G., & Rivarola, M. A. (2008). Adrenarche: Postnatal adrenal zonation and hormonal and metabolic regulation. *Hormone Research, 70*(5), 257–267.

Benkert, A. R., Young, M., Robinson, D., Hendrickson, C., Lee, P. A., & Strauss, K. A. (2015). Severe salt-losing 3β-hydroxysteroid dehydrogenase deficiency: Treatment and outcomes of *HSD3B2* c.35G > A homozygotes. *The Journal of Clinical Endocrinology & Metabolism, 100*(8), E1105–E1115. https://doi.org/10.1210/jc.2015-2098.

Bizzarri, C., Massimi, A., Federici, L., Cualbu, A., Loche, S., Bellincampi, L., et al. (2016). A new homozygous frameshift mutation in the HSD3B2 gene in an apparently non-consanguineous Italian family. *Hormone Research in Pædiatrics, 86*(1), 53–61. https://doi.org/10.1159/000444712.

Black, S. M., Harikrishna, J. A., Szklarz, G. D., & Miller, W. L. (1994). The mitochondrial environment is required for activity of the cholesterol side-chain cleavage enzyme, cytochrome P450scc. *Proceedings of the National Academy of Sciences of the United States of America, 91*(15), 7247–7251. Retrieved from http://www.ncbi.nlm.nih.gov/pubmed/8041774.

Burckhardt, M.-A., Udhane, S. S., Marti, N., Schnyder, I., Tapia, C., Nielsen, J. E., et al. (2015). Human 3β-hydroxysteroid dehydrogenase deficiency seems to affect fertility but may not harbor a tumor risk: Lesson from an experiment of nature. *European Journal of Endocrinology, 173*(5), K1–K12. https://doi.org/10.1530/EJE-15-0599.

Casey, M. L., & MacDonald, P. C. (1982). Extraadrenal formation of a mineralocorticosteroid: Deoxycorticosterone and deoxycorticosterone sulfate biosynthesis and metabolism. *Endocrine Reviews, 3*(4), 396–403. https://doi.org/10.1210/edrv-3-4-396.

Chang, Y. T., Kappy, M. S., Iwamoto, K., Wang, J., Yang, X., & Pang, S. (1993). Mutations in the type II 3 beta-hydroxysteroid dehydrogenase gene in a patient with classic salt-wasting 3 beta-hydroxysteroid dehydrogenase deficiency congenital adrenal hyperplasia. *Pediatric Research, 34*(5), 698–700. https://doi.org/10.1203/00006450-199311000-00026.

Chang, Y. T., Kulin, H. E., Garibaldi, L., Suriano, M. J., Bracki, K., & Pang, S. (1993). Hypothalamic-pituitary-gonadal axis function in pubertal male and female siblings with glucocorticoid-treated nonsalt-wasting 3 beta-hydroxysteroid dehydrogenase deficiency congenital adrenal hyperplasia. *The Journal of Clinical Endocrinology and Metabolism,* 77(5), 1251–1257. https://doi.org/10.1210/jcem.77.5.8077318.

Codner, E., Okuma, C., Iñiguez, G., Boric, M. A., Avila, A., Johnson, M. C., et al. (2004). Molecular study of the 3β-hydroxysteroid dehydrogenase gene type II in patients with hypospadias. *The Journal of Clinical Endocrinology & Metabolism, 89*(2), 957–964. https://doi.org/10.1210/jc.2002-020873.

Dardis, A., Saraco, N., Rivarola, M. A., & Belgorosky, A. (1999). Decrease in the expression of the 3beta-hydroxysteroid dehydrogenase gene in human adrenal tissue during prepuberty and early puberty: Implications for the mechanism of adrenarche. *Pediatric Research, 45*(3), 384–388. https://doi.org/10.1203/00006450-199903000-00016.

de Araújo, V. G. B., de Oliveira, R. S., Gameleira, K. P. D., Cruz, C. B., & Lofrano-Porto, A. (2014). 3β-Hydroxysteroid dehydrogenase type II deficiency on newborn screening test. *Arquivos Brasileiros de Endocrinologia e Metabologia*, *58*(6), 650–655. Retrieved from http://www.ncbi.nlm.nih.gov/pubmed/25211449.

de Peretti, E., & Forest, M. G. (1976). Unconjugated dehydroepiandrosterone plasma levels in normal subjects from birth to adolescence in human: The use of a sensitive radioimmunoassay 1. *The Journal of Clinical Endocrinology & Metabolism*, *43*(5), 982–991. https://doi.org/10.1210/jcem-43-5-982.

De Peretti, E., Forest, M. G., Feit, J. P., & David, M. (1980). Endocrine studies in two children with male pseudohermapbroditism due to to S/3-hydroxysteroid dehydrogenase defect. In A. R. Genazzani, J. H. H. Thijssen, & P. K. Siiteri (Eds.), *Adrenal androgens* (pp. 141–145). New York: Raven Press.

den Dunnen, J. T., Dalgleish, R., Maglott, D. R., Hart, R. K., Greenblatt, M. S., McGowan-Jordan, J., et al. (2016). HGVS recommendations for the description of sequence variants: 2016 update. *Human Mutation*, *37*(6), 564–569.

Doi, M., Satoh, F., Maekawa, T., Nakamura, Y., Fustin, J.-M., Tainaka, M., et al. (2014). Isoform-specific monoclonal antibodies against 3β-hydroxysteroid dehydrogenase/isomerase family provide markers for subclassification of human primary aldosteronism. *The Journal of Clinical Endocrinology & Metabolism*, *99*(2), E257–E262. https://doi.org/10.1210/jc.2013-3279.

Else, T., & Hammer, G. D. (2005). Genetic analysis of adrenal absence: Agenesis and aplasia. *Trends in Endocrinology and Metabolism*, *16*(10), 458–468. https://doi.org/10.1016/j.tem.2005.10.004.

Endoh, A., Kristiansen, S. B., Casson, P. R., Buster, J. E., & Hornsby, P. J. (1996). The zona reticularis is the site of biosynthesis of dehydroepiandrosterone and dehydroepiandrosterone sulfate in the adult human adrenal cortex resulting from its low expression of 3 beta-hydroxysteroid dehydrogenase. *The Journal of Clinical Endocrinology & Metabolism*, *81*(10), 3558–3565. https://doi.org/10.1210/jcem.81.10.8855801.

Flück, C. E., Meyer-Böni, M., Pandey, A. V., Kempná, P., Miller, W. L., Schoenle, E. J., et al. (2011). Why boys will be boys: Two pathways of fetal testicular androgen biosynthesis are needed for male sexual differentiation. *The American Journal of Human Genetics*, *89*(2), 201–218. https://doi.org/10.1016/j.ajhg.2011.06.009.

Gell, J. S., Carr, B. R., Sasano, H., Atkins, B., Margraf, L., Mason, J. I., et al. (1998). Adrenarche results from development of a 3beta-hydroxysteroid dehydrogenase-deficient adrenal reticularis. *The Journal of Clinical Endocrinology and Metabolism*, *83*(10), 3695–3701. https://doi.org/10.1210/jcem.83.10.5070.

Gendrel, D., Chaussain, J. L., Roger, M., & Job, J. C. (1979). Congenital adrenal hyperplasia due to blockade of 3-beta-hydroxysteroid dehydrogenase. *Archives Françaises de Pédiatrie*, *36*(7), 647–655. Retrieved from http://www.ncbi.nlm.nih.gov/pubmed/161161.

Ghayee, H. K., & Auchus, R. J. (2007). Basic concepts and recent developments in human steroid hormone biosynthesis. *Reviews in Endocrine & Metabolic Disorders*, *8*(4), 289–300. https://doi.org/10.1007/s11154-007-9052-2.

Goto, M., Piper Hanley, K., Marcos, J., Wood, P. J., Wright, S., Postle, A. D., et al. (2006). In humans, early cortisol biosynthesis provides a mechanism to safeguard female sexual development. *The Journal of Clinical Investigation*, *116*(4), 953–960. https://doi.org/10.1172/JCI25091.

Grumbach, M. M., & Auchus, R. J. (1999). Estrogen: Consequences and implications of human mutations in synthesis and action. *The Journal of Clinical Endocrinology and Metabolism*, *84*(12), 4677–4694. https://doi.org/10.1210/jcem.84.12.6290.

Gupta, M. K., Guryev, O. L., & Auchus, R. J. (2003). 5alpha-reduced C21 steroids are substrates for human cytochrome P450c17. *Archives of Biochemistry and Biophysics*, *418*(2), 151–160. Retrieved from http://www.ncbi.nlm.nih.gov/pubmed/14522586.

Hanley, N. A., & Arlt, W. (2006). The human fetal adrenal cortex and the window of sexual differentiation. *Trends in Endocrinology and Metabolism: TEM, 17*(10), 391–397. https://doi.org/10.1016/j.tem.2006.10.001.

Havelock, J. C., Auchus, R. J., & Rainey, W. E. (2004). The rise in adrenal androgen biosynthesis: Adrenarche. *Seminars in Reproductive Medicine, 22*(4), 337–347. https://doi.org/10.1055/s-2004-861550.

Homma, K., Hasegawa, T., Nagai, T., Adachi, M., Horikawa, R., Fujiwara, I., et al. (2006). Urine steroid hormone profile analysis in cytochrome P450 oxidoreductase deficiency: Implication for the backdoor pathway to dihydrotestosterone. *The Journal of Clinical Endocrinology & Metabolism, 91*(7), 2643–2649. https://doi.org/10.1210/jc.2005-2460.

Huang, Y., Zheng, J., Xie, T., Xiao, Q., Lu, S., Li, X., et al. (2014). A novel homozygous mutation p.E25X in the HSD3B2 gene causing salt wasting 3β-hydroxysteroid dehydrogenases deficiency in a Chinese pubertal girl: A delayed diagnosis until recurrent ovary cysts. *Zhonghua Er Ke Za Zhi = Chinese Journal of Pediatrics, 52*(12), 948–951. Retrieved from http://www.ncbi.nlm.nih.gov/pubmed/25619355.

Hui, X. G., Akahira, J. I., Suzuki, T., Nio, M., Nakamura, Y., Suzuki, H., et al. (2009). Development of the human adrenal zona reticularis: Morphometric and immunohistochemical studies from birth to adolescence. *Journal of Endocrinology, 203*(2), 241–252. https://doi.org/10.1677/JOE-09-0127.

Ishimoto, H., & Jaffe, R. B. (2011). Development and function of the human fetal adrenal cortex: A key component in the feto-placental unit. *Endocrine Reviews, 32*(3), 317–355. https://doi.org/10.1210/er.2010-0001.

Jeandron, D. D., & Sahakitrungruang, T. (2012). A novel homozygous Q334X mutation in the HSD3B2 gene causing classic 3β-hydroxysteroid dehydrogenase deficiency: An unexpected diagnosis after a positive newborn screen for 21-hydroxylase deficiency. *Hormone Research in Pædiatrics*, 77(5), 334–338. https://doi.org/10.1159/000336004.

Johannsen, T. H., Mallet, D., Dige-Petersen, H., Müller, J., Main, K. M., Morel, Y., et al. (2005). Delayed diagnosis of congenital adrenal hyperplasia with salt wasting due to type II 3beta-hydroxysteroid dehydrogenase deficiency. *The Journal of Clinical Endocrinology and Metabolism, 90*(4), 2076–2080. https://doi.org/10.1210/jc.2004-1374.

John, M. E., John, M. C., Boggaram, V., Simpson, E. R., & Waterman, M. R. (1986). Transcriptional regulation of steroid hydroxylase genes by corticotropin. *Proceedings of the National Academy of Sciences of the United States of America, 83*(13), 4715–4719. Retrieved from http://www.ncbi.nlm.nih.gov/pubmed/3014507.

Jones, C. M., Mallappa, A., Reisch, N., Nikolaou, N., Krone, N., Hughes, B. A., et al. (2017). Modified release and conventional glucocorticoids and diurnal androgen excretion in congenital adrenal hyperplasia. *The Journal of Clinical Endocrinology and Metabolism, 102*, 1797–1806. https://doi.org/10.1210/jc.2016-2855.

Kamrath, C., Hochberg, Z., Hartmann, M. F., Remer, T., & Wudy, S. A. (2012). Increased activation of the alternative "backdoor" pathway in patients with 21-hydroxylase deficiency: Evidence from urinary steroid hormone analysis. *The Journal of Clinical Endocrinology & Metabolism, 97*(3), E367–E375. https://doi.org/10.1210/jc.2011-1997.

Katsumata, N., Tanae, A., Yasunaga, T., Horikawa, R., Tanaka, T., & Hibi, I. (1995). A novel missense mutation in the type II 3 beta-hydroxysteroid dehydrogenase gene in a family with classical salt-wasting congenital adrenal hyperplasia due to 3 beta-hydroxysteroid dehydrogenase deficiency. *Human Molecular Genetics, 4*(4), 745–746. Retrieved from http://www.ncbi.nlm.nih.gov/pubmed/7633426.

Lee, T. C., Miller, W. L., & Auchus, R. J. (1999). Medroxyprogesterone acetate and dexamethasone are competitive inhibitors of different human steroidogenic enzymes. *The Journal of Clinical Endocrinology and Metabolism, 84*(6), 2104–2110. https://doi.org/10.1210/jcem.84.6.5646.

Lin, D., Sugawara, T., Strauss, J. F., Clark, B. J., Stocco, D. M., Saenger, P., et al. (1995). Role of steroidogenic acute regulatory protein in adrenal and gonadal steroidogenesis. *Science (New York, N.Y.)*, *267*(5205), 1828–1831. Retrieved from http://www.ncbi.nlm.nih.gov/pubmed/7892608.

Lorence, M. C., Corbin, C. J., Kamimura, N., Mahendroo, M. S., & Mason, J. I. (1990). Structural analysis of the gene encoding human 3 beta-hydroxysteroid dehydrogenase/delta 5----4-isomerase. *Molecular Endocrinology*, *4*(12), 1850–1855. https://doi.org/10.1210/mend.15.1.0585.

Lusa, L. G., de Lemos-Marini, S. H. V., Soardi, F. C., Ferraz, L. F. C., Guerra-Júnior, G., & de Mello, M. P. (2010). Structural aspects of the p.P222Q homozygous mutation of HSD3B2 gene in a patient with congenital adrenal hyperplasia. *Arquivos Brasileiros de Endocrinologia e Metabologia*, *54*(8), 768–774. Retrieved from http://www.ncbi.nlm.nih.gov/pubmed/21340167.

Lutfallah, C., Wang, W., Mason, J. I., Chang, Y. T., Haider, A., Rich, B., et al. (2002). Newly proposed hormonal criteria via genotypic proof for type II 3beta-hydroxysteroid dehydrogenase deficiency. *The Journal of Clinical Endocrinology and Metabolism*, *87*(6), 2611–2622. https://doi.org/10.1210/jcem.87.6.8615.

Luu, T. V., Lachance, Y., Labrie, C., Leblanc, G., Thomas, J. L., Strickler, R. C., et al. (1989). Full length cDNA structure and deduced amino acid sequence of human 3 beta-hydroxy-5-ene steroid dehydrogenase. *Molecular Endocrinology (Baltimore, Md.)*, *3*(8), 1310–1312. https://doi.org/10.1210/mend-3-8-1310.

Marti, N., Galván, J. A., Pandey, A. V., Trippel, M., Tapia, C., Müller, M., et al. (2017). Genes and proteins of the alternative steroid backdoor pathway for dihydrotestosterone synthesis are expressed in the human ovary and seem enhanced in the polycystic ovary syndrome. *Molecular and Cellular Endocrinology*, *441*, 116–123. https://doi.org/10.1016/j.mce.2016.07.029.

Marui, S., Castro, M., Latronico, A. C., Elias, L. L., Arnhold, I. J., Moreira, A. C., et al. (2000). Mutations in the type II 3beta-hydroxysteroid dehydrogenase (HSD3B2) gene can cause premature pubarche in girls. *Clinical Endocrinology*, *52*(1), 67–75. Retrieved from http://www.ncbi.nlm.nih.gov/pubmed/10651755.

Marui, S., Torrealba, I. M., Russell, A. J., Latronico, A. C., Sutcliffe, R. G., & Mendonca, B. B. (1998). A novel homozygous nonsense mutations E135* in the type II 3beta-hydroxysteroid dehydrogenase gene in a girl with salt-losing congenital adrenal hyperplasia. Mutations in brief no. 168. Online. *Human Mutation*, *12*(2), 139. https://doi.org/10.1002/(SICI)1098-1004(1998)12:2<139::AID-HUMU19>3.0.CO;2-4.

McCartin, S., Russell, A. J., Fisher, R. A., Wallace, A. M., Arnhold, I. J., Mason, J. I., et al. (2000). Phenotypic variability and origins of mutations in the gene encoding 3beta-hydroxysteroid dehydrogenase type II. *Journal of Molecular Endocrinology*, *24*(1), 75–82. Retrieved from http://www.ncbi.nlm.nih.gov/pubmed/10656999.

Mébarki, F., Sanchez, R., Rhéaume, E., Laflamme, N., Simard, J., Forest, M. G., et al. (1995). Nonsalt-losing male pseudohermaphroditism due to the novel homozygous N100S mutation in the type II 3 beta-hydroxysteroid dehydrogenase gene. *The Journal of Clinical Endocrinology & Metabolism*, *80*(7), 2127–2134. https://doi.org/10.1210/jcem.80.7.7608265.

Mendonca, B. B., Bloise, W., Arnhold, I. J., Batista, M. C., Toledo, S. P., Drummond, M. C., et al. (1987). Male pseudohermaphroditism due to nonsalt-losing 3 beta-hydroxysteroid dehydrogenase deficiency: Gender role change and absence of gynecomastia at puberty. *Journal of Steroid Biochemistry*, *28*(6), 669–675. Retrieved from http://www.ncbi.nlm.nih.gov/pubmed/2826919.

Mendonça, B. B., Russell, A. J., Vasconcelos-Leite, M., Arnhold, I. J., Bloise, W., Wajchenberg, B. L., et al. (1994). Mutation in 3 beta-hydroxysteroid dehydrogenase type II associated with pseudohermaphroditism in males and premature pubarche or cryptic expression in females. *Journal of Molecular Endocrinology, 12*(1), 119–122. Retrieved from http://www.ncbi.nlm.nih.gov/pubmed/8185809.

Mermejo, L. M., Elias, L. L. K., Marui, S., Moreira, A. C., Mendonca, B. B., & de Castro, M. (2005). Refining hormonal diagnosis of type II 3beta-hydroxysteroid dehydrogenase deficiency in patients with premature pubarche and hirsutism based on HSD3B2 genotyping. *The Journal of Clinical Endocrinology and Metabolism, 90*(3), 1287–1293. https://doi.org/10.1210/jc.2004-1552.

Miller, W. L. (2005). Minireview: Regulation of steroidogenesis by electron transfer. *Endocrinology, 146*(6), 2544–2550. https://doi.org/10.1210/en.2005-0096.

Miller, W. L., & Auchus, R. J. (2011). The molecular biology, biochemistry, and physiology of human steroidogenesis and its disorders. *Endocrine Reviews, 32*(1), 81–151. https://doi.org/10.1210/er.2010-0013.

Moisan, A. M., Ricketts, M. L., Tardy, V., Desrochers, M., Mébarki, F., Chaussain, J.-L., et al. (1999). New insight into the molecular basis of 3β-hydroxysteroid dehydrogenase deficiency: Identification of eight mutations in the *HSD3B2* gene in eleven patients from seven new families and comparison of the functional properties of twenty-five mutant enzymes. *The Journal of Clinical Endocrinology & Metabolism, 84*(12), 4410–4425. https://doi.org/10.1210/jcem.84.12.6288.

Nakamura, Y., Fujishima, F., Hui, X., Felizola, S. J. A., Shibahara, Y., Akahira, J., et al. (2015). 3βHSD and CYB5A double positive adrenocortical cells during adrenal development/aging. *Endocrine Research, 40*(1), 8–13. https://doi.org/10.3109/07435800.2014.895377.

Nakamura, Y., Gang, H. X., Suzuki, T., Sasano, H., & Rainey, W. E. (2009). Adrenal changes associated with adrenarche. *Reviews in Endocrine & Metabolic Disorders, 10*(1 spec. issue), 19–26. https://doi.org/10.1007/s11154-008-9092-2.

Nakamura, Y., Hornsby, P. J., Casson, P., Morimoto, R., Satoh, F., Xing, Y., et al. (2009). Type 5 17β-hydroxysteroid dehydrogenase (AKR1C3) contributes to testosterone production in the adrenal reticularis. *The Journal of Clinical Endocrinology & Metabolism, 94*(6), 2192–2198. https://doi.org/10.1210/jc.2008-2374.

Nakamura, Y., Xing, Y., Hui, X.-G., Kurotaki, Y., Ono, K., Cohen, T., et al. (2011). Human adrenal cells that express both 3β-hydroxysteroid dehydrogenase type 2 (HSD3B2) and cytochrome b5 (CYB5A) contribute to adrenal androstenedione production. *The Journal of Steroid Biochemistry and Molecular Biology, 123*(3–5), 122–126. https://doi.org/10.1016/j.jsbmb.2010.12.001.

Nayak, S., Lee, P. A., & Witchel, S. F. (1998). Variants of the type II 3beta-hydroxysteroid dehydrogenase gene in children with premature pubic hair and hyperandrogenic adolescents. *Molecular Genetics and Metabolism, 64*(3), 184–192. https://doi.org/10.1006/mgme.1998.2715.

Nordenström, A., Forest, M. G., & Wedell, A. (2007). A case of 3beta-hydroxysteroid dehydrogenase type II (HSD3B2) deficiency picked up by neonatal screening for 21-hydroxylase deficiency: Difficulties and delay in etiologic diagnosis. *Hormone Research, 68*(4), 204–208. https://doi.org/10.1159/000102593.

O'Reilly, M. W., Kempegowda, P., Jenkinson, C., Taylor, A. E., Quanson, J. L., Storbeck, K.-H., et al. (2017). 11-Oxygenated C19 steroids are the predominant androgens in polycystic ovary syndrome. *The Journal of Clinical Endocrinology and Metabolism, 102*(3), 840–848. https://doi.org/10.1210/jc.2016-3285.

Pang, S. (2001). Congenital adrenal hyperplasia owing to 3 beta-hydroxysteroid dehydrogenase deficiency. *Endocrinology and Metabolism Clinics of North America, 30*(1), 81–99. vi–vii. Retrieved from http://www.ncbi.nlm.nih.gov/pubmed/11344940.

Pang, S. Y., Lerner, A. J., Stoner, E., Levine, L. S., Oberfield, S. E., Engel, I., et al. (1985). Late-onset adrenal steroid 3 beta-hydroxysteroid dehydrogenase deficiency. I. A cause of hirsutism in pubertal and postpubertal women. *The Journal of Clinical Endocrinology and Metabolism, 60*(3), 428–439. https://doi.org/10.1210/jcem-60-3-428.

Pang, S., Levine, L. S., Stoner, E., Opitz, J. M., Pollack, M. S., Dupont, B., et al. (1983). Nonsalt-losing congenital adrenal hyperplasia due to 3β-hydroxysteroid dehydrogenase deficiency with normal glomerulosa function. *The Journal of Clinical Endocrinology & Metabolism*, *56*(4), 808–818. https://doi.org/10.1210/jcem-56-4-808.

Pang, S., Wang, W., Rich, B., David, R., Chang, Y. T., Carbunaru, G., et al. (2002). A novel nonstop mutation in the stop codon and a novel missense mutation in the type II 3β-hydroxysteroid dehydrogenase (3β-HSD) gene causing, respectively, nonclassic and classic 3β-HSD deficiency congenital adrenal hyperplasia. *The Journal of Clinical Endocrinology & Metabolism*, *87*(6), 2556–2563. https://doi.org/10.1210/jcem.87.6.8559.

Panzer, K., Ekhaguere, O. A., Darbro, B., Cook, J., & Shchelochkov, O. A. (2017). Uniparental isodisomy of chromosome 1 unmasking an autosomal recessive 3-beta hydroxysteroid dehydrogenase type II-related congenital adrenal hyperplasia. *Journal of Clinical Research in Pediatric Endocrinology*, *9*(1), 70–73. https://doi.org/10.4274/jcrpe.3680.

Parker, K. L., Rice, D. A., Lala, D. S., Ikeda, Y., Luo, X., Wong, M., et al. (2002). Steroidogenic factor 1: An essential mediator of endocrine development. *Recent Progress in Hormone Research*, *57*, 19–36.

Parks, G. A., Bermudez, J. A., Anast, C. S., Bongiovanni, A. M., & New, M. I. (1971). Pubertal boy with the 3β-hydroxy steroid dehydrogenase defect. *The Journal of Clinical Endocrinology & Metabolism*, *33*(2), 269–278. https://doi.org/10.1210/jcem-33-2-269.

Paula, F. J., Dick-de-Paula, I., Pontes, A., Schmitt, F. C., Mendonça, B. B., & Foss, M. C. (1994). Hyperandrogenism due to 3 beta-hydroxysteroid dehydrogenase deficiency with accessory adrenocortical tissue: A hormonal and metabolic evaluation. *Brazilian Journal of Medical and Biological Research = Revista Brasileira de Pesquisas Médicas e Biológicas*, *27*(5), 1149–1158. Retrieved from http://www.ncbi.nlm.nih.gov/pubmed/8000336.

Pretorius, E., Africander, D. J., Vlok, M., Perkins, M. S., Quanson, J., & Storbeck, K.-H. (2016). 11-Ketotestosterone and 11-ketodihydrotestosterone in castration resistant prostate cancer: Potent androgens which can no longer be ignored. *PLoS One*, *11*(7). e0159867. https://doi.org/10.1371/journal.pone.0159867.

Probst-Scheidegger, U., Udhane, S. S., l'Allemand, D., Flück, C. E., & Camats, N. (2016). Non-virilizing congenital adrenal hyperplasia in a female patient with a novel HSD3B2 mutation. *Sexual Development: Genetics, Molecular Biology, Evolution, Endocrinology, Embryology, and Pathology of Sex Determination and Differentiation*, *10*(4), 200–204. https://doi.org/10.1159/000448724.

Rabbani, B., Mahdieh, N., Haghi Ashtiani, M. T., Setoodeh, A., & Rabbani, A. (2012). In silico structural, functional and pathogenicity evaluation of a novel mutation: An overview of HSD3B2 gene mutations. *Gene*, *503*(2), 215–221. https://doi.org/10.1016/j.gene.2012.04.080.

Rege, J., Nakamura, Y., Satoh, F., Morimoto, R., Kennedy, M. R., Layman, L. C., et al. (2013). Liquid chromatography–tandem mass spectrometry analysis of human adrenal vein 19-carbon steroids before and after ACTH stimulation. *The Journal of Clinical Endocrinology & Metabolism*, *98*(3), 1182–1188. https://doi.org/10.1210/jc.2012-2912.

Remer, T., Boye, K. R., Hartmann, M. F., & Wudy, S. A. (2005). Urinary markers of adrenarche: Reference values in healthy subjects, aged 3-18 years. *The Journal of Clinical Endocrinology and Metabolism*, *90*(4), 2015–2021. https://doi.org/10.1210/jc.2004-1571.

Rhéaume, E., Lachance, Y., Zhao, H. F., Breton, N., Dumont, M., de Launoit, Y., et al. (1991). Structure and expression of a new complementary DNA encoding the almost exclusive 3 beta-hydroxysteroid dehydrogenase/delta 5-delta 4-isomerase in human adrenals and gonads. *Molecular Endocrinology (Baltimore, Md.)*, *5*(8), 1147–1157. https://doi.org/10.1210/mend-5-8-1147.

Rhéaume, E., Sanchez, R., Mébarki, F., Gagnon, E., Carel, J. C., Chaussain, J. L., et al. (1995). Identification and characterization of the G15D mutation found in a male patient with 3 beta-hydroxysteroid dehydrogenase (3 beta-HSD) deficiency: Alteration of the putative NAD-binding domain of type II 3 beta-HSD. *Biochemistry*, *34*(9), 2893–2900. Retrieved from http://www.ncbi.nlm.nih.gov/pubmed/7893703.

Rhéaume, E., Sanchez, R., Simard, J., Chang, Y. T., Wang, J., Pang, S., et al. (1994). Molecular basis of congenital adrenal hyperplasia in two siblings with classical nonsalt-losing 3 beta-hydroxysteroid dehydrogenase deficiency. *The Journal of Clinical Endocrinology and Metabolism*, *79*(4), 1012–1018. https://doi.org/10.1210/jcem.79.4.7962268.

Rhéaume, E., Simard, J., Morel, Y., Mebarki, F., Zachmann, M., Forest, M. G., et al. (1992). Congenital adrenal hyperplasia due to point mutations in the type II 3β-hydroxysteroid dehydrogenase gene. *Nature Genetics*, *1*(4), 239–245. https://doi.org/10.1038/ng0792-239.

Rone, M. B., Fan, J., & Papadopoulos, V. (2009). Cholesterol transport in steroid biosynthesis: Role of protein–protein interactions and implications in disease states. *Biochimica et Biophysica Acta (BBA)—Molecular and Cell Biology of Lipids*, *1791*(7), 646–658. https://doi.org/10.1016/j.bbalip.2009.03.001.

Rosenfield, R. L., Rich, B. H., Wolfsdorf, J. I., Cassorla, F., Parks, J. S., Bongiovanni, A. M., et al. (1980). Pubertal presentation of congenital delta 5-3 beta-hydroxysteroid dehydrogenase deficiency. *The Journal of Clinical Endocrinology and Metabolism*, *51*(2), 345–353. https://doi.org/10.1210/jcem-51-2-345.

Russell, A. J., Wallace, A. M., Forest, M. G., Donaldson, M. D., Edwards, C. R., & Sutcliffe, R. G. (1994). Mutation in the human gene for 3 beta-hydroxysteroid dehydrogenase type II leading to male pseudohermaphroditism without salt loss. *Journal of Molecular Endocrinology*, *12*(2), 225–237. Retrieved from http://www.ncbi.nlm.nih.gov/pubmed/8060486.

Sanchez, R., Mébarki, F., Rhéaume, E., Laflamme, N., Forest, M. G., Bey-Omard, F., et al. (1994). Functional characterization of the novel L108W and P186L mutations detected in the type II 3 beta-hydroxysteroid dehydrogenase gene of a male pseudohermaphrodite with congenital adrenal hyperplasia. *Human Molecular Genetics*, *3*(9), 1639–1645. Retrieved from http://www.ncbi.nlm.nih.gov/pubmed/7833923.

Sanchez, R., Rhéaume, E., Laflamme, N., Rosenfield, R. L., Labrie, F., & Simard, J. (1994). Detection and functional characterization of the novel missense mutation Y254D in type II 3 beta-hydroxysteroid dehydrogenase (3 beta HSD) gene of a female patient with nonsalt-losing 3 beta HSD deficiency. *The Journal of Clinical Endocrinology & Metabolism*, *78*(3), 561–567. https://doi.org/10.1210/jcem.78.3.8126127.

Shapiro, E., Huang, H. Y., & Wu, X. R. (2000). Uroplakin and androgen receptor expression in the human fetal genital tract: Insights into the development of the vagina. *The Journal of Urology*, *164*(3 Pt. 2), 1048–1051. Retrieved from http://www.ncbi.nlm.nih.gov/pubmed/10958738.

Simard, J., Moisan, A. M., & Morel, Y. (2002). Congenital adrenal hyperplasia due to 3β-hydroxysteroid dehydrogenase/Δ^5-Δ^4 isomerase deficiency. *Seminars in Reproductive Medicine*, *20*(3), 255–276. https://doi.org/10.1055/s-2002-35373.

Simard, J., Rhéaume, E., Leblanc, J. F., Wallis, S. C., Joplin, G. F., Gilbey, S., et al. (1994). Congenital adrenal hyperplasia caused by a novel homozygous frameshift mutation 273 delta AA in type II 3 beta-hydroxysteroid dehydrogenase gene (HSD3B2) in three male patients of Afghan/Pakistani origin. *Human Molecular Genetics*, *3*(2), 327–330. Retrieved from http://www.ncbi.nlm.nih.gov/pubmed/8004103.

Simard, J., Rhéaume, E., Sanchez, R., Laflamme, N., de Launoit, Y., Luu-The, V., et al. (1993). Molecular basis of congenital adrenal hyperplasia due to 3 beta-hydroxysteroid dehydrogenase deficiency. *Molecular Endocrinology*, 7(5), 716–728. https://doi.org/10.1210/mend.7.5.8316254.

Simard, J., Ricketts, M.-L., Gingras, S., Soucy, P., Feltus, F. A., & Melner, M. H. (2005). Molecular biology of the 3β-hydroxysteroid dehydrogenase/Δ^5-Δ^4 isomerase gene family. *Endocrine Reviews*, *26*(4), 525–582. https://doi.org/10.1210/er.2002-0050.

Sobel, V., Zhu, Y.-S., & Imperato-McGinley, J. (2004). Fetal hormones and sexual differentiation. *Obstetrics and Gynecology Clinics of North America*, *31*(4), 837–856. x–xi. https://doi.org/10.1016/j.ogc.2004.08.005.

Storbeck, K.-H., Bloem, L. M., Africander, D., Schloms, L., Swart, P., & Swart, A. C. (2013). 11β-Hydroxydihydrotestosterone and 11-ketodihydrotestosterone, novel C19 steroids with androgenic activity: A putative role in castration resistant prostate cancer? *Molecular and Cellular Endocrinology*, *377*(1–2), 135–146. https://doi.org/10.1016/j.mce.2013.07.006.

Suzuki, T., Sasano, H., Takeyama, J., Kaneko, C., Freije, W. A., Carr, B. R., et al. (2000). Developmental changes in steroidogenic enzymes in human postnatal adrenal cortex: Immunohistochemical studies. *Clinical Endocrinology*, *53*(6), 739–747. https://doi.org/10.1046/j.1365-2265.2000.01144.x.

Tajima, T., Fujieda, K., Nakae, J., Shinohara, N., Yoshimoto, M., Baba, T., et al. (1995). Molecular analysis of type II 3 beta-hydroxysteroid dehydrogenase gene in Japanese patients with classical 3 beta-hydroxysteroid dehydrogenase deficiency. *Human Molecular Genetics*, *4*(5), 969–971. Retrieved from http://www.ncbi.nlm.nih.gov/pubmed/7633460.

Takasawa, K., Ono, M., Hijikata, A., Matsubara, Y., Katsumata, N., Takagi, M., et al. (2014). Two novel *HSD3B2* missense mutations with diverse residual enzymatic activities for Δ5-steroids. *Clinical Endocrinology*, *80*(6), 782–789. https://doi.org/10.1111/cen.12394.

Thomas, J. L., Boswell, E. L., Scaccia, L. A., Pletnev, V., & Umland, T. C. (2005). Identification of key amino acids responsible for the substantially higher affinities of human type 1 3beta-hydroxysteroid dehydrogenase/isomerase (3beta-HSD1) for substrates, coenzymes, and inhibitors relative to human 3beta-HSD2. *The Journal of Biological Chemistry*, *280*(22), 21321–21328. https://doi.org/10.1074/jbc.M501269200.

Thomas, J. L., Frieden, C., Nash, W. E., & Strickler, R. C. (1995). An NADH-induced conformational change that mediates the sequential 3 beta-hydroxysteroid dehydrogenase/isomerase activities is supported by affinity labeling and the time-dependent activation of isomerase. *The Journal of Biological Chemistry*, *270*(36), 21003–21008. Retrieved from http://www.ncbi.nlm.nih.gov/pubmed/7673125.

Thomas, J. L., Mason, J. I., Blanco, G., & Veisaga, ML. (2001). The engineered, cytosolic form of human type I 3beta-hydroxysteroid dehydrogenase/isomerase: Purification, characterization and crystallization. *Journal of Molecular Endocrinology*, *27*(1), 77–83.

Turcu, A. F., Nanba, A. T., Chomic, R., Upadhyay, S. K., Giordano, T. J., Shields, J. J., et al. (2016). Adrenal-derived 11-oxygenated 19-carbon steroids are the dominant androgens in classic 21-hydroxylase deficiency. *European Journal of Endocrinology*, *174*(5), 601–609. https://doi.org/10.1530/EJE-15-1181.

Turcu, A., Smith, J. M., Auchus, R., & Rainey, W. E. (2014). Adrenal androgens and androgen precursors-definition, synthesis, regulation and physiologic actions. *Comprehensive Physiology*, *4*(4), 1369–1381. https:/doi.org/10.1002/cphy.c140006.

Voutilainen, R., Tapanainen, J., Chung, B.-C., Matteson, K. J., & Miller, W. L. (1986). Hormonal regulation of P450scc (20,22-desmolase) and P450cl7 (17α-hydroxylase/17,20-lyase) in cultured human granulosa cells. *The Journal of Clinical Endocrinology & Metabolism*, *63*(1), 202–207. https://doi.org/10.1210/jcem-63-1-202.

Welzel, M., Wüstemann, N., Simic-Schleicher, G., Dörr, H. G., Schulze, E., Shaikh, G., et al. (2008). Carboxyl-terminal mutations in 3beta-hydroxysteroid dehydrogenase type II cause severe salt-wasting congenital adrenal hyperplasia. *The Journal of Clinical Endocrinology and Metabolism*, *93*(4), 1418–1425. https://doi.org/10.1210/jc.2007-1874.

Wilhelm, D., Palmer, S., & Koopman, P. (2007). Sex determination and gonadal development in mammals. *Physiological Reviews*, *87*(1), 1–28. https://doi.org/10.1152/physrev.00009.2006.
Xing, Y., Lerario, A. M., Rainey, W., & Hammer, G. D. (2015). Development of adrenal cortex zonation. *Endocrinology and Metabolism Clinics of North America*, *44*(2), 243–274. https://doi.org/10.1016/j.ecl.2015.02.001.
Yoshimoto, M., Kawaguchi, T., Mori, R., Kinoshita, E., Baba, T., Tajima, T., et al. (1997). Pubertal changes in testicular 3 beta-hydroxysteroid dehydrogenase activity in a male with classical 3 beta-hydroxysteroid dehydrogenase deficiency showing spontaneous secondary sexual maturation. *Hormone Research*, *48*(2), 83–87. Retrieved from http://www.ncbi.nlm.nih.gov/pubmed/9251925.
Zachmann, M., Forest, M. G., & De Peretti, E. (1979). 3 beta-Hydroxysteroid dehydrogenase deficiency. Follow-up study in a girl with pubertal bone age. *Hormone Research*, *11*(6), 292–302. Retrieved from http://www.ncbi.nlm.nih.gov/pubmed/295036.
Zachmann, M., Vollmin, J. A., Murset, G., Curtius, H. C., & Prader, A. (1970). Unusual type of congenital adrenal hyperplasia probably due to deficiency of 3-beta-hydroxysteroid dehydrogenase. Case report of a surviving girl and steroid studies. *The Journal of Clinical Endocrinology and Metabolism*, *30*(6), 719–726. https://doi.org/10.1210/jcem-30-6-719.
Zhang, L., Mason, J. I., Naiki, Y., Copeland, K. C., Castro-Magana, M., Gordon-Walker, T. T., et al. (2000). Characterization of two novel homozygous missense mutations involving codon 6 and 259 of type II 3beta-hydroxysteroid dehydrogenase (3betaHSD) gene causing, respectively, nonsalt-wasting and salt-wasting 3betaHSD deficiency disorder. *The Journal of Clinical Endocrinology and Metabolism*, *85*(4), 1678–1685. https://doi.org/10.1210/jcem.85.4.6539.
Zhang, L., Sakkal-Alkaddour, H., Chang, Y. T., Yang, X., & Pang, S. (1996). A new compound heterozygous frameshift mutation in the type II 3 beta-hydroxysteroid dehydrogenase (3 beta-HSD) gene causes salt-wasting 3 beta-HSD deficiency congenital adrenal hyperplasia. *The Journal of Clinical Endocrinology & Metabolism*, *81*(1), 291–295. https://doi.org/10.1210/jcem.81.1.8550766.

FURTHER READING

Bongiovanni, A. M. (1962). The adrenogenital syndrome with deficiency of 3 beta-hydroxysteroid dehydrogenase. *The Journal of Clinical Investigation*, *41*(11), 2086–2092. https://doi.org/10.1172/JCI104666.
Schneider, G., Genel, M., Bongiovanni, A. M., Goldman, A. S., & Rosenfield, R. L. (1975). Persistent testicular delta5-isomerase-3beta-hydroxysteroid dehydrogenase (delta5-3beta-HSD) deficiency in the delta5-3beta-HSD form of congenital adrenal hyperplasia. *The Journal of Clinical Investigation*, *55*(4), 681–690. https://doi.org/10.1172/JCI107977.
Zachmann, M., Kempken, B., Anner, I., & Pezzoli, V. (1988). 160 age-related differences in the excretion of delta 5 steroids (D5S) in two related girls with 3b-hydroxy steroid dehydrogenase deficiency (3b0HD). *Pediatric Research*, *24*(4), 543. https://doi.org/10.1203/00006450-198810000-00181.

CHAPTER FOUR

DHEA Modulates Immune Function: A Review of Evidence

Sean P. Prall*,[1], **Michael P. Muehlenbein**[†]
*University of California, Los Angeles, CA, United States
[†]Baylor University, Waco, TX, United States
[1]Corresponding author: e-mail address: sprall@ucla.edu

Contents

Abstract

DHEA and DHEA-S have numerous associations with multiple aspects of immune function and are often characterized as beneficial and supportive of immunocompetence. However, closer inspection of these studies reveals confusion regarding the immunological components modified, the mechanisms of action, and degree of impact, and even whether these hormones even have direct action or are mediated by metabolites and interactions with other hormones and hormone receptors. Additionally, much of the research is conducted on rodent models using very high concentrations of hormone supplements, which may not be representative of the effects of these hormones in natural circulating concentrations, or may not translate to human physiology in a meaningful way. Here, we review the effects of DHEA and DHEA-S on immune function and examine the potential roles these hormones play on specific components of immune function. Drawing from the literature on hormone supplementation, as well as studies examining the natural circulating levels of DHEA and DHEA-S on specific immunological components and disease processes, we argue that DHEA has differential

Vitamins and Hormones, Volume 108
ISSN 0083-6729
https://doi.org/10.1016/bs.vh.2018.01.023

actions on human immune function, and that its effects are further shaped by concentrations of other hormones. Of particular interest is the role of DHEA as an anti-glucocorticoid, and for its actions on both androgen and estrogen receptors. With additional research, DHEA may be useful as a therapeutic, particularly in diseases with high levels of inflammation, or where adrenal production is altered. The convoluted nature of DHEA–immune interactions makes direct effects difficult to interpret, and future research needs to consider direct, intracrine, and downstream effects of these hormones.

1. INTRODUCTION

The adrenal androgen dehydroepiandrosterone (DHEA) has been implicated in a diverse array of physiological processes, despite that it is often considered a precursor hormone with a primary role as a storage pool for downstream androgen and estrogen synthesis (Labrie et al., 1998). Characterizations of the hormone as a "fountain of youth," as well as the availability of DHEA supplements to be bought over-the-counter as a dietary supplement, make research on DHEA particularly relevant and timely. DHEA has impacts on immune function and disease resistance, but its interactions with other hormones, lack of a clear independent mechanisms of action, discrepancies between human and animal physiologies that make direct conclusions complicated, and confusions about how to characterize the principal actions of this hormone make understanding these effects difficult. In this chapter, we review the current literature on DHEA and immunity, particularly in humans.

2. MECHANISMS OF ACTION IN IMMUNOMODULATION

As DHEA's numerous associations with immunological outcomes and biomarkers are investigated, the mechanism of action whereby DHEA moderates immunological components remains equivocal. Unlike other steroid hormones, DHEA has no known unique receptor, and its physiological role is described as a precursor for the synthesis of downstream hormones. Multiple avenues whereby DHEA may initiate immunological changes are being explored (Prough, Clark, & Klinge, 2016; see Table 1), although much more research is needed.

Table 1 Potential Mechanisms of Action on Immune Function

Mechanism	Explanation	Citations
Androgen and estrogen receptors	Concentration-dependent binding to androgen and estrogen receptors	Chen et al. (2005)
NF-κB	Inhibits transcription factor	Du, Khalil, et al. (2001)
Receptor in T-cells	Evidence of intracellular binding site in T cells	Meikle et al. (1992)
Receptor in monocytes	Evidence of intracellular binding site in monocytes	McLachlan et al. (1996)
IL-2 transcription enhanced	Evidence of enhanced transcription of IL2 mRNA in response to DHEA treatment	Suzuki, Suzuki, Daynes, and Engleman (1991)
Conversion to metabolites	DHEA metabolites show more potent protective responses to viral infections	Loria (2002)

One potential avenue of action is through conversion to downstream hormones in both endocrine and intracrine fashions. As a precursor hormone, DHEA concentrations are physiologically tied to downstream metabolites, and some of the purported actions of DHEA may be a result of modified synthesis of these other hormones. Some of these effects may be the result of conversion to androgens or estrogens, with consequential action on androgen and estrogen receptors via normal processes (Labrie et al., 1998). Other effects may be mediated by downstream metabolites. For example, DHEA enhances immunity in mice infected with lethal viral infections (Loria, Inge, Cook, Szakal, & Regelson, 1988), but further research illustrates that the downstream hormone androstenediol is substantially more effective in modulating immunological responses to infection (Loria & Padgett, 1992). Further, androstenediol is even more successful in limiting viral and bacterial infections and enhancing lymphocyte activity (Loria, 2002). Many other aspects of the immune–endocrine relationship may actually be modulated by downstream DHEA metabolites, instead of through direct actions of DHEA.

DHEA can also act directly on steroid hormone receptors. DHEA exhibits agnostic effects on ER-β and may in fact have greater effects on the receptor than estradiol itself (Chen et al., 2005). DHEA has some affinity for ER-α and exhibits antagonistic effects on the AR in high concentrations.

Since both ER-α and ER-β are present on many types of immune cells, including B cells, T cells, monocytes, and natural killer cells (Kovats, 2015), some action of DHEA on immunological parameters may be via ER binding and resultant transcription activities.

Beyond the role of steroid receptors, there is some evidence that DHEA exerts direct effects. DHEA's modulation of inflammatory responses appears to be mediated, at least in part, by its action on NF-κB. This transcription factor is responsible for activating cytokine expression and plays an important role in inflammatory responses. For example, NF-κB activation and translocation is decreased in the presence of DHEA in vitro, resulting in reduced inflammatory activity (Du, Khalil, & Sriram, 2001). Finally, DHEA may bind to receptors on T cells and monocytes (McLachlan, Serkin, & Bakouche, 1996; Meikle et al., 1992), although such binding is not yet fully described.

3. DHEA IN INFLAMMATION AND CYTOKINE MODULATION

DHEA potently modulates inflammation and cytokine responses to stimulation in a variety of cellular contexts (Table 2). For example, DHEA is very effective at blunting both Th-1 and Th-2 immunological responses, and further suppresses expression of various proinflammatory cytokines (Choi et al., 2008; Du, Guan, Khalil, & Sriram, 2001; Du, Khalil, et al., 2001). These effects appear to be mediated, in part, by regulation of NF-κB (Du, Khalil, et al., 2001). Other proinflammatory cytokines (e.g., IL-1β, TNF-α, IFN-γ, etc.) are suppressed by DHEA in a variety of experimental contexts. These cytokine modulations are not trivial, as suppression of TNF-α decreases mortality during severe sepsis (Oberbeck et al., 2001), and immunomodulation of DHEA as a response to supplementation may have beneficial outcomes in various disease states: in women with systematic lupus erythematosus (SLE), DHEA administration results in decreased disease activity and severity (Barry, McGuire, & van Vollenhoven, 1998; Petri et al., 2002; van Vollenhoven, Morabito, Engleman, & McGuire, 1998).

DHEA also plays an important role in regulation of IL-2 secretion, where immunological modulation has beneficial results in adaptive immune responses. DHEA stimulates IL-2 secretion from T cells, resulting in enhanced T-cell cytotoxicity (Suzuki et al., 1991). This effect may result

Table 2 Role of DHEA on Cytokine Release and Action

Cytokine	Effect	Citations
IL-1β	Inhibits production	Ben-Nathan, Padgett, and Loria (1999)
IL-2	Increases secretion	Suzuki et al. (1991)
IL-4	Increases secretion	Du, Guan, et al. (2001)
IL-5	Inhibits production	Choi et al. (2008)
IL-6	Inhibits production	Straub et al. (1998)
IL-10	Inhibits production	Choi et al. (2008) and Chang, Chu, Chen, Kuo, and Lai (2004)
IL-12	Inhibits production	Du, Khalil, et al. (2001)
TNF-α	Inhibits production	Ben-Nathan et al. (1999), Di Santo et al. (1996), Du, Khalil, et al. (2001), and Oberbeck et al. (2001)
IFN-γ	Inhibits production	Choi et al. (2008), Du, Khalil, et al. (2001), and Moynihan, Callahan, Kelley, and Campbell (1998)

from increased transcription of IL-2 in T-cells through an unknown mechanism (Meikle et al., 1992; Suzuki et al., 1991). Low IL-2 may be related to low levels of DHEA in patients with SLE (Suzuki, Suzuki, Engleman, Mizushima, & Sakane, 1995), and DHEA treatment results in increased IL-2 production (Suzuki, Suzuki, & Sakane, 1996).

4. COMPLEMENT PROTEIN ACTIVITY

DHEA may play a role in innate immunity via regulation of parts of the complement cascade. In vitro experimentation suggests that DHEA increases the expression of C1 inhibitor, a protein that inhibits activation of the complement cascade (Falus et al., 1990; Hidvégi, Fehér, Feher, Koó, & Füst, 1984; McLachlan et al., 1996). However, as concentrations used in in vitro studies may not reflect natural physiology in vivo, and studies such as live subject supplementation have not been conducted, it can be difficult to interpret these findings. Findings from a cross-sectional sample of young orangutans found that DHEA-S, but not DHEA, was higher in animals with reduced complement protein activity (Prall et al., 2015). Similar results were found in a cross-sectional study examining relationships between androgens and immunity in humans (Prall & Muehlenbein, 2015).

However, larger studies using natural variation in hormone concentrations, or using hormone supplementation, are needed to further support these findings.

5. LYMPHOCYTE PROLIFERATION AND CELLULAR CYTOTOXICITY

A number of studies have investigated DHEA's effects on human lymphocyte and murine splenocyte proliferation in response to mitogens, with highly variable results. Some evidence indicates that DHEA inhibits cellular proliferation in response to multiple mitogens (PHA, ConA, LPS) (Ben-Nathan et al., 1999; Padgett & Loria, 1994; Sakakura, Nakagawa, & Ohzeki, 2006), while other studies suggest that DHEA can result in proliferation only under very certain conditions (Catania et al., 1999; Sakakura et al., 2006; Zhang et al., 1999). These studies vary considerably by methodology, with different cell types, mitogens, DHEA concentration, and timing ofDHEA administration. DHEA inhibits proliferation when using T-cell mitogens, but increases proliferation when using B-cell mitogens, suggesting differential actions on adaptive immunity (Sakakura et al., 2006). In studies where physiological concentrations are used, DHEA tends to increase proliferation, but the opposite is true when supraphysiological concentrations are used (Hazeldine, Arlt, & Lord, 2010). These findings highlight the need for in vitro experimentation to use androgen concentrations that approximate natural variation found in vivo. More naturalistic studies may be of use here. In a small cross-sectional study examining the relationships among salivary androgens and mitogen-stimulated lymphocyte proliferation, there is a negative relationship between DHEA concentrations and lymphocyte responses, particularly in males (Prall & Muehlenbein, 2015), suggesting that the underlying endocrine profile may further modulate how immune responses are shaped by DHEA.

Evidence suggests that DHEA and DHEA-S further shape cytotoxic abilities of immunological components. Natural killer cell cytotoxicity is enhanced by DHEA-S through modulation of the production of IGF-1 in in vitro experimentation (Solerte, Fioravanti, & Vignati, 1999). Similar results are found in T-cell responses, where DHEA stimulation results in increased cytotoxicity in human cells in vitro (Suzuki et al., 1991). Additionally, monocytes stimulated with both DHEA and lipopolysaccharide, but not DHEA alone, result in increase monocyte toxicity and other immunological responses (McLachlan et al., 1996).

6. DHEA SUPPLEMENTATION RESULTS IN BENEFICIAL IMMUNOMODULATION AND DISEASE OUTCOMES

In vitro studies implicate specific cellular processes impacted by variation in DHEA concentrations. Further examination of DHEA supplementation in humans and other animals yields particular insight into how DHEA, along with naturally circulating concentrations of other hormones, can influence immunological components, and ultimately disease processes. Additionally, DHEA is available as an over-the-counter nutritional supplement, making studies of oral DHEA supplementation logistically easier than many other hormones in humans.

Numerous lab-based rodent studies illustrate the beneficial role of exogenous DHEA on disease outcomes and survival in experimental infections. DHEA treatment results in decreased mortality as a result of sepsis (Oberbeck et al., 2001) and *Escherichia coli* infection in mice (Gennari and Alexander, 1997). DHEA treatment in mice also increases survival to viral infection, increases time to disease onset, and decreases viral levels (Ben-Nathan, Lachmi, Lustig, and Feuerstein, 1991; Ben-Nathan et al., 1992; Loria et al., 1988). DHEA supplementation in mice infected with retroviruses also shows significant modulation of immunological components, including elevations of IL-2 and IFN-γ, and increased proliferation of T and B cells (Araghi-Niknam et al., 1997; Zhang et al., 1999). In rats experientially infected with *Trypanosoma cruzi* and treated with DHEA, parasitemia decreases compared to those not receiving DHEA (Brazão et al., 2010; Del Vecchio Filipin et al., 2010; Santos et al., 2007).

Supplementation research reveals further evidence that DHEA can serve as a vaccine adjuvant capable of bolstering immunological responses, particularly in elderly individuals. DHEA treatment in mice prior to vaccination increases splenic response to the pneumococcal vaccine and increases antibody responses from the tetanus toxoid vaccine (Araneo et al., 1995; Garg and Bondada, 1993). However, human trials have yielded contradictory results. DHEA and DHEA-S supplementation increase antibody titers to influenza vaccination in several studies of elderly subjects (Araneo et al., 1995; Degelau et al., 1997). Several other studies find no change, or even reduced responsiveness, to influenza and tetanus vaccines (Ben-Yehuda, Danenberg, Zakay-Rones, Gross, and Friedman, 1998; Danenberg et al., 1997; Evans et al., 1996). Again, variation in dosage concentration, length, timing, and immunological measurements may play some role in these discrepancies. Sorting out these discrepancies is necessary before DHEA can be considered as an effective vaccine adjuvant for the elderly.

Despite this ambiguity, many human DHEA supplementation studies report multiple immunological impacts. For example, DHEA supplementation elevates natural killer cell cytotoxicity and natural killer cell number (Casson et al., 1993; Khorram et al., 1997), and increases B- and T-cell mitogenic responses, although there are some inconsistencies in these results (Casson et al., 1993; Coles et al., 2005; Khorram et al., 1997; Kohut et al., 2003). There is little evidence of modification or altered production of cytokines or immunoglobulins, with only one study finding that DHEA supplementation in subjects with low basal DHEA-S resulted in elevated mitogen-stimulated IL-2 and IL-6 (Casson et al., 1993; Khorram et al., 1997; Kohut et al., 2003).

Supplementation studies are useful to show that variation in DHEA itself can impact immune function and health. However, some caution must be noted in interpretation of these results, as concentrations of hormones used can vary widely among studies. Supplementation often results in hormone concentrations much higher than what would be physiologically normal. Additionally, rodent studies must be interpreted cautiously, as rodents have different enzymatic pathways for hormone synthesis, making results from rodents difficult to generalize to human physiology (Maninger, Wolkowitz, Reus, Epel, & Mellon, 2009). More generally, while study outcomes may be muddled by systematic differences in supplementation concentration and duration, and disease and immune outcomes measured, results generally point to immunological benefits of elevated DHEA concentrations (Table 3).

7. DHEA AND INFECTION IN HUMAN POPULATIONS

To understand better the relationships among naturally circulating concentrations of DHEA and immunological measures, a number of studies have examined DHEA concentrations in relation to infection status using a cross-sectional sampling method (Table 4). While these studies cannot elucidate mechanistic relationships between DHEA and immunological responses, they nonetheless provide further evidence that DHEA variation can have important implications in health and disease. Some caution must be taken when interpreting these results, as various infections are known to impact androgen concentrations (Muehlenbein, Hirschtick, Bonner, & Swartz, 2010), including DHEA (Galindo-Sevilla et al., 2007; Libonati, de Mendonça, Maués, Quaresma, & de Souza, 2006; Prall & Muehlenbein, 2014).

Table 3 Studies of DHEA Supplementation and Immune Outcomes in Humans

Sex	Mean Age	Treatment	Length of Supplementation	Results	Citations
Females		DHEA supplementation	3 weeks	↑ Natural killer cell cytotoxicity, CD8+/CD5+ cells ↓ CD4+ T cells, lymphocyte proliferation No change in IL-6 production	Casson et al. (1993)
Males		DHEA and androstenedione supplementation	28 days	↑ PBMC proliferation to PHA No change in lymphocyte proliferation to LPS or ConA, production of IL-1β, IL-2, IL-4, IL-10, IFN-γ	Kohut et al. (2003)
Both	41	DHEA supplementation in patients with Addison's disease	12 weeks	↑ Treg cells, lymphocyte proliferation ↓ NK cells, NKT cells No change in IL-7	Coles et al. (2005)
Females	36.2/53.3	DHEA supplementation in patients with SLE	12 months	↓ SLE disease activity No change in erythrocyte sedimentation rate, CBC	van Vollenhoven et al. (1998)
Males		DHEA supplementation in symptomatic HIV-infected males	16 weeks	↑ Lymphocyte response to CMV ↓ CD4 count, neopterin No change in CD8 cells, % CD4 lymphocytes, β-2 microglobulin, delayed type hypersensitivity response, lymphocyte proliferation	Dyner et al. (1993)
Both	73	DHEA supplementation and influenza vaccination	4 days	↓ Antibody titer to select influenza strains	Ben-Yehuda et al. (1998)
Males	63	DHEA supplementation in subjects with low DHEA-S	20 weeks	↑ Monocyte number, B-cell number, B mitogenic response, T-cell mitogenic response, mitogen-stimulated IL-2 and IL-6, NK cell number, and cytotoxicity No change in IgG, IgA, IgM, total T lymphocyte number, T cell subset numbers	Khorram, Vu, and Yen (1997)

Continued

Table 3 Studies of DHEA Supplementation and Immune Outcomes in Humans—cont'd

Sex	Mean Age	Treatment	Length of Supplementation	Results	Citations
Both	73.4	DHEA supplementation and influenza vaccination	4 days	↓ Antibody titer to select influenza strains	Danenberg, Ben-Yehuda, Zakay-Rones, Gross, and Friedman (1997)
Both	73	DHEA-S supplementation and influenza or tetanus toxoid vaccination	4 days	↑ Antibody titer to vaccination (HAI assay) to influenza No changes in antibody response to tetanus toxoid	Araneo et al. (1995)
Both	78.61	Influenza vaccination with DHEA-S adjuvant	Single dose	↑ Antibody titer to vaccination (HAI assay) in select groups	Degelau, Guay, and Hallgren (1997)
Females		Systemic lupis erythematous patients supplemented with DHEA	6 months	↓ Disease activity	Barry et al. (1998)
Both	70.5	DHEA-S supplementation with influenza or tetanus vaccination	2 or 4 days	No change in antibody responses to tetanus or influenza	Evans et al. (1996)

Table 4 DHEA-S and Infection and Immune Function in Human Populations

Origin	Age/Sex	N	Immune Measures	Results	Citations
Kenya	12–25 Males	248	*Plasmodium falciparum* parasitemia	DHEA-S negatively associated with parasitemia	Kurtis et al. (2001)
Kenya	12–18 Females	648	*P. falciparum* parasitemia	DHEA-S negatively associated with parasitemia	Leenstra et al. (2003)
Brazil	15–47 Both	24	*P. falciparum* parasitemia	DHEA declined with parasitemia during treatment	Libonati et al. (2006)
Ethiopia	<9–40+ Both	135	*Schistosoma* infection intensity, IgG isotypes	Higher DHEA-S associated with decreased infection intensity, decreased IgG isotypes	Abebe et al. (2003)
Philippines	7–30 Both	727	*Schistosoma* infection intensity, reinfection following treatment	Higher DHEA-S associated with lower infection intensity, resistance to reinfection, and reinfection intensity	Kurtis et al. (2006)
Philippines	7–30 Both	731	C-reactive protein, IL-6 in heavily parasitized population	DHEA-S negatively related to C-reactive protein, IL-6	Coutinho et al. (2007)
Mexico	11–79 Both	40	IL-6 in individuals infected with leishmaniasis	DHEA negatively related to IL-6 and lower in infected individuals	Galindo-Sevilla et al. (2007)
Bolivia	8–23 Both	171	Secretory IgA	DHEA-S positively related to sIgA	Hodges-Simeon, Prall, Blackwell, Gurven, and Gaulin (2017)

As a potent modulator of inflammatory responses, DHEA-S has been associated with *Schistosoma* infection in several studies. For example, Abebe, Birkeland, Gaarder, Petros, and Gundersen (2003) evaluated the relationships among DHEA-S, immunity, and infection in Ethiopian adolescents and adults. While DHEA-S was negatively related to several immunoglobulins relevant to immunological responses against *Schistosoma*, DHEA-S concentration was significantly negatively related to infection intensity (Abebe et al., 2003). Similarly, adolescents and adults in the Philippines with high concentrations of DHEA-S showed lower *Schistosoma* infection intensity, and lower reinfection intensity following treatment (Kurtis et al., 2006). These findings mimic studies of DHEA and DHEA-S treatment in mice (Fallon, Richardson, Jones, & Dunne, 1998; Morales-Montor et al., 2001).

DHEA-S has been implicated as moderating malaria parasitemia in association with growth and development. For example, Kurtis, Mtalib, Onyango, and Duffy (2001) found that DHEA-S had effects on *Plasmodium falciparum* parasitemia in Kenyan males, again independent of age. Similar results were found in Kenyan females, where elevated DHEA-S predicted lower parasitemia and higher hemoglobin levels (Leenstra et al., 2003). These authors argue that age-independent associations between DHEA-S and immunity are mediated by the immunomodulatory properties of DHEA-S. However, more recent evidence suggests that DHEA can directly impact *P. falciparum* via inhibition of G6PD activity (Zhang et al., 2017). Whether DHEA and DHEA-S impact malarial infection via immunomodulation or through a direct effect, these results strongly implicate elevations in DHEA or DHEA-S as mediators of malaria resistance.

The modulation of inflammatory activity in response to chronic parasite infection can have important impacts beyond regulation of parasitemia. Prolonged exposure to proinflammatory responses can have detrimental effects on body condition, and elevations in DHEA-S during development may counteract these responses. To determine how DHEA-S mediates proinflammatory responses in populations with chronic parasite burden, Coutinho et al. (2007) examined the relationships among DHEA-S, nutritional status, CRP, and IL-6 in a sample from the Philippines. Inflammatory markers predicted undernutrition, although DHEA-S was negatively related to inflammation and positively associated with nutritional status, implicating DHEA-S as a potential molecule exerting protective effects on nutritional status. These findings explain further the age-related changes in morbidity associated with parasite infection. The ability of

DHEA and DHEA-S to exert some protective effect against parasite-induced wasting could have important implications for reproduction and survival in humans.

8. IMMUNOMODULATION VIA INTERACTIONS WITH GLUCOCORTICOIDS

Any review of the physiological effects of DHEA would be incomplete without some discussion of its interactions with glucocorticoids. As an adrenal androgen, DHEA responds to hypothalamic–pituitary–adrenal (HPA) stimulation and is in fact more sensitive to stimulation than is cortisol (Arvat et al., 2000). However, DHEA inhibits catecholamine release from the adrenal medulla (Liu & Wang, 2004), and its effects as an antiglucocorticoid have been characterized in numerous tissues and physiological systems (Browne, Wright, Porter, & Svec, 1992; Hu, Cardounel, Gursoy, Anderson, & Kalimi, 2000; Kimonides, Spillantini, Sofroniew, Fawcett, & Herbert, 1999; Shafagoj, Opoku, Qureshi, Regelson, & Kalimi, 1992). The mechanism behind these effects is unclear but appears to be unrelated to interference on binding on the glucocorticoid receptor (Mohan & Cleary, 1992). Within tissues, action may occur via modulation of 11β-HSD1, the enzyme responsible for synthesizing cortisol from cortisone, which is reduced in the presence of DHEA (Apostolova, Schweizer, Balazs, Kostadinova, & Odermatt, 2005). These actions, whether through DHEA directly or via downstream metabolites, are hypothesized to exert similar action on immune cells (Hazeldine et al., 2010).

Several studies involving rodents have explored the roles of DHEA supplementation on immunity or disease parameters in the presence of acute stress. For example, DHEA supplementation reverses suppression of IL-2 synthesis by glucocorticoids both in vitro and in vivo (Daynes, Dudley, & Araneo, 1990). Likewise, DHEA exposure protects against glucocorticoid-induced thymic involution and suppression of lymphocyte proliferation (Blauer, Poth, Rogers, & Bernton, 1991; May, Holmes, Rogers, & Poth, 1990). In comparing immune and disease outcomes in mice infected with lethal viruses and subjected to cold stress, Ben-Nathan et al. (1992) demonstrated that DHEA administration reduced mortality and viral levels compared with untreated mice.

Such work has not yet been extended to humans, but initial results from a small-cross-sectional study suggest that DHEA responses to acute stress do play a role in modulation of immunological responses during stress (Prall, Larson, & Muehlenbein, 2017). Whether DHEA has similar actions in reducing disease severity during infection remains to be seen. However, exploration of the relative concentrations of cortisol to DHEA, and corresponding health outcomes, can give some indication of this physiological relationship (Hechter, Grossman, & Chatterton, 1997). For example, cortisol/DHEA-S was related to mortality and metabolic syndrome in Vietnam era army veterans (Phillips et al., 2010a, 2010b). The ratio of DHEA to cortisol has also been linked with disease severity and mortality during septic shock (Arlt et al., 2006).

9. CONCLUSIONS

The studies outlined here suggest that DHEA, whether through direct action or through intracrine conversion to a metabolite, evokes potent responses from multiple immune components. DHEA appears to downregulate the complement cascade via increased expression of the C1 inhibitor, although few studies have examined this system specifically. DHEA has potent effects in cytokine production, downregulating inflammatory cytokines while directly upregulating IL-2 synthesis. DHEA generally acts to enhance lymphocyte proliferation, and increases T cell and NK cell cytotoxicity. Many of these results are derived from in vitro experimentation, but are supported by numerous studies showing the beneficial effects of DHEA supplementation on disease parameters and organism survival. Additional studies of supplementation as a vaccine adjuvant in humans are promising, but are compromised by differences in study design, and conflicting results suggest further study is required. Studies examining correlations between DHEA and some immune or disease parameter provide further evidence of the beneficial effect of DHEA and DHEA-S.

The mechanisms by which these actions occur remain equivocal. DHEA is a purported antiglucocorticoid, and it is clear that elevated DHEA is beneficial to immunocompetence in times of infection and acute HPA activation in mice. However, in a naturalistic context outside of supraphysiological dosing, the relationships among chronic and acute HPA activation, immunity, and hormone concentrations require further exploration.

REFERENCES

Abebe, F., Birkeland, K. I., Gaarder, P. I., Petros, B., & Gundersen, S. G. (2003). The relationships between dehydroepiandrosterone sulphate (DHEAS), the intensity of Schistosoma mansoni infection and parasite-specific antibody responses. A cross-sectional study in residents of endemic communities in north-east Ethiopia. *APMIS: Acta Pathologica, Microbiologica, et Immunologica Scandinavica, 111*(2), 319–328.

Apostolova, G., Schweizer, R. A. S., Balazs, Z., Kostadinova, R. M., & Odermatt, A. (2005). Dehydroepiandrosterone inhibits the amplification of glucocorticoid action in adipose tissue. *American Journal of Physiology—Endocrinology and Metabolism, 288*(5), E957–E964.

Araghi-Niknam, M., Zhang, Z., Jiang, S., Call, O., Eskelson, C. D., & Watson, R. R. (1997). Cytokine dysregulation and increased oxidation is prevented by dehydroepiandrosterone in mice infected with murine leukemia retrovirus. *Proceedings of the Society for Experimental Biology and Medicine. Society for Experimental Biology and Medicine (New York, N.Y.), 216*(3), 386–391.

Araneo, B., Dowell, T., Woods, M. L., Daynes, R., Judd, M., & Evans, T. (1995). DHEAS as an effective vaccine adjuvant in elderly humans. *Annals of the New York Academy of Sciences, 774*(1), 232–248.

Arlt, W., Hammer, F., Sanning, P., Butcher, S. K., Lord, J. M., Allolio, B., et al. (2006). Dissociation of serum dehydroepiandrosterone and dehydroepiandrosterone sulfate in septic shock. *The Journal of Clinical Endocrinology and Metabolism, 91*(7), 2548–2554.

Arvat, E., Di Vito, L., Lanfranco, F., Maccario, M., Baffoni, C., Rossetto, R., et al. (2000). Stimulatory effect of adrenocorticotropin on cortisol, aldosterone, and dehydroepiandrosterone secretion in normal humans: Dose-response study. *The Journal of Clinical Endocrinology and Metabolism, 85*(9), 3141–3146.

Barry, N. N., McGuire, J. L., & van Vollenhoven, R. F. (1998). Dehydroepiandrosterone in systemic lupus erythematosus: Relationship between dosage, serum levels, and clinical response. *The Journal of Rheumatology, 25*(12), 2352–2356.

Ben-Nathan, D., Lachmi, B., Lustig, S., & Feuerstein, G. (1991). Protection by dehydroepiandrosterone in mice infected with viral encephalitis. *Archives of Virology, 120*(3–4), 263–271.

Ben-Nathan, D., Lustig, S., Kobiler, D., Danenberg, H. D., Lupu, E., & Feuerstein, G. (1992). Dehydroepiandrosterone protects mice inoculated with West Nile virus and exposed to cold stress. *Journal of Medical Virology, 38*(3), 159–166.

Ben-Nathan, D., Padgett, D. A., & Loria, R. M. (1999). Androstenediol and dehydroepiandrosterone protect mice against lethal bacterial infections and lipopolysaccharide toxicity. *Journal of Medical Microbiology, 48*(5), 425–431.

Ben-Yehuda, A., Danenberg, H. D., Zakay-Rones, Z., Gross, D. J., & Friedman, G. (1998). The influence of sequential annual vaccination and of DHEA administration on the efficacy of the immune response to influenza vaccine in the elderly. *Mechanisms of Ageing and Development, 102*(2–3), 299–306.

Blauer, K. L., Poth, M., Rogers, W. M., & Bernton, E. W. (1991). Dehydroepiandrosterone antagonizes the suppressive effects of dexamethasone on lymphocyte proliferation. *Endocrinology, 129*(6), 3174–3179.

Brazão, V., Santello, F. H., Caetano, L. C., Del Vecchio Filipin, M., Toldo, M. P. A., & do Prado, J. C. (2010). Immunomodulatory effects of zinc and DHEA on the Th-1 immune response in rats infected with Trypanosoma cruzi. *Immunobiology, 215*(5), 427–434. https://doi.org/10.1016/j.imbio.2009.05.005.

Browne, E. S., Wright, B. E., Porter, J. R., & Svec, F. (1992). Dehydroepiandrosterone: Antiglucocorticoid action in mice. *The American Journal of the Medical Sciences, 303*(6), 366–371.

Casson, P. R., Andersen, R. N., Herrod, H. G., Stentz, F. B., Straughn, A. B., Abraham, G. E., et al. (1993). Oral dehydroepiandrosterone in physiologic doses modulates immune function in postmenopausal women. *American Journal of Obstetrics and Gynecology*, *169*(6), 1536–1539.

Catania, R. A., Angele, M. K., Ayala, A., Cioffi, W. G., Bland, K. I., & Chaudry, I. H. (1999). Dehydroepiandrosterone restores immune function following trauma-haemorrhage by a direct effect on T lymphocytes. *Cytokine*, *11*(6), 443–450.

Chang, D. M., Chu, S. J., Chen, H. C., Kuo, S. Y., & Lai, J. J. (2004). Dehydroepiandrosterone suppresses interleukin 10 synthesis in women with systemic lupus erythematosus. *Annals of the Rheumatic Diseases*, *63*, 1623–1626.

Chen, F., Knecht, K., Birzin, E., Fisher, J., Wilkinson, H., Mojena, M., et al. (2005). Direct agonist/antagonist functions of dehydroepiandrosterone. *Endocrinology*, *146*(11), 4568–4576.

Choi, I. S., Cui, Y., Koh, Y. A., Lee, H. C., Cho, Y. B., & Won, Y. H. (2008). Effects of dehydroepiandrosterone on Th2 cytokine production in peripheral blood mononuclear cells from asthmatics. *The Korean Journal of Internal Medicine*, *23*(4), 176–181.

Coles, A. J., Thompson, S., Cox, A. L., Curran, S., Gurnell, E. M., & Chatterjee, V. K. (2005). Dehydroepiandrosterone replacement in patients with Addison's disease has a bimodal effect on regulatory (CD4+CD25hi and CD4+FoxP3+) T cells. *European Journal of Immunology*, *35*(12), 3694–3703. https://doi.org/10.1002/eji.200526128.

Coutinho, H. M., Leenstra, T., Acosta, L. P., Olveda, R. M., McGarvey, S. T., Friedman, J. F., et al. (2007). Higher serum concentrations of DHEAS predict improved nutritional status in helminth-infected children, adolescents, and young adults in Leyte, the Philippines. *Journal of Nutrition*, *137*(2), 433–439.

Danenberg, H. D., Ben-Yehuda, A., Zakay-Rones, Z., Gross, D. J., & Friedman, G. (1997). Dehydroepiandrosterone treatment is not beneficial to the immune response to influenza in elderly subjects. *Journal of Clinical Endocrinology & Metabolism*, *82*(9), 2911–2914.

Daynes, R. A., Dudley, D. J., & Araneo, B. A. (1990). Regulation of murine lymphokine production in vivo. II. Dehydroepiandrosterone is a natural enhancer of interleukin 2 synthesis by helper T cells. *European Journal of Immunology*, *20*(4), 793–802.

Degelau, J., Guay, D., & Hallgren, H. (1997). The effect of DHEAS on influenza vaccination in aging adults. *Journal of the American Geriatrics Society*, *45*(6), 747–751.

Del Vecchio Filipin, M., Caetano, L. C., Brazão, V., Santello, F. H., Toldo, M. P. A., & do Prado, J. C. (2010). DHEA and testosterone therapies in Trypanosoma cruzi-infected rats are associated with thymic changes. *Research in Veterinary Science*, *89*(1), 98–103. https://doi.org/10.1016/j.rvsc.2010.01.016.

Di Santo, E., Sironi, M., Mennini, T., Zinetti, M., Savoldi, G., Di Lorenzo, D., et al. (1996). A glucocorticoid receptor-independent mechanism for neurosteroid inhibition of tumor necrosis factor production. *European Journal of Pharmacology*, *299*(1), 179–186.

Du, C., Guan, Q., Khalil, M. W., & Sriram, S. (2001). Stimulation of Th2 response by high doses of dehydroepiandrosterone in KLH-primed splenocytes. *Experimental Biology and Medicine*, *226*(11), 1051–1060.

Du, C., Khalil, M. W., & Sriram, S. (2001). Administration of dehydroepiandrosterone suppresses experimental allergic encephalomyelitis in SJL/J mice. *Journal of Immunology*, *167*(12), 7094–7101.

Dyner, T. S., Lang, W., Geaga, J., Golub, A., Stites, D., Winger, E., et al. (1993). An open-label dose-escalation trial of oral dehydroepiandrosterone tolerance and pharmacokinetics in patients with HIV disease. *Journal of Acquired Immune Deficiency Syndromes*, *6*(5), 459–465.

Evans, T. G., Judd, M. E., Dowell, T., Poe, S., Daynes, R. A., & Araneo, B. A. (1996). The use of oral dehydroepiandrosterone sulfate as an adjuvant in tetanus and influenza vaccination of the elderly. *Vaccine*, *14*(16), 1531–1537.

Fallon, P. G., Richardson, E. J., Jones, F. M., & Dunne, D. W. (1998). Dehydroepiandrosterone sulfate treatment of mice modulates infection with Schistosoma mansoni. *Clinical and Diagnostic Laboratory Immunology*, *5*(2), 251–253.

Falus, A., Fehér, K. G., Walcz, E., Brozik, M., Füst, G., Hidvégi, T., et al. (1990). Hormonal regulation of complement biosynthesis in human cell lines—I. Androgens and gamma-interferon stimulate the biosynthesis and gene expression of C1 inhibitor in human cell lines U937 and HepG2. *Molecular Immunology*, *27*(2), 191–195.

Galindo-Sevilla, N., Soto, N., Mancilla, J., Cerbulo, A., Zambrano, E., Chavira, R., et al. (2007). Low serum levels of dehydroepiandrosterone and cortisol in human diffuse cutaneous leishmaniasis by Leishmania mexicana. *American Journal of Tropical Medicine and Hygiene*, *76*(3), 566–572.

Garg, M., & Bondada, S. (1993). Reversal of age-associated decline in immune response to Pnu-imune vaccine by supplementation with the steroid hormone dehydroepiandrosterone. *Infection and Immunity*, *61*(5), 2238–2241.

Gennari, R., & Alexander, J. W. (1997). Arginine, glutamine, and dehydroepiandrosterone reverse the immunosuppressive effect of prednisone during gut-derived sepsis. *Critical Care Medicine*, *25*(7), 1207–1214.

Hazeldine, J., Arlt, W., & Lord, J. M. (2010). Dehydroepiandrosterone as a regulator of immune cell function. *The Journal of Steroid Biochemistry and Molecular Biology*, *120*(2–3), 127–136.

Hechter, O., Grossman, A., & Chatterton, R. T. (1997). Relationship of dehydroepiandrosterone and cortisol in disease. *Medical Hypotheses*, *49*(1), 85–91.

Hidvégi, T., Fehér, G. K., Feher, T., Koó, E., & Füst, G. (1984). Inhibition of the complement activation by an adrenal androgen, dehydroepiandrosterone. *Complement*, *1*(4), 201–206.

Hodges-Simeon, C. R., Prall, S. P., Blackwell, A. D., Gurven, M., & Gaulin, S. J. C. (2017). Adrenal maturation, nutritional status, and mucosal immunity in Bolivian youth. *American Journal of Human Biology*, *29*, e23025.

Hu, Y., Cardounel, A., Gursoy, E., Anderson, P., & Kalimi, M. (2000). Anti-stress effects of dehydroepiandrosterone: Protection of rats against repeated immobilization stress-induced weight loss, glucocorticoid receptor production, and lipid peroxidation. *Biochemical Pharmacology*, *59*(7), 753–762.

Khorram, O., Vu, L., & Yen, S. S. (1997). Activation of immune function by dehydroepiandrosterone (DHEA) in age-advanced men. *The Journals of Gerontology. Series A, Biological Sciences and Medical Sciences*, *52*(1), M1–M7.

Kimonides, V. G., Spillantini, M. G., Sofroniew, M. V., Fawcett, J. W., & Herbert, J. (1999). Dehydroepiandrosterone antagonizes the neurotoxic effects of corticosterone and translocation of stress-activated protein kinase 3 in hippocampal primary cultures. *Neuroscience*, *89*(2), 429–436.

Kohut, M. L., Thompson, J. R., Campbell, J., Brown, G. A., Vukovich, M. D., Jackson, D. A., et al. (2003). Ingestion of a dietary supplement containing dehydroepiandrosterone (DHEA) and androstenedione has minimal effect on immune function in middle-aged men. *Journal of the American College of Nutrition*, *22*(5), 363–371.

Kovats, S. (2015). Estrogen receptors regulate innate immune cells and signaling pathways. *Cellular Immunology*, *294*(2), 63–69.

Kurtis, J. D., Friedman, J. F., Leenstra, T., Langdon, G. C., Wu, H.-W., Manalo, D. L., et al. (2006). Pubertal development predicts resistance to infection and reinfection with Schistosoma japonicum. *Clinical Infectious Diseases*, *42*(12), 1692–1698.

Kurtis, J. D., Mtalib, R., Onyango, F. K., & Duffy, P. E. (2001). Human resistance to Plasmodium falciparum increases during puberty and is predicted by dehydroepiandrosterone sulfate levels. *Infection and Immunity*, *69*(1), 123–128.

Labrie, F., Bélanger, A., Luu-The, V., Labrie, C., Simard, J., Cusan, L., et al. (1998). DHEA and the intracrine formation of androgens and estrogens in peripheral target tissues: Its role during aging. *Steroids*, *63*(5–6), 322–328.

Leenstra, T., ter Kuile, F. O., Kariuki, S. K., Nixon, C. P., Oloo, A. J., Kager, P. A., et al. (2003). Dehydroepiandrosterone sulfate levels associated with decreased malaria parasite density and increased hemoglobin concentration in pubertal girls from western Kenya. *Journal of Infectious Diseases*, *188*(2), 297–304.

Libonati, R. M. F., de Mendonça, B. B., Maués, J. A., Quaresma, J. A. S., & de Souza, J. M. (2006). Some aspects of the behavior of the hypothalamus-pituitary-adrenal axis in patients with uncomplicated Plasmodium falciparum malaria: Cortisol and dehydroepiandrosterone levels. *Acta Tropica*, *98*(3), 270–276.

Liu, P.-S., & Wang, P.-Y. (2004). DHEA attenuates catecholamine secretion from bovine adrenal chromaffin cells. *Journal of Biomedical Science*, *11*(2), 200–205.

Loria, R. M. (2002). Immune up-regulation and tumor apoptosis by androstene steroids. *Steroids*, *67*, 953–966.

Loria, R. M., Inge, T. H., Cook, S. S., Szakal, A. K., & Regelson, W. (1988). Protection against acute lethal viral infections with the native steroid dehydroepiandrosterone (DHEA). *Journal of Medical Virology*, *26*(3), 301–314.

Loria, R. M., & Padgett, D. A. (1992). Androstenediol regulates systemic resistance against lethal infections in mice. *Archives of Virology*, *127*, 102–115.

Maninger, N., Wolkowitz, O. M., Reus, V. I., Epel, E. S., & Mellon, S. H. (2009). Neurobiological and neuropsychiatric effects of dehydroepiandrosterone (DHEA) and DHEA sulfate (DHEAS). *Frontiers in Neuroendocrinology*, *30*(1), 27.

May, M., Holmes, E., Rogers, W., & Poth, M. (1990). Protection from glucocorticoid induced thyme involution by dehydroepiandrosterone. *Life Sciences*, *46*(22), 1601–1609.

McLachlan, J. A., Serkin, C. D., & Bakouche, O. (1996). Dehydroepiandrosterone modulation of lipopolysaccharide-stimulated monocyte cytotoxicity. *Journal of Immunology*, *156*(1), 328–335.

Meikle, A., Dorchuck, R. W., Araneo, B. A., Stringham, J. D., Evans, T. G., Spruance, S. L., et al. (1992). The presence of a dehydroepiandrosterone-specific receptor binding complex in murine T cells. *Journal of Steroid Biochemistry and Molecular Biology*, *42*, 293–304.

Mohan, P. F., & Cleary, M. P. (1992). Studies on nuclear binding of dehydroepiandrosterone in hepatocytes. *Steroids*, *57*, 244–247.

Morales-Montor, J., Mohamed, F., Ghaleb, A. M., Baig, S., Hallal-Calleros, C., & Damian, R. T. (2001). In vitro effects of hypothalamic-pituitary-adrenal axis (HPA) hormones on Shistosoma mansoni. *The Journal of Parasitology*, *87*(5), 1132–1139.

Moynihan, J. A., Callahan, T. A., Kelley, S. P., & Campbell, L. M. (1998). Adrenal hormone modulation of type 1 and type 2 cytokine production by spleen cells: Dexamethasone and dehydroepiandrosterone suppress interleukin-2, interleukin-4, and interferon-γ production in vitro. *Cellular Immunology*, *184*(1), 58–64.

Muehlenbein, M. P., Hirschtick, J. L., Bonner, J. Z., & Swartz, A. M. (2010). Toward quantifying the usage costs of human immunity: Altered metabolic rates and hormone levels during acute immune activation in men. *American Journal of Human Biology*, *22*(4), 546–556.

Oberbeck, R., Dahlweid, M., Koch, R., van Griensven, M., Emmendörfer, A., Tscherne, H., et al. (2001). Dehydroepiandrosterone decreases mortality rate and improves cellular immune function during polymicrobial sepsis. *Critical Care Medicine*, *29*(2), 380–384.

Padgett, D. A., & Loria, R. M. (1994). In vitro potentiation of lymphocyte activation by dehydroepiandrosterone, and rostenediol, and androstenetriol. *Journal of Immunology*, *153*(4), 1544–1552.

Petri, M. A., Lahita, R. G., Van Vollenhoven, R. F., Merrill, J. T., Schiff, M., Ginzler, E. M., et al. (2002). Effects of prasterone on corticosteroid requirements of women with systemic lupus erythematosus: A double-blind, randomized, placebo-controlled trial. *Arthritis and Rheumatism*, *46*(7), 1820–1829.

Phillips, A. C., Carroll, D., Gale, C. R., Lord, J. M., Arlt, W., & Batty, G. D. (2010a). Cortisol, DHEAS, their ratio and the metabolic syndrome: Evidence from the Vietnam experience study. *European Journal of Endocrinology/European Federation of Endocrine Societies*, *162*(5), 919–923.

Phillips, A. C., Carroll, D., Gale, C. R., Lord, J. M., Arlt, W., & Batty, G. D. (2010b). Cortisol, DHEA sulphate, their ratio, and all-cause and cause-specific mortality in the Vietnam experience study. *European Journal of Endocrinology*, *163*(2), 285–292.

Prall, S. P., Ambu, L., Nathan, S., Alsisto, S., Ramirez, D., & Muehlenbein, M. P. (2015). Androgens and innate immunity in rehabilitated semi-captive orangutans (Pongo pygmaeus morio) from Malaysian Borneo. *American Journal of Primatology*, *77*, 642–650.

Prall, S. P., Larson, E. E., & Muehlenbein, M. P. (2017). The role of dehydroepiandrosterone on functional innate immune responses to acute stress. *Stress and Health*, *33*, 656–664. https://doi.org/10.1002/smi.2752.

Prall, S. P., & Muehlenbein, M. P. (2014). Testosterone and immune function in primates: A brief summary with methodological considerations. *International Journal of Primatology*, *35*, 805–824.

Prall, S. P., & Muehlenbein, M. P. (2015). Dehydroepiandrosterone and multiple measures of functional immunity in young adults. *American Journal of Human Biology*, *27*(6), 877–880.

Prough, R. A., Clark, B. J., & Klinge, C. M. (2016). Novel mechanisms for DHEA action. *Journal of Molecular Endocrinology*, *56*(3), R139–55.

Sakakura, Y., Nakagawa, Y., & Ohzeki, T. (2006). Differential effect of DHEA on mitogen-induced proliferation of T and B lymphocytes. *Journal of Steroid Biochemistry and Molecular Biology*, *99*(2–3), 115–120.

Santos, C. D., Toldo, M. P. A., Levy, A. M. A., Kawasse, L. M., Zucoloto, S., & do Prado, J. C., Jr. (2007). Dehydroepiandrosterone affects Trypanosoma cruzi tissue parasite burdens in rats. *Acta Tropica*, *102*(3), 143–150. https://doi.org/10.1016/j.actatropica.2007.04.010.

Shafagoj, Y., Opoku, J., Qureshi, D., Regelson, W., & Kalimi, M. (1992). Dehydroepiandrosterone prevents dexamethasone-induced hypertension in rats. *The American Journal of Physiology*, *263*(2 Pt. 1), E210–3.

Solerte, S. B., Fioravanti, M., & Vignati, G. (1999). Dehydroepiandrosterone sulfate enhances natural killer cell cytotoxicity in humans via locally generated immunoreactive insulin-like growth factor I. *The Journal of Clinical Endocrinology & Metabolism*, *84*, 3260–3267.

Straub, R. H., Konecna, L., Hrach, S., Rothe, G., Kreutz, M., Scholmerich, J., et al. (1998). Serum dehydroepiandrosterone (DHEA) and DHEA sulfate are negatively correlated with serum interleukin-6 (IL-6), and DHEA inhibits IL-6 secretion from mononuclear cells in man in vitro: Possible link between endocrinosenescence and immunosenescence. *The Journal of Clinical Endocrinology and Metabolism*, *83*, 2012–2017.

Suzuki, T., Suzuki, N., Daynes, R. A., & Engleman, E. G. (1991). Dehydroepiandrosterone enhances IL2 production and cytotoxic effector function of human T cells. *Clinical Immunology and Immunopathology*, *61*(2 Pt. 1), 202–211.

Suzuki, T., Suzuki, N., Engleman, E. G., Mizushima, Y., & Sakane, T. (1995). Low serum levels of dehydroepiandrosterone may cause deficient IL-2 production by lymphocytes in patients with systemic lupus erythematosus (SLE). *Clinical and Experimental Immunology*, *99*(2), 251–255.

Suzuki, N., Suzuki, T., & Sakane, T. (1996). Hormones and lupus: Defective dehydroepiandrosterone activity induces impaired interleukin-2 activity of T lymphocytes in patients with systemic lupus erythematosus. *Annales de Médecine Interne*, *147*(4), 248–252.

van Vollenhoven, R. F., Morabito, L. M., Engleman, E. G., & McGuire, J. L. (1998). Treatment of systemic lupus erythematosus with dehydroepiandrosterone: 50 patients treated up to 12 months. *The Journal of Rheumatology*, *25*(2), 285–289.

Zhang, Z., Araghi-Niknam, M., Liang, B., Inserra, P., Ardestani, S. K., Jiang, S., et al. (1999). Prevention of immune dysfunction and vitamin E loss by dehydroepiandrosterone and melatonin supplementation during murine retrovirus infection. *Immunology*, *96*(2), 291–297.

Zhang, Z., Che, X., Jiang, C., Fang, Z., Feng, Y., & Jiang, W. (2017). The effect and mechanism of inhibiting glucose-6-phosphate dehydrogenase activity on the proliferation of Plasmodium falciparum. *Biochimica et Biophysica Acta*, *1864*, 771–781.

CHAPTER FIVE

DHEA in Prenatal and Postnatal Life: Implications for Brain and Behavior

Tracey Quinn*, Ronda Greaves[†,‡], Emilio Badoer[†], David Walker[†,1]
*The Ritchie Centre, Hudson Institute of Medical Research, Monash Medical Centre, Clayton, VIC, Australia
[†]School of Health & Biomedical Sciences, RMIT University—Bundoora Campus, Melbourne, VIC, Australia
[‡]Murdoch Children's Research Institute, Parkville, Melbourne, VIC, Australia
[1]Corresponding author: e-mail address: david.walker@rmit.edu.au

Contents

Abstract

Dehydroepiandrosterone (DHEA) and its sulfated congener (DHEAS) are the principal C_{19} steroid produced by the adrenal gland in many mammals, including humans. It is secreted in high concentrations during fetal life, but synthesis decreases after birth until, in humans and some other primates, there is a prepubertal surge of DHEA production by the adrenal gland—a phenomenon known as adrenarche. There remains considerable uncertainty about the physiological role of DHEA and DHEAS. Moreover, the origin of the trophic drives that determine the waxing and waning of DHEA synthesis are poorly understood. These gaps in knowledge arise in some measure from

Vitamins and Hormones, Volume 108
ISSN 0083-6729
https://doi.org/10.1016/bs.vh.2018.03.001

the difficulty of understanding mechanistic determinants from observations made opportunistically in humans and primates, and have stimulated a search for other suitable species that exhibit adrenarche- and adrenopause-like changes of adrenal function. DHEA and DHEAS are clearly neuroactive steroids with actions at several neurotransmitter receptors; indeed, DHEA is now known to be also synthesized by many parts of the brain, and this capacity undergoes ontogenic changes, but whether this is dependent or independent of the changes in adrenal synthesis is unknown. In this chapter we review key contributions to this field over the last 50+ years, and speculate on the importance of DHEA for the brain, both during development and for maturation and aging of cerebral function and behavior.

1. INTRODUCTION

Dehydroepiandrosterone (3β-hydroxyandrost-5-ene-17-one (DHEA)) is the principal C_{19} steroid produced by the adrenal gland in humans and mammals (Parker, 1999). It is a 17-ketosteroid with an alcohol moiety at position three, and conjugation with sulfate reduces its hydrophobicity (Fig. 1). DHEA, and its sulfated derivative DHEA-sulfate (DHEAS), is multifunctional steroids with actions in a wide variety of physiological systems, including neurological (Kroboth, Salek, Pittenger, Fabian, & Frye, 1999), immune (Chen & Parker, 2004), and somatic growth and development (Arquitt, Stoecker, Hermann, & Winterfeldt, 1991; Zemel & Katz, 1986). Although DHEA and DHEAS were identified more than 50 years ago, there remains

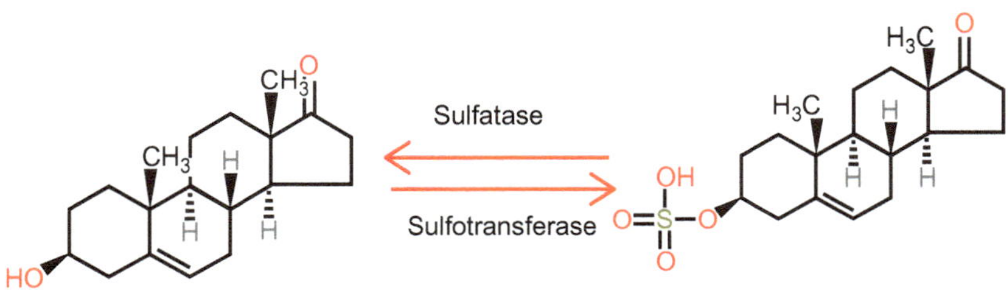

Fig. 1 DHEA is a carbon 19, 17-ketosteroid with the molecular formula of $C_{19}H_{28}O_2$ and IUPAC name of (3β)-3-hydroxyandrost-5-*en*-17-one. Average mass is 288.424 Da. The structure and naming of DHEA follows from the 1989 recommendations by the Joint Commission on Biochemical Nomenclature (JCBN) for *The Nomenclature of Steroids*. All steroids share the same basic 17 carbon structure with the presence of four linked rings (three six sided and one five sided) known as the cyclopentanophenanthrene (or cyclopentanoperhydrophenanthrene) ring. Cholesterol is recognized as the parent steroid and contains 27 carbon atoms, whereas the three main groups of steroids of interest in clinical endocrinology consist of 18, 19, or 21 carbon atoms, representing the estrane, androstane, and pregnane skeleton.

some uncertainty as to their physiological significance and mechanisms of action, and their roles in human disease are not yet completely understood (Allolio & Arlt, 2002; Eberling & Koivisto, 1994; Johnson, Bebb, & Sirrs, 2002).

In clinical obstetrics the fetoplacental unit is an important concept, as the synthesis and secretion of androstenedione, DHEA, and estriol by pregnant women depend heavily upon the functions of the fetal tissues. Therefore, measurement of the urinary estrogen excretion rates in the mother provides a means of assessing the health and well-being of the fetus, as well as placental function (Bolte, Mancuso, Eriksson, Wiqvist, & Diczfalusy, 1964a, 1964b). Of note, in babies born prematurely, despite separation from the maternal–placental unit, it is clearly demonstrated that the fetal adrenal gland does not mature into the adult adrenal until at least the equivalent of term (Midgley et al., 1998; Midgley, Russell, Oates, Shaw, & Honour, 1996). Because of the presence of the enzyme milieu that result in formation and release of DHEA, the concept of the fetoplacental unit is applicable to pregnant humans and primates, but not to rats or mice.

DHEA is a crucial precursor of sex steroid biosynthesis and exerts indirect endocrine and intracrine actions following conversion to other androgens (C_{19} steroids) and estrogens (C_{18} steroids). DHEA undergoes reversible sulfation to DHEAS, and both are active steroids, as distinct from the glucuronidation and similar processes of most steroids which renders them inactive metabolites. The level of DHEAS in the circulation is significantly higher than DHEA, and the levels are both age and gender dependent, but the ratio between these two steroids remains relatively constant with DHEAS being approximately 100 times higher than DHEA in blood (Guazzo, Kirkpatrick, Goodyer, Shiers, & Herbert, 1996). Importantly, human brain concentrations of DHEA and DHEAS are higher than plasma concentrations, although cerebrospinal fluid concentrations of DHEA are lower than in plasma, suggesting that the steroidal environment of the brain differs significantly from that in the blood (Guazzo et al., 1996).

Both DHEA and DHEAS are neuroactive steroids in the brain with direct interactions with neurotransmitter receptors, including sigma (σ), glutamate, and $GABA_A$ receptors (Baulieu & Robel, 1996; Kimonides, Khatibi, Svendsen, Sofroniew, & Herbert, 1998; Lapchak, Chapman, Nunez, & Zivin, 2000; Maurice, Grégoire, & Espallergues, 2006), and they have long been considered in the treatment of neuropsychiatric disorders, with published reports appearing as early as 1952 (Sands & Chamberlain, 1952; Strauss, Sands, Robinson, Tindall, & Stevenson, 1952). Since then, many studies have reported a deficiency in DHEA concentrations in patients with several major neurodevelopmental and neurodegenerative pathologies, including schizophrenia, bipolar affective disorder, depression, and Alzheimer's disease (Gallagher, Watson, Smith, Young, & Ferrier, 2007; Ritsner, Gibel, Ram, Maayan, & Weizman, 2006; Ritsner, Gibel, et al., 2007; Ritsner, Maayan, Gibel, & Weizman, 2007; Strous et al., 2003; Yanase et al., 1996). However, the hypothesis that alterations in normal DHEA biosynthesis in the brain and adrenal gland may contribute to the etiology of neurodevelopmental and age-related pathology is difficult to investigate in humans. This is partly owing to the inability to predict which individuals will go on to develop mental illness, and the need for retrospective studies to investigate fetal and/or neonatal complications in those individuals who later develop psychiatric pathologies as adolescents or adults.

2. DEVELOPMENT OF THE FETAL ADRENAL CORTEX WITH RESPECT TO DHEA BIOSYNTHESIS

Steroid hormones are produced by the fetus from the first trimester, with evidence of production at 8 weeks gestational age. At around 20 weeks gestation the fetal adrenal gland can potentially function to produce cortisol and aldosterone and by the third trimester most of the steroid enzymes, with the important exception of 3βHSD, are active. As a result the fetal adrenal has its own characteristic enzyme pattern, which includes very low activity of 3βHSD and an additional sulfatase step. At term the weight and size of the fetal adrenal glands are significantly greater than the adult adrenal.

The human fetal adrenal gland produces large amounts of DHEA and DHEAS as gestation proceeds, and at term produces more adrenal androgen per day than even an adult (Rainey, Rehman, & Carr, 2004). As shown in Fig. 2, the pathway leading to DHEA biosynthesis involves the enzymes—P450 side-chain cleavage (P450scc), Steroidogenic Acute Regulatory protein (StAR), 3β-hydroxysteroid-dehydrogenase (3βHSD), 17-α-hydroxylase/17,20-lyase (P450c17), and P450 sulfotransferase (SULT2A1), as well as the accessory hemoprotein Cytochrome *b*5 (Cyt*b*5). In this process, P450scc cleaves cholesterol to pregnenolone, the first committed intermediate in steroid biosynthesis. The enzyme P450c17 then mediates 17-α-hydroxylation of pregnenolone or progesterone to yield 17-α-OH-pregnenolone or 17-α-OH-progesterone, respectively. P450c17 also mediates cleavage of the c17,20 bond of 17-α-OH-pregnenolone to yield DHEA. It is principally the low

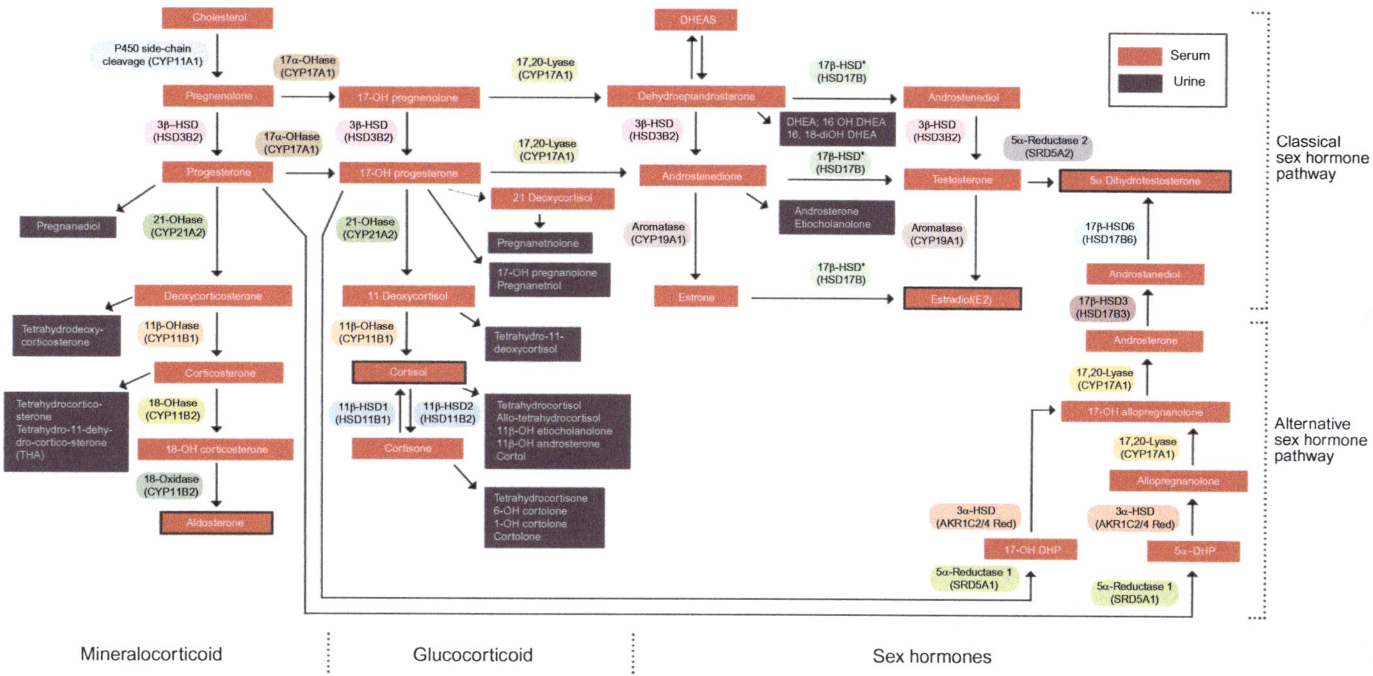

Fig. 2 A listing of the steroidogenic pathway and the enzymes involved in the biosynthesis of some key bioactive steroids. Key enzymes and cofactor proteins are shown near *arrows* indicating chemical reactions. P450scc cleaves cholesterol to pregnenolone, the first committed intermediate in steroid biosynthesis. The steroids pregnenolone, 17-OH-pregnenolone, DHEA, and androstenediol are Δ 5-steroids, which constitute the preferred pathway to C_{19} steroids in human beings. The enzyme P450c17 mediates 17-α-hydroxylation of pregnenolone or progesterone to yield 17-α-OH-pregnenolone or 17-α-OH-progesterone, respectively. P450c17 also mediates cleavage of the c17,20 bond of 17-α-OH-pregnenolone to yield DHEA. Interconversion of DHEA to its sulfated derivative, dehydroepiandrosterone sulfate (DHEAS), is mediated by sulfotransferase and sulfatase. The *dashed arrow* indicates alternative steroid hormone pathways. *From Greaves, R. F., Jolly, L., Hartmann, M., Ho, C. S., Kam, R., Joseph, J., et al. (2017). Harmonisation of serum dihydrotestosterone analysis: Establishment of an external quality assurance program.* Clinical Chemistry and Laboratory Medicine, 55, *522–529.*

3βHSD and high 17,20-lyase activity that accounts for the abundant production of DHEA and DHEAS by the fetal adrenal. The fetal adrenal gland can synthesize large quantities of cholesterol de novo from acetate, and it then uses both fetal and maternal cholesterol for the synthesis of the 3β-hydroxy-delta-5 forms of steroids such as pregnenolone, 17-α-hydroxypregnenolone, and DHEA (Mesiano & Jaffe, 1997). The movement of cholesterol from the outer mitochondrial membrane to the inner mitochondrial membrane by StAR represents the rate-limiting step in DHEA biosynthesis, and the relative expression level of the various steroidogenic enzymes determines further metabolism of pregnenolone in a zone-specific manner.

The fetal adrenal cortex consists of a small outer "definitive" zone (the neocortex), which appears steroidogenically quiescent until late gestation, and a larger inner "fetal" zone which is steroidogenically active throughout gestation. Steroidogenic enzymes have been detected by immunohistochemistry in the fetal zone 50–52 days after conception, and by 8 weeks of gestation the adrenal contains cortisol and responds to adrenocorticotropic hormone (ACTH) in primary culture systems (Goto et al., 2006). The outermost definitive zone, which is believed to give rise to the adult zona glomerulosa (ZG), is relatively inactive early in gestation because of low levels of the enzyme P450scc, but is quite active later in gestation when concentrations of mineralocorticoids have been observed to be high, including the months following birth (Miller, 1998). The small outer definitive zone expresses 3βHSD and aldosterone synthetase, but does not express P450c17 (Brosnan, 2001) or SULT2A1. Therefore, this zone is capable of synthesizing mineralocorticoids, but is not a major source of cortisol.

The smaller transitional zone of the fetal cortex, which gives rise to the zona fasciculata (ZF) in the adult pattern of cortical zonation, produces cortisol and expresses all enzymes of the mature steroid synthetic pathway, although not at adult levels (Mesiano, Coulter, & Jaffe, 1993). Fetal cortisol synthesis is regulated by pituitary ACTH and involves transient expression of adrenal 3βHSD2. In the human fetal adrenal the expression of 3βHSD2 and synthesis of cortisol decrease following the ninth week after conception; 3βHSD2 is barely detectable at 10–11 weeks and is absent at 14 weeks (Goto et al., 2006). However, during the second half of gestation, the transitional zone expresses both P450c17 and 3βHSD2.

In contrast, the huge innermost fetal zone has no 3βHSD, but high expression of P450c17, and so the principal products of the fetal adrenal are the Δ5-steroid precursors pregnenolone, 17-OH-pregnenolone, and large quantities of DHEA (Brosnan, 2001) and, compared to the ZR of the adult adrenal gland, extensive sulfokinase activity in the fetal zone ensures that the bulk of fetal steroids become conjugated with sulfates, including pregnenolone sulfate and DHEAS.

2.1 The Fetoplacental Unit

The biochemical synthesis of estrogens, including the 16α-hydroxylation, aromatization, and reduction of androstenedione and DHEA, requires coordinate processing in the fetal adrenal gland, liver, the placenta, and this interdependent relationship is known as the fetoplacental unit (Fig. 3), a concept proposed and advanced by Diczfalusy and Mancuso (1969). In this unit DHEA, which as mentioned earlier is synthesized and sulfated in the fetal zone of the adrenal gland, cannot be a substrate for adrenal 3βHSD2. Instead, it undergoes 16α-hydroxylation in the fetal liver by the enzyme CYP3A7 (Kitada, Kamataki, Itahashi, Rikihisa, & Kanakubo, 1987; Leeder et al., 2005) and is subsequently acted on by placental 3βHSD1, 17βHSD1, and P450aromatase to produce estriol.

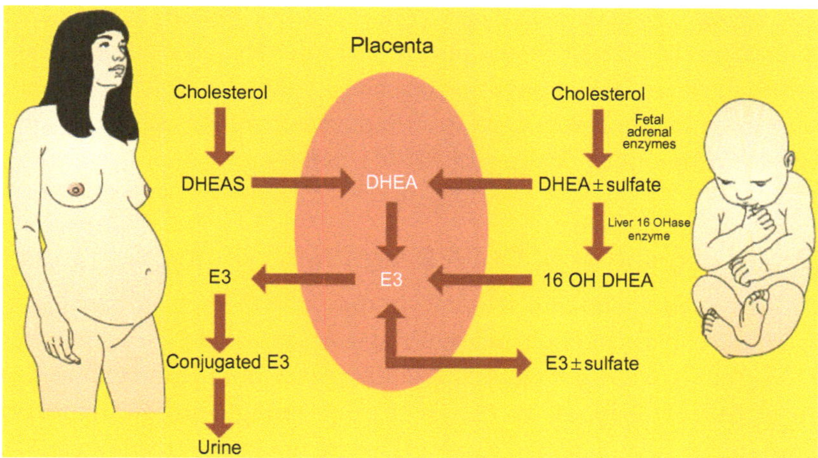

Fig. 3 The foetal placental unit and the production of steroids, represented by the foetal adrenal gland, foetal liver, and the placenta. Dehydroepiandrostane sulfate (DHEAS) is produced by the foetal adrenal gland and then in the absence of a significant amount of 3βHSD2 is metabolites by the liver to produce the 16 hydroxylated (OH) steroids. 16 OH DHEA is then utilized by the placenta to make oestriol (E3) which is then excreted in the maternal urine. *Note* 1: Liver sulfation of steroids predominates over glucuronidation in the foetal liver—e.g., Δ5 sulfated steroids (including pregnenolone sulfate and 17-α-OH-pregnenolone sulfate are produced). *Note* 2: Another metabolite of DHEA is 15,16-OH DHEA which will then lead to the production of oestetrol (E4) by the placenta. *From Greaves, R. F., Jawilakar, G., Hewitt, J. K., & Zacharin, M. R. (2014). A guide to understanding the steroid pathway: New insights and diagnostic implications.* Clinical Biochemistry, *47, 5–15.*

Small amounts of DHEA and DHEAS bypass the liver and are not 16α-hydroxylated, and are instead aromatized by the placenta to form estrone and oestradiol 17β, which subsequently pass into both the maternal and fetal circulations (Warren & Timberlake, 1964). Estrogens synthesized in this manner may then inhibit adrenal 3βHSD activity, providing an important feedback system to promote continual production of DHEAS (Fujieda, Faiman, Feyes, & Winter, 1982), but in the process resulting relatively low cortisol levels. Low fetal cortisol also results from high activity of placental 11βHSD2 converting maternal cortisol to cortisone. This is the situation until late in gestation when a rise in fetal cortisol is required for maturation of the fetal lungs and increased conversion of glucose to glycogen, particularly in the fetal liver, heart, and skeletal muscle (Fowden & Forhead, 2009; Liggins, 1994).

In contrast to the fetus, the placenta carries out very little de novo synthesis of cholesterol or other steroids from acetate (Jaffe, Pion, Eriksson, Wiqvist, & Diczfalu, 1965). However, the placenta does convert large amounts of maternal and fetal cholesterol to pregnenolone and progesterone (Bolte et al., 1964a, 1964b; Villee, Engel, & Villee, 1959; Warren & Timberlake, 1964). Steroids secreted by the placenta into the fetal circulation are again exposed to extensive hydroxylation reactions, and as a result, the fetal adrenal is able to synthesize corticosteroids, including aldosterone and cortisol, from placental progesterone (Bolte et al., 1964a, 1964b).

2.2 The Regulation of DHEA Biosynthesis During Gestation

In humans, corticotrophin-releasing hormone (CRH) synthesized in the hypothalamic paraventricular nucleus and secreted into the hypophyseal portal blood binds to anterior pituitary cell surface receptors to result in the secretion of ACTH (Tilbrook & Clarke, 2006). During gestation, human and animal studies have indicated that ACTH acts not only as the principal regulator of the fetal adrenal cortex and has a major role in adrenal growth, but it is not yet clear if ACTH is the principal driver of DHEA synthesis by the adrenal cortex. Studies in human anencephalic fetuses, or fetuses exposed to maternal synthetic glucocorticoid treatment, indicate that ACTH suppression during pregnancy can result in permanent alterations in hypothalamic–pituitary function, and failure of the fetal zone to develop beyond the size attained at approximately the 15th week of gestation (Benirschke, 1956; Easterling, Simmer, Dignam, Frankland, & Naftolin, 1966; Gray & Abramovich, 1980; Honnebier, Jöbsis, & Swaab, 1974). As maternal estrogen concentrations are dramatically reduced by these treatments, it is also likely that DHEA biosynthesis is significantly decreased in the adrenal gland of these fetuses (Milewich et al., 1978; Simpson & MacDonald, 1981). Hence, dexamethasone treatment of pregnant rhesus monkeys during late gestation results in significant atrophy of the fetal zone, as well as a decrease in maternal plasma concentrations of estradiol and estrone, and in fetal concentrations of cortisol (Challis, Davies, Benirschke, Hendrickx, & Ryan, 1974; Walsh, Norman, & Novy, 1979). Conversely, in utero administration of ACTH to normal and anencephalic human fetuses increases maternal estrogen levels, and can partially restore adrenal size (Honnebier et al., 1974). Furthermore, fetal adrenal cells respond to ACTH by secreting growth factors that appear to stimulate hyperplasia of the fetal zone (Mesiano & Jaffe, 1992), and incubation of human fetal adrenal cells in vitro with ACTH has been seen to promote the production of DHEA and cortisol by fetal adrenal cells (Smith, Mesiano, Chan, Brown, & Jaffe, 1998).

However, fetal adrenal development is also influenced by other factors acting independently or synergistically with ACTH, including growth, transcription, and placental factors (Parker, Stankovic, & Goland, 1999). Placental CRH has been shown to stimulate the production of proopiomelanocortin and ACTH in the fetal pituitary (Blumenfeld & Jaffe, 1986; Gibbs, Stewart, Vale, Rivier, & Yen, 1983) and placental syncytiotrophoblast cells (Petraglia, Sawchenko, Rivier, & Vale, 1987), thereby regulating growth and function of both pituitary corticotroph cells and the adrenal cortex. The importance of placental CRH in adrenal development is further supported by the fact that the fetal zone undergoes rapid involution immediately after birth when placenta-derived factors are no longer available, while there is relatively unchanged exposure to ACTH (Smith et al., 2009).

Together, these studies suggest that placental CRH may regulate the increase in fetal steroidogenesis that occurs during the last weeks of gestation. Certainly, it has been shown that placentally derived CRH can directly and preferentially stimulate human fetal adrenal cells in culture to produce DHEAS as well as cortisol (Chakravorty, Mesiano, & Jaffe, 1999; Smith et al., 1998). CRH is synthesized by the placenta in relatively large amounts (Campbell et al., 1987; Petraglia et al., 1987), and structurally and functionally appears to be identical to maternal and fetal hypothalamic CRH. Unlike hypothalamic CRH, which is under the control of glucocorticoid negative feedback, cortisol has been shown to stimulate placental CRH production in humans and in nonhuman primates, in both in vivo and in vitro studies (Campbell et al., 1987; Jones, Brooks, & Challis, 1989; Karalis & Majzoub, 1995; McLean et al., 1999; Petraglia et al., 1987; Robinson, Emanuel,

Frim, & Majzoub, 1998). The ability of cortisol to stimulate placental CRH makes it possible to create a feedforward endocrine cascade that does not end until parturition. As proposed by Smith (2007), this positive feedback cascade has been proposed as the driving the rise in CRH as well as fetal adrenal steroidogenesis in late gestation.

3. POSTNATAL DEVELOPMENT OF THE ADRENAL CORTEX WITH RESPECT TO DHEA BIOSYNTHESIS

Soon after birth, the human fetal adrenal gland undergoes involution due to the rapid disappearance of the ZF, accompanied by a significant decrease in adrenal androgen secretion (Johannisson, 1968; Mesiano et al., 1993; Parker, 1999; Spencer, Mesiano, Lee, & Jaffe, 1999). The size of the adrenal gland, as measured by ultrasonography, decreases by nearly 50% within the first 2 weeks after birth, and the large "fetal" zone completely disappears by 6 months of age (Rainey, Carr, Sasano, Suzuki, & Mason, 2002). The atrophy of the ZF appears to occur by apoptosis (Johannisson, 1968; Spencer et al., 1999), as the number of apoptotic nuclei in the fetal adrenal gland increases markedly during the postnatal period, whereas in mid-gestation it is very low. Concurrent with involution of the ZF, the outer adrenal cortex is remodeled to form the adult pattern of the ZF and glomerulosa. There is no evidence of adrenal cortical insufficiency during the perinatal period and the postnatal remodeling process (Parker, 1999); therefore, it is likely that these presumptive cortical zones are present and functional before birth.

4. STEROIDOGENESIS IN THE ADULT ADRENAL CORTEX

The three major pathways of zonal steroidogenesis in the adult human adrenal are shown in Fig. 4. All three zones of the adrenal cortex are able to produce steroids. Together, they can produce mineralocorticoids, glucocorticoids, and androgens; however, under normal conditions the major products are 11-hydroxylated C_{21} steroids (corticosteroids). Formation of active corticosteroids in each zone of the adrenal cortex involves a series of transformations of a common precursor, cholesterol, through a series of enzymatic reactions. In this process, the side chain of cholesterol is cleaved via the rate-limiting and hormonally regulated P450scc enzyme (Ritsner, Gibel, et al., 2007; Ritsner, Maayan, et al., 2007).

The two types of corticosteroids are the mineralocorticoids which regulate sodium and potassium levels, and the glucocorticoids, which regulate

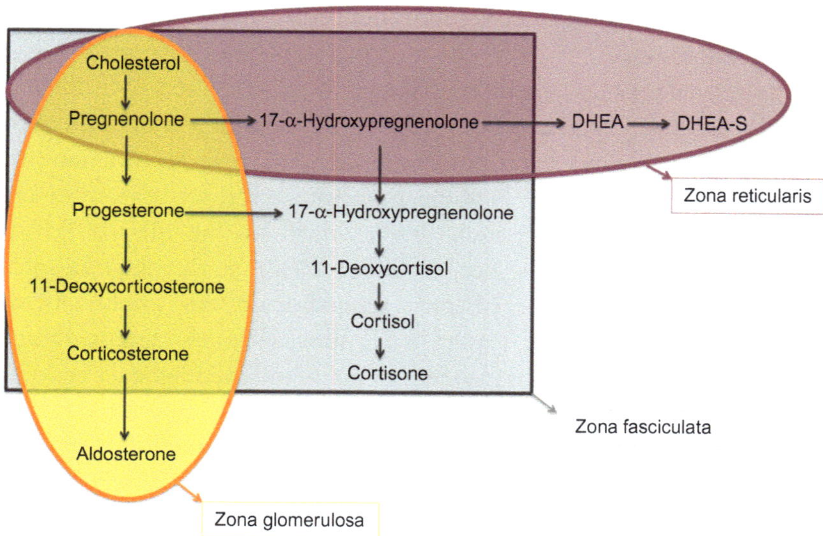

Fig. 4 Diagrammatic representation of zone-specific steroid synthesis in the adrenal cortex. All three zones can convert cholesterol into pregnenolone, but because of the presence or absence of certain enzymes, the flow of pregnenolone is preferentially directed to the synthesis of the steroid for which the particular zone is specific.

carbohydrate metabolism, among other functions. The major mineralocorticoid, aldosterone, is produced exclusively by the ZG, which is characterized by three distinct features: it expresses angiotensin II receptors, it expresses P450c11AS, but it does not express P450c17. As a result, the ZG produces aldosterone under regulation by the renin/angiotensin system.

By contrast, the adrenal ZF does not express angiotensin II receptors or P450c11AS, but instead expresses the ACTH receptor melanocortin type 2 (MC2R), and P450c11β, which cannot convert 18-hydroxycorticosterone to aldosterone (Mellon, Bair, & Monis, 1995; Miller & Auchus, 2011). As the ZF also expresses P450c17, this zone is the principle site of synthesis of the major glucocorticoid in humans, cortisol. This pattern of adrenal cortical specialization is similar in many other species, such as the nonhuman primate (Roohey, Raju, & Moustogiannis, 1997) and sheep (Dobbing, 1971); however, corticosterone (produced from the x-zone in rodents) is the major glucocorticoid produced in conventional rats and mice (Lamers et al., 1986). In the human, the ZF expresses little (if any) Cyt*b*5 (Suzuki et al., 2000) and consequently, P450c17 in the ZF catalyzes 17-α-hydroxylation but there is very little 17,20-lyase activity. Thus, the ZF produces two glucocorticoids (cortisol or corticosterone) under the influence of ACTH, but very little DHEA.

The adrenal zona reticularis also expresses MC2R, but very little 3βHSD2, P450c21, or P450c11β. As a result, the ZR produces minimal amounts of cortisol. In contrast to conventional rodent species such as the rat and mouse, the ZR of the human and some nonhuman primates expresses large amounts of P450c17 and Cyt*b*5 (Suzuki et al., 2000), which has the effect of favoring 17,20-lyase activity (Auchus, Lee, & Miller, 1998), so that large quantities of DHEA can be synthesized and secreted from this important zone (Endoh, Kristiansen, Casson, Buster, & Hornsby, 1996). As discussed further later, DHEA produced in this manner can go on to have biological actions of its own, or it is sulfated to DHEAS by SULT2A1. Small amounts of DHEA produced by the ZR can also be converted to androstenedione, and some of this is then metabolized to testosterone (Ghayee & Auchus, 2007).

5. ADRENARCHE AND ADRENOPAUSE

Both gender and age influence DHEA production. The most significant changes in steroid levels and pattern occur during the neonatal period, at adrenarche, puberty, and also menopause. As previously discussed, the human fetal adrenal can synthesize and release massive amounts of DHEA and DHEAS during the last months of gestation. Although the plasma concentrations of these C_{19} steroids remain high in the newborn, they decrease quickly as the fetal zone of the adrenal involutes after birth. From age 1 to 6 years, the adrenal glands secrete very low concentrations of DHEA, DHEAS, and androstenedione (de Peretti & Forest, 1978). However, at approximately 7–8 years of age, there is an increase in adrenal androgen production from the adrenal ZR, a process known as adrenarche, and a phenomenon apparently unique to humans and to some Old World primates, including the chimpanzee and orangutan (Campbell, 2006; Cutler et al., 1978; Smail, Faiman, Hobson, Fuller, & Winter, 1982). Plasma concentrations of DHEA(S) do not reach maximal levels until approximately 20–25 years of age, when it is seen that males have higher circulating levels of DHEAS compared with females (Fig. 5). It has been proposed that the higher concentration of DHEAS in men may be attributable to the location of the gene for steroid sulfatase on the X chromosome, and in having only one copy of the gene, men have less steroid sulfatase and consequently higher DHEAS concentrations (Miller, 2009).

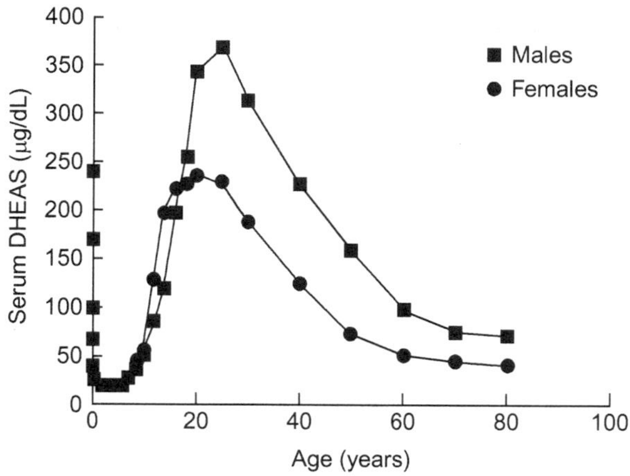

Fig. 5 Concentrations of serum DHEAS as a function of age in females and males. Values are high in cord blood and immediately after birth, before they fall rapidly in the first month of life as the fetal zone involutes. DHEAS concentrations remain low until the onset of adrenarche. Adrenarche generally begins at about age 7 years in girls and age 8 years in boys. Peak DHEAS concentrations in females do not reach values as high as those in males. In both sexes, the concentrations of DHEAS decline slowly during the adult years. *From Miller, W. L. (1999). The molecular basis of premature adrenarche: An hypothesis.* Acta Paediatrica, *88, 60–66.*

5.1 Adrenarche

Adrenarche occurs independently of an increase of estrogens, androgens, or gonadotropins associated with puberty (Ducharme et al., 1976; Sklar, Kaplan, & Grumbach, 1980). Furthermore, despite large increases in adrenal secretion of DHEA and DHEAS during adrenarche, circulating concentrations of ACTH and cortisol do not change significantly with age (Parker, 1993). However, ACTH is needed for normal adrenarche to occur, as illustrated by patients with familial glucocorticoid deficiency with mutations in MC2R (Weber et al., 1995). Thus, it has been suggested that ACTH plays a permissive role, but does not induce adrenarche. Hormones related to body mass, such as insulin and leptin, have been suggested as candidates for the trigger of adrenal growth and adrenarche (Biason-Lauber, Zachmann, & Schoenle, 2000).

The biochemical pathways that define adrenarche have been characterized in great detail, and it appears that alterations in key enzymes and cofactors in the adrenal ZR are closely related to the progression of adrenarche. Specifically, the abundance of 3βHSD protein in the ZR appears to decrease with the onset of adrenarche (Dardis, Saraco, Rivarola, & Belgorosky, 1999;

Endoh et al., 1996; Gell et al., 1999), and the increasing adrenal expression of Cyt*b*5 is almost exclusively confined to the ZR (Hui et al., 2009; Mapes, Corbin, Tarantal, & Conley, 1999; Yanase et al., 1998), two changes which together increase DHEA biosynthesis in adolescence.

5.2 Adrenopause

After ~35 years of age, there is evidence of a progressive decline in adrenal production of DHEA and DHEAS concentrations in both males and females, until concentrations return almost to preadrenarchal levels in persons over 80 years old (Bates, Harvey, Carruthers, & Martins, 2005). The magnitude of this deficiency with aging is such that serum levels of DHEA in elderly adults are only around 10%–20% of those in young adults (Havelock, Auchus, & Rainey, 2004; Parker, 1999). This diminution in adrenal androgens with age—often termed the "adrenopause"—is likely the result of a reduction in 17,20-lyase activity of P450c17 in the ZR (Dharia et al., 2005). Interestingly, it has been shown that the ZR of older men is reduced in size when compared adrenals of young men (Parker, Mixon, Brissie, & Grizzle, 1997), suggesting that at least part of the age-associated decrease in adrenal androgens might relate to a reduction in the number of DHEA-secreting cells in the ZR itself. In humans, serum DHEA concentrations decrease steadily with age, approaching lowest concentrations around the time at which many diseases of aging become apparent. In fact, DHEA and DHEAS have been widely publicized in both the lay press (Jaroff, 1995; Wick, 2002) and in the scientific literature (Morales, Nolan, Nelson, & Yen, 1994; Nawata, Yanase, Goto, Okabe, & Ashida, 2002) for their proposed antiaging and neuroprotective effects, and this has facilitated controversial speculation that DHEA treatment might be a remedy for a multitude of ailments (Watson, Huls, Araghinikuam, & Chung, 1996) and even, rather optimistically, having the potential to extend life (Olshansky, Hayflick, & Carnes, 2004). Low circulating levels of DHEA and DHEAS have been found in patients with breast and prostate cancers (Stahl, Schnorr, Pilz, & Dörner, 1992; Zumoff et al., 1981), and DHEA has been found to exert antioncogenic activity in animal models of various types of cancer (Gordon, Shantz, & Talalay, 1987; Li, Yan, Bélanger, & Labrie, 1994). DHEAS has also been shown to have immunomodulatory effects in vitro, including a role in the suppression of inflammatory genes (IL-1β, IL-6, TNF-α) associated with the human immune virus (Maingat et al., 2013).

6. ANIMAL MODELS

The difficulty in studying adrenarche and adrenopause over the long period of normal human life has stimulated the search for appropriate animal models of DHEA biosynthesis. The concentrations of DHEA, DHEAS, and androstenedione do not change significantly from before puberty until after sexual maturity in hamsters, sheep, pigs, goats, horses, cows, or chickens, although there are small changes in DHEA synthesis in rabbits and dogs which are suggestive of adrenarche (Cutler et al., 1978), as also in the rat, as seen by a transient increase in androstenedione levels in plasma before the onset of puberty itself (Pignatelli, Xiao, Gouveia, Ferreira, & Vinson, 2006). In contrast the spiny mouse shows that the enzymes and accessory proteins needed for DHEA synthesis, P450c17, Cyt*b*5, and 3βHSD, are expressed in the adrenal gland from at least 0.75 of gestation, and DHEA is present in fetal plasma from this time, and in fact increased fourfold in the week preceding birth (Quinn et al., 2013). Although DHEA was present in low concentrations in the prenatal and early postnatal period, there was a significant increase in the synthesis and secretion of DHEA in plasma from 8 to 20 days of age, a key developmental age which may be equivalent to early childhood and prepubescence in the human and nonhuman primate. During this time, there was increased immunoexpression of P450c17 and Cyt*b*5 expression in the ZR of the adrenal gland, while 3βHSD was found predominantly in the ZF and ZG. DHEA concentrations in both males and females decreased significantly after puberty, a change that was not seen in the concentrations of cortisol (T. Quinn and DW. Walker, unpublished findings). Notably, endogenous DHEA and the key enzymes for DHEA synthesis are found in the brain of the spiny mouse during both pre- and postnatal life, with a decrease in DHEA production associated with advancing age (Quinn, Ratnayake, Dickinson, Castillo-Melendez, & Walker, 2016). Some nonhuman primates, such as macaques and baboons, do synthesize high concentrations DHEA throughout life (Castracane, Cutler, & Loriaux, 1981; Cutler et al., 1978; Nguyen & Conley, 2008; Smail et al., 1982); however, these species do not exhibit the onset of adrenarche or the adrenopausal decline seen in humans. Only the chimpanzee shows the same biphasic pattern of DHEA and DHEAS secretion that is characteristic of human adrenarche (Bernstein, Sterner, & Wildman, 2012; Cutler et al., 1978; Smail et al., 1982). Any similarities between humans and other species, in terms of the morphological, endocrine, and biochemical changes that are associated with adrenarche, require further investigation. It has been proposed that the evolutionary roots for this developmental stage are of ancient origin (Bernstein et al., 2012), hypothesized to increase synaptogenesis and cortical maturation during the period of cortical and cognitive development (particularly gray matter reduction and synaptogenesis) that occurs from around age 6–20 in humans (Gogtay et al., 2004; Paus, 2005), also helping the brain to rewire in response to existing social environments (Campbell, 2006).

However, the evolutionary origin of DHEA synthesis in the ZR of some species and not others remains unexplained. It is unclear if an increase in DHEA synthesis at adrenarche is the result of "divergent" evolution (i.e., present in species with a common ancestor), or an example of "parallel" evolution (i.e., acquired or lost as a result of other adaptations). As production of DHEA is associated with the development of the ZR, which is a novel histological layer within the adrenal gland associated with primates more generally (Campbell, 2006), it is possible that DHEA is present only in species with a relatively long gestation and precocial fetal development, although this does not explain adrenarche, as such. The evolution of DHEA in these species could represent a mechanism that mediates maternal adaptations to pregnancy, as well as protect against maternal stress that otherwise would expose the fetus to the deleterious effects of excess glucocorticoids during gestation. In fact, it has been hypothesized that the primary role of DHEA is to function as a neuroactive steroid in the brain (Campbell, 2006).

7. DHEA(S) BIOSYNTHESIS AND ACTIONS IN THE BRAIN

During both prenatal and postnatal life, DHEA synthesis is increased following activation of the HPA axis (Genazzani et al., 1998; Rasmusson et al., 2004). This important system involves stimulation of parvocellular neurons from the paraventricular nucleus of the hypothalamus, and the subsequent release of the neuropeptide CRH and arginine vasopressin (AVP) into the hypophyseal portal blood system. The combined action of CRH and AVP on the corticotropes of the anterior pituitary gland then stimulates the release of ACTH, which subsequently acts on the cortex of the adrenal glands to stimulate the production and secretion of glucocorticoids (primarily cortisol in humans and corticosterone in rats and mice) from the ZF, and sex steroids (including DHEA) from the ZR. ACTH also stimulates the production of neuroactive steroid precursors and GABAergic neuroactive steroids in the brain (Majewska, 1995). Negative feedback mechanisms exist at various sites of the HPA axis to regulate axis activity, with mineralocorticoid and glucocorticoid receptors being the primary sites of regulation (Morrow, 2007). GABAergic neuroactive steroids can also inhibit CRF production and ACTH release and, in rodents at least, subsequently affect corticosterone levels (MacKenzie, Odontiadis, Le Melledo, Prior, & Baker, 2007).

Evidence that DHEA and DHEAS are neurosteroids (Baulieu & Robel, 1996), together with the fact that increases in DHEA(S) parallel patterns of cortical maturation from approximately age 6 years to the mid-20s (Gell et al., 1999; Hui et al., 2009), suggests that DHEA and DHEA(S) may play an important role in extended brain maturation among humans. As proposed by Campbell (2006), DHEA(S) may be important to human brain maturation by: (i) acting on the amygdala during prepubescence to reduce fearfulness and allow for the expression of an increased range of social interaction with unfamiliar individuals; (ii) acting on the hippocampus to promote memory, and in turn, both social and cognitive retention; and (iii) acting as an antagonist at the $GABA_A$ receptor (Campbell, 2006). In addition, the evolution of adrenarche is thought to increase the period of brain development in early childhood, so that there is increased synaptogenesis and cortical maturation during the period of cortical development from age 6 to the early 20s (Gogtay et al., 2004), also helping the brain to rewire in response to existing social environments (Campbell, 2006).

In the 1980s, Baulieu, Robel, and colleagues discovered that CNS concentrations of pregnenolone, DHEA, and their sulfate esters pregnenolone sulfate and DHEAS remained high long after adrenalectomy and castration in male rats (Corpechot, Robel, Axelson, Sjovall, & Baulieu, 1981; Le Goascogne et al., 1987; Robel & Baulieu, 1994). In addition to this, these studies found that circadian variations of these steroids in brain tissue were not synchronized with those of circulating steroids, suggesting that some steroids were produced in the brain. Residual skepticism that the brain was an actual site of steroidogenesis was dispelled by finding the mRNAs for P450scc, P450c11β, and ferredoxin in adult rat brain (Mellon & Deschepper, 1993). To date, most studies of neurosteroids have been done with rodents. At present, there is little evidence for the biosynthesis of steroids in the human brain, although such synthesis seems likely.

7.1 P450c17 Expression in the Developing Brain

During development, P450c17 may have an important role in the maintenance of neuronal connections and survival of specific neurons in the brain. Two steroid pathways seem to predominate in the rodent brain (Miller & Auchus, 2011). The first is the conventional pathway from cholesterol to DHEA and DHEAS, involving P450scc, P450c17, and SULT, whereas the second involves the synthesis of the pregnane steroid—allopregnanolone. Both pathways are initiated by P450scc; whether or not StAR also participates in this first step, or whether brain steroidogenesis, like that in the placenta, is independent of StAR, remains controversial (Miller, 2007).

Evidence in rodents suggests that P450c17 is expressed after embryonic day 9.5 in discrete areas of the developing brain, including the cortical subplate (Paul & Purdy, 1992), a region involved in guiding thalamic fibers to their cortical targets and in establishing appropriate thalamocortical connections (Compagnone, Bulfone, Rubenstein, & Mellon, 1995). Further studies have identified P450c17 in neuronal cell bodies in the hindbrain, mesencephalic nuclei, and the locus coeruleus (Compagnone et al., 1995). In the adult rat, P450c17 has been identified in the hippocampus (Hojo et al., 2004) and the spinal cord (Kibaly, Patte-Mensah, & Mensah-Nyagan, 2005). In the hippocampus, P450c17 was localized to pyramidal neurons in the CA1–CA3 regions and to granule cells of the dentate gyrus, and in these cells, P450c17 was localized in pre- and postsynaptic locations and in the endoplasmic reticulum as identified by immunoelectron microscope analysis (Hojo et al., 2004).

These studies further showed that the rat brain is capable of synthesis of DHEA from pregnenolone, suggesting functional activity of P450c17, and although basal P450c17 enzymatic activity was low in the hippocampus, function could be enhanced by exposing neurons to NMDA (Hojo et al., 2004). Similar findings have been reported for NMDA stimulation of pregnenolone synthesis from cholesterol in the hippocampus (Kimoto et al., 2001), suggesting that both P450scc and P450c17 activities are regulated by neurotransmitters. Interestingly, incubation of telencephalon, diencephalon, mesencephalon, metencephalon, or pituitary frog explants in culture has also been shown to synthesize several steroids from pregnenolone including DHEA, and P450c17 enzymatic activity has been seen to be reduced in a concentration-dependent manner by ketoconazole, an inhibitor of P450c17 (Do Rego et al., 2007). P450c17 expression has also been found in both neurons and glial cells adult rat spinal cord (Kibaly et al., 2005). Slices of spinal cord tissue containing P450c17 protein were also able to convert [^{3}H]-pregnenolone into [^{3}H]-DHEA, and again, this conversion was reduced by ketoconazole.

7.2 Function of DHEA in the Brain

In the developing brain, DHEA may have actions on the genesis and differentiation of neurons, as suggested by the potentiation of neurogenesis and neuronal survival in cultured human neural stem cells (Suzuki, Wright, Marwah, Lardy, & Svendsen, 2004). DHEA has also been shown to induce morphological changes in primary cultures of rodent embryonic neocortical neurons, such as increasing axonal length, presence of varicosities, and presence of basket-like formations around other cell bodies (Compagnone & Mellon, 1998). DHEA augmentation in rats has been shown to increase neurogenesis in the dentate gyrus (Karishma & Herbert, 2002).

DHEA and DHEAS may also be neuroprotective. In vivo studies have found DHEA can protect mouse striatal dopaminergic neurons against 1-methyl-4-phenyl-1,2,3,6-tetrahydropyridine (MPTP) toxicity and prevents kainic acid-induced cell death of rat hippocampal neurons (Karishma & Herbert, 2002). DHEA has also been shown to increase neuronal survival in cultured rat cerebellar granule cells against oxygen and glucose deprivation (Kaasik, Kalda, Jaako, & Zharkovsky, 2002; Marx et al., 2000), and in vivo DHEA augmentation prior to forebrain ischemia can reduce neuronal injury in the hippocampal CA1 region (Li, Klein, Sun, & Buchan, 2001).

Dehydroepiandrosterone (DHEA) is also protective after spinal cord injury (Guth, Zhang, & Roberts, 1994; Lapchak et al., 2000). Studies in female mice treated with DHEA showed more efficient locomotor recovery, left–right sensory coordination, and fine motor control than control animals of spinal cord injury treated with dimethyl sulfoxide (Fiore et al., 2004). In this study, mice treated with DHEA also had significantly decreased loss of myelin at the site of injury and reduced area of reactive gliosis surrounding the spinal lesion.

Other studies also suggest that DHEA may be neuroprotective by reducing excitotoxicity (Cardounel, Regelson, & Kalimi, 1999; Charalampopoulos et al., 2006; Kimonides et al., 1998; Kimonides, Spillantini, Sofroniew, Fawcett, & Herbert, 1999; Kurata, Takebayashi, Morinobu, & Yamawaki, 2004). For example, DHEA has been shown to be protect HT-22 cells (an immortalized mouse hippocampal cell line) against glutamate and amyloid β protein toxicity in a dose-dependent manner in vitro (Cardounel et al., 1999). Both DHEA and DHEAS have been shown to protect cultured neurons against NMDA toxicity in fetal rat hippocampal cultures (Kimonides et al., 1998), and DHEA is also protective against α-amino-3-hydroxy-5-methyl-4-isoxazolepropionic acid (AMPA) and kainic acid toxicity (Kimonides et al., 1998; Rupprecht, 1997).

The mechanism of DHEA protection against excitotoxicity is unclear, and it does not appear to involve direct interaction with glutamate receptors. The neuroprotective properties of DHEA may be facilitated via an alternative pathway, such as the σ_1 receptor (Maurice et al., 2006) or by protecting mitochondria against the effects of high intracellular Ca^{2+} (Kaasik, Safiulina, Kalda, & Zharkovsky, 2003). To this end, it has been found that DHEA can reduce cytoplasmic Ca^{2+}-induced loss of mitochondrial membrane potential by preventing Ca^{2+} influx into the mitochondrial matrix in primary cerebellar granule cell culture (Kaasik et al., 2003). The neuroprotective effect of DHEA against NMDA-induced excitotoxicity may be also mediated by the NMDA receptor-mediated modulation of the calcium/nitric oxide (NO) signaling pathway, where DHEA has been shown to inhibit NMDA-induced NO synthase activity and NO production in hippocampal cell culture (Kurata et al., 2004).

7.3 The Antiglucocorticoid Effects of DHEA in the Brain

Importantly, DHEA can also protect against the effects of glucocorticoid-mediated neurotoxicity (Hu, Cardounel, Gursoy, Anderson, & Kalimi, 2000; Kalimi, Shafagoj, Loria, Padgett, & Regelson, 1994). Previous studies have found that prolonged exposure to high levels of circulating corticosterone in rats can increase the age-related atrophy of apical dendrites and the rate at which hippocampal pyramidal neurons are lost (Wellman, 2001; Woolley, Gould, & McEwen, 1990). In addition, glucocorticoids have been shown to potentiate neurodegeneration induced by anoxia, kainic acid, or cholinergic agents (Hortnagl, Berger, Havelec, & Hornykiewicz, 1993; Tombaugh, Yang, Swanson, & Sapolsky, 1992). The corticoid-dependent mechanisms underlying these processes may include increased neuronal glutamate concentrations that eventually result in toxicity; however, the exact mechanism for the antiglucocorticoid action of DHEA remains unclear.

The antiglucocorticoid and hence neuroprotective actions of DHEA have been modeled in vivo where the toxic effects of corticosterone in the dentate gyrus of male rats were suppressed by even low concentrations of DHEA (Karishma & Herbert, 2002). Furthermore, this effect appeared to be specific for DHEA, since neither pregnenolone nor androstenediol had such antiglucocorticoid effects.

Similarly, DHEA has been shown to prevent neurotoxicity in vitro in cultured rat hippocampal tissue through the attenuation of the corticosterone-induced nuclear translocation of stress-activated protein kinase 3 (SAPK3), which might be important in the sequence of events leading to either neuronal death or cell survival (Kimonides et al., 1999). In cultured HT-22 cells DHEA augmentation can also induce suppression of the nuclear localization of the GR in response to glutamate toxicity, as assessed by immunohistochemistry (Cardounel et al., 1999). Thus, inhibition of GR translocation into the nucleus is a possible mechanism of DHEA's antiglucocorticoid effects.

8. CONCLUSION

In summary, previous studies have shown that in human and primate pregnancy, appropriate development and function of the fetal adrenal cortex are critical for fetal maturation and perinatal survival. Steroid hormones produced by the fetal adrenal cortex during gestation are important for the maturation of fetal organ systems necessary for extrauterine life. Importantly for the fetus, the production of cortisol and androgens by the fetal adrenal cortex also interacts with estrogen and progesterone produced by the placenta to influence uterine activity and, together with CRH, the onset of labor. In addition, the placenta and fetal adrenal gland also produce steroid precursors that form neuroactive metabolites in the fetal brain that affect both brain growth and activity, and can protect the brain against excitatory amino acid-induced neurotoxicity, various forms of oxidative stress, as well as the effects of excess glucocorticoids during important periods of brain development and maturation. Such relationships between the fetal adrenal gland, placenta, and brain are difficult to study in conventional rodents because the adrenal enzymes required for the formation androgens and subsequent aromatization in the placenta are not expressed. Due to ethical and financial restraints, there are only limited opportunities to fully study these systems in humans and primates during pregnancy. Therefore, an appropriate animal model is needed to further elucidate the relationship between the fetoplacental unit and the brain, and to further our knowledge of the possible role of adrenocortical hormones in the neurodevelopmental disorders that occur in postnatal life, and which are imputed to arise because of excess maternal glucocorticoids during pregnancy.

REFERENCES

Allolio, B., & Arlt, W. (2002). DHEA treatment: Myth or reality? *Trends in Endocrinology and Metabolism*, *13*, 288–294.

Arquitt, A. B., Stoecker, B. J., Hermann, J. S., & Winterfeldt, E. A. (1991). Dehydroepiandrosterone sulfate, cholesterol, hemoglobin, and anthropometric measures related to growth in male adolescents. *Journal of the American Dietetic Association*, *91*, 575–579.

Auchus, R. J., Lee, T. C., & Miller, W. L. (1998). Cytochrome b5 augments the 17, 20-lyase activity of human p450c17 without direct electron transfer. *Journal of Biological Chemistry, 273*, 3158–3165.

Bates, K. A., Harvey, A. R., Carruthers, M., & Martins, R. N. (2005). Androgens, andropause and neurodegeneration: Exploring the link between steroidogenesis, androgens and alzheimer's disease. *Cellular and Molecular Life Sciences, 62*, 281–292.

Baulieu, E. E., & Robel, P. (1996). Dehydroepiandrosterone and dehydroepiandrosterone sulfate as neuroactive neurosteroids. *The Journal of Endocrinology, 150*(Suppl), S221–S239.

Benirschke, K. (1956). Adrenals in anencephaly and hydrocephaly. *Obstetrics and Gynecology, 8*, 412–425.

Bernstein, R. M., Sterner, K. N., & Wildman, D. E. (2012). Adrenal androgen production in catarrhine primates and the evolution of adrenarche. *American Journal of Physical Anthropology, 147*, 389–400.

Biason-Lauber, A., Zachmann, M., & Schoenle, E. J. (2000). Effect of leptin on cyp17 enzymatic activities in human adrenal cells: New insight in the onset of adrenarche. *Endocrinology, 141*, 1446–1454.

Blumenfeld, Z., & Jaffe, R. B. (1986). Hypophysiotropic and neuromodulatory regulation of adrenocorticotropin in the human fetal pituitary gland. *Journal of Clinical Investigation, 78*, 288–294.

Bolte, E., Mancuso, S., Eriksson, G., Wiqvist, N., & Diczfalusy, E. (1964a). Studies on aromatisation of neutral steroids in pregnant women. 1. Aromatisation of c-19 steroids by placentas perfused in situ. *Acta Endocrinologica (Copenhagen), 45*, 535–539.

Bolte, E., Mancuso, S., Eriksson, G., Wiqvist, N., & Diczfalusy, E. (1964b). Studies on aromatisation of neutral steroids in pregnant women. 3. Over-all aromatisation of dehydroepiandrosterone sulphate circulating in foetal + maternal compartments. *Acta Endocrinologica (Copenhagen), 45*, 576–599.

Brosnan, P. G. (2001). The hypothalamic pituitary axis in the fetus and newborn. *Seminars in Perinatology, 25*, 371–384.

Campbell, B. (2006). Adrenarche and the evolution of human life history. *American Journal of Human Biology, 18*, 569–589.

Campbell, E., Linton, E., Wolfe, C., Scraggs, P., Jones, M., & Lowry, P. (1987). Plasma corticotropin-releasing hormone concentrations during pregnancy and parturition. *The Journal of Clinical Endocrinology and Metabolism, 64*, 1054–1059.

Cardounel, A., Regelson, W., & Kalimi, M. (1999). Dehydroepiandrosterone protects hippocampal neurons against neurotoxin-induced cell death: Mechanism of action. *Proceedings of the Society for Experimental Biology and Medicine, 222*, 145–149.

Castracane, V., Cutler, G., & Loriaux, D. (1981). Pubertal endocrinology of the baboon: Adrenarche. *American Journal of Physiology—Endocrinology and Metabolism, 241*, E305–E309.

Chakravorty, A., Mesiano, S., & Jaffe, R. B. (1999). Corticotropin-releasing hormone stimulates p450 17α-hydroxylase/17,20-lyase in human fetal adrenal cells via protein kinase c. *The Journal of Clinical Endocrinology and Metabolism, 84*, 3732–3738.

Challis, J. R., Davies, I. J., Benirschke, K., Hendrickx, A. G., & Ryan, K. J. (1974). The effects of dexamethasone on plasma steroid levels and fetal adrenal histology in the pregnant rhesus monkey. *Endocrinology, 95*, 1300–1305.

Charalampopoulos, I., Alexaki, V. I., Tsatsanis, C., Minas, V., Dermitzaki, E., Lasaridis, I., et al. (2006). Neurosteroids as endogenous inhibitors of neuronal cell apoptosis in aging. *Annals of the New York Academy of Sciences, 1088*, 139–152.

Chen CCG, Parker Jr CR (2004). Adrenal androgens and the immune system: Seminars in reproductive medicine. Thieme Medical Publishers, Inc., Vol. 22, 369–377.

Compagnone, N. A., Bulfone, A., Rubenstein, J. L., & Mellon, S. H. (1995). Expression of the steroidogenic enzyme p450scc in the central and peripheral nervous systems during rodent embryogenesis. *Endocrinology, 136*, 2689–2696.

Compagnone, N. A., & Mellon, S. H. (1998). Dehydroepiandrosterone: A potential signalling molecule for neocortical organization during development. *Proceedings of the National Academy of Sciences of the United States of America, 95*, 4678–4683.

Corpechot, C., Robel, P., Axelson, M., Sjovall, J., & Baulieu, E. E. (1981). Characterization and measurement of dehydroepiandrosterone sulfate in rat brain. *Proceedings of the National Academy of Sciences of the United States of America, 78*, 4704–4707.

Cutler, G. B., Glenn, M., Bush, M., Hodgen, G. D., Graham, C. E., & Loriaux, D. L. (1978). Adrenarche: A survey of rodents, domestic animals, and primates. *Endocrinology, 103*, 2112–2118.

Dardis, A., Saraco, N., Rivarola, M. A., & Belgorosky, A. (1999). Decrease in the expression of the 3β-hydroxysteroid dehydrogenase gene in human adrenal tissue during prepuberty and early puberty: Implications for the mechanism of adrenarche. *Pediatric Research, 45*, 384–388.

de Peretti, E., & Forest, M. G. (1978). Pattern of plasma dehydroepiandrosterone sulfate levels in humans from birth to adulthood: Evidence for testicular production. *The Journal of Clinical Endocrinology and Metabolism, 47*, 572–577.

Dharia, S., Slane, A., Jian, M., Conner, M., Conley, A. J., Brissie, R. M., et al. (2005). Effects of aging on cytochrome b5 expression in the human adrenal gland. *The Journal of Clinical Endocrinology and Metabolism, 90*, 4357–4361.

Diczfalusy, E., & Mancuso, S. (1969). *Oestrogen metabolism in pregnancy*. Oxford: Blackwell Publishers.

Do Rego, J. L., Tremblay, Y., Repetto, E., Castel, H., Vallarino, M., Bélanger, A., et al. (2007). Immunohistochemical localization and biological activity of the steroidogenic enzyme cytochrome p450 17α-hydroxylase/c17, 20-lyase (p450c17) in the frog brain and pituitary. *Journal of Neurochemistry, 100*, 251–268.

Dobbing, J. (1971). Undernutrition and the developing brain—Use of animal models to elucidate the human problem. *Psychiatria Neurologia Neurochirurgia, 74*, 433.

Ducharme, J. R., Forest, M. G., Deperetti, E., Sempe, M., Collu, R., & Bertrand, J. (1976). Plasma adrenal and gonadal sex steroids in human pubertal development. *The Journal of Clinical Endocrinology and Metabolism, 42*, 468–476.

Easterling, W. K., Simmer, H. H., Dignam, W. J., Frankland, M. V., & Naftolin, F. (1966). Neutral C_{19}-steroids and steroid sulfates in human pregnancy II. Dehydroepiandrosterone sulfate, 16α-hydroxydehydroepi-androsterone, and 16α-hydroxydehydroepiandrosterone sulfate in maternal and fetal blood of pregnancies with anenphalic and normal fetuses. *Steroids, 8*, 157–178.

Eberling, P., & Koivisto, V. (1994). Physiological importance of dehydroepiandrosterone. *The Lancet, 343*, 1479–1481.

Endoh, A., Kristiansen, S., Casson, P., Buster, J., & Hornsby, P. (1996). The zona reticularis is the site of biosynthesis of dehydroepiandrosterone and dehydroepiandrosterone sulfate in the adult human adrenal cortex resulting from its low expression of 3 beta-hydroxysteroid dehydrogenase. *The Journal of Clinical Endocrinology and Metabolism, 81*, 3558–3565.

Fiore, C., Inman, D. M., Hirose, S., Noble, L. J., Igarashi, T., & Compagnone, N. A. (2004). Treatment with the neurosteroid dehydroepiandrosterone promotes recovery of motor behavior after moderate contusive spinal cord injury in the mouse. *Journal of Neuroscience Research, 75*, 391–400.

Fowden, A. L., & Forhead, A. J. (2009). Hormones as epigenetic signals in developmental programming. *Experimental Physiology, 94*, 607–625.

Fujieda, K., Faiman, C., Feyes, F., & Winter, J. (1982). The control of steroidogenesis by human fetal adrenal cells in tissue culture. IV. The effects of exposure to placental steroids. *The Journal of Clinical Endocrinology and Metabolism, 54*, 89–94.

Gallagher, P., Watson, S., Smith, M. S., Young, A. H., & Ferrier, I. N. (2007). Plasma cortisol-dehydroepiandrosterone (DHEA) ratios in schizophrenia and bipolar disorder. *Schizophrenia Research, 90*, 258–265.

Gell, J. S., Carr, B. R., Sasano, H., Atkins, B., Margraf, L., Mason, J. I., et al. (1999). Adrenarche results from development of a 3 beta-hydroxysteroid dehydrogenase-deficient adrenal reticularis. *Obstetric and Gynecology Survey, 54*, 387–388.

Genazzani, A. R., Petraglia, F., Bernardi, F., Casarosa, E., Salvestroni, C., Tonetti, A., et al. (1998). Circulating levels of allopregnanolone in humans: Gender, age, and endocrine influences. *The Journal of Clinical Endocrinology and Metabolism, 83*, 2099–2103.

Ghayee, H. K., & Auchus, R. J. (2007). Basic concepts and recent developments in human steroid hormone biosynthesis. *Reviews in Endocrine & Metabolic Disorders, 8*, 289–300.

Gibbs, D., Stewart, R., Vale, W., Rivier, J., & Yen, S. (1983). Synthetic corticotropin-releasing factor stimulates secretion of immunoreactive β-endorphin/β-lipotropin and acth by human fetal pituitaries in vitro. *Life Sciences, 32*, 547–550.

Gogtay, N., Giedd, J. N., Lusk, L., Hayashi, K. M., Greenstein, D., Vaituzis, A. C., et al. (2004). Dynamic mapping of human cortical development during childhood through early adulthood. *Proceedings of the National Academy of Sciences of the United States of America, 101*, 8174–8179.

Gordon, G. B., Shantz, L. M., & Talalay, P. (1987). Modulation of growth, differentiation and carcinogenesis by dehydroepiandrosterone. *Advances in Enzyme Regulation, 26*, 355–382.

Goto, M., Hanley, K. P., Marcos, J., Wood, P. J., Wright, S., Postle, A. D., et al. (2006). In humans, early cortisol biosynthesis provides a mechanism to safeguard female sexual development. *Journal of Clinical Investigation, 116*, 953–960.

Gray, E. S., & Abramovich, D. (1980). Morphologic features of the anencephalic adrenal gland in early pregnancy. *American Journal of Obstetrics and Gynecology, 137*, 491–495.

Guazzo, E. P., Kirkpatrick, P. J., Goodyer, I. M., Shiers, H. M., & Herbert, J. (1996). Cortisol, dehydroepiandrosterone (DHEA), and DHEA sulfate in the cerebrospinal fluid of man: Relation to blood levels and the effects of age. *The Journal of Clinical Endocrinology and Metabolism, 81*, 3951–3960.

Guth, L., Zhang, Z., & Roberts, E. (1994). Key role for pregnenolone in combination therapy that promotes recovery after spinal cord injury. *Proceedings of the National Academy of Sciences of the United States of America, 91*, 12308–12312.

Havelock, J. C., Auchus, R. J., & Rainey, W. E. (2004). The rise in adrenal androgen biosynthesis: Adrenarche. *Seminars in Reproductive Medicine, 22*, 337–347.

Hojo, Y., Hattori, T.-A., Enami, T., Furukawa, A., Suzuki, K., Ishii, H.-T., et al. (2004). Adult male rat hippocampus synthesizes estradiol from pregnenolone by cytochromes p45017α and p450 aromatase localized in neurons. *Proceedings of the National Academy of Sciences of the United States of America, 101*, 865–870.

Honnebier, W., Jöbsis, A., & Swaab, D. (1974). The effect of hypophysial hormones and human chorionic gonadotrophin (hcg) on the anencephalic fetal adrenal cortex and on parturition in the human. *British Journal of Obstetrics and Gynaecology, 81*, 423–438.

Hortnagl, H., Berger, M. L., Havelec, L., & Hornykiewicz, O. (1993). Role of glucocorticoids in the cholinergic degeneration in rat hippocampus induced by ethylcholine aziridinium (af64a). *Journal of Neuroscience, 13*, 2939–2945.

Hu, Y., Cardounel, A., Gursoy, E., Anderson, P., & Kalimi, M. (2000). Anti-stress effects of dehydroepiandrosterone. Protection of rats against repeated immobilization stress-induced weight loss, glucocorticoid receptor production, and lipid peroxidation. *Biochemical Pharmacology, 59*, 753–762.

Hui, X.-G., Akahira, J.-I., Suzuki, T., Nio, M., Nakamura, Y., Suzuki, H., et al. (2009). Development of the human adrenal zona reticularis: Morphometric and immunohistochemical studies from birth to adolescence. *Journal of Endocrinology, 203*, 241–252.

Jaffe, R., Pion, R., Eriksson, G., Wiqvist, N., & Diczfalu, E. (1965). Studies on aromatisation of neutral steroids in pregnant women. 4. Lack of oestrogen formation from progesterone. *Acta Endocrinologica (Copenhagen), 48*, 413–422.

Jaroff, L. (1995). New age therapy. *Time, 23*, 52.

Johannisson, E. (1968). The foetal adrenal cortex in the human. Its ultrastructure at different stages of development and in different functional states. *Acta Endocrinology (Copenhagen), 130*, S1–107.

Johnson, M., Bebb, R., & Sirrs, S. (2002). Uses of DHEA in aging and other disease states. *Ageing Research Reviews, 1*, 29–41.

Jones, S., Brooks, A., & Challis, J. (1989). Steroids modulate corticotropin-releasing hormone production in human fetal membranes and placenta. *The Journal of Clinical Endocrinology and Metabolism, 68*, 825–830.

Kaasik, A., Kalda, A., Jaako, K., & Zharkovsky, A. (2002). Dehydroepiandrosterone sulphate prevents oxygen-glucose deprivation-induced injury in cerebellar granule cell culture. *Neuroscience, 102*, 427–432.

Kaasik, A., Safiulina, D., Kalda, A., & Zharkovsky, A. (2003). Dehydroepiandrosterone with other neurosteroids preserve neuronal mitochondria from calcium overload. *Journal of Steroid Biochemistry and Molecular Biology, 87*, 97–103.

Kalimi, M., Shafagoj, Y., Loria, R., Padgett, D., & Regelson, W. (1994). Antiglucocorticoid effects of dehydroepiandrosterone (DHEA). *Molecular and Cellular Biochemistry, 131*, 99–104.

Karalis, K., & Majzoub, J. A. (1995). Regulation of placental corticotropin-releasing hormone by steroids. *Annals of New York Academy of Sciences, 771*, 551–555.

Karishma, K. K., & Herbert, J. (2002). Dehydroepiandrosterone (DHEA) stimulates neurogenesis in the hippocampus of the rat, promotes survival of newly formed neurons and prevents corticosterone-induced suppression. *European Journal of Neuroscience, 16*, 445–453.

Kibaly, C., Patte-Mensah, C., & Mensah-Nyagan, A. G. (2005). Molecular and neurochemical evidence for the biosynthesis of dehydroepiandrosterone in the adult rat spinal cord. *Journal of Neurochemistry, 93*, 1220–1230.

Kimonides, V. G., Khatibi, N. H., Svendsen, C. N., Sofroniew, M. V., & Herbert, J. (1998). Dehydroepiandrosterone (DHEA) and dhea-sulfate (DHEAS) protect hippocampal neurons against excitatory amino acid-induced neurotoxicity. *Proceedings of the National Academy of Sciences of the United States of America, 95*, 1852–1857.

Kimonides, V. G., Spillantini, M. G., Sofroniew, M. V., Fawcett, J. W., & Herbert, J. (1999). Dehydroepiandrosterone antagonizes the neurotoxic effects of corticosterone and translocation of stress-activated protein kinase 3 in hippocampal primary cultures. *Neuroscience, 89*, 429–436.

Kimoto, T., Tsurugizawa, T., Ohta, Y., Makino, J., Tamura, H., Hojo, Y., et al. (2001). Neurosteroid synthesis by cytochrome p450-containing systems localized in the rat brain hippocampal neurons: N-methyl-d-aspartate and calcium-dependent synthesis. *Endocrinology, 142*, 3578–3589.

Kitada, M., Kamataki, T., Itahashi, K., Rikihisa, T., & Kanakubo, Y. (1987). P-450 hfla, a form of cytochrome p-450 purified from human fetal livers, is the 16 alpha-hydroxylase of dehydroepiandrosterone 3-sulfate. *Journal of Biological Chemistry, 262*, 13534–13537.

Kroboth, P. D., Salek, F. S., Pittenger, A. L., Fabian, T. J., & Frye, R. F. (1999). DHEA and DHEA-S: A review. *Journal of Clinical Pharmacology, 39*, 327–348.

Kurata, K., Takebayashi, M., Morinobu, S., & Yamawaki, S. (2004). Beta-estradiol, dehydroepiandrosterone, and dehydroepiandrosterone sulfate protect against n-methyl-d-aspartate-induced neurotoxicity in rat hippocampal neurons by different mechanisms. *Journal of Pharmacology and Experimental Therapeutics, 311*, 237–245.

Lamers, W. H., Mooren, P. G., Griep, H., Endert, E., Degenhart, H. J., & Charles, R. (1986). Hormones in perinatal rat and spiny mouse—Relation to altricial and precocial timing of birth. *American Journal of Physiology—Endocrinology and Metabolism, 251*, E78–E85.

Lapchak, P. A., Chapman, D. F., Nunez, S. Y., & Zivin, J. A. (2000). Dehydroepiandrosterone sulfate is neuroprotective in a reversible spinal cord ischemia model possible involvement of gabaa receptors. *Stroke, 31*, 1953–1957.

Le Goascogne, C., Robel, P., Gouezou, M., Sananes, N., Baulieu, E.-E., & Waterman, M. (1987). Neurosteroids: Cytochrome p-450scc in rat brain. *Science, 237*, 1212–1215.

Leeder, J. S., Gaedigk, R., Marcucci, K. A., Gaedigk, A., Vyhlidal, C. A., Schindel, B. P., et al. (2005). Variability of cyp3a7 expression in human fetal liver. *Journal of Pharmacology and Experimental Therapeutics, 314*, 626–635.

Li, H., Klein, G., Sun, P., & Buchan, A. M. (2001). Dehydroepiandrosterone (DHEA) reduces neuronal injury in a rat model of global cerebral ischemia. *Brain Research, 888*, 263–266.

Li, S., Yan, X., Bélanger, A., & Labrie, F. (1994). Prevention by dehydroepiandrosterone of the development of mammary carcinoma induced by 7, 12-dimethylbenz(a)anthracene (DMBA) in the rat. *Breast Cancer Research and Treatment, 29*, 203–217.

Liggins, G. (1994). The role of cortisol in preparing the fetus for birth. *Reproduction, Fertility and Development, 6*, 141–150.

MacKenzie, E. M., Odontiadis, J., Le Melledo, J. M., Prior, T. I., & Baker, G. B. (2007). The relevance of neuroactive steroids in schizophrenia, depression, and anxiety disorders. *Cellular and Molecular Neurobiology, 27*, 541–574.

Maingat, F. G., Polyak, M. J., Paul, A. M., Vivithanaporn, P., Noorbakhsh, F., Ahboucha, S., et al. (2013). Neurosteroid-mediated regulation of brain innate immunity in hiv/aids: DHEA-S suppresses neurovirulence. *The FASEB Journal, 27*, 725–737.

Majewska, M. D. (1995). Neuronal actions of dehydroepiandrosterone possible roles in brain development, aging, memory, and affect. *Annals of the New York Academy of Sciences, 774*, 111–120.

Mapes, S., Corbin, C. J., Tarantal, A., & Conley, A. (1999). The primate adrenal zona reticularis is defined by expression of cytochrome b5, 17 alpha-hydroxylase/17,20-lyase cytochrome p450 (p450c17) and nadph-cytochrome p450 reductase (reductase) but not 3 beta-hydroxysteroid dehydrogenase/delta 5-4 isomerase (3 beta-HSD). *The Journal of Clinical Endocrinology and Metabolism, 84*, 3382–3385.

Marx, C. E., Jarskog, L. F., Lauder, J. M., Gilmore, J. H., Lieberman, J. A., & Morrow, A. L. (2000). Neurosteroid modulation of embryonic neuronal survival in vitro following anoxia. *Brain Research, 871*, 104–112.

Maurice, T., Grégoire, C., & Espallergues, J. (2006). Neuro (active) steroids actions at the neuromodulatory sigma$_{\sigma 1}$ receptor: Biochemical and physiological evidences, consequences in neuroprotection. *Pharmacology Biochemistry and Behavior, 84*, 581–597.

McLean, M., Bisits, A., Davies, J., Walters, W., Hackshaw, A., De Voss, K., et al. (1999). Predicting risk of preterm delivery by second-trimester measurement of maternal plasma corticotropin-releasing hormone and α-fetoprotein concentrations. *American Journal of Obstetrics and Gynecology, 181*, 207–215.

Mellon, S. H., Bair, S. R., & Monis, H. (1995). P450c11b3 mRNA, transcribed from a third p450c11 gene, is expressed in a tissue-specific, developmentally, and hormonally regulated fashion in the rodent adrenal and encodes a protein with both 11-hydroxylase and 18-hydroxylase activities. *Journal of Biological Chemistry, 270*, 1643–1649.

Mellon, S. H., & Deschepper, C. F. (1993). Neurosteroid biosynthesis: Genes for adrenal steroidogenic enzymes are expressed in the brain. *Brain Research, 629*, 283–292.

Mesiano, S., Coulter, C. L., & Jaffe, R. B. (1993). Localization of cytochrome p450 cholesterol side-chain cleavage, cytochrome p450 17 alpha-hydroxylase/17, 204yase, and 3 - beta-hydroxysteroid dehydrogenase isomerase steroidogenic enzymes in human and rhesus monkey fetal adrenal glands: Reappraisal of functional zonation. *The Journal of Clinical Endocrinology and Metabolism,* 77, 1184–1189.

Mesiano, S., & Jaffe, R. B. (1992). In *Regulation of growth and differentiated function in the human fetal adrenal* (p. 235). *Colloques-institut national de la sante et de la recherche medicale colloques et seminaires*.

Mesiano, S., & Jaffe, R. B. (1997). Developmental and functional biology of the primate fetal adrenal cortex. *Endocrine Reviews, 18*, 378–403.

Midgley, P. C., Russell, K., Oates, N., Holownia, P., Shaw, J. C., & Honour, J. W. (1998). Adrenal function in preterm infants: Acth may not be the sole regulator of the fetal zone. *Pediatric Research*, *44*, 887–893.
Midgley, P. C., Russell, K., Oates, N., Shaw, J. C., & Honour, J. W. (1996). Activity of the adrenal fetal zone in preterm infants continues to term. *Endocrine Research*, *22*, 729–733.
Milewich, L., Gomez-Sanchez, C., Madden, J. D., Bradfield, D. J., Parker, P. M., Smith, S. L., et al. (1978). Dehydroisoandrosterone sulfate in peripheral blood of premenopausal, pregnant and postmenopausal women and men. *Journal of Steroid Biochemistry*, *9*, 1159–1164.
Miller, W. L. (1998). Steroid hormone biosynthesis and actions in the materno-feto-placental unit. *Clinics in Perinatology*, *25*, 799–817.
Miller, W. L. (2007). Steroidogenic acute regulatory protein (StAR), a novel mitochondrial cholesterol transporter. *Biochimica et Biophysica Acta (BBA)—Molecular and Cell Biology of Lipids*, *1771*, 663–676.
Miller, W. L. (2009). Androgen synthesis in adrenarche. *Reviews in Endocrine and Metabolic Disorders*, *10*, 3–17.
Miller, W. L., & Auchus, R. J. (2011). The molecular biology, biochemistry, and physiology of human steroidogenesis and its disorders. *Endocrine Reviews*, *32*, 81–151.
Morales, A. J., Nolan, J. J., Nelson, J. C., & Yen, S. (1994). Effects of replacement dose of dehydroepiandrosterone in men and women of advancing age. *Journal of Clinical Endocrinology and Metabolism*, *78*, 1360–1367.
Morrow, A. L. (2007). Recent developments in the significance and therapeutic relevance of neuroactive steroids—Introduction to the special issue. *Pharmacology and Therapeutics*, *116*, 1–6.
Nawata, H., Yanase, T., Goto, K., Okabe, T., & Ashida, K. (2002). Mechanism of action of anti-aging DHEA-S and the replacement of DHEA-S. *Mechanisms of Ageing and Development*, *123*, 1101–1106.
Nguyen, A., & Conley, A. (2008). Adrenal androgens in humans and nonhuman primates: Production, zonation and regulation. *Endocrine Development*, *13*, 33–54.
Olshansky, S. J., Hayflick, L., & Carnes, B. A. (2004). No truth to the fountain of youth. *Scientific American*, *14*, 98–102.
Parker, L. N. (1993). Adrenarche. *The Endocrinologist*, *3*, 385–391.
Parker, C. R. (1999). Dehydroepiandrosterone and dehydroepiandrosterone sulfate production in the human adrenal during development and aging. *Steroids*, *64*, 640–647.
Parker, C. R., Jr., Stankovic, A. M., & Goland, R. S. (1999). Corticotropin-releasing hormone stimulates steroidogenesis in cultured human adrenal cells. *Molecular and Cellular Endocrinology*, *155*, 19–25.
Parker, C., Mixon, R., Brissie, R., & Grizzle, W. (1997). Aging alters zonation in the adrenal cortex of men. *Journal of Clinical Endocrinology and Metabolism*, *82*, 3898–3901.
Paul, S. M., & Purdy, R. H. (1992). Neuroactive steroids. *The FASEB Journal*, *6*, 2311–2322.
Paus, T. (2005). Mapping brain maturation and cognitive development during adolescence. *Trends in Cognitive Sciences*, *9*, 60–68.
Petraglia, F., Sawchenko, P. E., Rivier, J., & Vale, W. (1987). Evidence for local stimulation of acth secretion by corticotropin-releasing factor in human placenta. *Nature*, *328*, 717–719.
Pignatelli, D., Xiao, F., Gouveia, A. M., Ferreira, J. G., & Vinson, G. P. (2006). Adrenarche in the rat. *Journal of Endocrinology*, *191*, 301–308.
Quinn, T. A., Ratnayake, U., Dickinson, H., Castillo-Melendez, M., & Walker, D. W. (2016). Ontogenetic change in the regional distribution of dehydroepiandrosterone-synthesizing enzyme and the glucocorticoid receptor in the brain of the spiny mouse (acomys cahirinus). *Developmental Neuroscience*, *38*, 54–73.

Quinn, T. A., Ratnayake, U., Dickinson, H., Nguyen, T.-H., McIntosh, M., Castillo-Melendez, M., et al. (2013). Ontogeny of the adrenal gland in the spiny mouse, with particular reference to production of the steroids cortisol and dehydroepiandrosterone. *Endocrinology, 154*, 1190–1201.
Rainey, W. E., Carr, B. R., Sasano, H., Suzuki, T., & Mason, J. I. (2002). Dissecting human adrenal androgen production. *Trends in Endocrinology and Metabolism, 13*, 234–239.
Rainey, W. E., Rehman, K. S., & Carr, B. R. (2004). The human fetal adrenal: Making adrenal androgens for placental estrogens. *Seminars in Reproductive Medicine, 22*, 327–336.
Rasmusson, A. M., Vasek, J., Lipschitz, D. S., Vojvodal, D., Mustone, M. E., Shi, Q. H., et al. (2004). An increased capacity for adrenal DHEA release is associated with decreased avoidance and negative mood symptoms in women with PTSD. *Neuropsychopharmacology, 29*, 1546–1557.
Ritsner, M., Gibel, A., Maayan, R., Ratner, Y., Ram, E., Modai, I., et al. (2007). State and trait related predictors of serum cortisol to DHEA(S) molar ratios and hormone concentrations in schizophrenia patients. *European Neuropsychopharmacology, 17*, 257–264.
Ritsner, M., Gibel, A., Ram, E., Maayan, R., & Weizman, A. (2006). Alterations in DHEA metabolism in schizophrenia: Two-month case-control study. *European Neuropsychopharmacology, 16*, 137–146.
Ritsner, M., Maayan, R., Gibel, A., & Weizman, A. (2007). Differences in blood pregnenolone and dehydroepiandrosterone levels between schizophrenia patients and healthy subjects. *European Neuropsychopharmacology, 17*, 358–365.
Robel, P., & Baulieu, E.-E. (1994). Neurosteroids: Biosynthesis and function. *Trends in Endocrinology and Metabolism, 5*, 1–8.
Robinson, B. G., Emanuel, R. L., Frim, D. M., & Majzoub, J. A. (1998). Glucocorticoid stimulates expression of corticotropin-releasing hormone gene in human placenta. *Proceedings of the National Academy of Sciences of the United States of America, 85*, 5244–5248.
Roohey, T., Raju, T. N. K., & Moustogiannis, A. N. (1997). Animal models for the study of perinatal hypoxic-ischemic encephalopathy: A critical analysis. *Early Human Development, 47*, 115–146.
Rupprecht, R. (1997). The neuropsychopharmacological potential of neuroactive steroids. *Journal of Psychiatric Research, 31*, 297–314.
Sands, D. E., & Chamberlain, G. H. (1952). Treatment of inadequate personality in juveniles by dehydroisoandrosterone; preliminary report. *British Medical Journal, 2*, 66–68.
Simpson, E. R., & MacDonald, P. C. (1981). Endocrine physiology of the placenta. *Annual Review of Physiology, 43*, 163–188.
Sklar, C. A., Kaplan, S. L., & Grumbach, M. M. (1980). Evidence for dissociation between adrenarche and gonadarche: Studies in patients with idiopathic precocious puberty, gonadal dysgenesis, isolated gonadotropin deficiency, and constitutionally delayed growth and adolescence. *The Journal of Clinical Endocrinology and Metabolism, 51*, 548–556.
Smail, P. J., Faiman, C., Hobson, W. C., Fuller, G. B., & Winter, J. S. D. (1982). Further studies on adrenarche in nonhuman primates. *Endocrinology, 111*, 844–848.
Smith, R. (2007). Parturition. *New England Journal of Medicine, 356*, 271–283.
Smith, R., Mesiano, S., Chan, E.-C., Brown, S., & Jaffe, R. B. (1998). Corticotropin-releasing hormone directly and preferentially stimulates dehydroepiandrosterone sulfate secretion by human fetal adrenal cortical cells. *The Journal of Clinical Endocrinology and Metabolism, 83*, 2916–2920.
Smith, R., Smith, J. I., Shen, X., Engel, P. J., Bowman, M. E., McGrath, S. A., et al. (2009). Patterns of plasma corticotropin-releasing hormone, progesterone, estradiol, and estriol change and the onset of human labor. *The Journal of Clinical Endocrinology and Metabolism, 94*, 2066–2074.

Spencer, S. J., Mesiano, S., Lee, J. Y., & Jaffe, R. B. (1999). Proliferation and apoptosis in the human adrenal cortex during the fetal and perinatal periods: Implications for growth and remodeling. *Journal of Clinical Endocrinology and Metabolism*, *84*, 1110–1115.
Stahl, F., Schnorr, D., Pilz, C., & Dörner, G. (1992). Dehydroepiandrosterone (DHEA) levels in patients with prostatic cancer, heart diseases and under surgery stress. *Experimental and Clinical Endocrinology and Diabetes*, *99*, 68–70.
Strauss, E. B., Sands, D. E., Robinson, A. M., Tindall, W. J., & Stevenson, W. A. (1952). Use of dehydroisoandrosterone in psychiatric treatment: A preliminary survey. *British Medical Journal*, *2*, 64–66.
Strous, R. D., Maayan, R., Lapidus, R., Stryjer, R., Lustig, M., Kotler, M., et al. (2003). Dehydroepiandrosterone augmentation in the management of negative, depressive, and anxiety symptoms in schizophrenia. *Archives of General Psychiatry*, *60*, 133–141.
Suzuki, T., Sasano, H., Takeyama, J., Kaneko, C., Freije, W. A., Carr, B. R., et al. (2000). Developmental changes in steroidogenic enzymes in human postnatal adrenal cortex: Immunohistochemical studies. *Clinical Endocrinology (Oxford)*, *53*, 739–747.
Suzuki, M., Wright, L. S., Marwah, P., Lardy, H. A., & Svendsen, C. N. (2004). Mitotic and neurogenic effects of dehydroepiandrosterone (DHEA) on human neural stem cell cultures derived from the fetal cortex. *Proceedings of the National Academy of Sciences of the United States of America*, *101*, 3202–3207.
Tilbrook, A. J., & Clarke, I. J. (2006). Neuroendocrine mechanisms of innate states of attenuated responsiveness of the hypothalamo-pituitary adrenal axis to stress. *Frontiers in Neuroendocrinology*, *27*, 285–307.
Tombaugh, G. C., Yang, S. H., Swanson, R. A., & Sapolsky, R. M. (1992). Glucocorticoids exacerbate hypoxic and hypoglycemic hippocampal injury in vitro: Biochemical correlates and a role for astrocytes. *Journal of Neurochemistry*, *59*, 137–146.
Villee, D. B., Engel, L. L., & Villee, C. A. (1959). Steroid hydroxylation in human fetal adrenals. *Endocrinology*, *65*, 465–474.
Walsh, S. W., Norman, R. L., & Novy, M. J. (1979). In utero regulation of rhesus monkey fetal adrenals: Effects of dexamethasone, adrenocorticotropin, thyrotropin-releasing hormone, prolactin, human chorionic gonadotropin, and α-melanocyte-stimulating hormone on fetal and maternal plasma steroids. *Endocrinology*, *104*, 1805–1813.
Warren, J. C., & Timberlake, C. E. (1964). Biosynthesis of estrogens in pregnancy—Precursor role of plasma dehydroisoandrosterone. *Obstetrics and Gynecology*, *23*, 689–698.
Watson, R. R., Huls, A., Araghinikuam, M., & Chung, S. (1996). Dehydroepiandrosterone and diseases of aging. *Drugs & Aging*, *9*, 274–291.
Weber, A., Toppari, J., Harvey, R., Klann, R., Shaw, N., Ricker, A., et al. (1995). Adrenocorticotropin receptor gene mutations in familial glucocorticoid deficiency: Relationships with clinical features in four families. *The Journal of Clinical Endocrinology and Metabolism*, *80*, 65–71.
Wellman, C. L. (2001). Dendritic reorganization in pyramidal neurons in medial prefrontal cortex after chronic corticosterone administration. *Journal of Neurobiology*, *49*, 245–253.
Wick, G. (2002). 'Anti-aging' medicine: Does it exist? A critical discussion of 'anti-aging' health products. *Experimental Gerontology*, *37*, 1137–1140.
Woolley, C. S., Gould, E., & McEwen, B. S. (1990). Exposure to excess glucocorticoids alters dendritic morphology of adult hippocampal pyramidal neurons. *Brain Research*, *531*, 225–231.
Yanase, T., Fukahori, M., Taniguchi, S., Nishi, Y., Sakai, Y., Takayanagi, R., et al. (1996). Serum dehydroepiandrosterone (DHEA) and DHEA-sulfate (DHEA-S) in Alzheimer's disease and in cerebrovascular dementia. *Endocrine Journal*, *43*, 119–123.

Yanase, T., Sasano, H., Yubisui, T., Sakai, Y., Takayanagi, R., & Nawata, H. (1998). Immunohistochemical study of cytochrome b5 in human adrenal gland and in adrenocortical adenomas from patients with cushing's syndrome. *Endocrine Journal*, *45*, 89–95.

Zemel, B. S., & Katz, S. H. (1986). The contribution of adrenal and gonadal androgens to the growth in height of adolescent males. *American Journal of Physical Anthropology*, *71*, 459–466.

Zumoff, B., Levin, J., Rosenfeld, R. S., Markham, M., Strain, G. W., & Fukushima, D. K. (1981). Abnormal 24-hr mean plasma concentrations of dehydroisoandrosterone and dehydroisoandrosterone sulfate in women with primary operable breast cancer. *Cancer Research*, *41*, 3360–3363.

FURTHER READING

Widstrom, R. L., & Dillon, J. S. (2004). Is there a receptor for dehydroepiandrosterone or dehydroepiandrosterone sulfate? *Seminars in Reproductive Medicine*, *22*, 289–298.

CHAPTER SIX

Astrocyte Neuroprotection and Dehydroepiandrosterone

Bruno D. Arbo[*,†,1], **Felipe S. Ribeiro**[‡], **Maria F. Ribeiro**[‡]
[*]Instituto de Ciências Biológicas, Universidade Federal do Rio Grande (FURG), Rio Grande, Brazil
[†]Programa de Pós-Graduação em Ciências Biológicas: Farmacologia e Terapêutica, ICBS, Universidade Federal do Rio Grande do Sul (UFRGS), Porto Alegre, Brazil
[‡]Laboratório de Interação Neuro-Humoral, ICBS, Universidade Federal do Rio Grande do Sul (UFRGS), Porto Alegre, Brazil
[1]Corresponding author: e-mail address: brunoarbo@gmail.com

Contents

Abstract

Dehydroepiandrosterone (DHEA) and its sulfate ester (DHEAS) are the most abundant steroid hormones in the systemic circulation of humans. Due to their abundance and reduced production during aging, these hormones have been suggested to play a role in many aspects of health and have been used as drugs for a multiple range of therapeutic actions, including hormonal replacement and the improvement of aging-related diseases. In addition, several studies have shown that DHEA and DHEAS are neuroprotective under different experimental conditions, including models of ischemia, traumatic brain injury, spinal cord injury, glutamate excitotoxicity, and neurodegenerative diseases. Since astrocytes are responsible for the maintenance of neural tissue homeostasis and the control of neuronal energy supply, changes in astrocytic function have been associated with neuronal damage and the progression of different pathologies.

Vitamins and Hormones, Volume 108
ISSN 0083-6729
https://doi.org/10.1016/bs.vh.2018.01.004

Therefore, the aim of this chapter is to discuss the neuroprotective effects of DHEA against different types of brain and spinal cord injuries and how the modulation of astrocytic function by DHEA could represent an interesting therapeutic approach for the treatment of these conditions.

1. INTRODUCTION

Dehydroepiandrosterone (DHEA) and its sulfate ester (DHEAS) are the most abundant steroid hormones in the systemic circulation of humans. These hormones, which are mainly synthesized in the adrenals, may be converted into androgens and estrogens, being able to act alone or through their metabolites through multiple mechanisms of action. Due to their abundance and reduced production during aging, these hormones have been suggested to play a role in many aspects of health, being used as drugs for a multiple range of therapeutic actions, including hormonal replacement (Bramwell, 2010; Panjari & Davis, 2010; Raven & Hinson, 2007), weight loss (Gómez-Santos, Hernández-Morante, Tébar, Granero, & Garaulet, 2012; Kurzman, MacEwen, & Haffa, 1990), and for the improvement of aging-related diseases (Aly, Metwally, & Ahmed, 2011; Johnson, Bebb, & Sirrs, 2002; Weiss, Villareal, Ehsani, Fontana, & Holloszy, 2012; Weiss, Villareal, Fontana, Han, & Holloszy, 2011; Wolkowitz et al., 2003). Because of those effects, DHEA has been considered to be the "fountain of youth" during the 90s (Baulieu, 1996), being also used for people seeking for increased muscle mass and adaptation to high-intensity interval training (Labrie et al., 2006; Liu, Lin, Huang, Ivy, & Kuo, 2013; Villareal & Holloszy, 2006).

In addition to their synthesis in the adrenals, DHEA and DHEAS can be synthesized in the brain, directly by neurons and glial cells, being classified as neurosteroids (Baulieu, 1981; Baulieu, Robel, & Schumacher, 2001). The discovery that DHEA and DHEAS could be also produced by the brain has led to several studies showing that DHEA and DHEAS are also neuroactive steroids, being able to modulate several neural functions (Baulieu & Robel, 1998). In this context, numerous studies have shown that these hormones are neuroprotective under different experimental conditions, including models of excitotoxicity (Kokona, Charalampopoulos, Pediaditakis, Gravanis, & Thermos, 2012; Mao & Barger, 1998; Veiga, Garcia-Segura, & Azcoitia, 2003), ischemia (Aragno et al., 2000; Lapchak, Chapman, Nunez, & Zivin, 2000; Li et al., 2009, 2006; Yabuki et al., 2015), oxygen and glucose deprivation (Kaasik, Kalda, Jaako, & Zharkovsky, 2001;

Vieira-Marques et al., 2016), neurodegenerative diseases (Aly et al., 2011; Lapchak & Araujo, 2001; Li et al., 2010; Litim, Morissette, & Di Paolo, 2016), traumatic brain injury (TBI) (Hoffman, Virmani, Simkins, & Stein, 2003; Juhász-Vedres et al., 2006; Lür et al., 2006; Milman, Zohar, Maayan, Weizman, & Pick, 2008), spinal cord injury (Compagnone, 2008; Fiore et al., 2004), and neuroinflammation (Barger, Chavis, & Drew, 2000; Wang, Huang, Chen, Kuo, & Jeng, 2001). As astrocytes play critical roles in the development, function, and survival of neurons in many of these conditions, it has been suggested that they could be an important cell target for DHEA- and DHEAS-mediated neuroprotection (Arbo, Benetti, & Ribeiro, 2016). In this chapter, we review our current knowledge on the neuroprotective actions of DHEA and DHEAS in different experimental models, focusing on the modulation of glial cells functioning.

2. NEUROBIOLOGY OF ASTROCYTES

Astroglia is represented by a highly heterogeneous population of neural cells, located both at the gray and white matter of the CNS, which are responsible for the homeostasis of the neural tissue among other functions (Verkhratsky, Steardo, Parpura, & Montana, 2016). Astrocytes are generated from astrocytic lineage cells, which appear early during the histogenesis of the CNS. These cells shift from a bipolar to a monopolar radial form, before finally becoming mature and multipolar astrocytes (Mission, Takahashi, & Caviness, 1991; Morest & Silver, 2003). Functionally, astrocytes have been viewed for a long time as "supportive cells," due to its functions related to the maintenance of neuronal homeostasis, including the regulation of ion homeostasis (Walz, 1989), the control of neuronal metabolism (Dringen, Gebhardt, & Hamprecht, 1993; Pellerin et al., 2007; Pellerin & Magistretti, 1994), the clearance of glutamate (Rothstein et al., 1996), and antioxidant protection (Dringen & Hamprecht, 1997; Dringen, Kussmaul, Gutterer, Hirrlinger, & Hamprecht, 1999). Later, however, after the establishment of the concept of the tripartite synapse, whereby astrocytic processes envelop the pre- and postsynaptic terminals, forming a physical barrier that limits the diffusion of the neurotransmitters away from the synapses (Araque, Parpura, Sanzgiri, & Haydon, 1999), it has been shown that astrocytes also contribute to information processing in the brain, sensing and integrating synaptic activity, and releasing gliotransmitters such as glutamate, GABA, D-serine, and ATP among others in response to neuronal activation (Araque, 2008; Araque et al., 2014). Regarding this context, it has been shown that changes in gliotransmission may play a role in the pathophysiology of different disorders, including neurodevelopment disorders and neuropsychiatric and neurodegenerative diseases, and that the modulation of the astrocytes and the tripartite synapse could be an interesting therapeutic target for the treatment of these conditions (Halassa, Fellin, & Haydon, 2007; Machado-Vieira, Manji, & Zarate, 2009; Parpura et al., 2012; Verkhratsky et al., 2016; Wierońska & Pilc, 2009). In addition, astrocytic end feet of the basal process cover almost the entire surface of intraparenchymal capillaries in the adult brain, being involved in the formation of the blood brain barrier (BBB), a diffusion barrier which regulates the transport of different molecules between blood and brain (Liu & Chopp, 2016) (Fig. 1). The astrocytic end feet express surface proteins, including glucose transporters, regulating glucose uptake from the blood (Kacem, Lacombe, Seylaz, & Bonvento, 1998). Astrocytes uptake glucose and store it as glycogen, which is synthesized by glycogen synthase. During greater energy demand, astrocytes transform glycogen into glucose-6-phosphate for glycolysis, resulting in increased lactate production, which is transported to the neurons, where it can be converted into pyruvate for oxidative phosphorylation and ATP production (Stobart & Anderson, 2013) (Fig. 1). Therefore, since astrocytes regulate neuronal energy supply (Stobart & Anderson, 2013) and BBB disruption is a hallmark of TBI and cerebral ischemia (Li et al., 2016; Liu & Chopp, 2016), they could be also an important cellular target for the treatment of these injuries, as we will see later in this chapter.

3. ASTROGLIA RESPONSES TO BRAIN INJURY

Glial cells are responsible for the maintenance of neural tissue homeostasis. Therefore, after any kind of brain damage, they respond in a coordinated way to restore the equilibrium in this tissue. This response, aimed to protect stressed neurons and brain in general, is called "reactive gliosis" and is specific for the damage that triggered it, involving changes in cellular morphology, proliferation, migration, and in the expression of numerous genes related to inflammation and cell survival regulation (Acaz-Fonseca, Avila-Rodriguez, Garcia-Segura, & Barreto, 2016; Burda & Sofroniew, 2014). In general, reactive gliosis is divided into numerous stages, involving the isolation of the injured area, the elimination of pathogens, dead cells, and cellular debris, and the tissue remodeling after the resolution of the lesion (Peng, Parpura, & Verkhratsky, 2014). After a focal brain damage, multiple

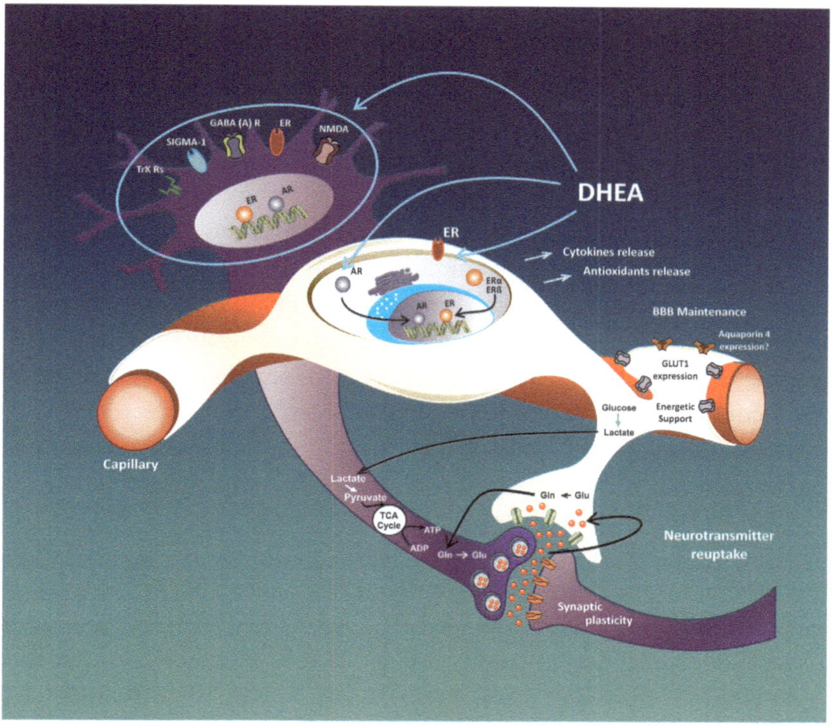

Fig. 1 DHEA, astrocyte function, and neuroprotection. DHEA binds to several intracellular receptors, including the classical androgen (AR) and estrogen (ER) receptors, regulating gene expression in both neurons and astrocytes. Moreover, DHEA is also able to bind to membrane receptors, including GABA(A), NMDA, membrane ER, sigma-1, and Trk (A, B, and C) receptors, among others. The neuroprotective actions of DHEA have been associated with the modulation of astrocytic function, including the stimulation of antioxidants production, the inhibition of proinflammatory cytokines release, the regulation of glutamate levels, and the modulation of the neuronal energy supply. Other neuroprotective actions may involve a protective effect on blood–brain barrier (BBB) function and the modulation of aquaporin-4 expression and edema formation after brain injury. *Adapted from Acaz-Fonseca, E., Avila-Rodriguez, M., Garcia-Segura, L. M., & Barreto, G. E. (2016). Regulation of astroglia by gonadal steroid hormones under physiological and pathological conditions.* Progress in Neurobiology, 144, *5–26. doi:10.1016/j.pneurobio.2016.06.002.*

molecular and cellular events begin and progress during several weeks. The responses of the CNS to acute focal damage can be divided into three overlapping but distinct phases: (1) cell death and inflammation, (2) cell proliferation for tissue replacement, and (3) tissue remodeling. First, rapid events occur, including the activation of the coagulation cascade, platelet aggregation, and clot formation, followed by the recruitment of inflammatory and immune cells, subacute death of parenchymal cells, and initiation of debris removal. Microglial cells are the first glial cells to be activated, migrating from other brain regions to the injured area, while astrocytes remain in situ and do not migrate to or from injury sites (Bardehle et al., 2013; Burda & Sofroniew, 2014). At this point, astrocytes can swell osmotically and die in the center of severe injuries, or become reactive and hypertrophy or even proliferate depending on the degree of damage (Bardehle et al., 2013; Burda & Sofroniew, 2014). This first phase of the inflammatory response aims the elimination of the pathogens (depending on the injury type) and cellular debris, but can also affect the integrity of some neurons if not controlled (secondary cell death). The second phase of reactive gliosis starts later, aiming the repair of the lesioned area, and being characterized by the proliferation and migration of cells that will promote the tissue repair and replacement, including fibroblast lineage cells, endothelial progenitors, inflammatory cells, and several types of glia and neural-lineage progenitor cells, including scar-forming astrocytes and their progenitors (Burda & Sofroniew, 2014). Initially, the glial scar is formed by local astroglial progenitors that assemble around the lesioned area containing inflammatory and fibroblast-lineage cells, isolating the core of the injury (Pekny, Wilhelmsson, & Pekna, 2014; Wanner et al., 2013). Finally, the last phase of the inflammatory response to injury involves the tissue remodeling, including glial scar organization and BBB repair, as well as other chronic events that can continue over several months. At this phase, the lesion core is surrounded by a well-organized astrocyte scar that acts as a protective barrier, blocking the migration of inflammatory cells from the fibrous lesion core to the adjacent viable neural tissue (Burda & Sofroniew, 2014).

Although reactive gliosis has been traditionally seen as detrimental, since astrocyte scar form a physical barrier that inhibit the axonal regeneration and the recovery of the neuronal connectivity (Cregg et al., 2014; Voskuhl et al., 2009), the blockade of glial scar formation in some different conditions was associated with increased lesion size, neuronal death, and demyelination, and decreased functional recovery after injury (Bush et al., 1999; Faulkner et al., 2004; Li et al., 2008). It is important to mention that astrocyte activation is related with changes in its morphology (hypertrophy and stellation) and in the expression of several molecules, including cytokines, growth factors, adhesion molecules, and different enzymes, depending on the type of injury that initiated this response (Acaz-Fonseca et al., 2016; Khakh & Sofroniew, 2015). The dysregulation of the astroglial response to brain injuries may contribute to the development of neurodegenerative diseases and to the outcome of other injuries such as stroke and TBI (Burda, Bernstein, & Sofroniew, 2016; Correale & Farez, 2015; Heneka et al., 2015; Liu & Chopp, 2016). Therefore, reactive gliosis is a complex and heterogeneous response, involving a combination of positive and negative effects that will affect neuronal survival and regeneration. Regarding this context, the identification of regulatory factors of this process is under intensive research, since astrogliosis modulation could be an important therapeutic strategy for the treatment of different types of brain injury (Acaz-Fonseca et al., 2016; Arbo et al., 2016; Liu & Chopp, 2016; Verkhratsky et al., 2016).

4. STEROIDOGENESIS AND DHEA PRODUCTION

DHEA is synthesized from cholesterol, which is the precursor molecule for the synthesis of all steroid hormones, in a multistep process involving several enzymes. In the periphery, most of steroidogenic cholesterol is derived from circulating lipoproteins (primarily low-density lipoproteins (LDL) in humans and high-density lipoproteins (HDL) in rodents) (Miller, 2013), while in the CNS, most of the cholesterol is locally synthesized by astrocytes and oligodendrocytes, since circulating cholesterol does not cross the BBB (Giudetti, Romano, Lavecchia, & Gaetani, 2016). The rate-limiting step in steroidogenesis is the cholesterol transport into the mitochondria, which is regulated by a multiprotein complex, including the steroidogenic acute regulatory protein (StAR), among others (Rone et al., 2012). Steroid hormones biosynthesis initiates with the enzymatic side chain cleavage of cholesterol by the cytochrome P450 side chain cleavage (CYP11A1), which forms pregnenolone (Miller & Auchus, 2011). After, pregnenolone can be transformed in DHEA, through the action of the cytochrome P450 side-chain cleavage at carbon 17 (CYP17), that converts pregnenolone to 17-hydroxy-pregnenolone and finally to DHEA (Charalampopoulos, Remboutsika, Margioris, & Gravanis, 2008). DHEA can be transformed into its more stable sulfate ester (DHEAS) through the action of the hydroxysteroid sulfotransferase (HST, SULT2A1), also known as DHEA sulfotransferase. On the other hand, DHEAS can be converted back to DHEA by the steroid sulfatase (Maninger, Wolkowitz, Reus, Epel, & Mellon, 2009).

In addition to its conversion into DHEAS, DHEA can be converted into other steroids, including androstenedione via 3β-HSD or androstenediol via 17β-hydroxysteroid dehydrogenase (17β-HSD). Both androstenedione and androstenediol are precursors of testosterone that can be converted into 5α-dihydrotestosterone (5α-DHT) by the action of the 5α-reductase, or aromatized into estradiol. Alternatively, DHEA can be transformed into 7-oxygenated derivatives, such as 7α-hydroxydehydroepiandrosterone (7α-OH-DHEA) and 7β-hydroxydehydroepiandrosterone (7β-OH-DHEA) (Baulieu et al., 2001).

As previously mentioned, DHEA and DHEAS are the most abundant steroid hormones in humans. In humans from both sexes, circulating levels of DHEA start to rise during the adrenarche, when DHEA is produced and secreted by the adrenal cortex (Havelock, Auchus, & Rainey, 2004). During the adult life, in addition to its synthesis from the adrenal cortex, humans also synthesize DHEA in their gonads (Maninger et al., 2009; Nieschlag et al., 1973). The peripheral levels of DHEA reach their peak early in adulthood and gradually decline with age (Baulieu, 1996; Parker, 1999). In comparison with humans, rodents present lower levels of DHEA and DHEAS, as they only synthesize DHEA in their gonads, since they do not express CYP17 in the adrenals (Le Goascogne et al., 1991; van Weerden, Bierings, van Steenbrugge, de Jong, & Schröder, 1992). DHEA synthesis is regulated by the hypothalamic–pituitary–adrenal axis. Corticotropin-releasing hormone (CRH) is released from the hypothalamus and stimulates the anterior pituitary to produce and secrete adrenocorticotropin hormone (ACTH). In turn, ACTH is released in the circulation and reaches the adrenals, binding to melanocortin-2 plasma membrane G-protein-coupled receptors, increasing cholesterol delivery to the mitochondria and the expression of CYP11 and CYP17, stimulating the synthesis of DHEA. The levels of DHEA reach their peak of the day in the morning, following the circadian pattern of ACTH secretion (Prough, Clark, & Klinge, 2016).

In addition to its synthesis in peripheral glands, DHEA is also synthesized in the CNS. The first clue that DHEA could be synthesized in the CNS came from the study of Corpéchot, Robel, Axelson, Sjövall, and Baulieu (1981), which showed the presence of DHEAS in the brain of castrated and adrenalectomized rats. Although later it would be discovered that the rodent adult brain, unlike the human brain, does not express the enzyme CYP17, which is necessary for the synthesis of DHEA from pregnenolone (Compagnone, Bulfone, Rubenstein, & Mellon, 1995a; Le Goascogne et al., 1991; Mellon, 2007), and does not present significant levels of DHEA or DHEAS (Liere et al., 2009, 2004; Schumacher et al., 2015), that study was important as it stimulated other researchers to investigate the synthesis of DHEA and other steroid hormones in the brain. In this context, in contrast with what was reported regarding the absence of the expression of CYP17 in the rodent adult brain, it has been shown that the embryonic rodent brain expresses this enzyme (Compagnone, Bulfone, Rubenstein, & Mellon, 1995b; Hammer, Compagnone, Vigne, Bair, & Mellon, 2004), suggesting that DHEA could be locally produced in the CNS during the embryogenic period. Moreover, Liere et al. (2009) showed that the human brain, unlike the rodent brain, presents considerable levels of DHEA and DHEAS, suggesting that there are important differences in the synthesis of steroid hormones between human and rodent brains. Interestingly, Zwain and Yen (1999) demonstrated that astrocytes are the major producers of DHEA in the CNS, pointing to a close relationship between astrocytes and DHEA.

5. MECHANISMS OF ACTION OF DHEA AND DHEAS

Despite being the most abundant steroids in the human body, little is known about the mechanisms of action of DHEA and DHEAS. In contrast with what have been shown regarding the mechanisms of action of other steroid hormones, a specific cellular receptor for DHEA hormone remains to be identified. However, studies have reported that DHEA is able to bind and activate some intracellular receptors (Fig. 1), including the pregnane X receptor (Ripp, Fitzpatrick, Peters, & Prough, 2002), androgen receptor (AR) (Chen et al., 2005; Lu, Mo, Hu, Garippa, & Simon, 2003; Mo, Lu, & Simon, 2006), estrogen receptors α (ERα) and β (ERβ) (Chen et al., 2005), and peroxisome proliferator-activated receptors (Webb, Geoghegan, Prough, & Michael Miller, 2006), although with a lower affinity than the cognate ligands of these receptors. In addition, DHEA can elicit genomic actions through the activation of ER and AR after its conversion into testosterone or dihydrotestosterone (DHT) and estradiol, respectively (Stárka, Dušková, & Hill, 2015). Moreover, DHEA modulates several membrane receptors (Fig. 1), including the GABA(A) receptor, glycine receptors, *N*-methyl-D-aspartate (NMDA) receptor, kainate receptors, ionotropic glutamate receptors, σ1 receptors, neurotrophin receptors, nicotinic acetylcholine receptors, muscarinic receptors, and ionic channels such as calcium, potassium, and sodium channels (Hill, Dušková, & Stárka, 2015). Studies have shown that DHEAS act as an antagonist of GABA(A) receptors when administered at low/medium micromolar concentrations, exerting multiple prosurvival effects (Lapchak et al., 2000; Park-Chung, Malayev, Purdy, Gibbs, & Farb, 1999; Sachidanandan & Bera, 2015; Svob Štrac,

Jazvinšćak Jembrek, Erhardt, Mirković Kos, & Peričić, 2012; Waters, Miller, Aleo, & Schnellmann, 1997). Moreover, in addition to the modulation of GABA(A) receptor, DHEAS also decreases the permeability of glycine receptor, regulating chloride influx into the cells (Fodor, Boros, Dezso, & Maksay, 2006; Maksay, Laube, & Betz, 2001). On the other hand, studies show that DHEA and DHEAS may act as positive allosteric modulators of NMDA receptors, increasing the intracellular levels of calcium and stimulating axon development and dendrite outgrowth (Compagnone & Mellon, 1998; Irwin, Lin, Rogawski, Purdy, & Paul, 1994). However, in contrast with the idea that DHEA activates NMDA receptors, Kimonides, Khatibi, Svendsen, Sofroniew, and Herbert (1998) showed that DHEA is neuroprotective against NMDA, α-amino-3-hydroxy-5-methyl-4-isoxazolepropionic acid (AMPA), and kainate excitotoxicity in hippocampal neurons. Furthermore, recent data show that DHEA may bind to σ1 and neurotrophin receptors TrkA, TrkB, and TrkC, activating Akt and Erk, which are involved with prosurvival signaling through PI3K and MAPK pathways, respectively, eliciting neuroprotective actions and enhancing neurogenesis and synaptic efficacy (Janner, Jacob, Jahn, Kucharski, & Ribeiro, 2010; Lazaridis et al., 2011; Moriguchi et al., 2013; Pediaditakis et al., 2015; Yabuki et al., 2015). In addition, the activation of σ1 receptors have also been related with the activation of NMDA receptors (Maninger et al., 2009) and calcium/calmodulin-dependent protein kinase II (CaMKII) (Moriguchi, Yamamoto, Ikuno, & Fukunaga, 2011; Yabuki et al., 2015). Interestingly, in addition to the activation of neurotrophin receptors, DHEA also stimulates the synthesis of neurotrophins and growth factors, including BDNF, nerve growth factor, and insulin-like growth factor 1 which are associated with cell survival and proliferation (Rahmani, Shoae-Hassani, Keyhanvar, Kheradmand, & Darbandi-Azar, 2013; Ribeiro & Garcia-Segura, 2002). Finally, studies have demonstrated that DHEA enhances mitochondrial oxidative capacity and modulates CNS metabolism, interacting with lactate and controlling glucose uptake (de Souza, Ribeiro, & Kucharski, 2012; Patel & Katyare, 2006).

In addition to its direct actions, DHEA is also able to act through its metabolites. Studies show that DHEA is metabolized by cells from the hippocampus, frontal cortex, striatum, amygdala, and cerebellum into 7-oxygenated derivatives, including 7α-OH-DHEA and 7β-OH-DHEA, which may mediate the neuroprotective actions normally attributed to DHEA (Akwa, Morfin, Robel, & Baulieu, 1992; Schumacher et al., 2003). Moreover, DHEA-reduced metabolites, including estradiol, testosterone, 5α-DHT, and androstenediol, are also neuroactive steroids that act through different mechanisms, regulating reactive gliosis and eliciting neuroprotection among other actions (Arevalo, Azcoitia, & Garcia-Segura, 2015; Arevalo, Santos-Galindo, Acaz-Fonseca, Azcoitia, & Garcia-Segura, 2013; Białek, Zaremba, Borowicz, & Czuczwar, 2004; Elkabes & Nicot, 2014; Zup, Edwards, & McCarthy, 2014).

6. ASTROCYTES AS A TARGET FOR NEUROPROTECTION: MODULATION BY DHEA AND DHEAS

Several studies have reported that DHEA and DHEAS are neuroprotective in different experimental models, being potential candidates for the development of new therapeutic strategies against different types of injury in the CNS. In this section, we will discuss the use of DHEA and DHEAS as neuroprotective agents, focusing on its actions on astrocytes, since the modulation of these cells could represent a relevant strategy for neuroprotection.

6.1 Stroke

Stroke is the second leading cause of death in the world, killing every year almost 6 million people (Feigin et al., 2014). In spite of its high prevalence, few treatments for stroke are currently available, and their efficacy is generally low. Currently, there are more than 6000 clinical trials registered with patients victims of stroke (clinicaltrials.gov), indicating a high number of studies trying to identify new therapeutic approaches for the treatment of this injury. In this context, studies have shown that DHEA and DHEAS may be neuroprotective in experimental models of stroke. Li, Klein, Sun, and Buchan (2001) showed that rats that received implants of DHEA pellets (100 mg, 13 days before the injury) presented a decreased damage at the CA1 region of the hippocampus after a forebrain ischemia in comparison with controls. In another study, male rats submitted to a model of transient cerebral ischemia were treated with DHEAS in different protocols, and it was demonstrated that repeated administration of DHEAS (20 mg/kg for 3 days), starting from the first 3 h of reperfusion, produced a neuroprotective effect, while acute DHEAS administration was not effective (Li et al., 2006). Moreover, another study performed by the same group demonstrated that a single injection of DHEA (20 mg/kg) from 3 to 48 h after lesion exerted neuroprotective effects in male rats submitted to a transient global cerebral ischemia, while the administration of DHEA 1 h before or after injury exacerbated the neuronal death (Li et al., 2009). More recently, it was demonstrated that DHEA administration decreases neuronal damage and improves the memory deficits after a transient brain ischemia (Yabuki et al., 2015).

Since diabetes is a risk factor for stroke, Aragno et al. (2000) have investigated if DHEA could protect diabetic rats against cerebral ischemia. In this study, normoglycemic or diabetic rats were subjected to cerebral ischemia/reperfusion and treated with vehicle or DHEA (2 or 4 mg/kg/day) for 7, 14, and 21 days, and it was shown that DHEA reduced the neuronal damage induced by ischemia/reperfusion in both normoglycemic and diabetic rats (Aragno et al., 2000). In addition to its effects in experimental models of cerebral ischemia, one study demonstrated that DHEAS is also neuroprotective in an experimental model of spinal cord ischemia, produced by the occlusion of the infrarenal aorta (Lapchak et al., 2000).

Ischemic stroke is frequently caused by cerebral artery occlusion and is characterized by a pathologic cascade that includes loss of cerebral blood flow, BBB disruption, and edema formation (Liu & Chopp, 2016). The neuroinflammatory response after cerebral ischemia is characterized by the activation of glial cells and increased production of proinflammatory cytokines, such as tumoral necrosis factor (TNF)-α and several interleukins (IL) (McColl, Allan, & Rothwell, 2007; Nilupul Perera et al., 2006). In particular, TNF-α and IL-6 potentiate the postischemic neuroinflammatory response and have been related to BBB breakdown (Ballabh, Braun, & Nedergaard, 2004; Nilupul Perera et al., 2006). In this context, it has been shown that DHEA and DHEAS are able to inhibit TNF and IL-6 production in astrocytes in vitro and in vivo in different experimental conditions (Di Santo, Foddi, Ricciardi-Castagnoli, Mennini, & Ghezzi, 1996; Di Santo, Sironi, et al., 1996; Kipper-Galperin, Galilly, Danenberg, & Brenner, 1999). Therefore, one of the possible mechanisms involved in the neuroprotective effects of DHEA(S) against cerebral ischemia could be a protective effect on BBB function through the inhibition of TNF and IL-6 production by astrocytes after injury.

In addition to the neuroinflammatory response, ischemic stroke is also characterized by an important oxidative damage after injury. It is known that after acute ischemic stroke, a rapid increase in ROS production occurs, overwhelming antioxidant defenses and causing tissue damage (Rodrigo et al., 2013). Regarding this context, Aragno et al. (2000) showed that DHEA is also able to prevent oxidative damage caused by transient ischemia/reperfusion in both normoglycemic and diabetic rats, decreasing the levels of H_2O_2, hydroxyl radical, and ROS after injury. Since astrocytes produce different antioxidants, such as ascorbate, SODs 1, 2, and 3, and glutathione (Barreto, Gonzalez, Torres, & Morales, 2011; Dringen, Gutterer, & Hirrlinger, 2000; Slemmer, Shacka, Sweeney, & Weber, 2008), the neuroprotective effects of DHEA against experimental stroke may be related with an increased astrocytic production of antioxidants after injury. Furthermore, it was demonstrated that DHEA protects immunostimulated rat C6 glioma cells from glucose deprivation by inhibiting the production of superoxide anion (Shin et al., 2001), suggesting that DHEA may also modulate the production of free radicals by glial cells after injury.

Cytoplasmic Ca^{2+} overload and the following Ca^{2+} accumulation in the mitochondrial matrix are other important mediators of cell damage after ischemia. The accumulation of Ca^{2+} in the mitochondrial matrix leads to the loss of the mitochondrial membrane potential, to an increase in the opening of the mPTP and to a decreased energy production, which will lead to the activation of apoptosis signaling pathways (Leist & Nicotera, 1998). Ca^{2+} homeostasis in brain cells depends on the activity of several Ca^{2+} -permeable receptors and channels, and store-operated channels and pumps. Astrocytes communicate with neighboring neurons through Ca^{2+} signals that are able to modulate synaptic interactions (Alberdi, Sánchez-Gómez, & Matute, 2005). Therefore, Ca^{2+} overload resulting from dysregulation of channels and pumps can promote neuronal and glial damage. In this context, Kaasik, Safiulina, Kalda, and Zharkovsky (2003) showed that DHEA, but not DHEAS, is able to protect neuronal mitochondria from calcium overload. Although these data have been obtained using cerebellar granule cell cultures, DHEA may also protect astrocytic mitochondria from calcium overload after ischemia, and this could be another mechanism related with its neuroprotective effects.

Finally, it is important to mention that most of the studies regarding the actions of DHEA have been performed in male young-adult rats only. Therefore, other studies evaluating the effects of DHEA in female rats, as well as in aged rats, are needed. In this context, it was recently demonstrated that the administration of DHEA protects male and female hippocampal neurons from glucose deprivation in vitro, suggesting that DHEA could be a promising candidate to be used for the treatment of patients from both sexes after stroke (Vieira-Marques et al., 2016).

6.2 Traumatic Brain Injury

In addition to its neuroprotective effects against experimental stroke, numerous studies have shown that DHEA and DHEAS are also neuroprotective against TBI. Studies show that DHEAS is neuroprotective in different experimental models of TBI, including bilateral penetrating lesion and close-head injury, decreasing cognitive-deficits and behavioral changes after injury (Hoffman et al., 2003; Milman et al., 2008). Furthermore, Juhász-Vedres et al. (2006) showed that DHEAS is neuroprotective against focal cortical cold lesion when administered either before or after injury, protecting BBB integrity and reducing edema formation after injury. Interestingly, the administration of letrozole, an aromatase inhibitor that blocks the conversion of testosterone into estradiol, abolished the neuroprotective effects of DHEAS, suggesting that estradiol could mediate the neuroprotective effects of DHEAS in this model (Juhász-Vedres et al., 2006). Moreover, it has been shown that fluasterone, a DHEA analog, also improves functional recovery after TBI (Malik et al., 2003). On the other hand, it has been recently shown that TBI induces several changes in brain levels of neuroactive steroids in mice, and that the levels of progesterone and DHEA are positively correlated with a better functional recovery after injury (Lopez-Rodriguez et al., 2015). Therefore, a pharmacological increase in the brain levels of DHEA may represent an interesting therapeutic strategy for the treatment of TBI.

TBI is characterized by the direct damage related to the injury and by a secondary injury cascade, including brain edema, oxidative damage, excitotoxicity, and apoptosis. Brain edema occurs due to the disruption of fluid homeostasis in neurons and glial cells affected by the injury and due to the vasogenic edema caused by the disruption of the BBB, leading to brain swelling and increased intracranial pressure. In addition to the edema formation, TBI triggers an inflammatory response characterized by the migration of neutrophils and macrophages, by the induction of gliosis around the injured region, and by the activation of NF-κB signaling pathway and proinflammatory cytokines release, such as IL-1, IL-6, and TNF-α (Pearn et al., 2017). In this context, it has been shown that DHEA attenuates reactive gliosis, astrocyte proliferation, and its accumulation in the proximity of the wound after a penetrating brain injury (García-Estrada, Luquín, Fernández, & Garcia-Segura, 1999). These findings could be with the inhibition of proinflammatory cytokine release, as some studies showed that DHEA and DHEAS inhibit the production of TNF-α and IL-6 in astrocytes under different experimental conditions (Di Santo, Foddi, et al., 1996; Di Santo, Sironi, et al., 1996; Kipper-Galperin et al., 1999). Moreover, it has been shown that DHEA is able to decrease the protein expression of TNF-α, IL-1β, and iNOS, as well as the nitrite production in microglial cells treated with LPS (Barger et al., 2000; Wang et al., 2001; Wu et al., 2015), corroborating the idea that DHEA could act as an antiinflammatory agent.

In addition to brain edema formation and inflammation, TBI is also associated with oxidative stress, represented by increased levels of NO, superoxide, and peroxynitrite, which inhibit the ATP production and affect the oxidative metabolism in neurons and astrocytes (Bentzer, Davidsson, & Grände, 2000; Cherian, Goodman, & Robertson, 2000; Hüttemann et al., 2008). Also, TBI is associated with an increased expression of NOS, which leads to cytochrome-*c* release from the mitochondria and apoptosis (Cherian, Hlatky, & Robertson, 2004). In this context, the modulation of the rapidly increase in the levels of free radicals in the brain is an important target for the neuroprotective actions of steroid hormones against brain injuries and requires the cooperation between astroglia and microglia (Johann & Beyer, 2013). Therefore, since some studies have shown that DHEA and DHEAS present antioxidant effects under different experimental conditions, including experimental models of cerebral ischemia, oxidative damage, glucose deprivation, neurodegenerative diseases, aging, and diabetes (Aly et al., 2011; Aragno et al., 2002, 2000, 1999; Bastianetto, Ramassamy, Poirier, & Quirion, 1999; Kumar, Taha, Sharma, Kale, & Baquer, 2008; Leskiewicz et al., 2008; Shin et al., 2001; Túnez, Muñoz, & Montilla, 2005), the neuroprotective actions of DHEA and DHEAS against TBI may involve an antioxidant effect related to the modulation of astrocytes, which are involved in the regulation of the levels of free radicals and antioxidants after injury.

6.3 Spinal Cord Injury

Similar to TBI, spinal cord injury (SCI) is followed by reactive gliosis and inflammation, which result in the loss of neurons and oligodendrocytes and demyelination (Haan, Zhu, Wang, Wei, & Song, 2015). As previously mentioned, astrocyte activation and glial scar formation at the epicenter of the injury represent a barrier for axon regeneration and for the remyelination of intact axons (Elkabes & Nicot, 2014). On the other hand, myelin breakdown is usually associated with increased levels of myelin basic protein (MBP)-related peptides, which induce astrogliosis and glial scar formation (South et al., 2000). In this context, studies have shown that DHEA is also neuroprotective against SCI, improving functional recovery after injury (Compagnone, 2008; Fiore et al., 2004). The neuroprotective effects of

DHEA were associated with increased white matter spared area at the epicenter and reduced reactive gliosis around the injury (Fiore et al., 2004). These effects may involve the modulation of astrogliosis that usually follows myelin breakdown, since it has been demonstrated that both DHEA and DHEAS inhibit the mitogenic effect of MBP on astrocytes (Muntwyler & Bologa, 1989). As previously mentioned, since it has been shown that DHEA decreases TNF-α and IL-6 production in astrocytes in vitro (Di Santo, Foddi, et al., 1996; Di Santo, Sironi, et al., 1996; Kipper-Galperin et al., 1999), the neuroprotective effects of DHEA against SCI could be also related with its antiinflammatory actions after injury. Interestingly, it has been shown that DHEA levels decrease after SCI (Rouleau, Ung, Lapointe, & Guertin, 2007). Therefore, a pharmacological increase in the levels of DHEA could be a potential therapeutic strategy for the treatment of SCI, decreasing reactive gliosis and the inflammatory process after injury.

6.4 Glutamate Excitotoxicity

Glutamate excitotoxicity is a key component of the secondary injury cascade that occurs after ischemia and TBI. It is known that astrocytes play an important role in the regulation of glutamate levels in the CNS due to their ability to synthesize glutamine, the precursor of glutamate, and to uptake glutamate through glutamate transporters (Anderson & Swanson, 2000; Zou et al., 2010). In the astrocytes, the glutamate reuptaked from the synaptic cleft is recycled for neurotransmission through a process known as the glutamate–glutamine shuttle. Glutamate is converted into glutamine by astrocytic glutamine synthase, and then glutamine is transported through the neurons, where it is converted back into glutamate by phosphate-activated glutaminase and repackaged into vesicles for synaptic release (Stobart & Anderson, 2013) (Fig. 1). Regarding this context, it has been demonstrated that DHEA and DHEAS are also neuroprotective against excitotoxic cell death under different conditions. Kimonides et al. (1998) showed that DHEA and DHEAS protect hippocampal neurons against NMDA, AMPA, and kainic acid excitotoxicity. These data were corroborated by other study showing that DHEA is neuroprotective against kainic acid excitotoxicity in the hippocampus of gonadectomized male rats (Veiga et al., 2003). The neuroprotective effect of DHEA seems to involve its conversion into estradiol, since the blockade of aromatase abolished its neuroprotective action (Veiga et al., 2003). Interestingly, although astrocytes does not express aromatase in healthy states, kainic acid administration stimulates the expression of this enzyme (Garcia-Segura et al., 1999), suggesting that astrocytes could play a pivotal role in the neuroprotective actions of DHEA against excitotoxic cell death. Moreover, Cardounel, Regelson, and Kalimi (1999) showed that glutamate excitotoxicity is associated with an increased neuronal expression of glucocorticoid receptor (GR), and that DHEA is able to protect hippocampal neurons against glutamate excitotoxicity by decreasing the expression of GR. Furthermore, Kokona et al. (2012) showed that DHEA protects the retina against AMPA-induced excitotoxicity through the activation of TrkA receptors. In addition, Lhullier et al. (2004) showed that DHEA inhibits the glutamate release from synaptosomes of rats, corroborating the idea that DHEA could be neuroprotective against excitotoxic cell death due to its inhibitory effect on glutamatergic system overstimulation after injury. These results are supported by other study showing that DHEA administration increases the expression of different glutamate transporters in the rat hippocampus (Mishra, Singh, Mukherjee, & Sharma, 2013). Therefore, the neuroprotective actions of DHEA against glutamate excitotoxicity seem to depend on the overexpression of aromatase and on the regulation of glutamate levels by astrocytes.

6.5 Neurodegenerative Diseases

Studies show that changes in astrocytic function are involved in the pathogenesis and progression of several neurodegenerative diseases, including Alzheimer's disease (AD), prion disease, and amyotrophic lateral sclerosis (Brambilla, Martorana, & Rossi, 2013). In addition, since aging and several neurodegenerative diseases are associated with changes in the serum and cerebrospinal fluid levels of DHEA and DHEAS (Azuma et al., 1993; Foroughipour, Norbakhsh, Najafabadi, & Meamar, 2012; Rammouz, Lecanu, & Papadopoulos, 2011; Raven & Hinson, 2007; Téllez et al., 2006), studies have suggested that DHEA and DHEAS may also play a role in the pathophysiology of these diseases. Therefore, several studies have investigated if DHEA administration could promote neuroprotective actions in different experimental models of neurodegenerative diseases.

Most of the studies regarding the neuroprotective actions of DHEA against neurodegenerative diseases have been performed in experimental models of AD. This disease, which is the most common neurodegenerative disease and the major cause of dementia in people over 60 years old, is characterized by the accumulation of the amyloid-beta (Aβ) peptide and by the presence of high levels of intracellular neurofibrillary tangles formed by hyperphosphorylated aggregates of Tau, which lead to the activation of several proapoptotic signaling pathways (Cavallucci, D'Amelio, & Cecconi, 2012; Querfurth & LaFerla, 2010). In this context, Li et al. (2010) showed that DHEA prevents Aβ-impaired survival and dendritic growth of newborn neurons in the dentate gyrus of the hippocampus, in a mechanism involving the activation of σ1 receptors and the modulation of PI3K/Akt signaling. These results are corroborated by other studies showing that DHEA is neuroprotective in different experimental models of AD in vivo, improving the performance of the animals in learning and memory tasks (Aly et al., 2011; Farr, Banks, Uezu, Gaskin, & Morley, 2004). Astrocytes may have a pivotal role in the neuroprotective actions of DHEA, since it has been demonstrated that DHEA is able to protect the brain from the oxidative damage associated with AD, decreasing lipid peroxidation and increasing the activity of antioxidant enzymes (Aly et al., 2011; Ramassamy et al., 1999). This hypothesis is corroborated by other study showing that DHEA is able to modulate redox homeostasis by increasing antioxidant activity in neurons, and to improve bioenergetic activity in these cells by increasing ATP levels, mitochondrial membrane potential, and basal mitochondrial respiration (Grimm, Schmitt, Lang, Mensah-Nyagan, & Eckert, 2014). Furthermore, the neuroprotective effects of DHEA were associated with increased levels of Bcl-2 and BDNF, as well as with decreased brain acetylcholinesterase activity and increased levels of acetylcholine (Aly et al., 2011).

In addition to its neuroprotective effects in experimental models AD, studies show that DHEA is also neuroprotective in experimental models of Parkinson's disease (PD). PD, which is the second most common neurodegenerative disease just behind AD, is associated with the loss of dopaminergic neurons in the substantia nigra of the midbrain and the development of neuronal Lewy bodies, leading to rest tremor, rigidity, bradykinesia, stooping posture, neurobehavioral disorders, and cognitive impairment (Beitz, 2014). Regarding this context, studies show that DHEA is neuroprotective in experimental models of PD in mouse and monkeys, protecting dopaminergic neurons against 1-methyl-4-phenyl-1,2,3,6-tetrahydropyridine (MPTP) administration and improving the mean parkinsonian score (Bélanger, Grégoire, Bédard, & Di Paolo, 2006; Litim et al., 2016). The neuroprotective effects of DHEA may involve an interaction with dopamine and serotonin systems, since it is known that DHEA regulates the turnover ratios of these neurotransmitters in the striatum and nucleus accumbens (Pérez-Neri, Méndez-Sánchez, Montes, & Ríos, 2008).

7. CONCLUSION AND FUTURE DIRECTIONS

In summary, several studies show that DHEA and DHEAS are neuroprotective in different experimental models of stroke, TBI, SCI, and neurodegenerative diseases. The neuroprotective actions of DHEA seem to involve the modulation of astrocytic function, including the modulation of reactive gliosis, inflammation, BBB function, and redox and glutamate homeostasis (Fig. 1). Therefore, the development of DHEA-based therapies could represent a promising therapeutic strategy for the treatment of those conditions. Further studies should be focused on deepening the identification of the mechanisms of action of DHEA in the brain, in particularly its actions on glial cells. Moreover, the identification of sex and age-related differences regarding the neuroprotective effects of DHEA, providing evidence if DHEA could be neuroprotective in individuals of both sexes and different ages, is crucial to establish possible clinical trials for the evaluation of the neuroprotective effects of DHEA in humans.

ACKNOWLEDGMENTS

B.D.A. received a PNPD scholarship from Coordenação de Aperfeiçoamento de Pessoal de Nível Superior (CAPES). M.F.R. received a CNPq 1D Researcher Productivity Grant.

REFERENCES

Acaz-Fonseca, E., Avila-Rodriguez, M., Garcia-Segura, L. M., & Barreto, G. E. (2016). Regulation of astroglia by gonadal steroid hormones under physiological and pathological conditions. *Progress in Neurobiology*, *144*, 5–26. https://doi.org/10.1016/j.pneurobio.2016.06.002.

Akwa, Y., Morfin, R. F., Robel, P., & Baulieu, E. E. (1992). Neurosteroid metabolism. 7 Alpha-hydroxylation of dehydroepiandrosterone and pregnenolone by rat brain microsomes. *Biochemical Journal*, *288*, 959–964.

Alberdi, E., Sánchez-Gómez, M. V., & Matute, C. (2005). Calcium and glial cell death. *Cell Calcium*, *38*, 417–425.

Aly, H. F., Metwally, F. M., & Ahmed, H. H. (2011). Neuroprotective effects of dehydroepiandrosterone (DHEA) in rat model of Alzheimer's disease. *Acta Biochimica Polonica*, *58*, 513–520.

Anderson, C. M., & Swanson, R. A. (2000). Astrocyte glutamate transport: Review of properties, regulation, and physiological functions. *Glia*, *32*, 1–14.

Aragno, M., Mastrocola, R., Brignardello, E., Catalano, M., Robino, G., Manti, R., et al. (2002). Dehydroepiandrosterone modulates nuclear factor-kappaB activation in hippocampus of diabetic rats. *Endocrinology*, *143*, 3250–3258.

Aragno, M., Parola, S., Brignardello, E., Mauro, A., Tamagno, E., Manti, R., et al. (2000). Dehydroepiandrosterone prevents oxidative injury induced by transient ischemia/reperfusion in the brain of diabetic rats. *Diabetes*, *49*, 1924–1931.

Aragno, M., Tamagno, E., Gatto, V., Brignardello, E., Parola, S., Danni, O., et al. (1999). Dehydroepiandrosterone protects tissues of streptozotocin-treated rats against oxidative stress. *Free Radical Biololgy & Medicine, 26*, 1467–1474.

Araque, A. (2008). Astrocytes process synaptic information. *Neuron Glia Biology, 4*, 3–10.

Araque, A., Carmignoto, G., Haydon, P. G., Oliet, S. H., Robitaille, R., & Volterra, A. (2014). Gliotransmitters travel in time and space. *Neuron, 81*, 728–739.

Araque, A., Parpura, V., Sanzgiri, R. P., & Haydon, P. G. (1999). Tripartite synapses: Glia, the unacknowledged partner. *Trends in Neurosciences, 22*, 208–215.

Arbo, B. D., Benetti, F., & Ribeiro, M. F. (2016). Astrocytes as a target for neuroprotection: Modulation by progesterone and dehydroepiandrosterone. *Progress in Neurobiology, 144*, 27–47. https://doi.org/10.1016/j.pneurobio.2016.03.010.

Arevalo, M. A., Azcoitia, I., & Garcia-Segura, L. M. (2015). The neuroprotective actions of oestradiol and oestrogen receptors. *Nature Reviews. Neuroscience, 16*, 17–29.

Arevalo, M. A., Santos-Galindo, M., Acaz-Fonseca, E., Azcoitia, I., & Garcia-Segura, L. M. (2013). Gonadal hormones and the control of reactive gliosis. *Hormones and Behavior, 63*, 216–221.

Azuma, T., Matsubara, T., Shima, Y., Haeno, S., Fujimoto, T., Tone, K., et al. (1993). Neurosteroids in cerebrospinal fluid in neurologic disorders. *Journal of the Neurological Sciences, 120*, 87–92.

Ballabh, P., Braun, A., & Nedergaard, M. (2004). The blood-brain barrier: An overview: Structure, regulation, and clinical implications. *Neurobiology of Disease, 16*, 1–13.

Bardehle, S., Krüger, M., Buggenthin, F., Schwausch, J., Ninkovic, J., Clevers, H., et al. (2013). Live imaging of astrocyte responses to acute injury reveals selective juxtavascular proliferation. *Nature Neuroscience, 16*, 580–586.

Barger, S. W., Chavis, J. A., & Drew, P. D. (2000). Dehydroepiandrosterone inhibits microglial nitric oxide production in a stimulus-specific manner. *Journal of Neuroscience Research, 62*, 503–509.

Barreto, G. E., Gonzalez, J., Torres, Y., & Morales, L. (2011). Astrocytic-neuronal crosstalk: Implications for neuroprotection from brain injury. *Neuroscience Research, 71*, 107–113.

Bastianetto, S., Ramassamy, C., Poirier, J., & Quirion, R. (1999). Dehydroepiandrosterone (DHEA) protects hippocampal cells from oxidative stress-induced damage. *Brain Research. Molecular Brain Research, 66*, 35–41.

Baulieu, E. E. (1981). Steroid hormones in the brain: Several mechanisms? In K. Fuxe, J. A. Gustafsson, & L. Wetterberg (Eds.), *Steroid hormone regulation of the brain* (pp. 3–14). Oxford: Pergamon Press.

Baulieu, E. E. (1996). Dehydroepiandrosterone (DHEA): A fountain of youth? *The Journal of Clinical Endocrinology and Metabolism, 81*, 3147–3151.

Baulieu, E. E., & Robel, P. (1998). Dehydroepiandrosterone (DHEA) and dehydroepiandrosterone sulfate (DHEAS) as neuroactive neurosteroids. *Proceedings of the National Academy of Sciences of the United States of America, 95*, 4089–4091.

Baulieu, E. E., Robel, P., & Schumacher, M. (2001). Neurosteroids: Beginning of the story. *International Review of Neurobiology, 46*, 1–32.

Beitz, J. M. (2014). Parkinson's disease: A review. *Frontiers in Biosciences (Scholar Edition), 6*, 65–74.

Bélanger, N., Grégoire, L., Bédard, P. J., & Di Paolo, T. (2006). DHEA improves symptomatic treatment of moderately and severely impaired MPTP monkeys. *Neurobiology of Aging, 27*, 1684–1693.

Bentzer, P., Davidsson, H., & Grände, P. O. (2000). Microdialysis-based long-term measurements of energy-related metabolites in the rat brain following a fluid percussion trauma. *Journal of Neurotrauma, 17*, 441–447.

Białek, M., Zaremba, P., Borowicz, K. K., & Czuczwar, S. J. (2004). Neuroprotective role of testosterone in the nervous system. *Polish Journal of Pharmacology, 56*, 509–518.

Brambilla, L., Martorana, F., & Rossi, D. (2013). Astrocyte signaling and neurodegeneration: New insights into CNS disorders. *Prion*, *7*, 28–36.
Bramwell, B. L. (2010). The clinical use of dehydroepiandrosterone in postmenopausal women. *International Journal of Pharmaceutical Compounding*, *14*, 465–471.
Burda, J. E., Bernstein, A. M., & Sofroniew, M. V. (2016). Astrocyte roles in traumatic brain injury. *Experimental Neurology*, *275*, 305–315.
Burda, J. E., & Sofroniew, M. V. (2014). Reactive gliosis and the multicellular response to CNS damage and disease. *Neuron*, *81*, 229–248.
Bush, T. G., Puvanachandra, N., Horner, C. H., Polito, A., Ostenfeld, T., Svendsen, C. N., et al. (1999). Leukocyte infiltration, neuronal degeneration, and neurite outgrowth after ablation of scar-forming, reactive astrocytes in adult transgenic mice. *Neuron*, *23*, 297–308.
Cardounel, A., Regelson, W., & Kalimi, M. (1999). Dehydroepiandrosterone protects hippocampal neurons against neurotoxin-induced cell death: Mechanism of action. *Proceedings of the Society for Experimental Biology and Medicine*, *222*, 145–149.
Cavallucci, V., D'Amelio, M., & Cecconi, F. (2012). Aβ toxicity in Alzheimer's disease. *Molecular Neurobiology*, *45*, 366–378.
Charalampopoulos, I., Remboutsika, E., Margioris, A. N., & Gravanis, A. (2008). Neurosteroids as modulators of neurogenesis and neuronal survival. *Trends in Endocrinology and Metabolism*, *19*, 300–307.
Chen, F., Knecht, K., Birzin, E., Fisher, J., Wilkinson, H., Mojena, M., et al. (2005). Direct agonist/antagonist functions of dehydroepiandrosterone. *Endocrinology*, *146*, 4568–4576.
Cherian, L., Goodman, J. C., & Robertson, C. S. (2000). Brain nitric oxide changes after controlled cortical impact injury in rats. *Journal of Neurophysiology*, *83*, 2171–2178.
Cherian, L., Hlatky, R., & Robertson, C. S. (2004). Nitric oxide in traumatic brain injury. *Brain Pathology*, *14*, 195–201.
Compagnone, N. A. (2008). Treatments for spinal cord injury: Is there hope in neurosteroids? *The Journal of Steroid Biochemistry and Molecular Biology*, *109*, 307–313.
Compagnone, N. A., Bulfone, A., Rubenstein, J. L., & Mellon, S. H. (1995a). Expression of the steroidogenic enzyme P450scc in the central and peripheral nervous systems during rodent embryogenesis. *Endocrinology*, *136*, 2689–2696.
Compagnone, N. A., Bulfone, A., Rubenstein, J. L., & Mellon, S. H. (1995b). Steroidogenic enzyme P450c17 is expressed in the embryonic central nervous system. *Endocrinology*, *136*, 5212–5223.
Compagnone, N. A., & Mellon, S. H. (1998). Dehydroepiandrosterone: A potential signalling molecule for neocortical organization during development. *Proceedings of the National Academy of Sciences of the United States of America*, *95*, 4678–4683.
Corpéchot, C., Robel, P., Axelson, M., Sjövall, J., & Baulieu, E. E. (1981). Characterization and measurement of dehydroepiandrosterone sulfate in rat brain. *Proceedings of the National Academy of Sciences of the United States of America*, *78*, 4704–4707.
Correale, J., & Farez, M. F. (2015). The role of astrocytes in multiple sclerosis progression. *Frontiers in Neurology*, *6*, 180.
Cregg, J. M., DePaul, M. A., Filous, A. R., Lang, B. T., Tran, A., & Silver, J. (2014). Functional regeneration beyond the glial scar. *Experimental Neurology*, *253*, 197–207.
de Souza, D. K., Ribeiro, M. F., & Kucharski, L. C. (2012). Effects of dehydroepiandrosterone (DHEA) and lactate on glucose uptake in the central nervous system. *Neuroscience Letters*, *507*, 62–66.
Di Santo, E., Foddi, M. C., Ricciardi-Castagnoli, P., Mennini, T., & Ghezzi, P. (1996). DHEAS inhibits TNF production in monocytes, astrocytes and microglial cells. *Neuroimmunomodulation*, *3*, 285–288.
Di Santo, E., Sironi, M., Mennini, T., Zinetti, M., Savoldi, G., Di Lorenzo, D., et al. (1996). A glucocorticoid receptor-independent mechanism for neurosteroid inhibition of tumor necrosis factor production. *European Journal of Pharmacology*, *299*, 179–186.

Dringen, R., Gebhardt, R., & Hamprecht, B. (1993). Glycogen in astrocytes: Possible function as lactate supply for neighboring cells. *Brain Research*, *623*, 208–214.
Dringen, R., Gutterer, J. M., & Hirrlinger, J. (2000). Glutathione metabolism in brain metabolic interaction between astrocytes and neurons in the defense against reactive oxygen species. *European Journal of Biochemistry*, *267*, 4912–4916.
Dringen, R., & Hamprecht, B. (1997). Involvement of glutathione peroxidase and catalase in the disposal of exogenous hydrogen peroxide by cultured astroglial cells. *Brain Research*, *759*, 67–75.
Dringen, R., Kussmaul, L., Gutterer, J. M., Hirrlinger, J., & Hamprecht, B. (1999). The glutathione system of peroxide detoxification is less efficient in neurons than in astroglial cells. *Journal of Neurochemistry*, *72*, 2523–2530.
Elkabes, S., & Nicot, A. B. (2014). Sex steroids and neuroprotection in spinal cord injury: A review of preclinical investigations. *Experimental Neurology*, *259*, 28–37.
Farr, S. A., Banks, W. A., Uezu, K., Gaskin, F. S., & Morley, J. E. (2004). DHEAS improves learning and memory in aged SAMP8 mice but not in diabetic mice. *Life Sciences*, *75*, 2775–2785.
Faulkner, J. R., Herrmann, J. E., Woo, M. J., Tansey, K. E., Doan, N. B., & Sofroniew, M. V. (2004). Reactive astrocytes protect tissue and preserve function after spinal cord injury. *The Journal of Neuroscience*, *24*, 2143–2155.
Feigin, V. L., Forouzanfar, M. H., Krishnamurthi, R., Mensah, G. A., Connor, M., Bennett, D. A., et al. (2014). Global and regional burden of stroke during 1990–2010: Findings from the Global Burden of Disease Study 2010. *The Lancet*, *383*, 245–254.
Fiore, C., Inman, D. M., Hirose, S., Noble, L. J., Igarashi, T., & Compagnone, N. A. (2004). Treatment with the neurosteroid dehydroepiandrosterone promotes recovery of motor behavior after moderate contusive spinal cord injury in the mouse. *Journal of Neuroscience Research*, *75*, 391–400.
Fodor, L., Boros, A., Dezso, P., & Maksay, G. (2006). Expression of heteromeric glycine receptor-channels in rat spinal cultures and inhibition by neuroactive steroids. *Neurochemistry International*, *49*, 577–583.
Foroughipour, A., Norbakhsh, V., Najafabadi, S. H., & Meamar, R. (2012). Evaluating sex hormone levels in reproductive age women with multiple sclerosis and their relationship with disease severity. *Journal of Research in Medical Sciences*, *17*, 882–885.
García-Estrada, J., Luquín, S., Fernández, A. M., & Garcia-Segura, L. M. (1999). Dehydroepiandrosterone, pregnenolone and sex steroids down-regulate reactive astroglia in the male rat brain after a penetrating brain injury. *International Journal of Developmental Neuroscience*, *17*, 145–151.
Garcia-Segura, L. M., Wozniak, A., Azcoitia, I., Rodriguez, J. R., Hutchison, R. E., & Hutchison, J. B. (1999). Aromatase expression by astrocytes after brain injury: Implications for local estrogen formation in brain repair. *Neuroscience*, *89*, 567–578.
Giudetti, A. M., Romano, A., Lavecchia, A. M., & Gaetani, S. (2016). The role of brain cholesterol and its oxidized products in Alzheimer's disease. *Current Alzheimer Research*, *13*, 198–205.
Gómez-Santos, C., Hernández-Morante, J. J., Tébar, F. J., Granero, E., & Garaulet, M. (2012). Differential effect of oral dehydroepiandrosterone-sulphate on metabolic syndrome features in pre- and postmenopausal obese women. *Clinical Endocrinology*, *77*, 548–554.
Grimm, A., Schmitt, K., Lang, U. E., Mensah-Nyagan, A. G., & Eckert, A. (2014). Improvement of neuronal bioenergetics by neurosteroids: Implications for age-related neurodegenerative disorders. *Biochimica et Biophysica Acta*, *1842*, 2427–2438.
Haan, N., Zhu, B., Wang, J., Wei, X., & Song, B. (2015). Crosstalk between macrophages and astrocytes affects proliferation, reactive phenotype and inflammatory response,

suggesting a role during reactive gliosis following spinal cord injury. *Journal of Neuroinflammation*, *12*, 109.

Halassa, M. M., Fellin, T., & Haydon, P. G. (2007). The tripartite synapse: Roles for gliotransmission in health and disease. *Trends in Molecular Medicine*, *13*, 54–63.

Hammer, F., Compagnone, N. A., Vigne, J. L., Bair, S. R., & Mellon, S. H. (2004). Transcriptional regulation of P450scc gene expression in the embryonic rodent nervous system. *Endocrinology*, *145*, 901–912.

Havelock, J. C., Auchus, R. J., & Rainey, W. E. (2004). The rise in adrenal androgen biosynthesis: Adrenarche. *Seminars in Reproductive Medicine*, *22*, 337–347.

Heneka, M. T., Carson, M. J., El Khoury, J., Landreth, G. E., Brosseron, F., Feinstein, D. L., et al. (2015). Neuroinflammation in Alzheimer's disease. *The Lancet. Neurology*, *14*, 388–405.

Hill, M., Dušková, M., & Stárka, L. (2015). Dehydroepiandrosterone, its metabolites and ion channels. *The Journal of Steroid Biochemistry and Molecular Biology*, *145*, 293–314.

Hoffman, S. W., Virmani, S., Simkins, R. M., & Stein, D. G. (2003). The delayed administration of dehydroepiandrosterone sulfate improves recovery of function after traumatic brain injury in rats. *Journal of Neurotrauma*, *20*, 859–870.

Hüttemann, M., Lee, I., Pecinova, A., Pecina, P., Przyklenk, K., & Doan, J. W. (2008). Regulation of oxidative phosphorylation, the mitochondrial membrane potential, and their role in human disease. *Journal of Bioenergetics and Biomembranes*, *40*, 445–456.

Irwin, R. P., Lin, S. Z., Rogawski, M. A., Purdy, R. H., & Paul, S. M. (1994). Steroid potentiation and inhibition of N-methyl-D-aspartate receptor-mediated intracellular Ca++ responses: Structure-activity studies. *The Journal of Pharmacology and Experimental Therapeutics*, *271*, 677–682.

Janner, D. R., Jacob, M. H., Jahn, M. P., Kucharski, L. C., & Ribeiro, M. F. (2010). Dehydroepiandrosterone effects on Akt signaling modulation in central nervous system of young and aged healthy rats. *The Journal of Steroid Biochemistry and Molecular Biology*, *122*, 142–148.

Johann, S., & Beyer, C. (2013). Neuroprotection by gonadal steroid hormones in acute brain damage requires cooperation with astroglia and microglia. *The Journal of Steroid Biochemistry and Molecular Biology*, *137*, 71–81.

Johnson, M. D., Bebb, R. A., & Sirrs, S. M. (2002). Uses of DHEA in aging and other disease states. *Ageing Research Reviews*, *1*, 29–41.

Juhász-Vedres, G., Rózsa, E., Rákos, G., Dobszay, M. B., Kis, Z., Wölfling, J., et al. (2006). Dehydroepiandrosterone sulfate is neuroprotective when administered either before or after injury in a focal cortical cold lesion model. *Endocrinology*, *147*, 683–686.

Kaasik, A., Kalda, A., Jaako, K., & Zharkovsky, A. (2001). Dehydroepiandrosterone sulphate prevents oxygen-glucose deprivation-induced injury in cerebellar granule cell culture. *Neuroscience*, *102*, 427–432.

Kaasik, A., Safiulina, D., Kalda, A., & Zharkovsky, A. (2003). Dehydroepiandrosterone with other neurosteroids preserve neuronal mitochondria from calcium overload. *The Journal of Steroid Biochemistry and Molecular Biology*, *87*, 97–103.

Kacem, K., Lacombe, P., Seylaz, J., & Bonvento, G. (1998). Structural organization of the perivascular astrocyte endfeet and their relationship with the endothelial glucose transporter: A confocal microscopy study. *Glia*, *23*, 1–10.

Khakh, B. S., & Sofroniew, M. V. (2015). Diversity of astrocyte functions and phenotypes in neural circuits. *Nature Neuroscience*, *18*, 942–952.

Kimonides, V. G., Khatibi, N. H., Svendsen, C. N., Sofroniew, M. V., & Herbert, J. (1998). Dehydroepiandrosterone (DHEA) and DHEA-sulfate (DHEAS) protect hippocampal neurons against excitatory amino acid-induced neurotoxicity. *Proceedings of the National Academy of Sciences of the United States of America*, *95*, 1852–1857.

Kipper-Galperin, M., Galilly, R., Danenberg, H. D., & Brenner, T. (1999). Dehydroepiandrosterone selectively inhibits production of tumor necrosis factor alpha and interleukin-6 in astrocytes. *International Journal of Developmental Neuroscience*, *17*, 765–775.

Kokona, D., Charalampopoulos, I., Pediaditakis, I., Gravanis, A., & Thermos, K. (2012). The neurosteroid dehydroepiandrosterone (DHEA) protects the retina from AMPA-induced excitotoxicity: NGF TrkA receptor involvement. *Neuropharmacology*, *62*, 2106–2117.

Kumar, P., Taha, A., Sharma, D., Kale, R. K., & Baquer, N. Z. (2008). Effect of dehydroepiandrosterone (DHEA) on monoamine oxidase activity, lipid peroxidation and lipofuscin accumulation in aging rat brain regions. *Biogerontology*, *9*, 235–246.

Kurzman, I. D., MacEwen, E. G., & Haffa, A. L. (1990). Reduction in body weight and cholesterol in spontaneously obese dogs by dehydroepiandrosterone. *International Journal of Obesity*, *14*, 95–104.

Labrie, F., Luu-The, V., Martel, C., Chernomoretz, A., Calvo, E., Morissette, J., et al. (2006). Dehydroepiandrosterone (DHEA) is an anabolic steroid like dihydrotestosterone (DHT), the most potent natural androgen, and tetrahydrogestrinone (THG). *The Journal of Steroid Biochemistry and Molecular Biology*, *100*, 52–58.

Lapchak, P. A., & Araujo, D. M. (2001). Preclinical development of neurosteroids as neuroprotective agents for the treatment of neurodegenerative diseases. *International Review of Neurobiology*, *46*, 379–397.

Lapchak, P. A., Chapman, D. F., Nunez, S. Y., & Zivin, J. A. (2000). Dehydroepiandrosterone sulfate is neuroprotective in a reversible spinal cord ischemia model: Possible involvement of GABA(A) receptors. *Stroke*, *31*, 1953–1956.

Lazaridis, I., Charalampopoulos, I., Alexaki, V. I., Avlonitis, N., Pediaditakis, I., Efstathopoulos, P., et al. (2011). Neurosteroid dehydroepiandrosterone interacts with nerve growth factor (NGF) receptors, preventing neuronal apoptosis. *PLoS Biology*, *9*, e1001051.

Le Goascogne, C., Sananès, N., Gouézou, M., Takemori, S., Kominami, S., Baulieu, E. E., et al. (1991). Immunoreactive cytochrome P-450(17 alpha) in rat and guinea-pig gonads, adrenal glands and brain. *Journal of Reproduction and Fertility*, *93*, 609–622.

Leist, M., & Nicotera, P. (1998). Calcium and neuronal death. *Reviews of Physiology, Biochemistry and Pharmacology*, *132*, 79–125.

Leskiewicz, M., Regulska, M., Budziszewska, B., Jantas, D., Jaworska-Feil, L., Basta-Kaim, A., et al. (2008). Effects of neurosteroids on hydrogen peroxide- and staurosporine-induced damage of human neuroblastoma SH-SY5Y cells. *Journal of Neuroscience Research*, *86*, 1361–1370.

Lhullier, F. L., Riera, N. G., Nicolaidis, R., Junqueira, D., Dahm, K. C., Cipriani, F., et al. (2004). Effect of DHEA glutamate release from synaptosomes of rats at different ages. *Neurochemical Research*, *29*, 335–339.

Li, Z., Cui, S., Zhang, Z., Zhou, R., Ge, Y., Sokabe, M., et al. (2009). DHEA-neuroprotection and -neurotoxicity after transient cerebral ischemia in rats. *Journal of Cerebral Blood Flow and Metabolism*, *29*, 287–296.

Li, H., Klein, G., Sun, P., & Buchan, A. M. (2001). Dehydroepiandrosterone (DHEA) reduces neuronal injury in a rat model of global cerebral ischemia. *Brain Research*, *888*, 263–266.

Li, L., Lundkvist, A., Andersson, D., Wilhelmsson, U., Nagai, N., Pardo, A. C., et al. (2008). Protective role of reactive astrocytes in brain ischemia. *Journal of Cerebral Blood Flow and Metabolism*, *28*, 468–481.

Li, W., Watts, L., Long, J., Zhou, W., Shen, Q., Jiang, Z., et al. (2016). Spatiotemporal changes in blood-brain barrier permeability, cerebral blood flow, T2 and diffusion following mild traumatic brain injury. *Brain Research*, *1646*, 53–61. https://doi.org/10.1016/j.brainres.2016.05.036.

Li, L., Xu, B., Zhu, Y., Chen, L., Sokabe, M., & Chen, L. (2010). DHEA prevents Aβ25-35-impaired survival of newborn neurons in the dentate gyrus through a modulation of PI3K-Akt-mTOR signaling. *Neuropharmacology, 59*, 323–333.

Li, Z., Zhou, R., Cui, S., Xie, G., Cai, W., Sokabe, M., et al. (2006). Dehydroepiandrosterone sulfate prevents ischemia-induced impairment of long-term potentiation in rat hippocampal CA1 by up-regulating tyrosine phosphorylation of NMDA receptor. *Neuropharmacology, 51*, 958–966.

Liere, P., Pianos, A., Eychenne, B., Cambourg, A., Bodin, K., Griffiths, W., et al. (2009). Analysis of pregnenolone and dehydroepiandrosterone in rodent brain: Cholesterol autoxidation is the key. *Journal of Lipid Research, 50*, 2430–2444.

Liere, P., Pianos, A., Eychenne, B., Cambourg, A., Liu, S., Griffiths, W., et al. (2004). Novel lipoidal derivatives of pregnenolone and dehydroepiandrosterone and absence of their sulfated counterparts in rodent brain. *Journal of Lipid Research, 45*, 2287–2302.

Litim, N., Morissette, M., & Di Paolo, T. (2016). Neuroactive gonadal drugs for neuroprotection in male and female models of Parkinson's disease. *Neuroscience and Biobehavioral Reviews, 67*, 79–88.

Liu, Z., & Chopp, M. (2016). Astrocytes, therapeutic targets for neuroprotection and neurorestoration in ischemic stroke. *Progress in Neurobiology, 144*, 103–120. https://doi.org/10.1016/j.pneurobio.2015.09.008.

Liu, T. C., Lin, C. H., Huang, C. Y., Ivy, J. L., & Kuo, C. H. (2013). Effect of acute DHEA administration on free testosterone in middle-aged and young men following high-intensity interval training. *European Journal of Applied Physiology, 113*, 1783–1792.

Lopez-Rodriguez, A. B., Acaz-Fonseca, E., Giatti, S., Caruso, D., Viveros, M. P., Melcangi, R. C., et al. (2015). Correlation of brain levels of progesterone and dehydroepiandrosterone with neurological recovery after traumatic brain injury in female mice. *Psychoneuroendocrinology, 56*, 1–11.

Lu, S. F., Mo, Q., Hu, S., Garippa, C., & Simon, N. G. (2003). Dehydroepiandrosterone upregulates neural androgen receptor level and transcriptional activity. *Journal of Neurobiology, 57*, 163–171.

Lür, G., Rákos, G., Juhász-Vedres, G., Farkas, T., Kis, Z., & Toldi, J. (2006). Effects of dehydroepiandrosterone sulfate on the evoked cortical activity of controls and of brain-injured rats. *Cellular and Molecular Neurobiology, 26*, 1505–1519.

Machado-Vieira, R., Manji, H. K., & Zarate, C. A. (2009). The role of the tripartite glutamatergic synapse in the pathophysiology and therapeutics of mood disorders. *The Neuroscientist, 15*, 525–539.

Maksay, G., Laube, B., & Betz, H. (2001). Subunit-specific modulation of glycine receptors by neurosteroids. *Neuropharmacology, 41*, 369–376.

Malik, A. S., Narayan, R. K., Wendling, W. W., Cole, R. W., Pashko, L. L., Schwartz, A. G., et al. (2003). A novel dehydroepiandrosterone analog improves functional recovery in a rat traumatic brain injury model. *Journal of Neurotrauma, 20*, 463–476.

Maninger, N., Wolkowitz, O. M., Reus, V. I., Epel, E. S., & Mellon, S. H. (2009). Neurobiological and neuropsychiatric effects of dehydroepiandrosterone (DHEA) and DHEA sulfate (DHEAS). *Frontiers in Neuroendocrinology, 30*, 65–91.

Mao, X., & Barger, S. W. (1998). Neuroprotection by dehydroepiandrosterone-sulfate: Role of an NFkappaB-like factor. *Neuroreport, 9*, 759–763.

McColl, B. W., Allan, S. M., & Rothwell, N. J. (2007). Systemic inflammation and stroke: Aetiology, pathology and targets for therapy. *Biochemical Society Transactions, 35*, 1163–1165.

Mellon, S. H. (2007). Neurosteroid regulation of central nervous system development. *Pharmacology & Therapeutics, 116*, 107–124.

Miller, W. L. (2013). Steroid hormone synthesis in mitochondria. *Molecular and Cellular Endocrinology, 379*, 62–73.

Miller, W. L., & Auchus, R. J. (2011). The molecular biology, biochemistry, and physiology of human steroidogenesis and its disorders. *Endocrine Reviews*, *32*, 81–151.
Milman, A., Zohar, O., Maayan, R., Weizman, R., & Pick, C. G. (2008). DHEAS repeated treatment improves cognitive and behavioral deficits after mild traumatic brain injury. *European Neuropsychopharmacology*, *18*, 181–187.
Mishra, M., Singh, R., Mukherjee, S., & Sharma, D. (2013). Dehydroepiandrosterone's anti-epileptic action in FeCl3-induced epileptogenesis involves upregulation of glutamate transporters. *Epilepsy Research*, *106*, 83–91.
Mission, J. P., Takahashi, T., & Caviness, V. S., Jr. (1991). Ontogeny of radial and other astroglial cells in murine cerebral cortex. *Glia*, *4*, 138–148.
Mo, Q., Lu, S. F., & Simon, N. G. (2006). Dehydroepiandrosterone and its metabolites: Differential effects on androgen receptor trafficking and transcriptional activity. *The Journal of Steroid Biochemistry and Molecular Biology*, *99*, 50–58.
Morest, D. K., & Silver, J. (2003). Precursors of neurons, neuroglia, and ependymal cells in the CNS: What are they? Where are they from? How do they get where they are going? *Glia*, *43*, 6–18.
Moriguchi, S., Shinoda, Y., Yamamoto, Y., Sasaki, Y., Miyajima, K., Tagashira, H., et al. (2013). Stimulation of the sigma-1 receptor by DHEA enhances synaptic efficacy and neurogenesis in the hippocampal dentate gyrus of olfactory bulbectomized mice. *PLoS One*, *8*, e60863.
Moriguchi, S., Yamamoto, Y., Ikuno, T., & Fukunaga, K. (2011). Sigma-1 receptor stimulation by dehydroepiandrosterone ameliorates cognitive impairment through activation of CaM kinase II, protein kinase C and extracellular signal-regulated kinase in olfactory bulbectomized mice. *Journal of Neurochemistry*, *117*, 879–891.
Muntwyler, R., & Bologa, L. (1989). In vitro hormonal regulation of astrocyte proliferation. *Schweizer Archiv für Neurologie und Psychiatrie*, *140*, 29–33.
Nieschlag, E., Loriaux, D. L., Ruder, H. J., Zucker, I. R., Kirschner, M. A., & Lipsett, M. B. (1973). The secretion of dehydroepiandrosterone and dehydroepiandrosterone sulphate in man. *The Journal of Endocrinology*, *57*, 123–134.
Nilupul Perera, M., Ma, H. K., Arakawa, S., Howells, D. W., Markus, R., Rowe, C. C., et al. (2006). Inflammation following stroke. *Journal of Clinical Neuroscience*, *13*, 1–8.
Panjari, M., & Davis, S. R. (2010). DHEA for postmenopausal women: A review of the evidence. *Maturitas*, *66*, 172–179.
Park-Chung, M., Malayev, A., Purdy, R. H., Gibbs, T. T., & Farb, D. H. (1999). Sulfated and unsulfated steroids modulate gamma-aminobutyric acidA receptor function through distinct sites. *Brain Research*, *830*, 72–87.
Parker, C. R., Jr. (1999). Dehydroepiandrosterone and dehydroepiandrosterone sulfate production in the human adrenal during development and aging. *Steroids*, *64*, 640–647.
Parpura, V., Heneka, M. T., Montana, V., Oliet, S. H., Schousboe, A., Haydon, P. G., et al. (2012). Glial cells in (patho)physiology. *Journal of Neurochemistry*, *121*, 4–27.
Patel, M. A., & Katyare, S. S. (2006). Treatment with dehydroepiandrosterone (DHEA) stimulates oxidative energy metabolism in the cerebral mitochondria. A comparative study of effects in old and young adult rats. *Neuroscience Letters*, *402*, 131–136.
Pearn, M. L., Niesman, I. R., Egawa, J., Sawada, A., Almenar-Queralt, A., Shah, S. B., et al. (2017). Pathophysiology associated with traumatic brain injury: Current treatments and potential novel therapeutics. *Cellular and Molecular Neurobiology*, *37*, 571–585. https://doi.org/10.1007/s10571-016-0400-1.
Pediaditakis, I., Iliopoulos, I., Theologidis, I., Delivanoglou, N., Margioris, A. N., Charalampopoulos, I., et al. (2015). Dehydroepiandrosterone: An ancestral ligand of neurotrophin receptors. *Endocrinology*, *156*, 16–23.
Pekny, M., Wilhelmsson, U., & Pekna, M. (2014). The dual role of astrocyte activation and reactive gliosis. *Neuroscience Letters*, *565*, 30–38.

Pellerin, L., Bouzier-Sore, A. K., Aubert, A., Serres, S., Merle, M., Costalat, R., et al. (2007). Activity-dependent regulation of energy metabolism by astrocytes: An update. *Glia, 55*, 1251–1262.

Pellerin, L., & Magistretti, P. J. (1994). Glutamate uptake into astrocytes stimulates aerobic glycolysis: A mechanism coupling neuronal activity to glucose utilization. *Proceedings of the National Academy of Sciences of the United States of America, 91*, 10625–10629.

Peng, L., Parpura, V., & Verkhratsky, A. (2014). EDITORIAL neuroglia as a central element of neurological diseases: An underappreciated target for therapeutic intervention. *Current Neuropharmacology, 12*, 303–307.

Pérez-Neri, I., Méndez-Sánchez, I., Montes, S., & Ríos, C. (2008). Acute dehydroepiandrosterone treatment exerts different effects on dopamine and serotonin turnover ratios in the rat corpus striatum and nucleus accumbens. *Progress in Neuropsychopharmacology & Biological Psychiatry, 32*, 1584–1589.

Prough, R. A., Clark, B. J., & Klinge, C. M. (2016). Novel mechanisms for DHEA action. *Journal of Molecular Endocrinology, 56*, R139–155.

Querfurth, H. W., & LaFerla, F. M. (2010). Alzheimer's disease. *The New England Journal of Medicine, 362*, 329–344.

Rahmani, A., Shoae-Hassani, A., Keyhanvar, P., Kheradmand, D., & Darbandi-Azar, A. (2013). Dehydroepiandrosterone stimulates nerve growth factor and brain derived neurotrophic factor in cortical neurons. *Advances in Pharmacological Sciences, 2013*, 506191.

Ramassamy, C., Averill, D., Beffert, U., Bastianetto, S., Theroux, L., Lussier-Cacan, S., et al. (1999). Oxidative damage and protection by antioxidants in the frontal cortex of Alzheimer's disease is related to the apolipoprotein E genotype. *Free Radical Biology & Medicine, 27*, 544–553.

Rammouz, G., Lecanu, L., & Papadopoulos, V. (2011). Oxidative stress-mediated brain dehydroepiandrosterone (DHEA) formation in Alzheimer's disease diagnosis. *Frontiers in Endocrinology, 2*, 69.

Raven, P. W., & Hinson, J. P. (2007). Dehydroepiandrosterone (DHEA) and the menopause: An update. *Menopause International, 13*, 75–78.

Ribeiro, M. F., & Garcia-Segura, L. M. (2002). Dehydroepiandrosterone regulates insulin-like growth factor-1 system in adult rat hypothalamus. *Endocrine, 17*, 129–134.

Ripp, S. L., Fitzpatrick, J. L., Peters, J. M., & Prough, R. A. (2002). Induction of CYP3A expression by dehydroepiandrosterone: Involvement of the pregnane X receptor. *Drug Metabolism and Disposition, 30*, 570–575.

Rodrigo, R., Fernández-Gajardo, R., Gutiérrez, R., Matamala, J. M., Carrasco, R., Miranda-Merchak, A., et al. (2013). Oxidative stress and pathophysiology of ischemic stroke: Novel therapeutic opportunities. *CNS & Neurological Disorders Drug Targets, 12*, 698–714.

Rone, M. B., Midzak, A. S., Issop, L., Rammouz, G., Jagannathan, S., Fan, J., et al. (2012). Identification of a dynamic mitochondrial protein complex driving cholesterol import, trafficking, and metabolism to steroid hormones. *Molecular Endocrinology, 26*, 1868–1882.

Rothstein, J. D., Dykes-Hoberg, M., Pardo, C. A., Bristol, L. A., Jin, L., Kuncl, R. W., et al. (1996). Knockout of glutamate transporters reveals a major role for astroglial transport in excitotoxicity and clearance of glutamate. *Neuron, 16*, 675–686.

Rouleau, P., Ung, R. V., Lapointe, N. P., & Guertin, P. A. (2007). Hormonal and immunological changes in mice after spinal cord injury. *Journal of Neurotrauma, 24*, 367–378.

Sachidanandan, D., & Bera, A. K. (2015). Inhibition of the GABAA receptor by sulfated neurosteroids: A mechanistic comparison study between pregnenolone sulfate and dehydroepiandrosterone sulfate. *Journal of Molecular Neuroscience, 56*, 868–877.

Schumacher, M., Guennoun, R., Mattern, C., Oudinet, J. P., Labombarda, F., De Nicola, A. F., et al. (2015). Analytical challenges for measuring steroid responses to stress, neurodegeneration and injury in the central nervous system. *Steroids*, *103*, 42–57.
Schumacher, M., Weill-Engerer, S., Liere, P., Robert, F., Franklin, R. J., Garcia-Segura, L. M., et al. (2003). Steroid hormones and neurosteroids in normal and pathological aging of the nervous system. *Progress in Neurobiology*, *71*, 3–29.
Shin, C. Y., Choi, J. W., Jang, E. S., Ju, C., Kim, W. K., Kim, H. C., et al. (2001). Dehydroepiandrosterone inhibits the death of immunostimulated rat C6 glioma cells deprived of glucose. *Brain Research*, *922*, 267–275.
Slemmer, J. E., Shacka, J. J., Sweeney, M. I., & Weber, J. T. (2008). Antioxidants and free radical scavengers for the treatment of stroke, traumatic brain injury and aging. *Current Medicinal Chemistry*, *15*, 404–414.
South, S. A., Deibler, G. E., Tzeng, S. F., Badache, A., Kirchner, M. G., Muja, N., et al. (2000). Myelin basic protein (MBP) and MBP peptides are mitogens for cultured astrocytes. *Glia*, *29*, 81–90.
Stárka, L., Dušková, M., & Hill, M. (2015). Dehydroepiandrosterone: A neuroactive steroid. *The Journal of Steroid Biochemistry and Molecular Biology*, *145*, 254–260.
Stobart, J. L., & Anderson, C. M. (2013). Multifunctional role of astrocytes as gatekeepers of neuronal energy supply. *Frontiers in Cellular Neuroscience*, *7*, 38.
Svob Štrac, D., Jazvinšćak Jembrek, M., Erhardt, J., Mirković Kos, K., & Peričić, D. (2012). Modulation of recombinant GABA(A) receptors by neurosteroid dehydroepiandrosterone sulfate. *Pharmacology*, *89*, 163–171.
Téllez, N., Comabella, M., Julià, E., Río, J., Tintoré, M., Brieva, L., et al. (2006). Fatigue in progressive multiple sclerosis is associated with low levels of dehydroepiandrosterone. *Multiple Sclerosis*, *12*, 487–494.
Túnez, I., Muñoz, M. C., & Montilla, P. (2005). Treatment with dehydroepiandrosterone prevents oxidative stress induced by 3-nitropropionic acid in synaptosomes. *Pharmacology*, *74*, 113–118.
van Weerden, W. M., Bierings, H. G., van Steenbrugge, G. J., de Jong, F. H., & Schröder, F. H. (1992). Adrenal glands of mouse and rat do not synthesize androgens. *Life Sciences*, *50*, 857–861.
Veiga, S., Garcia-Segura, L. M., & Azcoitia, I. (2003). Neuroprotection by the steroids pregnenolone and dehydroepiandrosterone is mediated by the enzyme aromatase. *Journal of Neurobiology*, *56*, 398–406.
Verkhratsky, A., Steardo, L., Parpura, V., & Montana, V. (2016). Translational potential of astrocytes in brain disorders. *Progress in Neurobiology*, *144*, 188–205. https://doi.org/10.1016/j.pneurobio.2015.09.003.
Vieira-Marques, C., Arbo, B. D., Ruiz-Palmero, I., Ortiz-Rodriguez, A., Ghorbanpoor, S., Kucharski, L. C., et al. (2016). Dehydroepiandrosterone protects male and female hippocampal neurons and neuroblastoma cells from glucose deprivation. *Brain Research*, *1644*, 176–182.
Villareal, D. T., & Holloszy, J. O. (2006). DHEA enhances effects of weight training on muscle mass and strength in elderly women and men. *American Journal of Physiology. Endocrinology and Metabolism*, *291*, E1003–1008.
Voskuhl, R. R., Peterson, R. S., Song, B., Ao, Y., Morales, L. B., Tiwari-Woodruff, S., et al. (2009). Reactive astrocytes form scar-like perivascular barriers to leukocytes during adaptive immune inflammation of the CNS. *The Journal of Neuroscience*, *29*, 11511–11522.
Walz, W. (1989). Role of glial cells in the regulation of the brain ion microenvironment. *Progress in Neurobiology*, *33*, 309–333.

Wang, M. J., Huang, H. M., Chen, H. L., Kuo, J. S., & Jeng, K. C. (2001). Dehydroepiandrosterone inhibits lipopolysaccharide-induced nitric oxide production in BV-2 microglia. *Journal of Neurochemistry*, *77*, 830–838.

Wanner, I. B., Anderson, M. A., Song, B., Levine, J., Fernandez, A., Gray-Thompson, Z., et al. (2013). Glial scar borders are formed by newly proliferated, elongated astrocytes that interact to corral inflammatory and fibrotic cells via STAT3-dependent mechanisms after spinal cord injury. *The Journal of Neuroscience*, *33*, 12870–12886.

Waters, S. L., Miller, G. W., Aleo, M. D., & Schnellmann, R. G. (1997). Neurosteroid inhibition of cell death. *The American Journal of Physiology*, *273*, F869–876.

Webb, S. J., Geoghegan, T. E., Prough, R. A., & Michael Miller, K. K. (2006). The biological actions of dehydroepiandrosterone involves multiple receptors. *Drug Metabolism Reviews*, *38*, 89–116.

Weiss, E. P., Villareal, D. T., Ehsani, A. A., Fontana, L., & Holloszy, J. O. (2012). Dehydroepiandrosterone replacement therapy in older adults improves indices of arterial stiffness. *Aging Cell*, *11*, 876–884.

Weiss, E. P., Villareal, D. T., Fontana, L., Han, D. H., & Holloszy, J. O. (2011). Dehydroepiandrosterone (DHEA) replacement decreases insulin resistance and lowers inflammatory cytokines in aging humans. *Aging*, *3*, 533–542.

Wierońska, J. M., & Pilc, A. (2009). Metabotropic glutamate receptors in the tripartite synapse as a target for new psychotropic drugs. *Neurochemistry International*, *55*, 85–97.

Wolkowitz, O. M., Kramer, J. H., Reus, V. I., Costa, M. M., Yaffe, K., Walton, P., et al. (2003). DHEA treatment of Alzheimer's disease: A randomized, double-blind, placebo-controlled study. *Neurology*, *60*, 1071–1076.

Wu, Z., Li, L., Zheng, L. T., Xu, Z., Guo, L., & Zhen, X. (2015). Allosteric modulation of sigma-1 receptors by SKF83959 inhibits microglia-mediated inflammation. *Journal of Neurochemistry*, *134*, 904–914.

Yabuki, Y., Shinoda, Y., Izumi, H., Ikuno, T., Shioda, N., & Fukunaga, K. (2015). Dehydroepiandrosterone administration improves memory deficits following transient brain ischemia through sigma-1 receptor stimulation. *Brain Research*, *1622*, 102–113.

Zou, J., Wang, Y. X., Dou, F. F., Lü, H. Z., Ma, Z. W., Lu, P. H., et al. (2010). Glutamine synthetase down-regulation reduces astrocyte protection against glutamate excitotoxicity to neurons. *Neurochemistry International*, *56*, 577–584.

Zup, S. L., Edwards, N. S., & McCarthy, M. M. (2014). Sex- and age-dependent effects of androgens on glutamate-induced cell death and intracellular calcium regulation in the developing hippocampus. *Neuroscience*, *281*, 77–87.

Zwain, I. H., & Yen, S. S. (1999). Neurosteroidogenesis in astrocytes, oligodendrocytes, and neurons of cerebral cortex of rat brain. *Endocrinology*, *140*, 3843–3852.

CHAPTER SEVEN

The Role of Dehydroepiandrosterone (DHEA) in Skeletal Muscle

Koji Sato*, Motoyuki Iemitsu†,1

*Graduate School of Human Development and Environment, Kobe University, Kobe, Japan
†Faculty of Sport and Health Science, Ritsumeikan University, Kusatsu, Japan
[1]Corresponding author: e-mail address: iemitsu@fc.ritsumei.ac.jp

Contents

Abstract

Dehydroepiandrosterone (DHEA) is a precursor of sex steroid hormones and is converted to testosterone and estradiol. Normally, androgens and estrogens produced adrenal cortex, testis, and ovary; however, recent studies revealed androgens and estrogens are synthesized by peripheral tissues such as brain, skin, liver, kidney, bone, etc. We found skeletal muscles are also capable of synthesizing androgens and estrogens from DHEA. Circulating DHEA provides substrates required for conversion into potent androgens and estrogens in peripheral tissues.

Sex steroid hormone administration has important roles: one is that the enhancement of protein synthesis and anabolism, resulting in muscle growth and increased muscle strength. The other is improvement of hyperglycemia through the activation of glucose signaling pathway in skeletal muscle as well as acceleration of muscle lipid metabolism that increase peroxisome proliferator-activated receptor alpha (PPARα) and PPAR delta (PPARδ). We introduce the effect of DHEA and sex steroid hormones administration on muscle glucose and lipid metabolisms as well as the effect of sex steroid hormone on the muscle hypertrophy.

Vitamins and Hormones, Volume 108
ISSN 0083-6729
https://doi.org/10.1016/bs.vh.2018.03.002

1. INTRODUCTION

Sex steroid hormones are generally secreted from the testis, ovary, and adrenal cortex and regulate diverse physiological processes of various target tissues, including those of the reproductive tissues, bone, liver, cardiovascular tissues, brain, and skeletal muscles (Aizawa et al., 2007; Labrie, Simard, Luu-The, Belanger, & Pelletier, 1992, Labrie, Sugimoto, et al., 1992; Martel, Melner, Gagne, Simard, & Labrie, 1994; Milewich et al., 1993; Pelletier et al., 1992; Sato, Iemitsu, Aizawa, & Ajisaka, 2008). Dehydroepiandrosterone (DHEA) and its sulfate-bound form (DHEA-S), are mainly secreted by the adrenal gland, ovary, and testis. These are the most abundant hormones in humans and other mammals and are important precursors of sex steroid hormones. Although ovaries and testis are the main source of androgens and estrogens, active sex steroids are in large part or entirely synthesized locally in peripheral tissues in human (Labrie et al., 1995; Labrie, Simard, et al., 1992; Labrie, Sugimoto, et al., 1992; Luu-The, Zhang, Poirier, & Labrie, 1995). Adrenals secrete large amount of DHEA, which are converted into androstenedione (4-dione) then into potent androgens and/or estrogens in peripheral tissues (Labrie, Luu-The, Labrie, & Simard, 2001; Labrie et al., 1997). In peripheral tissues, the biosynthesis of active androgens and estrogens is synthesized from DHEA by three steroidogenesis-related enzymes. Testosterone is synthesized by 3β-hydroxysteroid dehydrogenase (3β-HSD) and 17β-HSD, and in turn, estrogens are converted from androgens catalyzed by aromatase cytochrome P450 (P450arom; Fig. 1). The physiological balance between sex steroid hormones is largely controlled by aromatase (Conley et al., 2001). There

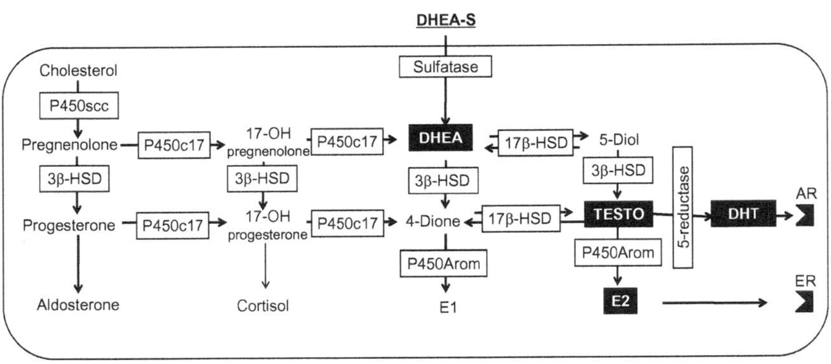

Fig. 1 The schema of pathway in steroidogenesis.

are several types of 17β-HSDs, which are each products of a distinct gene. Both androgenic and estrogenic 17β-HSD activities were found in all 25 rhesus monkey and 15 human peripheral intracrine tissues examined; so far, 10 different 17β-HSDs have been cloned, 1–5, 7, 8, 10, 11, and 12 (Lukacik et al., 2007). The 17β-HSDs differ in tissue distribution, catalytic preferences, substrate specificity, subcellular localization, and mechanisms of regulation. Therefore, 17β-HSDs gene family are responsible for the interconversion of DHEA and 5-diol, 4-dione and testosterone, estrone, and estradiol (E_2), additionally, androstanedione and 5α-dihydrotestosterone (DHT), androsterone and androstane 3α, 17β-diol, and epiandrosterone and androstane 3β, 17β-diol. The enzymatic activity is required for the synthesis of all active androgens and estrogens, and also their inactivation (Labrie et al., 1995).

3β-HSD is the membrane-bound enzyme, which catalyzes an essential step in the transformation of all 5-pregnen-3β-ol and 5-androsten-3β-ol steroids into progesterone, and precursors of all androgens, estrogens, glucocorticoids, and mineralocorticoids (Labrie et al., 2001). Additionally, 3β-HSD is responsible for the interconversion of 3β-hydroxy- and 3-keto-5α-androstane steroids. Thus, the role of 3β-HSD is catalyzing the first step in the intracrine transformation of DHEA into 4-dione, the precursor of both androgens and estrogens (Labrie et al., 2001). Furthermore, the enzyme 5α-reductase has a role of transformation of testosterone into DHT, the most potent androgen, which is responsible for the differentiation of the male prostate and virilization at puberty. The type of 5α-reductase in human is identified as type 1 and 2. The structure of type 1 5α-reductase gene has been elucidated (Jenkins et al., 1991) as well as it is relatively intensive to the inhibitor finasteride (Andersson, Bergman, Jenkins, & Russel, 1991).

Estrogens are produced by the irreversible aromatization of androgens by P450 aromatase, they work by their nuclear and cell surface receptors (Carreau et al., 2003, 2007; O'Donnell, Robertson, Jones, & Simpson, 2001). P450 aromatase is an enzymatic complex, composed of two proteins, a ubiquitous NADPH–cytochrome P450 reductase and a cytochrome P450 aromatase, which contains the steroid-binding site (Carreau, Bouraima-Lelong, & Delalande, 2011). P450 aromatase is localized in the cellular endoplasmic reticulum of different areas of the body, including the testis (Carreau et al., 2001). In humans, P450 aromatase is the product of a single gene and is encoded by the CYP19 gene, which maps to chromosome 15q21.2 in humans (Carreau et al., 2011). The structure and hormonal regulation of CYP19 are complex: the gene spans 123 kb, with a coding region of 30 kb comprising nine translated exons. In humans, aromatase is expressed in both the granulosa and luteal cells of the ovary, and also in various extraglandular sites including the placenta, brain, bone, testis, and adipose tissue.

As we stated earlier, all four genes are found not only in classic steroidogenic tissues (placenta, adrenal cortex, ovary, and testis) but also in several peripheral tissues, such as adipose tissue, breast, lung endometrium, skin, prostate, liver, kidney, brain, as well as in skeletal muscles (Aizawa et al., 2007; Labrie, Simard, et al., 1992; Labrie, Sugimoto, et al., 1992; Martel et al., 1994; Milewich et al., 1993; Pelletier et al., 1992; Sato et al., 2008). This chapter provides the information for the role of DHEA for glucose and lipid metabolism in skeletal muscle as well as introduces the effect of the elevation of DHEA level on muscle hypertrophy especially for type 2 diabetic and older aged adults.

2. STEROIDOGENESIS AND ITS FUNCTIONS IN SKELETAL MUSCLE

The skeletal muscle is a sex steroid-sensitive tissue, and both androgen and estrogen receptors are expressed in skeletal muscles (Glenmark et al., 2004). Synthesis of testosterone is regulated by P450 side chain clevage, 17α-hydroxylase cytochrome P450, 17β-HSD, and 3β-HSD enzymes (Labrie et al., 2005). In our previous studies, steroidogenesis-related enzymes such as 17β-HSD, 3β-HSD, and P450arom were expressed in skeletal muscle cells, and DHEA administration increased muscle testosterone concentration (Aizawa et al., 2007); as well as in our early study, we revealed that 5α-reductase was expressed in skeletal muscle cells, and testosterone administration increased DHT concentration. Thus, skeletal muscles were capable of synthesizing and metabolizing sex steroid hormones from DHEA (Aizawa et al., 2007; Sato et al., 2008; Fig. 2). Generally, the function of testosterone and estradiol is related to growth, strength, and antioxidant action in muscles (Ihemelandu, 1980) as well as the administration of testosterone induces increase in muscle strength through acceleration of protein synthesis (Urban et al., 1995). Additionally, a long-term DHEA administration increases muscle strength and mass in humans, whereas a deficiency of sex steroid hormones leads to retardation in skeletal muscle development (Morales, Haubrich, Hwang, Asakura, & Yen, 1998). Other experiments show that lipid metabolic enzymes in muscle are enhanced in obese rats by DHEA administration (Horii, Sato, Mesaki, & Iemitsu, 2016a, 2016b), also DHEA administration improves hyperglycemia through enhances glucose metabolism signaling pathway; however, administration of DHT synthesizing blocker, the activations of glucose metabolism signaling pathway are blocked (Sato, Iemitsu, Aizawa, & Ajisaka, 2009;

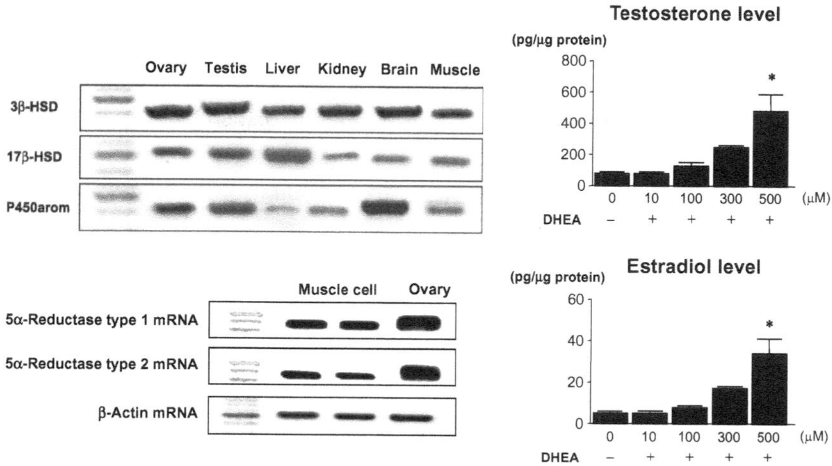

Fig. 2 mRNA expressions of 3β-HSD, 17β-HSD, P450arom, 5α-reductase type 1 and 2, and testosterone and estradiol concentrations with DHEA addition to culture skeletal muscle cell. *$P < 0.01$ compared with 0 μM DHEA addition.

Sato, Iemitsu, Aizawa, Mesaki, & Fujita, 2011), so metabolite active androgen, DHT, would influence skeletal muscle metabolisms. Therefore, DHEA administration induces acceleration of steroidogenesis in skeletal muscle and affects not only muscle strength and mass but also glucose and lipid metabolism in skeletal muscles.

3. DHEA AND GLUCOSE METABOLISM IN SKELETAL MUSCLE

Individuals with obese and type 2 diabetes have lower serum DHEA levels (Yamaguchi et al., 1998). Sex steroid hormones normally decrease with aging, and it has been reported that a deficiency of sex steroid hormones in both men and women has a high risk for developing obesity and type 2 diabetes (Ding et al., 2007). Thus, lower sex steroid hormone concentrations may be related to the pathogenesis of obesity and type 2 diabetes. In the 1980s, Coleman, Leiter, and Schwizer (1982) and Coleman, Schwizer, and Leiter (1984) reported DHEA administration induced the remission of hyperglycemia and was strongly correlated with improved insulin sensitivity in db/db mouse. Moreover, DHEA administration also has the role to protect against the development of visceral obesity and muscle insulin resistance in rats with high-fat fed (Hansen, Han, Nolte, Chen, & Holloszy, 1997). According to Kang et al. (2004), the principal steroidal androgens, testosterone and DHT are thought to predominantly mediate the biological effect of DHEA as well as DHT activated the insulin signal pathway such as protein kinase B (Akt) via stimulation of phosphatidylinositol 3-kinase (PI3-kinase) in osteoblast cells. However, the mechanism by which sex steroid hormone administration improves hyperglycemia has not been identified. In our earlier study (Sato et al., 2008), DHEA or testosterone administration activated Akt and PKC ζ/λ phosphorylations, and it results in the enhancement of glucose transporter type 4 (GLUT4) translocation in muscle cell. Moreover, DHEA and testosterone administration-induced increases in both Akt and PKC ζ/λ phosphorylations and GLUT4 translocation were partly blocked by DHT synthesizing blocker administration, also acute DHEA injection (2 mg/kg body weight) enhanced the glucose metabolism signaling pathway (Akt/PKC ζ/λ-GLUT4) in skeletal muscle, and it induced the improvement of hyperglycemia in streptozotocin (STZ)-induced type 1 diabetes rat. In the study, acute DHEA injection increased muscle GLUT4 content approximately 2.7-fold compared to STZ rats without DHEA injection and decreased blood glucose level from 22.8 mmol/L at baseline to 16.7 mmol/L at 90 min after DHEA injection (Sato et al., 2009; Fig. 3). Although DHEA has similar effect as insulin, acute DHEA injection (2 mg/kg body weight)

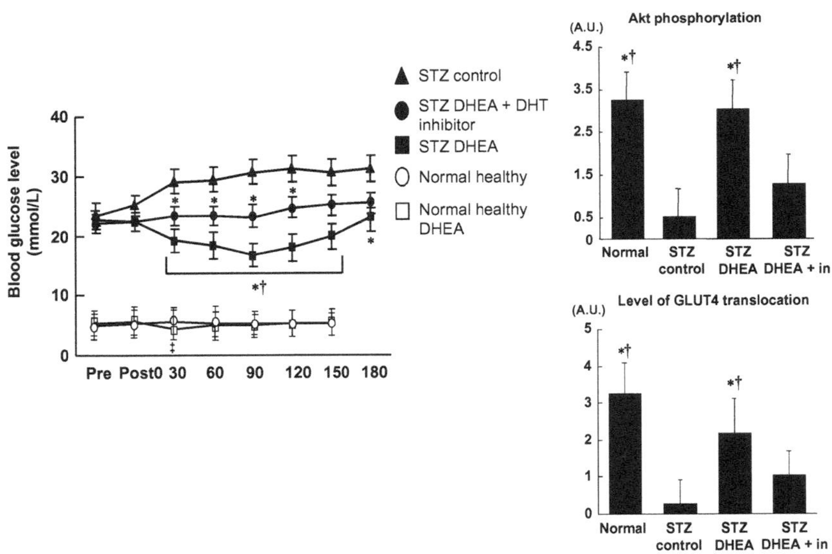

Fig. 3 Effect of dehydroepiandrosterone (DHEA) injection on blood glucose levels in normal healthy rats and type 1 diabetes mellitus model rats. Akt phosphorylation level and glucose transporter-4 (GLUT4) translocation in the gastrocnemius muscle. Data are means ± SE. *$P < 0.01$ compared with the control; †$P < 0.05$ compared with DHT inhibitor injection. ‡$P < 0.05$ compared with normal-control.

had a weaker effect on reducing blood glucose levels than on insulin injection; however, the study revealed that DHEA injection decreased blood glucose level, and it was achieved without changing blood insulin levels. In addition, the DHEA-induced Akt/PKC ζ/λ-GLUT4 activations were blocked by DHT synthesizing inhibitor. Therefore, DHEA effects might be achieved partly by DHT, which converted from DHEA in the skeletal muscle.

Several studies reported that the effect of DHEA administration for patients with obese and type 2 diabetes. In our previous study that administered DHEA for 6 weeks in high-sucrose hyperglycemic obese rats, induced significant increase in Akt/PKC ζ/λ-GLUT4 signaling activations and cleared the negative correlation between muscle DHEA and DHT levels and fasting glucose levels (Sato et al., 2011; Fig. 4), so DHEA administration-induced increase in sex steroid hormones improved higher fasting blood glucose level

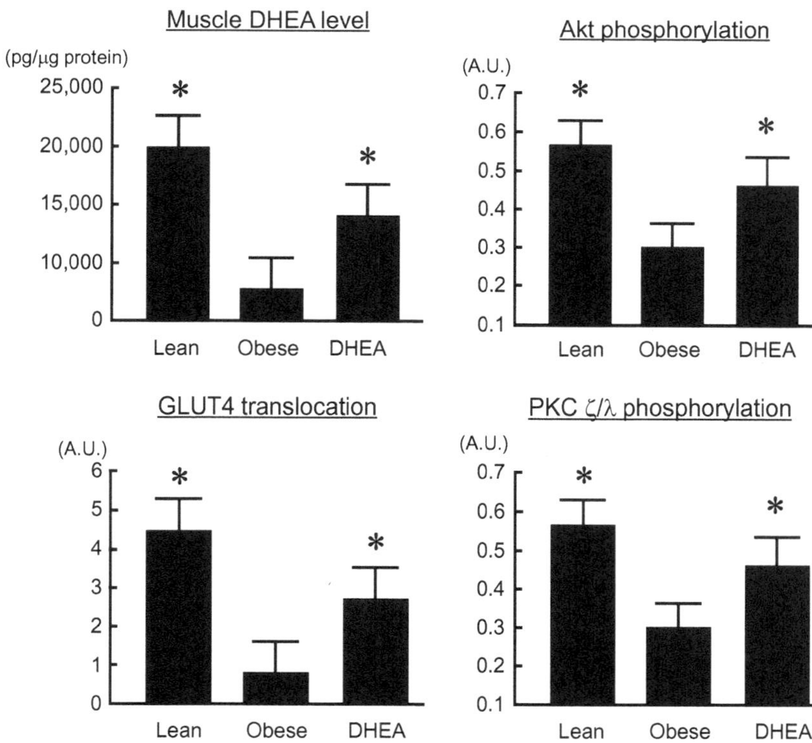

Fig. 4 Effect of 6-week DHEA administration in high-sucrose-induced obese rats on muscle DHEA levels and glucose metabolism signaling pathway (Akt/PKC ζ/λ-GLUT4). *$P < 0.01$ compared with the obese control.

in high-sucrose-induced obese rats. Moreover, chronic DHEA administration also improved quantitative insulin sensitivity check index (QUICKI), and it was positively correlated with muscle sex steroid hormones concentrations (Sato et al., 2012). Thus, chronic DHEA administration improved not only fasting blood glucose level but also insulin resistance in high-sucrose obese rats. In short-term (2 weeks) DHEA administration studies, DHEA induced an acute decrease in blood glucose levels in diabetic mice (Coleman et al., 1982, 1984) and enhanced the activities of enzyme-related hepatic glucose metabolism in diet-induced or Zucker obese rats (Mohan & Cleary, 1988; Mohan, Ihnen, Levin, & Cleary, 1990). In addition, short-term DHEA administration demonstrated that DHEA enhanced glucose metabolism such as PI3-kinase and Akt phosphorylation in liver and adipose tissues (Ishizuka et al., 2007; Mohan et al., 1990). Taken together, DHEA administration increased lower sex steroid hormone levels in obesity and type 2 diabetes model rats, and enhanced glucose metabolism signaling pathway in skeletal muscle.

4. DHEA AND LIPID METABOLISM IN SKELETAL MUSCLE

In human studies, Villareal and Holloszy (2004) reported that 6 months DHEA treatment for elderly men and women, DHEA administration induced significant decreases in both visceral and subcutaneous fat, and DHEA administration resulted in a significant improvement in insulin action that correlated with the reduction in visceral fat. The previous epidemiologic studies of the relationship between DHEA and abdominal fat have been conflicting. In another human studies by Barrett-Connor and Ferrara (1996), in postmenopausal women, DHEA-S levels were positively correlated with waist–hip ratio, and authors conclude that DHEA does not protect against obesity. In contrast, Haffner, Valdez, Mykkanen, Stern, and Katz (1994), DHEA-S level is significantly inversely related to abdominal obesity and insulin concentration in middle-aged men. This discrepancy probably explained by the difference in DHEA-S levels. The study by Barrett-Connor and Ferrara, the women with the highest quartile of waist–hip ratio had about 420 ng/mL in a mean serum DHEA-S level. Therefore, DHEA administration induced increase in circulating DHEA-S and DHEA levels and might cause significant decrease in visceral and subcutaneous fat in human.

In people with obesity, adiposity is caused the impairment of fatty acid metabolism in skeletal muscles (Houmard, 2008). To activate proliferator-activated receptor alpha (PPARα) and PPAR delta (PPARδ), stimulate mitochondrial enzyme expressions, which involve fat metabolism and synthesis (Lefebvre, Chinetti, Fruchart, & Steals, 2006; Vega, Huss, & Kelly, 2000; Wang et al., 2003). PPARα and PPARδ cooperate with PGC-1α and retinoid X receptor, and promote transcription of nuclear genes encoding mitochondrial fatty acid oxidation enzymes (Vega et al., 2000; Wang et al., 2003). Therefore, the enhancement of PPARα and PPARδ activation in skeletal muscle would suppress fat accumulation. In previous studies reported that DHEA administration reduced total and visceral fat accumulation in high fat/sucrose diet-induced (Corona et al., 2013; Hansen et al., 1997; Mohan et al., 1990; Sato et al., 2011) and genetically obese rats (Kimura et al., 1998). The mechanism which DHEA administration induced decrease in total and visceral fat accumulation is upregulation of adiponectin gene expression through activation of PPAR gamma (PPARγ) and enhancement of adiponectin secretion in adipose tissue or adipocytes (Karbowska & Kochan, 2005, 2011; Perez-de-Heredia et al., 2008). Adiponectin binds to adiponectin receptor (AdipoR1) in skeletal muscle, then stimulates fatty acid oxidation in muscle cells, which activates PPARs and PGC-1α via 5′-AMP-activated protein kinase (AMPK) phosphorylation (Barnea, Madar, & Froy, 2010; Okada-Iwabu et al., 2013; Yoon et al., 2006) as well as DHEA-induced increase in PGC-1α expression and AMPK phosphorylation has been shown in C2C12 myotubes (Yokokawa et al., 2015). In our recent study, the molecular mechanisms of the long-term DHEA-induced activation in muscle PPAR levels in obesity have been cleared. Long-term (6 weeks) DHEA administration in high-sucrose fed-induced obese rats significantly increased PPARα, PPARδ, and PGC-1α. Moreover, circulating DHEA level was significantly correlated with total adiponectin level and the levels of PPARα, PPARδ, and PGC-1α protein expression (Horii et al., 2016a, 2016b; Fig. 5). In this study, DHEA

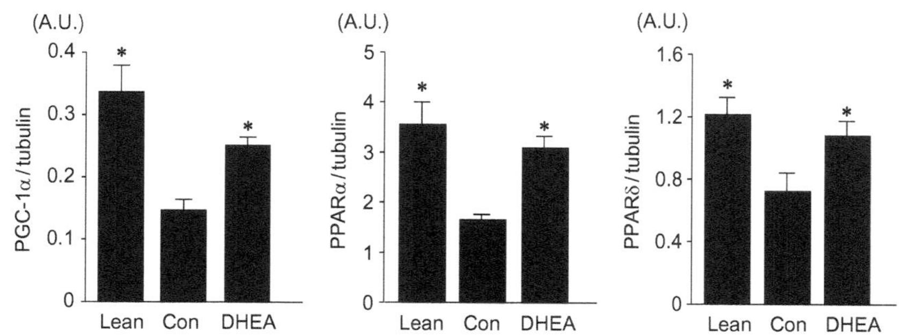

Fig. 5 Effects of 6 weeks of DHEA administration on PGC-1α, PPARα, and PPARδ protein expressions in high-sucrose-induced obesity rats. *$P < 0.01$ compared with obese control.

administration in obese rats also reduced epididymal fat mass, and it caused weight loss. The other study revealed that 17 days of DHEA administration ameliorates the decrease in PGC-1α and β3 adrenergic receptor gene expression in adipose tissue in obese rats, which leaded to weight loss in obese (Ryu et al., 2003). DHEA administration for 2 weeks in rats also activated PPARα gene expression, and it induced acceleration of fatty acid metabolism in adipose tissue (Kochan & Karbowska, 2004). Thus, DHEA administration may promote fatty acid utilization in skeletal muscle and lipolysis in adipose tissue that resulting in a reduction in visceral fat accumulation. In addition, long-term DHEA administration induced the enhancement of adiponectin secretion and would be associated with transcriptional regulation of lipid metabolic enzymes through activation of PPARα and PPARδ.

5. DHEA AND MUSCLE HYPERTROPHY

The secretion of DHEA and DHEA-S increases during childhood, which around 6–8 years, and maximal levels of circulating DHEA and DHEA-S are peaked between the ages of 20 and 30 years. Thereafter, circulating DHEA and DHEA-S levels remarkably decrease (Belanger et al., 1994; Migeon, Keller, Lawrence, & Shepart, 1957; Orentreich, Brind, Rizer, & Vogelman, 1984). Circulating DHEA-S levels are at approximately 20% of peak values at 70 years of age, while it can decrease by 95% until the age of 85–90 years. The 70%–95% reduction in the formation of DHEA by the adrenals during aging results in a remarkable reduction in the formation of androgens and estrogens in peripheral target tissues, which may be involved in the pathogenesis of age-related diseases such as type 2 diabetes (Schwartz, Pashko, & Whitcomb, 1986) and obesity (MacEwen & Kurzman, 1991). In fact, our latest study revealed that protein expression of steroidogenic enzymes such as 3β-HSD, 17β-HSD, 5α-reductase, and androgen receptor in skeletal muscle were lower in older men compared with younger counterparts. In addition, both serum and muscle DHEA, free testosterone, and DHT levels were also approximately 30%–50% lower in older men compared to young (Sato et al., 2014; Fig. 6). In previous studies, testosterone administration induced increase in muscle strength through an acceleration of protein synthesis in skeletal muscle (Urban et al., 1995), and deficiency of testosterone levels leads to suppress muscle development (Morales et al., 1998). Moreover, long-term DHEA administration for humans induced increase in muscle mass (Morales et al., 1998). Additionally, a significant correlation was found between serum DHEA levels and muscle force per cross-sectional area in

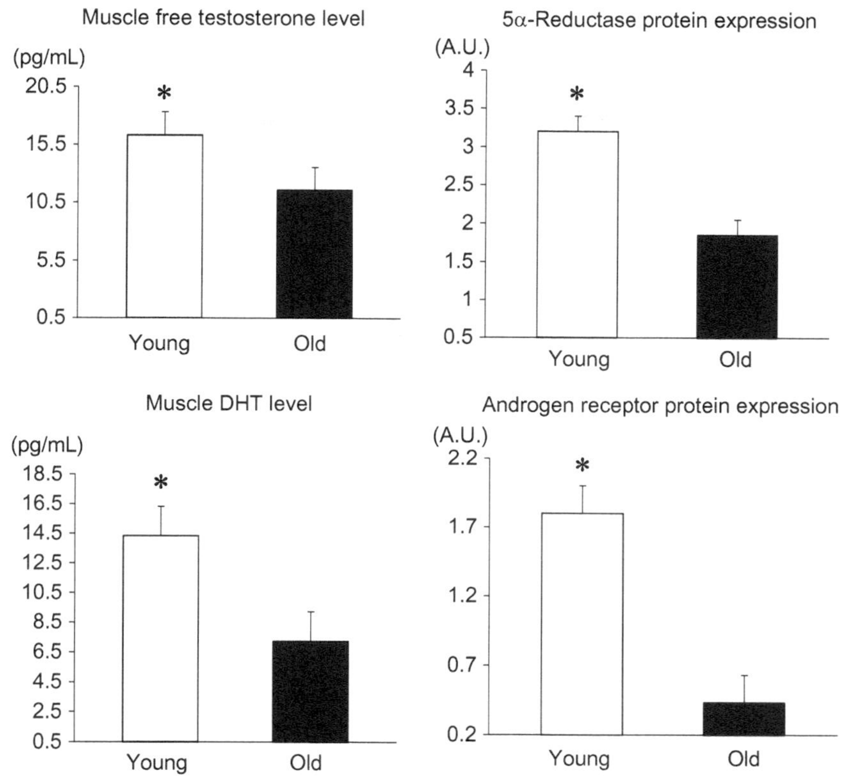

Fig. 6 Muscle free testosterone and DHT concentrations and 5α-reductase and androgen receptor protein expression in comparison between young and old in vastus lateralis muscle. $*P < 0.01$ compared with old subjects.

postmenopausal women (Pöllänen et al., 2010). On the contrast, Villareal and Holloszy (2006) revealed that 6 months DHEA administration did not significantly increase strength or thigh muscle volume in elderly women and men; however, DHEA administration potentiated the effect of 4 months of weightlifting training (resistance training) on muscle strength, evaluated by means of one-repetition maximum measurement, and on thigh muscle volume, measured by magnetic resonance imaging (Villareal & Holloszy, 2006), and 12 weeks resistance training in older men restores muscle DHEA, free testosterone, and DHT level, also increased steroidogenesis-related enzymes such as 3β-HSD, 17β-HSD, 5α-reductase protein expressions (Sato et al., 2014). In younger subjects, acute resistance exercise did not change steroidogenesis-related enzymes. Although sex steroid hormone administration increased muscle mass and strength in elderly subjects, it is only effective in individuals with lower sex steroid hormone as well as it would be difficult to improve muscle mass and strength only by DHEA treatment, may need to combine DHEA administration and resistance exercise to gain muscle mass and strength more effectively.

Patients with obesity and type 2 diabetes also have lower sex steroid hormones, and patients with bad glycemic control lead to muscle atrophy (Bianchi & Volpato, 2016). In fact, type 2 diabetic rats showed significant decrease in muscle DHEA, free testosterone, DHT levels, and protein expression of steroidogenic enzymes with loss of skeletal muscle mass and hyperglycemia. However, 8 weeks resistance training in type 2 diabetes model rats significantly increased the levels of muscle sex steroid hormones and steroidogenic enzyme protein expressions with a concomitant increase in skeletal muscle mass. However, resistance training-induced improvements of muscle mass were suppressed by chronic inhibition of 5α-reductase synthesizing. Furthermore, muscle DHT levels were significantly correlated with muscle mass (Horii et al., 2016a, 2016b). Thus, resistance training-induced elevations in muscle DHT may necessary to induce muscle hypertrophy in type 2 diabetic rats.

6. SUMMARY

- Ovaries and testis are the main source of androgens and estrogens; active sex steroids are in large part or entirely synthesized locally in peripheral tissues even in skeletal muscle.
- DHEA administration for 6 weeks in obesity rats induced significant increase in Akt/PKC ζ/λ-GLUT4 signaling activations in skeletal muscle, and it causes decrease in blood glucose level.
- DHEA administration in obesity rats induced increase in PPARα, PPARδ, and PGC-1α, and circulating DHEA level was correlated with total adiponectin level and the levels of PPARα, PPARδ, and PGC-1α protein expression. Thus, DHEA administration-induced increase in DHEA has a role of protecting against fat accumulation.
- Sex steroid administration increased muscle mass and strength in elderly subjects, and it is only effective in individuals with lower sex steroid hormone levels, but the combination between DHEA administration and resistance exercise may effectively gain the muscle mass and strength.
- 12 weeks resistance training in older men restores muscle DHEA, free testosterone and DHT level, also increased steroidogenesis-related enzymes such as 3β-HSD, 17β-HSD, 5α-reductase protein expressions.

- 8 weeks resistance training in type 2 diabetes model rats significantly increased the levels of muscle sex steroid hormones and steroidogenic enzyme protein expressions with a concomitant increase in skeletal muscle mass.

7. CONCLUSION AND FUTURE DIRECTIONS

DHEA and its sulfate are secretory products of adrenal glands. These are the most abundant hormones in the systemic circulation of humans, convertible into androgens and estrogens in the peripheral tissues. Impairment of steroidogenesis and reduced production, it causes frequent diseases with aging, and it should be critical to many aspects of health. In this chapter, we reviewed that DHEA administration improved fasting blood glucose level and hyperglycemia both obesity and type 2 diabetes model rats through activation of glucose metabolism-related signaling pathway, and also that administration increased muscle lipid metabolism in obese rats. Furthermore, DHEA or testosterone administration enhanced muscle protein synthesis and induced increase in muscle mass and strength in both elderly human and type 2 diabetes rats. Although DHEA administration-induced increase in sex steroid hormones induced various beneficial effects, only DHEA administration may be limited to gain beneficial effects for individuals with lower sex steroid hormones level such as elderly people, obese, and type 2 diabetic patients. Our previous studies revealed that the combination of DHEA administration and exercise training dramatically increased both circulating and muscle sex steroid hormones, and more successfully improved insulin resistance and increased muscle mass and strength. Therefore, DHEA administration combined with exercise training may be one of the new therapeutic candidates for treatment of obesity, type 2 diabetes, and sarcopenia. However, further studies should be needed which focuses on the sex differences as administration of sex steroid hormone and any other functions of sex steroid hormone in different tissues.

REFERENCES

Aizawa, K., Iemitsu, M., Maeda, S., Jesmin, S., Otsuki, T., Mowa, C. N., et al. (2007). Expression of steroidogenic enzymes and synthesis of sex steroid hormones from DHEA in skeletal muscle of rats. *American Journal of Physiology. Endocrinology and Metabolism*, *292*, E577–E584.

Andersson, S., Bergman, D. M., Jenkins, E. P., & Russel, D. W. (1991). Deletion of steroid 5α-reductase 2 gene in male pseudophermaphroditism. *Nature*, *354*, 159–161.

Barnea, M., Madar, Z., & Froy, O. (2010). High-fat diet followed by fasting disrupts circadian expression of adiponectin signaling pathway in muscle and adipose tissue. *Obesity*, *18*, 230–238.

Barrett-Connor, E., & Ferrara, A. (1996). Dehydroepiandrosterone, dehydroepiandrosterone sulfate, obesity, waist-hip ratio, and noninsulin-dependent diabetes in postmenopausal women: The Rancho Bernardo study. *The Journal of Clinical Endocrinology and Metabolism*, *81*, 59–64.

Belanger, A., Candas, B., Dupont, A., Cusan, L., Diamond, P., Gomez, J. L., et al. (1994). Changes in serum concentrations of conjugated and unconjugated steroids in 40- to 80-year-old men. *The Journal of Clinical Endocrinology and Metabolism*, *79*, 1086–1090.

Bianchi, L., & Volpato, S. (2016). Muscle dysfunction in type 2 diabetes: A major threat to patient's mobility and independence. *Acta Diabetologica*, *53*, 879–889.

Carreau, S., Bouraima-Lelong, H., & Delalande, C. (2011). Estrogens in male germ cells. *Spermatogenesis*, *1*, 90–94.

Carreau, S., Bourguiba, S., Lambard, S., Galeraud-Denis, I., Genissel, C., Bilinska, B., et al. (2001). Aromatase expression in male germ cells. *The Journal of Steroid Biochemistry and Molecular Biology*, *79*, 203–208.

Carreau, S., Lambard, S., Delalande, C., Denis-Galeraud, I., Bilinska, B., & Bourguiba, S. (2003). Aromatase expression and role of estrogens in male gonad: A review. *Reproductive Biology and Endocrinology*, *11*, 1–35.

Carreau, S., Silandre, D., Bourguiba, S., Hamden, K., Said, L., Lambard, S., et al. (2007). Estrogens and male reproduction: A new concept. *Brazilian Journal of Medical and Biological Research*, *40*, 761–768.

Coleman, D. L., Leiter, E. H., & Schwizer, R. W. (1982). Therapeutic effects of dehydroepiandrosterone (DHEA) in diabetic mice. *Diabetes*, *31*, 830–833.

Coleman, D. L., Schwizer, R. W., & Leiter, E. H. (1984). Effect of genetic background on the therapeutic effect of dehydroepiandrosterone (DHEA) in diabetic-obesity mutants and in aged normal mice. *Diabetes*, *33*, 26–32.

Conley, A., Mapes, S., Corbin, C. J., Greger, D., Walters, K., Trant, J., et al. (2001). A comparative approach to structure-function studies of mammalian aromatases. *The Journal of Steroid Biochemistry and Molecular Biology*, *79*, 289–297.

Corona, G., Rastrelli, G., Giagulli, V. A., Sila, A., Sforza, A., Forti, G., et al. (2013). Dehydroepiandrosterone supplementation in elderly men: A meta-analysis study of placebo-controlled trials. *The Journal of Clinical Endocrinology and Metabolism*, *98*, 3615–3626.

Ding, E. L., Song, Y., Manson, J. E., Rifai, N., Buring, J. E., & Liu, S. (2007). Plasma sex steroid hormones and risk of developing type 2 diabetes in women: A prospective study. *Diabetologia*, *50*, 2076–2084.

Glenmark, B., Nilsson, M., Gao, H., Gustafsson, J. A., Dahlman-Wright, K., & Westerbland, H. (2004). Difference in skeletal muscle function in males vs. females: Role of estrogen receptor-beta. *American Journal of Physiology. Endocrinology and Metabolism*, *287*, E1125–1131.

Haffner, S. M., Valdez, R. A., Mykkanen, L., Stern, M. P., & Katz, M. S. (1994). Decreased testosterone and dehydroepiandrosterone sulfate concentrations are associated with increased insulin and glucose concentrations in nondiabetic men. *Metabolism*, *43*, 599–603.

Hansen, P. A., Han, D. H., Nolte, L. A., Chen, M., & Holloszy, J. O. (1997). DHEA protects against visceral obesity and muscle insulin resistance in rats fed a high-fat diet. *American Journal of Physiology. Regulatory, Integrative and Comparative Physiology*, *273*, R1704–1708.

Horii, N., Sato, K., Mesaki, N., & Iemitsu, M. (2016a). DHEA administration activates transcription of muscular lipid metabolic enzymes via PPARα and PPARδ in obese rats. *Hormone and Metabolic Research*, *48*, 207–212.

Horii, N., Sato, K., Mesaki, N., & Iemitsu, M. (2016b). Increased muscular 5α-dihydrotestosterone in response to resistance training relates to skeletal muscle mass and glucose metabolism in type 2 diabetic rats. *PLoS One*, *11*, e0165689.

Houmard, J. A. (2008). Intramuscular lipid oxidation and obesity. *American Journal of Physiology. Regulatory, Integrative and Comparative Physiology, 294*, R1111–R1116.
Ihemelandu, E. C. (1980). Effect of oestrogen on muscle development of female rabbits. *Acta Anatomica, 108*, 310–315.
Ishizuka, T., Miura, A., Kajita, K., Matsumoto, M., Sugiyama, C., Matsubara, K., et al. (2007). Effect of dehydroepiandrosterone on insulin sensitivity in Otsuka Long-Evans Tokushima-fatty rats. *Acta Diabetologica, 44*, 219–226.
Jenkins, E. P., Hsieh, C. L., Milatovich, A., Normington, K., Berman, D. M., Francke, U., et al. (1991). Characterization and chromosomal mapping of human steroid 5α-reductase gene and pseudogene and mapping of the mouse homologue. *Genomics, 11*, 1102–1112.
Kang, H. Y., Cho, C. L., Huang, K. L., Wang, J. C., Hu, Y. C., Lin, H. K., et al. (2004). Nongenomic androgen activation of phosphatidylinositol 3-kinase/Akt signalling pathway in MC3T3-E1 osteoblasts. *Journal of Bone and Mineral Research, 19*, 1181–1190.
Karbowska, J., & Kochan, Z. (2005). Effect of DHEA on endocrine functions of adipose tissue, the involvement of PPAR gamma. *Biochemical Pharmacology, 70*, 249–257.
Karbowska, J., & Kochan, Z. (2011). Effects of DHEA on metabolic and endocrine functions of adipose tissue. *Hormone Molecular Biology and Clinical Investigation, 14*, 65–74.
Kimura, M., Tanaka, S., Yamada, Y., Kiuchi, Y., Yamakawa, T., & Sekihara, H. (1998). Dehydroepiandrosterone decreases serum tumor necrosis factor-alpha and restores insulin sensitivity: Independent effect from secondary weight reduction in genetically obese Zucker fatty rats. *Endocrinology, 139*, 3249–3253.
Kochan, Z., & Karbowska, J. (2004). Dehydroepiandrosterone up-regulates resistin gene expression in white adipose tissue. *Molecular and Cellular Endocrinology, 218*, 57–64.
Labrie, Y., Durocher, F., Lachance, Y., Turgeon, C., Simard, J., Labrie, C., et al. (1995). The human type II 17β-hydroxysteroid dehydrogenase gene encodes two alternatively-spliced messenger RNA species. *DNA and Cell Biology, 14*, 849–861.
Labrie, F., Luu-The, V., Bélanger, A., Lin, S. X., Simard, J., Pelletier, G., et al. (2005). Is dehydroepiandrosterone a hormone? *The Journal of Endocrinology, 187*, 169–196.
Labrie, F., Luu-The, V., Labrie, C., & Simard, J. (2001). DHEA and its transformation into androgens and estrogens in peripheral tissues: Intracrinology. *Frontiers in Neuroendocrinology, 22*, 185–212.
Labrie, F., Luu-The, V., Lin, S. X., Labrie, C., Simard, J., Breton, R., et al. (1997). The key role of 17β-hydroxysteroid dehydrogenases in sex steroid biology. *Steroids, 62*, 148–158.
Labrie, F., Simard, J., Luu-The, V., Belanger, A., & Pelletier, G. (1992). Structure, function and tissue specific gene expression of 3β-hydroxysteroid dehydrogenase/5-ene-4-ene isomerase enzymes in classical and peripheral intracrine steroidogenic tissues. *The Journal of Steroid Biochemistry and Molecular Biology, 43*, 805–826.
Labrie, F., Sugimoto, Y., Luu-The, V., Simard, J., Lachance, Y., Bachvarov, D., et al. (1992). Structure of human typeII5α-reductase. *Endocrinology, 131*, 1571–1573.
Lefebvre, P., Chinetti, G., Fruchart, J. C., & Steals, B. (2006). Sorting out the roles of PPAR alpha in energy metabolism and vascular homeostasis. *The Journal of Clinical Investigation, 116*, 571–580.
Lukacik, P., Keller, B., Bunkoczi, G., Kavanagh, K., Lee, W. H., Adamski, J., et al. (2007). Structural and biological characterization of human orphan DHRA10 reveals a novel cytosolic enzyme with steroid dehydrogenase activity. *Biochemical Journal, 402*, 419–427.
Luu-The, V., Zhang, Y., Poirier, D., & Labrie, F. (1995). Characteristics of human type 1, 2, and 3 17β-hydroxysteroid dehydrogenase activities: Oxidation-reduction and inhibition. *The Journal of Steroid Biochemistry and Molecular Biology, 55*, 581–587.
MacEwen, E. G., & Kurzman, I. D. (1991). Obesity in the dog: Role of the adrenal steroid dehydroepiandrosterone (DHEA). *The Journal of Nutrition, 121*, S51–S55.

Martel, C., Melner, M. H., Gagne, D., Simard, J., & Labrie, F. (1994). Widespread tissue distribution of steroid sulfatase, 3 beta-hydroxysteroid dehydrogenase/delta 5-delta 4 isomerase (3 beta-HSD), 17 beta-HSD 5 alpha-reductase and aromatase activities in the rhesus monkey. *Molecular and Cellular Endocrinology, 104*, 103–111.

Migeon, C. J., Keller, A. R., Lawrence, B., & Shepart, T. H. (1957). Dehydroepiandrosterone and androsterone levels in human placenta. Effect of age and sex: Day-to-day and diurnal variations. *The Journal of Clinical Endocrinology and Metabolism, 17*, 1051–1062.

Milewich, L., Shaw, C. E., Mason, J. I., Carr, B. R., Blomquist, C. H., & Thomas, J. L. (1993). 3β-Hydroxysteroid dehydrogenase activity in tissues of human fetus determined 5α-androstane-3β, 17β-diol and dehydroepiandrosterone as substrates. *The Journal of Steroid Biochemistry and Molecular Biology, 45*, 525–537.

Mohan, P. F., & Cleary, M. P. (1988). Effect of short-term DHEA administration on liver metabolism of lean and obese rats. *American Journal of Physiology. Endocrinology and Metabolism, 255*, E2–E8.

Mohan, P. F., Ihnen, J. S., Levin, B. E., & Cleary, M. P. (1990). Effects of dehydroepiandrosterone treatment in rats with diet-induced obesity. *The Journal of Nutrition, 120*, 1103–1114.

Morales, A. J., Haubrich, R. H., Hwang, J. Y., Asakura, H., & Yen, S. S. (1998). The effect of six months treatment with a 100 mg daily dose of dehydroepiandrosterone (DHEA) on circulating sex steroids, body composition and muscle strength in age-advanced men and women. *Clinical Endocrinology, 49*, 421–432.

O'Donnell, L., Robertson, K. M., Jones, M. E., & Simpson, E. R. (2001). Estrogens and spermatogenesis. *Endocrine Reviews, 22*, 289–318.

Okada-Iwabu, M., Yamauchi, T., Iwabu, M., Honma, T., Hamaguchi, K., Matsuda, K., et al. (2013). A small-molecule AdipoR agonist for type 2 diabetes and short life in obesity. *Nature, 503*, 493–499.

Orentreich, N., Brind, J. L., Rizer, R. L., & Vogelman, J. H. (1984). Age changes and sex differences in serum dehydroepiandrosterone sulfate concentrations throughout adulthood. *The Journal of Clinical Endocrinology and Metabolism, 59*, 551–555.

Pelletier, G., Dupont, E., Simard, J., Luu-The, V., Belanger, A., & Labrie, F. (1992). Ontogeny and subcellular localization of 3β- hydroxysteroid dehydrogenases (3β-HSD) in the human and rat adrenal, ovary and testis. *The Journal of Steroid Biochemistry and Molecular Biology, 43*, 451–467.

Perez-de-Heredia, F., Sanchez, J., Priego, T., Nicolas, F., Portillo Mdel, P., Palou, A., et al. (2008). Adiponectin is involved in the protective effect of DHEA against metabolic risk in aged rats. *Steroids, 73*, 1128–1136.

Pöllänen, E., Ronkainen, P. H., Horttanainen, M., Takala, T., Puolakka, J., Suominen, H., et al. (2010). Effects of combined hormone replacement therapy or its effective agents on the IGF-1 pathway in skeletal muscle. *Growth Hormone & IGF Research, 20*, 372–379.

Ryu, J. W., Kim, M. S., Kim, C. H., Song, K. H., Park, J. Y., Lee, J. D., et al. (2003). DHEA administration increases brown fat uncoupling protein 1 levels in obese OLETF rats. *Biochemical and Biophysical Research Communications, 303*, 726–731.

Sato, K., Iemitsu, M., Aizawa, K., & Ajisaka, R. (2008). Testosterone and DHEA activate the glucose metabolism-related signaling pathway in skeletal muscle. *American Journal of Physiology. Endocrinology and Metabolism, 294*, E961–968.

Sato, K., Iemitsu, M., Aizawa, K., & Ajisaka, R. (2009). DHEA improves impaired activation of Akt and PKC ζ/λ-GLUT4 pathway in skeletal muscle and improves hyperglycemia in streptozotocin-induced diabetes rats. *Acta Physiologica, 197*, 217–225.

Sato, K., Iemitsu, M., Aizawa, K., Mesaki, N., Ajisaka, R., & Fujita, S. (2012). DHEA administration and exercise training improves insulin resistance in obese rats. *Nutrition & Metabolism, 9*, 47.

Sato, K., Iemitsu, M., Aizawa, K., Mesaki, N., & Fujita, S. (2011). Increased muscular dehydroepiandrosterone levels are associated with improved hyperglycemia in obese rats. *American Journal of Physiology. Endocrinology and Metabolism*, *301*, E274–E280.

Sato, K., Iemitsu, M., Matsutani, K., Kurihara, T., Hamaoka, T., & Fujita, S. (2014). Resistance training restores muscle sex steroid hormone steroidogenesis in older men. *The FASEB Journal*, *28*, 1891–1897.

Schwartz, A. G., Pashko, L., & Whitcomb. (1986). Inhibition of tumor development by dehydroepiandrosterone and related steroids. *Toxicologic Pathology*, *14*, 357–362.

Urban, R. J., Bodenburg, Y. H., Gilkison, C., Foxworth, J., Coggan, A. R., Wolfe, R. R., et al. (1995). Testosterone administration to elderly men increases skeletal muscle strength and protein synthesis. *American Journal of Physiology. Heart and Circulatory Physiology*, *269*, H820–826.

Vega, R. B., Huss, J. M., & Kelly, D. P. (2000). The coactivator PGC-1 cooperates with peroxisome proliferator-activated receptor alpha in transcriptional control of nuclear genes encoding mitochondrial fatty acid oxidation enzymes. *Molecular and Cellular Biology*, *20*, 1868–1876.

Villareal, D. T., & Holloszy, J. O. (2004). Effect of DHEA on abdominal fat and insulin action in elderly women and men. *JAMA*, *292*, 2243–2248.

Villareal, D. T., & Holloszy, J. O. (2006). DHEA enhances effects of weight training on muscle mass and strength in elderly women and men. *American Journal of Physiology. Endocrinology and Metabolism*, *291*, E1003–E1008.

Wang, Y. X., Lee, C. H., Tiep, S., Yu, R. T., Ham, J., & Kang, H. (2003). Peroxisome proliferator-activated receptor delta activates fat metabolism to prevent obesity. *Cell*, *113*, 159–170.

Yamaguchi, Y., tanaka, S., Yamakawa, T., Kimura, M., Ukawa, K., Yamada, Y., et al. (1998). Reduced serum dehydroepiandrosterone levels in diabetic patients with hyperinsulinaemia. *Clinical Endocrinology*, *49*, 377–383.

Yokokawa, T., Sato, K., Iwanaka, N., Honda, H., Higashida, K., Iemitsu, M., et al. (2015). Dehydroepiandrosterone activates AMP kinase and regulates GLUT4 and PGC-1α expression in C2C12 myotubes. *Biochemical and Biophysical Research Communications*, *463*, 42–47.

Yoon, M. J., Lee, G. Y., Chung, J. J., Ahn, Y. H., Hong, S. H., & Kim, J. B. (2006). Adiponectin increases fatty acid oxidation in skeletal muscle cells by sequential activation of AMP-activated protein kinase, p38 mitogen-activated protein kinase, and peroxisome proliferator-activated receptor alpha. *Diabetes*, *55*, 2562–2570.

FURTHER READING

Wilson, J. D., Griffin, J. E., & Russel, D. W. (1993). Steroid 5α-reductase 2 deficiency. *Endocrine Reviews*, *14*, 577–593.

CHAPTER EIGHT

Dehydroepiandrosterone (DHEA) and Its Sulfate (DHEA-S) in Mammalian Reproduction: Known Roles and Novel Paradigms

Bindu N. Chimote[1], Natchandra M. Chimote
Vaunshdhara Clinic and Assisted Conception Centre, Nagpur, Maharashtra, India
[1]Corresponding author: e-mail address: bindunm10@yahoo.com

Contents

Abstract

Steroid hormones form an integral part of normal development in mammalian organisms. Cholesterol is the parent compound from which all steroid hormones are synthesized. The product pregnenolone formed from cholesterol serves as precursor for mineralocorticoids, glucocorticoids, as well as dehydroepiandrosterone (DHEA) and its derived sexual hormones. DHEA assumes the prohormone status of a predominant endogenous precursor and a metabolic intermediate in ovarian follicular steroidogenesis. DHEA supplementation has been used to enhance ovarian reserve. Steroids like estradiol and testosterone have long been contemplated to play important roles in regulating meiotic maturation of oocytes in conjunction with gonadotropins. It is known that oocyte priming with estrogen is necessary to develop calcium (Ca^{2+}) oscillations during maturation. Accruing evidence from diverse studies suggests that DHEA and

Vitamins and Hormones, Volume 108
ISSN 0083-6729
https://doi.org/10.1016/bs.vh.2018.02.001

its sulfate (dehydroepiandrosterone sulfate, DHEA-S) play significantly vital role not only as intermediates in androgen and estrogen formation, but may also be the probable 'oocyte factor' and behave as endogenous agonists triggering calcium oscillations for oocyte activation. DHEA/DHEA-S have been reported to regulate calcium channels for the passage of Ca^{2+} through the oocyte cytoplasm and for maintaining required threshold of Ca^{2+} oscillations. This role of DHEA/DHEA-S assumes critical significance in assisted reproductive technology and in-vitro fertilization treatment cycles where physical, chemical, and mechanical methods are employed for artificial oocyte activation to enhance fertilization rates. However, since these methods are invasive and may also cause adverse epigenetic modifications; oral or culture-media supplementation with DHEA/DHEA-S provides a noninvasive innate mechanism of in-vitro oocyte activation based on physiological metabolic pathway.

1. INTRODUCTION

Steroid hormones form an integral part of normal development in mammalian organisms. For, e.g., mineralocorticoids (aldosterone) play an essential role in regulating blood pressure; glucocorticoids (cortisol) provide an energy source for gluconeogenesis, whereas the sex hormones (androgens and estrogens) are indispensable for development of sexual characteristics and reproductive functions (Fig. 1A and B).

Cholesterol is the parent compound from which all steroid hormones are synthesized. Formation of pregnenolone by the side chain cleavage of cholesterol initiated by the enzyme CYP11A is the rate-limiting step of steroid hormone biosynthesis (Bernhardt & Waterman, 2007; Lisurek & Bernhardt, 2004). The product pregnenolone serves as precursor for mineralocorticoids, glucocorticoids, as well as dehydroepiandrosterone (DHEA) and its derived sexual hormones.

Steroid hormones, in an "unconjugated" form, have been believed to exercise their biological functions via interaction with their corresponding receptors. On the other hand, "conjugated" steroids with sulfate groups attached have been cogitated to be exclusively designated for excretion (Strott, 1996). However, since mammals produce a substantial concentration of sulfonated steroids especially during pregnancy, the physiological role of conjugated steroids, besides being inactive reservoirs for steroid hormones (Claus & Hoffmann, 1980; Hoffmann & Landeck, 1999; Janowski, Zdunczyk, Ras, & Mwaanga, 1999; Pasqualini & Chetrite, 2005), still needs to be elucidated.

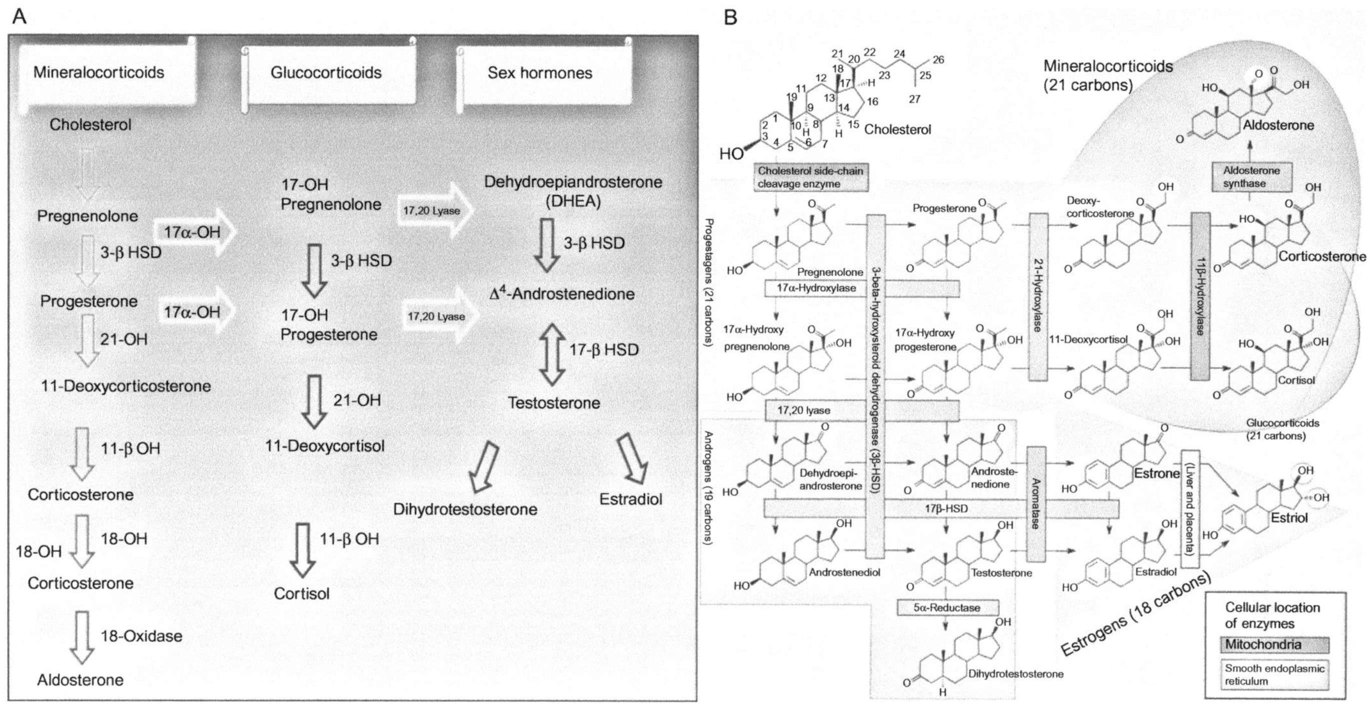

Fig. 1 Biosynthesis of steroid hormones from cholesterol. (A) Schematic layout. (B) Structural representation. *Panel (A) and (B) reproduced from Erhuma, A. M. (2012). Glucocorticoids: Biochemical group that play key role in fetal programming of adult disease. In X. Oian (Ed.),* Medicine glucocorticoids-new recognition of our familiar friend. *ISBN:978-953-51-0872-6 and Häggström, M., Stannered, Hoffmeier, Settersr, Richfield, D. (2014). Diagram of the pathways of human steroidogenesis.* WikiJournal of Medicine, 1*(1), 5. doi:10.15347/wim/2014.005. ISSN 2002-4436, respectively.*

Cholesterol sulfate (CS) is the only sulfonated steroid that has been investigated in detail, and its role in regulating activity of serine proteases has been elucidated. The side chain cleavage of cholesterol is catalyzed by the enzyme CYP11A1, and CS has been shown to have an inhibitory effect on this enzyme (Lambeth, Xu, & Glover, 1987; Tsutsumi et al., 2008). This implies that sulfonated steroids play a more direct role in steroid hormone biosynthesis than earlier believed. Moreover, discovery and the presence of estrogen receptors, steroid sulfatases and estrogen sulfotransferases in the same tissue (Schular et al., 2008), suggest a putative system that controls the availability of free steroids.

2. DEHYDROEPIANDROSTERONE (DHEA) AND DEHYDROEPIANDROSTERONE SULFATE (DHEAS)

DHEA is produced from cholesterol in the adrenal cortex, gonads, and the brain (Schulman et al., 2007). DHEA is also called dehydroisoandrosterone or dehydroandrosterone. It is the most abundant circulating steroid hormone in humans (William & Ganong, 2005). It is the primary precursor of natural estrogens. DHEA is converted to testosterone in ovarian connective tissue (theca/stroma) and is subsequently processed by the granulosa cells to form estradiol. It therefore assumes the prohormone status of a predominant endogenous precursor and a metabolic intermediate in ovarian follicular steroidogenesis (Mo, Lu, & Simon, 2006; Scott, 1996).

DHEA is reversibly catalyzed by sulfotransferase (SULT2A1) to its sulfate ester dehydroepiandrosterone sulfate DHEA-S) (Fig. 2). In humans, 99% of DHEA circulates in its sulfonated form, DHEA-S (Leowattana, 2004). The concentration of circulating DHEA-S reaches its peak in young adults and subsequently decreases with age with concentrations of DHEA-S between 20% and 30% in 70- to 80-year-old people compared with the value of young adults (Kroboth, Salek, Pittenger, Fabian, & Frye, 1999). The naturally occurring concentrations of DHEA-S and DHEA have been determined to be

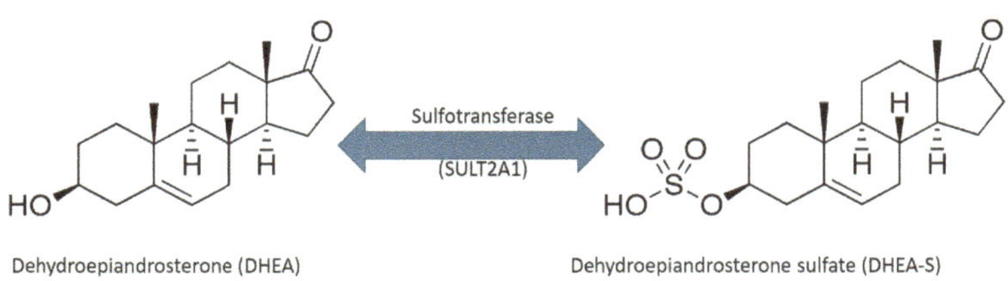

Fig. 2 Conversion of DHEA to DHEA-S.

around 10 mM and 10 nM, respectively, in young adults (Baulieu, 1996). Most DHEA in blood is found as DHEA-S with levels that are about 300 times higher than those of free DHEA. Whereas DHEA levels naturally reach their peak in the early morning hours, DHEA-S levels show no diurnal variation. DHEA-S is also the most abundant steroid in follicular fluid (FF) of preovulatory follicles being present at micromolar levels (Dehenin, Jondet, & Scholler, 1987). In contrast to the short terminal half-life of DHEA which is pegged at only 15–30 min, DHEA-S has a much longer half-life of 7–10 h (White & Porterfield, 2013). Thus from a practical point of view, measurement of DHEA-S seems more preferable than DHEA since its level remains stable.

DHEA-S has been speculated to play a rather secondary role in the generally accepted pathway for steroidogenesis, as being the inactive sulfonated form which gets reconverted into DHEA for the production of androstenedione, the precursor for estrogen biosynthesis (Fig. 1).

However, since DHEA-S is the most abundantly circulating steroid in humans (Neunzig & Bernhardt, 2014) and has a longer half-life, Neunzig and Bernhardt (2014) reconstructed an in-vitro system to study the effect of sulfonated DHEA on CYP11A1 activity at molecular level.

Their contention was based on an earlier finding that as against unconjugated steroids, sulfonated steroids need to be actively transported through cell membranes and that sodium-dependent organic anion transporters were found to possess a high sensitivity toward sulfonated steroids (Geyer, Godoy, & Petzinger, 2004). Therefore, sulfonated steroids can easily cross the mitochondrial membranes and have direct contact with CYP11A1. In conformity with their contention, Neunzig and Bernhardt (2014) indeed concluded that DHEA-S but not DHEA increases the CYP11A1-dependent pregnenolone formation caused by improved cholesterol and redox partner binding. This was a path-breaking finding so far as role of sulfonated steroids in steroidogenesis was concerned. However, detailed investigations about the interplay between steroid biosynthesis and steroid conjugation had still been missing.

Several workers continued with their experimentation on role of DHEA and DHEA-S. Liu et al. (2002) identified a putative membrane receptor for DHEA/DHEA-S, suggesting that although the molecular mechanisms of the effects of DHEA/DHEA-S still remain obscure, they may possess some physiological role. Recently, Neunzig and Bernhardt (2014) demonstrated higher binding of the enzyme CYP11A1 to substrate cholesterol in the presence of DHEA-S and showed that the activity of CYP11A1 displayed a 26%

increase in pregnenolone formation in the presence of DHEA-S as compared to that in the presence of DHEA. They also observed a 75% upregulation in the catalytic efficiency of the side-chain cleavage of cholesterol in the presence of DHEA-S but not DHEA. They surmised that the negatively charged sulfate group of DHEA-S may be essential for the enzyme activity. They also noted that DHEA-S stringently regulated the first step of steroidogenesis and prompted interaction with the redox partner Adx (Neunzig & Bernhardt, 2014).

Indeed, there has been a rather incriminating evidence of involvement of DHEA-S utilization as precursor for estrogen production (Fig. 3) since a sulfatase enzyme that accepts DHEA-S as a substrate has been reported in the ovarian follicles (Jimena et al., 1992). Later, Bonser et al. (2000) also demonstrated the activity of sulfatase in granulosa cells and concluded that DHEA-S is not only used as a substrate but also its supplementation to granulosa cells results in an increased production of estradiol and androstenedione in a dose-dependent manner.

All these findings lay credence to the speculation that DHEA-S may actually bear a more significant role than just being a reservoir for DHEA molecule. Yet surprisingly, none of the studies till date have made any attempt to seriously examine if DHEA-S by itself could pose as an active substrate eventually leading to the formation of estrogen and contribute as a primary entity in the overall process.

Nevertheless, it would be worthwhile to discuss here the roles of both DHEA and DHEA-S in the mammalian reproduction process and more elaborately in the mammalian fertilization process. But, before we begin our exploratory journey toward understanding this complicated phenomenon, it is essential to understand the basic concepts of the fertilization process involving fusion of the male and female gametes.

3. MAMMALIAN FERTILIZATION PROCESS

3.1 Brief Outline

In each menstrual cycle, every normally ovulating woman ovulates, i.e., releases a mature oocyte by the rupture of her follicle. A "mature" oocyte is one that has attained "meiotic competence." It is pertinent to note that the preovulatory oocyte is retained in the Prophase I stage of Meiosis I of the cell cycle. This stage is also called the germinal vesicle (GV) arrested stage. The cumulus cells which surround the preovulatory oocyte send inhibitory signals, chiefly in the form of cyclic adenosine monophosphate (c-AMP), to

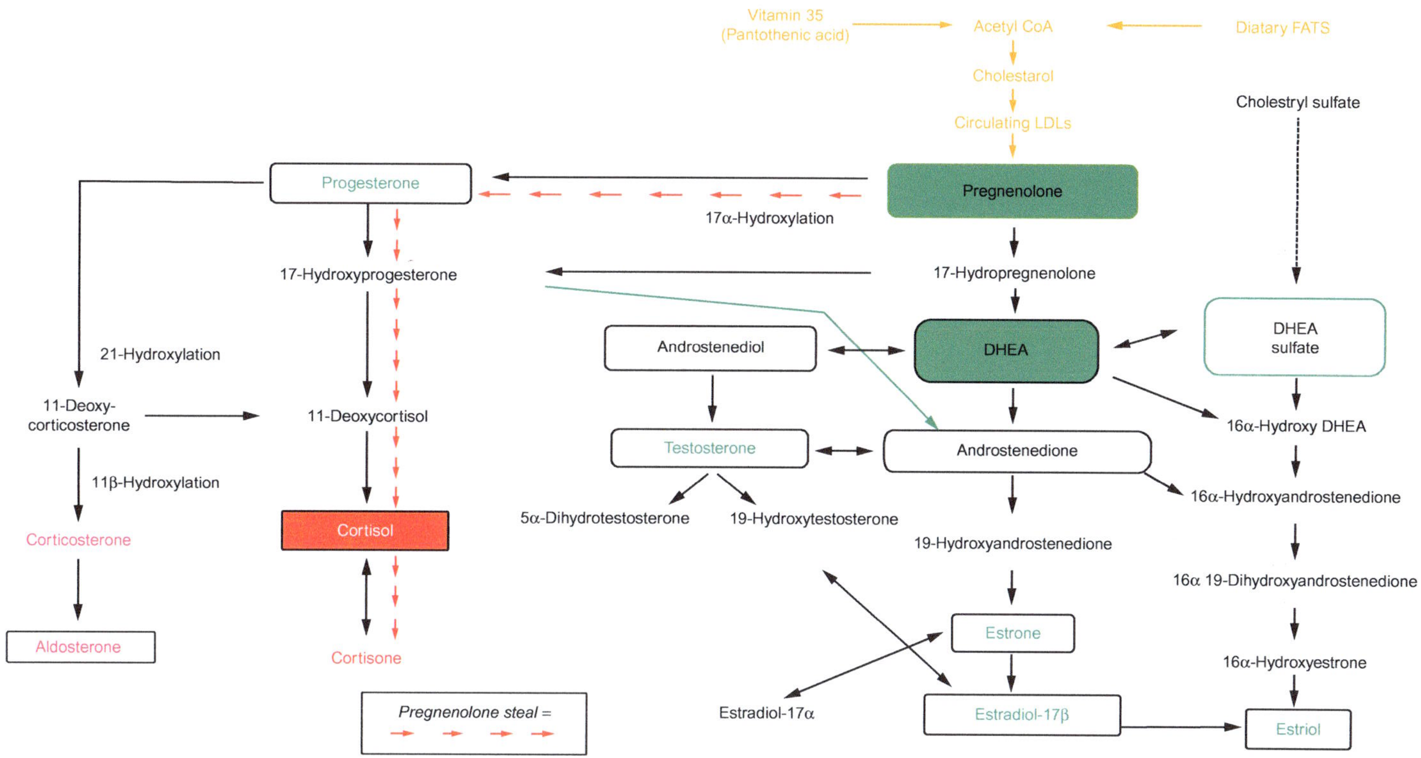

Fig. 3 Pathway showing DHEA-S as an important intermediate in the formation of estriol. *Reproduced from McEvoy, M. (2012). The endocrine journey: Steroidal hormones & their pathways, January 12, 2012.*

retain the oocyte in this stage (Norris et al., 2009). However, in response to the ovulatory surge in luteinizing hormone (LH), GV breaks down and causes resumption of meiosis at Metaphase I (Park et al., 2004). The "nuclear maturation" process begins; the oocyte divides asymmetrically; and whereas the DNA content remains same (i.e., 2C DNA), the chromosome number is halved, and the egg cell becomes haploid. The excess chromosome is extruded in the form of first polar body (PB) in perivitelline space. Concurrently, cytoplasmic maturation also occurs, involving tremendous flux in cytoplasmic activities whereby maternal organelles like mitochondria, endoplasmic reticulum, etc., undergo or elicit significant changes (Albertini, 1987; Van Blerkom, 1991). Cell cycle continues in preparation for fertilization, and the oocyte is again arrested at Metaphase II of Meiosis II. Subsequent to nuclear and cytoplasmic maturation, the now "mature" oocyte, that has attained cortical polarity and is arrested at Metaphase II, awaits the sperm in anticipation of fertilization. The ground is laid for the initiation of the process of fertilization in which two haploid cells (i.e., the oocyte and the sperm) will come together to form a diploid cell or zygote.

Absolutely unpretentious as it may seem, this entire process follows a palpably complicated path, with checkpoints at discrete intervals. Understandably so, because fertilization is arguably the most important step in the larger canvass of procreation and production of next-generation progeny. So although the oocyte has attained maturation, fertilization is fomented only after it is further activated by the presence of the sperm at its surface. Rise in calcium levels by binding of sperm is the key feature for oocyte activation from its arrest (Hörmanseder, Tischer, & Mayer, 2013; Von Stetina & Orr-Weaver, 2011). Thus, oocyte maturation and activation are almost indispensable events leading to successful fertilization.

3.2 Oocyte Activation: The Sperm Factor

Several theories have been put forward for the sperm-induced oocyte activation (Tosti & Ménézo, 2010). However, the one that is largely accepted involves a sperm factor called phospholipase C-zeta (PLCζ) (Kashir, Nomikos, Lai, & Swann, 2014; Swann & Lai, 2016) (Fig. 4). After binding to the oocyte zona pellucida, sperm releases the sperm factor (PLCζ), following which, in mammals, an outward ion current is generated due to electrical modification of oocyte plasma membrane. This results in release of calcium from the intracellular stores and generation of calcium oscillations.

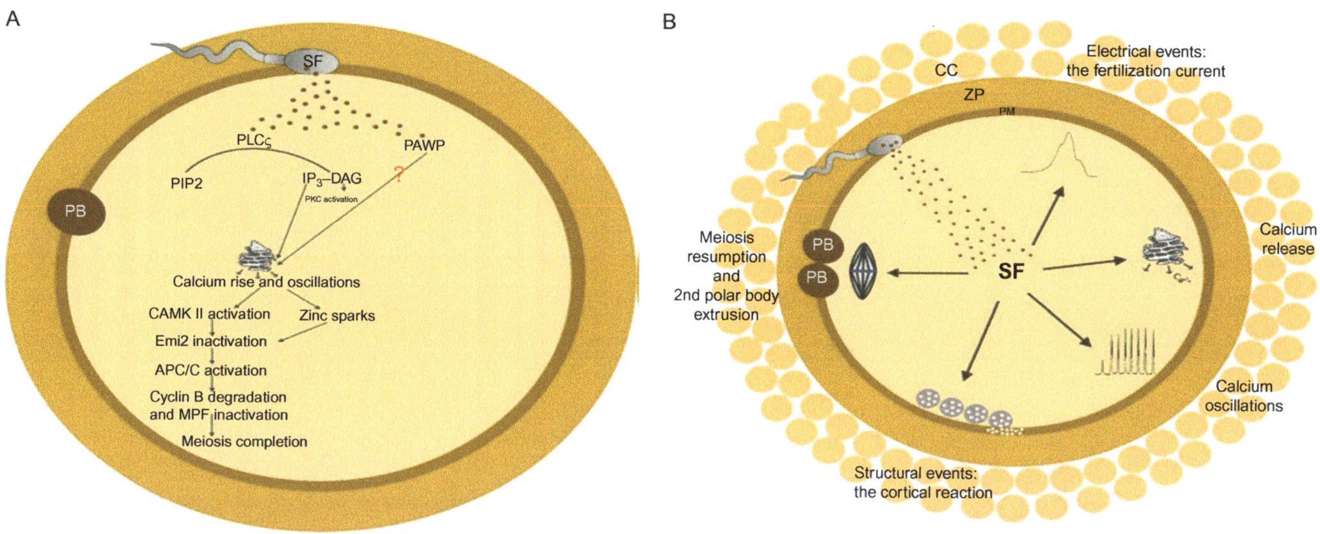

Fig. 4 See legend on next page.

The presence of zinc spikes have also been implicated in regulating mammalian oocyte meiotic cycle (Que et al., 2015).

Once sperm penetrates the inner vestments of oocyte, it activates the oocyte in preparation for fusion. Oocyte undergoes physical changes, and meiosis is resumed allowing completion of cell cycle, extrusion of second polar body, and triggering formation of zygote (oocyte fertilized by sperm).

Interaction of sperm with the oocyte elicits a signaling event involving mobilization of Ca^{2+} from its endoplasmic reticulum stores that trigger completion of meiosis and initiates embryo development (Stricker, 1999). The subsequent rise in intracellular Ca^{2+} oscillations is an obligatory step toward oocyte activation and for the mammalian gamete–zygote–embryo transition (Ducibella et al., 2002). Various studies (Ducibella et al., 2002; Ozil, Banrezes, Toth, Pan, & Schultz, 2006; Schultz &

Fig. 4 Schematic representation of sperm-induced oocyte activation. (A) How PLCζ acts: the hypothetical mechanism of action of PLCζ and downstream events in mammalian oocyte activation. Following gamete plasma membrane fusion, the SF diffuses into the ooplasm, binds and hydrolyzes PIP2 generating 2-s messengers, namely, 1P3 and DAG. IP3 induces calcium release from the endoplasmic reticulum stores. The calcium rise and oscillations may induce a cascade of translational events, such as CAMKII activation, Emi2 inactivation, APC/C activation, cyclin B degradation, and MPF inactivation. Altogether, these events result in the resumption and completion of meiosis. In alternative, calcium rise and oscillations may induce zinc sparks which in turn generate Emi2 inactivation. PAWP is potential SF, which induces calcium rise and oscillations by a still unknown mechanism. (B) Image representing events occurring during sperm-induced oocyte activation: upon release of the SF, electrical modification of oocyte plasma membrane properties generates an outward (in mammals) or inward ion current (nonmammals). Release of calcium from the intracellular stores generates calcium oscillations. Physical changes of the oocyte occur by the release of CG contents, and finally meiosis is resumed, allowing completion of the cell cycle, extrusion of the polar body, and triggering of zygote formation. *APC/C*: anaphase promoting complex/cyclosome; *cADPr*: cyclic adenosine diphosphoribose; *CAMKII*: calcium calmodulin-dependent protein kinase; *CC*: cumulus complex; *CdkI*: cyclin-dependent kinase; *CG*: conical granule; *DAG*: diacylglycerol; *EMI/ERP*: early mitotic inhibitors; *IP3*: inositol 1,4,5-trisphosphate; *MAPK*: mitogen-activated protein kinase; *Mos*: serine/threonine kinase; *MPF*: maturation-promoting factor; *NAADP*: nicotinic acid adenine dinucleotide phosphate; *NSF*: *N*-ethilmaleimide sensitive factor; *PAWP*: postacrosomal sheath WW domain-binding protein; *PIP2*: phosphatidylinositol (4,5)-bisphosphate; *PKC*: protein kinase C; *PLCC*: phospholipase C zeta; *SF*: sperm factor; PB: polar body; *SFEI, SFE9*: structural matrix proteins; *SNAP*: *N*-ethylmaleimide-sensitive factor attachment protein alpha; *TPCs*: two-pore channels; *ZP*: zona pellucida. *Figure reproduced from Tosti, E., & Menezo, Y. (2016). Gamete activation: Basic knowledge and clinical applications. Human Reproduction Update, 22(4), 420–439. doi:10.1093/humupd/dmw014. Epub 2016 Jun 8.*

Kopf, 1995) have surmised that the rise as well as pattern of Ca^{2+} oscillation is thus the decisive factor for:

(i) release of oocyte from its meiotic arrest
(ii) recruitment of oocyte mRNAs
(iii) proper development of maternal and paternal pronuclei
(iv) embryonic gene expression
(v) embryonic development, conception, and sustenance till term.

Although indispensable to fertilization and further embryonic development, the molecular mechanism underlying this phenomenon is hitherto obscure. In the most accepted model, inositol 1,4,5-triphosphate (IP3) has been implicated in release of Ca^{2+} from its ER stores.

However, the ER store of Ca^{2+} must be replenished by an influx of Ca^{2+} into the ooplasm from external environment through calcium pumps. At the same time, since a continued disproportionate rise in Ca^{2+} oscillations may be detrimental to embryo development (Gordo et al., 2002; Ozil et al., 2005), the excess Ca^{2+} needs to be swiftly pumped out so that a proper intracellular homeostasis of Ca^{2+} concentration is maintained. Any anomaly in mitochondrial function or distortion in the Ca^{2+} ATPase pumps abrogates the entire pattern of Ca^{2+} oscillations. Indeed, Wakai and Fissore (2013) emphasized the obligatory role of mitochondria and the active contribution of the PMCA and SERCA pumps in maintaining Ca^{2+} oscillations in mouse eggs. Regulation of Ca^{2+} concentration within the ER stores therefore seems to be the key criterion in maintaining the level of oscillations (Takahashi, Kikuchi, Kidokoro, & Shirakawa, 2013; Wakai & Fissore, 2013).

4. IN-VITRO FERTILIZATION: BRIEF CONCEPT

In-vitro fertilization (IVF) comprises of retrieval of oocytes from the follicles present in both ovaries of the female. Development of several follicles to recover supernumery eggs involves superovulation which is achieved hormonally by exogenously supplementing follicle-stimulating hormone (FSH) and stimulating the ovaries to produce multiple follicles. Human chorionic gonadotropin injection is administered to attain final oocyte maturation. Several factors govern the ovarian response to stimulation elicited in the female. As will be explained later in the chapter, DHEA/DHEA-S levels share a significant correlation with ovarian response. Nevertheless, the women may be categorized into three broad groups on the basis of ovarian response to standard stimulation protocol using 200 IU of recombinant follicle-stimulating hormone (r-FSH):

(i) *Poor responders*: Less than five oocytes retrieved owing to advanced maternal age (>37 years) or abnormal ovarian reserve tests (like antral follicle count, baseline serum levels of antimüllerian hormone, and FSH)
(ii) *Average/normal responders*: About 5–15 oocytes retrieved after stimulation
(iii) *Hyper/exaggerated responders*: More than 15 oocytes retrieved. Usually such women exhibit polycystic ovaries and suffer from polycystic ovarian syndrome (PCOS)

The oocytes that are so retrieved are still in the periovulatory stage. Postretrieval, communication between cumulus cells and oocytes allows metabolic transfer and regulates meiotic maturation (Zuke & Brackett, 1990). Several pathways have been postulated for the gonadotropin-induced meiotic maturation of oocytes (Downs & Hunzicker-Dunn, 1995; Su, Xia, Byskov, Fu, & Yang, 1999; Su, Xia, He, Fu, & Yng, 1998). Whatever pathway the gonadotropins may follow, the ultimate function is to modulate c-AMP levels which in turn affect cumulus cells–oocyte communication and enhance synthesis of stimulatory factor(s) for induction of nuclear and cytoplasmic maturation of oocytes (Hegele Hartung et al., 1999).

Involvement of steroids like estradiol and testosterone in the regulation of meiotic maturation in conjunction with gonadotropins, although contemplated, is still unclear (Kikuchi et al., 2000). An earlier study had cited that oocyte priming with estrogen is necessary to develop Ca^{2+} oscillations during maturation (Tesarik & Mendoza, 1997). Based on the corollary that oocytes are particularly sensitive to steroids while still at the GV stage, it was speculated that any imbalance in the estrogen-to-androgen ratios in-vivo may cause an abnormal Ca^{2+} response thus influencing fertilization.

5. DHEA/DHEA-S IN INFERTILITY TREATMENT AND IVF

5.1 DHEA Supplementation to Poor Responders

Advanced maternal age has been associated with high risk of miscarriages, and the causes have largely been attributed to increased rates of chromosomal abnormalities with aging. However, genetic studies have revealed that both chromosomally normal and abnormal embryos display the same developmental defects and developmental arrest pattern (Ford, Wilkin, Thomas, & McCarthy, 1996; Poland, Miller, Harris, & Livingston, 1981). These findings suggest that rather than nuclear origin being the sole cause, there may be involvement of cytoplasmic factors for the observed disruptions. Several studies have implicated role of the cytoplasmic organelle mitochondria in oocyte aging, reduced respiratory capacity/adenosine triphosphate (ATP) content as well as its correlation with oocyte developmental potential and IVF outcome (Kushnir et al., 2012; Simsek-Duran et al., 2013; Sugimura et al., 2012; Van Blerkom, Davis, & Lee, 1995).

The zona reticularis of adrenal cortex is the primary site for synthesis of DHEA and DHEA-S in humans and primates undergoing adrenarche. It is hypothesized that with advancing age, the fall in number of functional zona reticularis cells in the adrenal cortex might result in decreased synthesis of DHEA and DHEA-S (Hornsby, 2004). The reduction in size of adrenal cortex with age is known to be governed by the hypothalamus and pituitary (Hornsby, 2004). The hormones inhibin B (which plays a major role in the recruitment of primordial follicles and is a marker for follicle number) and activin are both expressed by the zona reticularis and are involved in important signaling mechanisms in the adrenal cortex (Arola et al., 2000; Vanttinen, Liu, Kuulasmaa, Kivinen, & Voutilainen, 2003). Therefore, decline and/or cessation in production of DHEA from ZR of adrenal cortex is a major determinant in the rapid age-linked depreciation in follicle number postage 37. In this context, it is pertinent to note that among all mammals, adrenal synthesis of DHEA is seen only in humans. It may be inferred from available experimental data (Haning et al., 1993) that whereas DHEA/DHEA-S of adrenal origin may serve as a prehormone for testosterone production in FF, the synthesis of DHEA/DHEA-S by follicular theca cells may be an essential prerequisite for ovarian estrogen biosynthesis. Lower levels of ovarian DHEA have been reported (Ford, 2013) to influence activity of several important metabolic enzymes including peroxisome proliferator-activated receptor alpha (PPARα) which in turn affects oxidative phosphorylation within mitochondria. This sequence of events may explain the rapid age-related decline in number and quality of follicles among most mammals (refer to Fig. 5). Thus, by virtue of its additional role in maintaining mitochondrial activity, DHEA supplementation acts via a two pronged mechanisms: (i) enhances number and quality of follicles and (ii) reduces oxidative stress and further atresia of ovarian follicles.

Within the ovary, DHEA and androgen are solely produced by the theca cells. DHEA is first produced by the theca cells of secondary follicles where it leads to the formation of estrone and 17β estradiol (Ethun et al., 2012; Young & McNeilly, 2010). DHEA levels decline with age and number of follicles (Ford, 2013). DHEA treatment has been shown to enhance

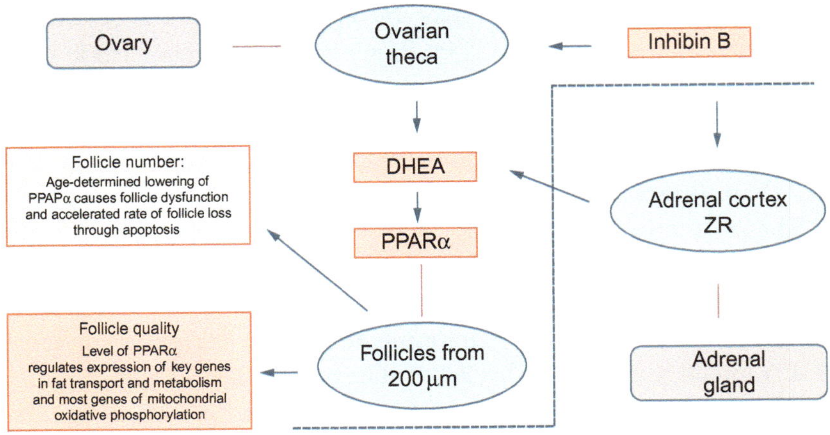

Fig. 5 The model (*left and above dotted line*) shows the response of the preantral theca to inhibin B, which results in production of DHEA and PPARα in follicles from about 200 μm. It is proposed that reduced levels of PPARα primarily account for both the decline in follicle number and the loss of oocyte quality that ultimately leads to errors of cell division and arrested development. The defects that occur in cytoplasmic organelles, especially mitochondria, are largely responsible for decline in follicle quality. These can be attributed to key changes in fat metabolism and transport and mitochondrial function that are directly caused by dysfunction of PPARα secondary to low DHEA. These same changes also lead to increased production of ceramide, apoptosis and hence accelerated decline in the size of the follicle pool. The area to the right of the *dotted line* indicates that in humans and those primates where the adrenal ZR produces an alternative source of DHEA, that age-related loss of ZR cells accelerates the rate of aging in the late 30s age group. *Reproduced from Ford, J.H. (2013). Reduced quality and accelerated follicle loss with female reproductive aging—Does decline in theca dehydroepiandrosterone (DHEA) underlie the problem?* Journal of Biomedical Science, 20:*93.*

mitochondrial function. The subsequent upregulation in the energy metabolism pathways (Patel & Katyare, 2006) may be the principle underlying DHEA supplementation to "poor responder women" (Artini et al., 2012).

Accordingly, several studies have reported enhancement in number and quality of oocytes after oral DHEA supplementation to elderly women or to women with poor ovarian reserve undergoing IVF (Barad & Gleicher, 2005, 2006; Casson, Lindsay, Pisarska, Carson, & Buster, 2000; Gleicher & Barad, 2011; Gleicher, Weghofer, & Barad, 2011). In a case study, the authors (Barad & Gleicher, 2005) could salvage ovarian function after DHEA supplementation in an elderly, 43-year-old single woman who wanted to bank her embryos. The same authors later (Gleicher & Barad, 2011) further reported enhancement in ovarian function, ovarian reserve, oocyte/embryo quality, probability of pregnancy accompanied by lower rates of miscarriages, and genetic abnormalities after DHEA therapy. Mamas and Mamas (2009) also depicted improvement in ovarian response to stimulation after oral DHEA administration.

However, a recent systematic review (Fouany & Sharara, 2013) emphasizes that there has to be a consensus among various studies on the dose of DHEA to be supplemented and that more prospective randomized control trials need to be carried out to evaluate the effect of DHEA on increase in pregnancy rates. Also, the patient population that will benefit from DHEA supplementation needs to be selected with care and caution.

Thus, it would be unwise to empirically provide DHEA therapy to all poor responder subjects since although it may apparently seem to be safe and beneficial for women with normal or low adrenal and ovarian DHEA or testosterone content; such a therapy could in fact be counterproductive and highly detrimental for women who are susceptible to "testosterone overload." Also, it is known that sustained exposure to high LH bioactivity leads to overgrowth of ovarian connective tissue (hyperthecosis/stromal hyperplasia), and that LH stimulates conversion of DHEA to testosterone. Consequently, especially in IVF cycles, it is highly probable that under the influence of LH, excessive ovarian testosterone would enter the FF, causing exhaustion of estradiol producing granulosa cells thus compromising not only with the development of follicle and egg but also with the egg quality.

Thus, DHEA supplementation can actually backfire and have negative consequences in women who are prone to ovarian hyperthecosis (elderly women, women with diminished ovarian reserve, and those having PCOS), thereby reducing fertility potential. DHEA supplementation therefore has to be done discreetly and with caution. To obviate the repercussions of empirical DHEA supplementation, it would be advisable to first study and confirm the function of endogenously present DHEA/DHEA-S and then to decide on whether DHEA/DHEA-S supplementation is required and in what doses.

6. DHEA-S IN OOCYTE MATURATION AND ACTIVATION: THE OOCYTE FACTOR?

It has widely been accepted that oocytes need to be primed with estrogen in order to develop Ca^{2+} oscillations during maturation (Tesarik & Mendoza, 1997). In yet another study (Watson, Jeng, & Kochukov, 2008), all estrogens viz estrone E_1, estradiol E_2, and estriol E_3 were found to enhance the oscillation frequency of calcium (Ca) spikes, with the same time delay (~200 s) at all levels (10^{-15}–10^{-9} M). At some concentrations, E_1 and E_3 provoked more Ca-responding cells than E_2. DHEA-S, by its conversion to E_3, may aid in the generation of CA^{2+} wave thereby facilitating the nongenomic maturation of oocytes.

It has been documented that the action of estrogen can be counteracted by androgen, and in-vivo exposure of oocytes to inadequate estrogen-to-androgen ratios may impel an abnormal Ca^{2+} response to spermatozoa during fertilization and consequently impair developmental potential of the resulting embryo (Tesarik & Mendoza, 1997). The same contention that the developmental potential of human oocytes appears to be better related to the estrogen-to-androgen ratio in FF than to absolute concentrations of follicular estrogen was also made earlier (Yding Andersen, 1993).

In a significant study, Chimote, Nath, Chimote, and Chimote (2015) contended that DHEA-S in conjunction with DHEA actively participates toward oocyte maturation. They measured DHEA-S levels in FF of normal responder women undergoing IVF. FF represents the dynamic microenvironmental milieu in which the oocyte undergoes its growth and development. FF is aspirated during IVF when the follicle is punctured to retrieve the oocyte/egg and can be used to assess various biomarkers of follicular and/or oocyte function. In their study, these authors tried to expound the effect of DHEA-S on oocyte maturation vis-à-vis its fertilization potential. Their hypothesis was based on and explored the chemical properties of DHEA/DHEA-S.

Chemical dynamics suggest that DHEA acts as a weak partial agonist of the androgen receptor albeit with low affinity ($K_i = 1\ \mu M$). Its intrinsic activity at the receptor is thus almost completely negligible since it competes with the full agonists like testosterone for receptor binding. Piquantly, it actually acts more like an antagonist there, and hence behaves like an antiandrogen. Remarkably, in addition to its affinity for the androgen receptor, DHEA acts as a full agonist for ERβ estrogen receptor with a maximal response similar to or slightly greater than that of estradiol. Hence, DHEA has been considered to be a potentially major endogenous estrogen in the body (Chen, Knecht, Birzin, et al., 2005; Webb, Geoghegan, Prough, & Michael Miller, 2006). Thus on one hand, DHEA, by virtue of its antiandrogenic property and ability to bind with ERβ estrogen receptor, mimics estrogen for granulosa cell activity, DHEA-S on the other aids in actually synthesizing requisite quantities of estrogens as to maintain a stable androgen-to-estrogen ratio so necessary for oocyte maturation. This might be the reason why previous studies with FF DHEA were found to be either contradictory or inconclusive and a reliable marker still remained elusive. Although, DHEA appears to play the role of a clasp by holding on to the ERβ estrogen receptor to give a momentary illusion of estrogen rise and keeping androgen levels at bay; DHEA-S seems to be the more effective biomolecule that actually maneuvers the process of oocyte maturation. These authors obtained high number of the meiotically mature MII oocytes and observed that higher fertilization rates correlated more significantly with FF DHEA-S than with FF E_2 levels thus emphasizing the role of DHEA-S as a primary regulator of estrogen formation. Indeed, these results are amply substantiated by other studies and pathways that have also reported direct utilization of DHEA-S as precursor for estrogen production (Bonser et al., 2000; McEvoy, 2012). Another interesting finding was the observed correlation of DHEA-S levels in FF with embryo development rates. These promising results explicate DHEA-S in FF as a reliable and robust marker for oocyte maturation as well as for the developmental competence of oocyte to blastocyst stage.

However, the authors also reiterated that a concomitant measurement of FF DHEA levels may facilitate a more thorough and in-depth understanding of the overall process. The results obtained in that study are represented in Fig. 6.

In yet another abstract of an unpublished study (Chimote, Nath, & Chimote, 2017), in lieu with their earlier study in FF, these same authors recruited various patient populations and evaluated DHEA-S levels in baseline serum samples of non-PCOS as well as PCOS women. DHEA-S levels have been known to be reduced among poor responder women, whereas they have been reported to be elevated in about 20%–70% of women with PCOS.

In IVF, lower fertilization rates have been reported in elderly/poor responder women as well as in PCOS women. However, it is pertinent to note that neither all poor responders have low levels nor all PCOS women have raised levels of DHEA-S. Therefore, these authors tried to establish threshold levels of DHEA-S for fertilization among these patient populations in order to endorse the role of DHEA-S in oocyte activation. They observed that the lower threshold of day 3 DHEA-S for fertilization was $>95\,\mu g/dL$, whereas upper threshold limit was $<270\,\mu g/dL$ with a high degree of sensitivity and specificity. These results suggest that while normal average DHEA-S levels (95–270 μg/dL: in PCOS and/or non-PCOS) are conducive, very low ($<95\,\mu g/dL$: in PCOS and/or non-PCOS) and very

A Parameter	Low FF DHEA-S <760 ng/mL (n = 115)	Medium FF DHEA-S 760-1850 ng/mL (n = 243)	High FF DHEA-S >1850 ng/mL (n = 134)	One Way ANOVA (Kruskal–Wallis test) P Value
Age (years)	31.36 ± 0.29	30.93 ± 0.21	30.73 ± 0.30	0.18
Day of hCG E2 (pg/mL)	1950 ± 66.31	2100 ± 75.43	2320 ± 77.58	0.29
FF E2 (pg/mL)	80193 ± 6819	102777 ± 6012	186713 ± 20709	0.01
No. eggs retrieved	1002	2469	1424	0.64
No. of MII oocytes (%)	796 (79.44)	2112 (85.54)	1303 (91.5)	<0.0001
No. of eggs fertilized (%)	714/1002 (71.26%)	1831/2469 (74.16%)	1123/1424 (78.86%)	<0.0001
No. of embryos cleaved (%)	663/1002 (66.17%)	1706/2469 (69.1%)	1049/1424 (73.67%)	0.0002
No. of blastocysts formed (%)	270/1002 (26.95%)	815/2469 (33.01%)	529/1424 (37.15%)	<0.0001
Mean no. of d5 blastocysts transferred	1.82 ± 0.04	1.85 ± 0.07	1.90 ± 0.05	0.23

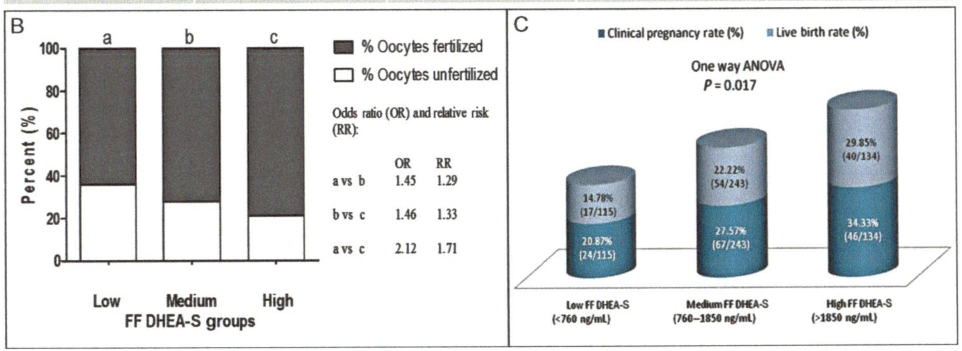

Fig. 6 Results obtained from the study of Chimote et al. (2015). (A) Table showing intercomparison of data. (B) Odds ratio (OR) and relative risk (RR). (C) Clinical pregnancy and live birth rates. All (A), (B), and (C) in low, medium, and high FF DHEA-S groups.

high levels (>270 μg/dL: in PCOS) are detrimental for oocyte activation, fertilization, and embryonic development.

Thus, their futuristic hypothesis is that those women with lower baseline serum levels of DHEA-S, who are more likely to face fertilization failure or lower fertilization rates after IVF, may be given oral supplementation of DHEA/DHEA-S instead of taking them for IVF treatment in that cycle. Alternatively, they may undergo IVF but the culture media may be supplemented with DHEA-S. This study did not however arrive at a dose of DHEA-S for supplementation. Similarly, the authors hypothesized that PCOS women exhibiting higher threshold levels of DHEA-S may undergo treatment with metformin before being evaluated again for the serum DHEA-S levels and undergoing IVF treatment. This conjecture is in conformity with other studies where metformin treatment has been suggested/provided to PCOS patients exhibiting higher levels of DHEA/DHEA-S. Although preliminary, the findings of this study are quite significant. Also, since their study involved human subjects, it was not possible to actually measure Ca^{2+} oscillations in oocytes and corroborate it with the observed DHEA-S levels. It would be interesting to design such a study involving mouse or some other mammalian oocytes where variations in Ca^{2+} oscillation pattern can be measured and investigated.

7. DHEA/DHEA-S-INDUCED OOCYTE ACTIVATION IN-VIVO AND IN-VITRO: SIGNIFICANCE AND PROBABLE MECHANISM OF ACTION

Inspite of elaborate knowledge regarding the absolute necessity of controlled Ca^{2+} oscillations in oocyte maturation and activation process leading to fertilization, little is yet known about the mechanism/s that trigger and govern the calcium channels. In assisted reproductive technology, it was believed that the observed instances of lower fertilization rates or fertilization failure would be obliterated with the advent of intracytoplasmic sperm injection (ICSI) technique. ICSI involves manual injection of a single sperm within each oocyte. It was believed that the mechanical agitation of the oocyte due to aspiration of its cytoplasm during ICSI would lead to oocyte activation and hence better fertilization rates (Ebner, Moser, Sommergruber, et al., 2004). However, although minimized to some extent, the problem of fertilization failure persisted even after ensuring sperm entry within the inner vestments of the oocyte. Infact, in a study carried out in pigs (Nakai et al., 2016) reduced oocyte activation was demonstrated after ICSI due to lack of Ca^{2+} oscillations or abnormal pattern of oscillations.

Low or no fertilization observed even after ICSI with treated/activated sperm has led to a shift in focus on oocyte activation. Sperm activation methods to improve its functionality and increase fertilization rates have been in vogue for quite some time in IVF cycles. The aim of artificial oocyte activation is to mimic physiological mechanisms, based mainly on calcium changes. One of the earliest methods of oocyte activation employed electrical stimulation/electrical pulses (Manipalviratn et al., 2006; Yanagida et al., 1999) to create pores for increased calcium permeability with physical agitation. Few later studies reported use of the chemical strontium chloride ($SrCl_2$) in oocyte activation (Chen, Qian, Tan, & Mima, 2010; Kim et al., 2012). Calcium ionophores such as ionomycin and A23187 have also been shown to be effective artificial oocyte activation agents in enhancing fertilization rates (Montag, Koster, van der Ven, Bohlen, & van der Ven, 2012; Tesarik & Sousa, 1995; Yeste, Jones, Amdani, Patel, & Coward, 2016). The more recent methodology exploits the role of the sperm factor PLCζ in oocyte activation. Several authors have advocated the use of human recombinant PLCζ injections to rescue the oocyte from activation deficiency (Chithiwala et al., 2015; Ramadan, Kashir, Jones, & Coward, 2012; Sanusi, Yu, Nomikos, Lai, & Swann, 2015; Yoon et al., 2012). Although it deviates from its physiological mode of action in interacting with intracellular vesicles containing PIP2 rather than binding to membrane bound PIP2, use of PLCζ is the current novel strategy of interest for oocyte activation (Amdani, Yeste, Jones, & Coward, 2015; Nomikos, 2015).

However, the efficiency and safety of most of these methods is under question due to conflicting data on the benefits and hazards involved in use of these techniques. The major concerns raised can be summarized as:

- **(i)** Early exposure and manipulation of oocytes with external agents may interfere with normal embryo development and pose risks toward healthy live births (Vanden Meerschaut, Nikiforaki, Heindryckx, & De Sutter, 2014).
- **(ii)** The unforeseen downstream cascade of events that the changes in cell homoeostasis may entail following spatial and temporal changes in calcium concentrations due to artificial mechanisms (Santella & Dale, 2015).
- **(iii)** Not all artificial methods are capable of activating oocytes of all animal species (Bos-Mikich, Bressan, Ruggeri, Watanabe, & Meirelles, 2015).

(iv) Most importantly, there may be alterations and disruptions in the basic physiologic endocrine pathways that may affect the tertiary structure of the nucleus.

(v) Finally, possibility of adverse epigenetic modifications resulting in abnormal offsprings cannot be ruled out.

(vi) These factors desist from the routine application of artificial oocyte activation methods.

Investigations in human erythrocytes and neuronal cells (Zylinska, Kowalska, & Ferene, 2009; Zylinska, Kowalska, & Kozaczuk, 2008) have revealed that the steroid DHEA and its sulfate DHEA-S have a bimodal effect on the regulation of the plasma membrane calcium pump activity (PMCA). This action involves the nongenomic regulation of calcium homeostasis in different excitable cell types by modulating the ATP-dependent transport across membranes. Thus, these steroids facilitate in maintaining threshold levels of calcium oscillation within the cell cytoplasm. The critical levels of DHEA/DHEA-S may trigger a signaling cascade of events involving several receptor proteins and enzyme systems for, e.g., the Ca^{2+}-ATPase present on membrane of endoplasmic reticulum. The property of steroids to remodel the membrane structure by altering its fluidity may be an important contributory factor toward its mode of action. Incidentally, these authors (Zylinska et al., 2008) also elucidated that at very high/higher than threshold levels (as may be observed in PCOS women), DHEA-S may have a reduced stimulatory effect probably owing to the presence of the negatively charged sulfate group which may confer conformational modifications, altering its action. This seems to be a convincing explanation for reduced oocyte activation as observed vis-à-vis fertilization potential in poor responder women having lower threshold DHEA-S levels and in PCOS women having higher threshold levels of DHEA-S.

A more in-depth understanding of the chemical structure and function of these steroids may go a long way in devising strategies for use of DHEA/DHEA-S as endogenous agonist/s for oocyte activation in-vivo in natural cycles as well as in-vitro in IVF cycles.

8. CONCLUSION

Any intervention method employed for inducing oocyte maturation in-vitro may affect fertilization and subsequently early embryonic development. Because eventually, it is the meiotically competent oocytes that store all macromolecular organelles and information to be utilized during fertilization and post fertilization events which decides the future of a developing embryo (Chaube, 2001). During IVF, several parthenogenetic agonists like $SrCl_2$, thiomersal have been employed to activate intracellular Ca^{2+} oscillations. These components sensitize the IP3 receptors but do not bring about the release of IP3 per se (Cheek et al., 1993; Galione et al., 1993). Thus, since the natural biochemical pathway of activation is bypassed, the physiological consequences of such artificial methods on embryonic development and postnatal effects need to be thoroughly evaluated.

Devising noninvasive molecular stratagems is still an arduous task (McRae, Sharma, & Fisher, 2013; Nel-Themaat & Nagy, 2011). Understanding the biochemical and molecular mechanisms by which the steroid sulfate combination of DHEA/DHEA-S brings about oocyte activation may help establish it as an innate molecule that will not only be instrumental in regulating Ca^{2+} homeostasis in-vivo but can also serve as an in-vitro agonist for inducing oocyte activation. The greatest advantage in its favor is that it is a natural intermediate in the steroid biosynthesis pathway and therefore will avert any adverse effects associated with the use of artificial activators. Avenues for research in this area hold tremendous promise. Sustained experiments involving DHEA/DHEA-S supplementation and corresponding measurements of dose-dependent changes in calcium oscillations, if fructified, may pave the way for novel paradigms where the emphasis for oocyte activation will not only be focused on the sperm factor PLCζ but also on the "oocyte factor" DHEA/DHEA-S.

REFERENCES

Albertini, D. F. (1987). Cytoplasmic reorganization during the resumption of meiosis in cultured preovulatory rat oocytes. *Developmental Biology*, *120*, 121–131.

Amdani, S. N., Yeste, M., Jones, C., & Coward, K. (2015). Phospholipase C zeta (PLCζ) and male infertility: Clinical update and topical developments. *Advances in Biological Regulation*, *61*, 58–67.

Arola, J., Liu, J., Heikkila, P., Ilvesmaki, V., Salmenkivi, K., Voutilainen, R., et al. (2000). Expression of inhibin alpha in adrenocortical tumours reflects the hormonal status of the neoplasm. *The Journal of Endocrinology*, *165*(2), 223–229.

Artini, P. G., Simi, G., Ruggiero, M., Pinelli, S., Di Berardino, O. M., Papini, F., et al. (2012). DHEA supplementation improves follicular microenvironment in poor responder patients. *Gynecological Endocrinology*, *28*(9), 669–673.

Barad, D. H., & Gleicher, N. (2005). Increased oocyte production after treatment with dehydroepiandrosterone. *Fertility and Sterility*, *84*. 756.e1–3.

Barad, D. H., & Gleicher, N. (2006). Effect of dehydroepiandrosterone on oocyte and embryo yields, embryo grade and cell number in IVF. *Human Reproduction*, *21*(11), 2845–2849.

Baulieu, E. E. (1996). Dehydroepiandrosterone (DHEA): A fountain of youth? *The Journal of Clinical Endocrinology and Metabolism*, *81*, 3147–3151.

Bernhardt, R., & Waterman, M. R. (2007). *Cytochrome P450 and steroid hormone biosynthesis. The ubiquitous roles of cytochrome P450 proteins* (pp. 361–396). John Wiley & Sons, Ltd.

Bonser, J., Walker, J., Purohit, A., Reed, M. J., Potter, B. V., Willis, D. S., et al. (2000). Human granulosa cells are a site of sulphatase activity and are able to utilize dehydroepiandrosterone sulphate as a precursor for oestradiol production. *The Journal of Endocrinology, 167*, 465–471.

Bos-Mikich, A., Bressan, F. F., Ruggeri, R. R., Watanabe, Y., & Meirelles, F. V. (2015). Parthenogenesis and human assisted reproduction. *Stem Cells International, 2016*. 1970843. 8 pp.

Casson, P. R., Lindsay, M. S., Pisarska, M. D., Carson, S. A., & Buster, J. E. (2000). Dehydroepiandrosterone supplementation augments ovarian stimulation in poor responders: A case series. *Human Reproduction, 15*(10), 2129–2132.

Chaube, S. K. (2001). Role of meiotic maturation regulatory factors in developmental competence of mammalian oocytes. *Health and Population; Perspectives and Issues, 24*(4), 218–231.

Cheek, T. R., McGuinness, O. M., Vincent, C., Morton, R. B., Berridge, M. J., & Johnson, M. H. (1993). Fertilization and thimerosal stimulate similar calcium spiking patterns in mouse oocytes but by separate mechanisms. *Development, 119*, 179–189.

Chen, F., Knecht, K., Birzin, E., et al. (2005). Direct agonist/antagonist functions of dehydroepiandrosterone. *Endocrinology, 146*(11), 4568–4576.

Chen, J., Qian, Y., Tan, Y., & Mima, H. (2010). Successful pregnancy following oocyte activation by stronyium in normozoospermic patients of unexplained infertility with fertilization failure during previous intracytoplasmic sperm injection treatment. *Reproduction, Fertility, and Development, 22*, 852–855.

Chimote, B. N., Nath, N., & Chimote, N. M. (2017). Dehydroepiandrosterone sulphate (DHEAs) supplementation: A promising non-invasive futuristic strategy for oocyte activation in IVF cycles. *Human Reproduction, 32*(1), i190. Abstracts of the 33rd annual meeting of ESHRE 2017 Geneva, Switzerland, 2-5 July 2017.Poster P-117.

Chimote, N. M., Nath, N. M., Chimote, N. N., & Chimote, B. N. (2015). Follicular fluid dehydroepiandrosterone sulfate is a credible marker of oocyte maturity and pregnancy outcome in conventional in vitro fertilization cycles. *Journal Human Reproduction Science, 8*, 209–213.

Chithiwala, Z. H., Lee, H. C., Hill, D. L., Jellerette-Nolan, T., Fissore, R., Grow, D., et al. (2015). Phospholipase C zeta deficiency as a cause for repetitive oocyte fertilization failure during ovarian stimulation for in vitro fertilization with ICSI: A case report. *Journal of Assisted Reproduction and Genetics, 32*, 1415–1419.

Claus, R., & Hoffmann, B. (1980). Oestrogens, compared to other steroids of testicular origin, in blood plasma of boars. *Acta Endocrinologica, 94*, 404–411.

Dehenin, L., Jondet, M., & Scholler, R. (1987). Androgen and 19-norsteroid profiles in human preovulatory follicles from stimulated cycles: An isotope dilution mass spectrometric study. *Journal of Steroid Biochemistry, 26*, 399–405.

Downs, S. M., & Hunzicker-Dunn, M. (1995). Differential regulation of oocyte maturation and cumulus expansion in the mouse oocyte-cumulus cells complex by site-selective analog of cyclic adenosine monophosphate. *Developmental Biology, 172*, 72–85.

Ducibella, T., Huneau, D., Angelichio, E., Xu, Z., Schultz, R. M., Kopf, G. S., et al. (2002). Egg-to-embryo transition is driven by differential responses to Ca(2+) oscillation number. *Developmental Biology, 250*, 280–291.

Ebner, T., Moser, M., Sommergruber, M., et al. (2004). Complete oocyte activation failure after ICSI can be overcome by a modified injection technique. *Human Reproduction, 19*, 1837–1841.

Ethun, K. F., Wood, C. E., Parker, C. R., Kaplan, J. R., Chen, H., & Appt, S. E. (2012). Effect of ovarian aging on androgen biosynthesis in a cynomolgus macaque model. *Climacteric, 15*(1), 82–92.

Ford, J. H. (2013). Reduced quality and accelerated follicle loss with female reproductive aging—Does decline in theca dehydroepiandrosterone (DHEA) underlie the problem? *Journal of Biomedical Science*, *20*, 93.

Ford, J. H., Wilkin, H. Z., Thomas, P., & McCarthy, C. (1996). A 13-year cytogenetic study of spontaneous abortion: Clinical applications of testing. *The Australian & New Zealand Journal of Obstetrics & Gynaecology*, *36*(3), 314–318.

Fouany, M. R., & Sharara, F. I. (2013). Is there a role for DHEA supplementation in women with diminished ovarian reserve? *Journal of Assisted Reproduction and Genetics*, *30*(9), 1239–1244. https://doi.org/10.1007/s10815-013-0018-x.

Galione, A., McDougall, A., Busa, W. B., Willmott, N., Gillot, I., & Whitaker, M. (1993). Redundant mechanisms of calcium-induced calcium release underlying calcium waves during fertilization of sea urchin eggs. *Science*, *261*, 348–352.

Geyer, J., Godoy, J. R., & Petzinger, E. (2004). Identification of a sodium-dependent organic anion transporter from rat adrenal gland. *Biochemical and Biophysical Research Communications*, *316*, 300–306.

Gleicher, N., & Barad, D. H. (2011). Dehydroepiandrosterone (DHEA) supplementation in diminished ovarian reserve (DOR). *Reproductive Biology and Endocrinology*, *9*, 67.

Gleicher, N., Weghofer, A., & Barad, D. H. (2011). The role of androgens in follicle maturation and ovulation induction: Friend or foe of infertility treatment? *Reproductive Biology and Endocrinology*, *9*, 116.

Gordo, A. C., Rodrigues, P., Kurokawa, M., Jellerette, T., Exley, G. E., Warner, C., et al. (2002). Intracellular calcium oscillations signal apoptosis rather than activation in in-vitro aged mouse eggs. *Biology of Reproduction*, *66*, 1828–1837.

Haning, R. V., Hackett, R. J., Flood, C. A., Loughlin, J. S., Zhao, Q. Y., & Longcope, C. (1993). Plasma dehydroepiandrosterone sulfate serves as a prehormone for 48% of follicular fluid testosterone during treatment with menotropins. *The Journal of Clinical Endocrinology and Metabolism*, *76*(5), 1301–1307.

Hegele Hartung, C., Kuhnke, J., Lessl, M., Grondahl, C., Ottesen, C., Beier, H. M., et al. (1999). Nuclear and cytoplasmic maturation of mouse oocytes after treatment with synthetic meiosis activating sterol in-vitro. *Biology of Reproduction*, *61*, 1362–1372.

Hoffmann, B., & Landeck, A. (1999). Testicular endocrine function, seasonality and semen quality of the stallion. *Animal Reproduction Science*, *57*, 89–98.

Hörmanseder, E., Tischer, T., & Mayer, T. U. (2013). Modulation of cell cycle control during oocyte-to-embryo transitions. *The EMBO Journal*, *32*, 2191–2203.

Hornsby, P. J. (2004). Aging of the human adrenal cortex. *Science of Aging Knowledge Environment*, *2004*(35), re6.

Janowski, T., Zdunczyk, S., Ras, A., & Mwaanga, E. S. (1999). Use of estrone sulfate determination in goat blood for the detection of pregnancy and prediction of fetal number. *Tierärztliche Praxis. Ausgabe G, Grosstiere/Nutztiere*, *27*, 107–109.

Jimena, P., Castilla, J. A., Peran, F., Ramirez, J. P., Vergara, F., Jr., Molina, R., et al. (1992). Adrenal hormones in human follicular fluid. *Acta Endocrinologica*, *127*, 403–406.

Kashir, J., Nomikos, M., Lai, F. A., & Swann, K. (2014). Sperm-induced Ca2+ release during egg activation in mammals. *Biochemical and Biophysical Research Communications*, *450*, 1204–1211.

Kikuchi, K., Naito, K., Noguchi, J., Shimada, A., Kaneko, H., Yamashita, M., et al. (2000). Maturation/M-phase promoting factor: A regulator of aging in porcine oocytes. *Biology of Reproduction*, *63*, 715–722.

Kim, J. W., Choi, J. L., Yang, S. H., Yoon, S. H., Jung, J. H., & Lim, J. H. (2012). Live birth after SrCl(2) oocyte activation in previous repeated failed or low fertilization rates after ICSI of frozen thawed testicular spermatozoa: Case report. *Journal of Assisted Reproduction and Genetics*, *29*, 1393–1396.

Kroboth, P. D., Salek, F. S., Pittenger, A. L., Fabian, T. J., & Frye, R. F. (1999). DHEA and DHEA-S: A review. *Journal of Clinical Pharmacology, 39*, 327–348.

Kushnir, V. A., Ludaway, T., Russ, R. B., Fields, E. J., Koczor, C., & Lewis, W. (2012). Reproductive aging is associated with decreased mitochondrial abundance and altered structure in murine oocytes. *Journal of Assisted Reproduction and Genetics, 29*(7), 637–642.

Lambeth, J. D., Xu, X. X., & Glover, M. (1987). Cholesterol sulfate inhibits adrenal mitochondrial cholesterol side chain cleavage at a site distinct from cytochrome P450scc. Evidence for an intramitochondrial cholesterol translocator. *The Journal of Biological Chemistry, 262*, 9181–9188.

Leowattana, W. (2004). DHEAS as a new diagnostic tool. *Clinica Chimica Acta; International Journal of Clinical Chemistry, 341*, 1–15.

Lisurek, M., & Bernhardt, R. (2004). Modulation of aldosterone and cortisol synthesis on the molecular level. *Molecular and Cellular Endocrinology, 215*, 149–159.

Liu, D., & Dillon, J. S. (2002). Dehydroepiandrosterone activates endothelial cell nitric-oxide synthase by a specific plasma membrane receptor coupled to Gαi2,3. *The Journal of Biological Chemistry, 277*, 21379–21388.

Mamas, L., & Mamas, E. (2009). Dehydroepiandrosterone supplementation in assisted reproduction: Rationale and results. *Current Opinion in Obstetrics & Gynecology, 21*, 306–308.

Manipalviratn, S., Ahnonkitpanit, V., Numchaisrika, P., Chompurat, D., Pansatha, J., & Suwajanakorn, S. (2006). Results of direct current electrical activation of failed to fertilize oocytes after intracytoplasmic sperm injection. *The Journal of Reproductive Medicine, 51*, 493–499.

McEvoy, M. (2012). *The endocrine journey: Steroidal hormones & their pathways*. January 12.

McRae, C., Sharma, V., & Fisher, J. (2013). Metabolite profiling in the pursuit of biomarkers for IVF outcome: The case for metabolomics studies. *International Journal of Reproductive Medicine, 2013*. 603167.

Mo, Q., Lu, S. F., & Simon, N. G. (2006). Dehydroepiandrosterone and its metabolites: Differential effects on androgen receptor trafficking and transcriptional activity. *The Journal of Steroid Biochemistry and Molecular Biology, 99*(1), 50–58.

Montag, M., Koster, M., van der Ven, K., Bohlen, U., & van der Ven, H. (2012). The benefit of artificial oocyte activation is dependent on the fertilization rate in a previous treatment cycle. *Reproductive Biomedicine Online, 24*, 521–526.

Nakai, M., Ito, J., Suzuki, S., Fuchimoto, D., Sembon, S., Suzuki, M., et al. (2016). Lack of calcium oscillation causes failure of oocyte activation after intracytoplasmic sperm injection in pigs. *The Journal of Reproduction and Development, 62*, 615–621.

Nel-Themaat, L., & Nagy, Z. P. (2011). A review of the promises and pitfalls of oocyte and embryo metabolomics. *Placenta, 32*, 257–263.

Neunzig, J., & Bernhardt, R. (2014). Dehydroepiandrosterone sulfate (DHEAS) stimulates the first step in the biosynthesis of steroid hormones. *PLoS One 92*, e89727. https://doi.org/10.1371/journal.pone.0089727.

Nomikos, M. (2015). Novel signaling mechanism and clinical applications of sperm specific PLCζ. *Biochemical Society Transactions, 43*, 371–376.

Norris, R. P., et al. (2009). Cyclic GMP from the surrounding somatic cells regulates cyclic AMP and meiosis in the mouse oocyte. *Development, 136*, 1869–1878.

Ozil, J. P., Banrezes, B., Toth, S., Pan, H., & Schultz, R. M. (2006). Ca2+ oscillatory pattern in fertilized mouse eggs affects gene expression and development to term. *Developmental Biology, 300*, 534–544.

Ozil, J. P., Markoulaki, S., Toth, S., Matson, S., Banzeres, B., Knott, J. G., et al. (2005). Egg activation events are regulated by the duration of a sustained (Ca2+) cyt signal in the mouse. *Developmental Biology, 282*, 39–54.

Park, J. Y., et al. (2004). EGF-like growth factors as mediators of LH action in the ovulatory follicle. *Science, 30*, 682–684.
Pasqualini, J. R., & Chetrite, G. S. (2005). Recent insight on the control of enzymes involved in estrogen formation and transformation in human breast cancer. *The Journal of Steroid Biochemistry and Molecular Biology, 93*, 221–236.
Patel, M. A., & Katyare, S. S. (2006). Dehydroepiandrosterone (DHEA) treatment stimulates oxidative energy metabolism in the cerebral mitochondria from developing rats. *International Journal of Developmental Neuroscience, 24*(5), 327–334.
Poland, B. J., Miller, J. R., Harris, M., & Livingston, J. (1981). Spontaneous abortion. A study of 1961 women and their conceptuses. *Acta Obstetricia et Gynecologica Scandinavica. Supplement, 102*, 1–32.
Que, E. L., Bleher, R., Duncan, F. E., Kong, B. Y., Gleber, S. C., Vogt, S., et al. (2015). Quantitative mapping of zinc fluxes in the mammalian egg reveals the origin of fertilization-induced zinc sparks. *Nature Chemistry*, 7, 130–139.
Ramadan, W. M., Kashir, J., Jones, C., & Coward, K. (2012). Oocyte activation and phospholipase C zeta (PLCζ): Diagnostic and therapeutic implications for assisted reproductive technology. *Cell Communication and Signaling: CCS, 10*, 12.
Santella, L., & Dale, B. (2015). Assisted yes, but where do we draw the line? *Reproductive Biomedicine Online, 31*, 476–478.
Sanusi, R., Yu, Y., Nomikos, M., Lai, F. A., & Swann, K. (2015). Rescue of failed oocyte activation after ICSI in a mouse model of male factor infertility by recombinant phospholipase C ζ. *Molecular Human Reproduction, 21*, 783–791.
Schular, G., Greven, H., Kowalewski, M. P., Doring, B., Ozalp, G. R., et al. (2008). Placental steroids in cattle: Hormones, placental growth factors or by-products of trophoblast giant cell differentiation? *Experimental and Clinical Endocrinology & Diabetes, 116*, 429–436.
Schulman, et al. (2007). *The Merck index* (13th ed., p. 7798).
Schultz, R. M., & Kopf, G. S. (1995). Molecular basis of mammalian egg activation. *Current Topics in Developmental Biology, 30*, 21–62.
Scott, T. (1996). *Concise encyclopedia biology* (p. 49). Walter de Gruyter.
Simsek-Duran, F., Li, F., Ford, W., Swanson, R. J., Jones, H. W., & Castora, F. J. (2013). Age associated metabolic and morphologic changes in mitochondria of individual mouse and hamster oocytes. *PLoS One, 8*(5), e64955.
Stricker, S. A. (1999). Comparative biology of calcium signaling during fertilization and egg activation in animals. *Developmental Biology, 211*, 157–176.
Strott, C. A. (1996). Steroid sulfo-transferases. *Endocrine Reviews, 17*, 670–697.
Su, Y. Q., Xia, G. L., Byskov, A. G., Fu, G. D., & Yang, C. R. (1999). Protein kinase C and intracellular calcium are involved in follicle stimulating hormone-mediated meiotic resumption of cumulus cell-enclosed porcine oocytes in hypoxanthine-supplemented medium. *Molecular Reproduction and Development, 53*(1), 51–58.
Su, Y. Q., Xia, G. L., He, C., Fu, G. D., & Yng, C. R. (1998). Studies on the effects of gonadotropins on the meiotic maturation of cumulus-enclosed porcine oocytes in-vitro. *Chi Journal Applied Physiology, 14*, 264–267.
Sugimura, S., Matoba, S., Hashiyada, Y., Aikawa, Y., Ohtake, M., Matsuda, H., et al. (2012). Oxidative phosphorylation-linked respiration in individual bovine oocytes. *The Journal of Reproduction and Development, 58*(6), 636–641.
Swann, K., & Lai, F. A. (2016). Egg activation at fertilization by a soluble sperm protein. *Physiological Reviews, 96*, 127–149.
Takahashi, T., Kikuchi, T., Kidokoro, Y., & Shirakawa, H. (2013). Ca2+ influx dependent refilling of intracellular Ca2+ stores determines the frequency of Ca2+ oscillations in fertilized mouse eggs. *Biochemical and Biophysical Research Communications, 430*, 60–65.

Tesarik, J., & Mendoza, C. (1997). Direct non-genomic effects of follicular steroids on maturing human oocytes: Oestrogen versus androgen antagonism. *Human Reproduction Update, 3*(2), 95–100.

Tesarik, J., & Sousa, M. (1995). More than 90% fertilization rates after intracytoplasmic sperm injection and artificial induction of oocyte activation with calcium ionophore. *Fertility and Sterility, 63*, 343–349.

Tosti, E., & Ménézo, Y. (2010). Sperm induced oocyte activation. In T. Lejeune & P. Delvaux (Eds.), *2010. Human spermatozoa: Maturation, capacitation and abnormalities* (pp. 379–397). New York: Nova Biomedical Books, Science Publishers Inc.

Tsutsumi, R., Hiroi, H., Momoeda, M., hosokawa, Y., Nakazava, F., et al. (2008). Inhibitory effects of cholesterol sulfate on progesterone production in human granulosa-like tumor cell line, KGN. *Endocrine Journal, 55*, 575–581.

Van Blerkom, J. (1991). Microtubule mediation of cytoplasmic and nuclear maturation during the early stages of resumed meiosis in cultured mouse oocytes. *Proceedings of the National Academy of Sciences of the United States of America, 88*, 5031–5035.

Van Blerkom, J., Davis, P. W., & Lee, J. (1995). ATP content of human oocytes and developmental potential and outcome after in-vitro fertilization and embryo transfer. *Human Reproduction, 10*(2), 415–424.

Vanden Meerschaut, F., Nikiforaki, D., Heindryckx, B., & De Sutter, P. (2014). Assisted oocyte activation following ICSI fertilization failure. *Reproductive Biomedicine Online, 28*, 560–571.

Vanttinen, T., Liu, J., Kuulasmaa, T., Kivinen, P., & Voutilainen, R. (2003). Expression of activin/inhibin signaling components in the human adrenal gland and the effects of activins and inhibins on adrenocortical steroidogenesis and apoptosis. *The Journal of Endocrinology, 1783*, 479–489.

Von Stetina, J. R., & Orr-Weaver, T. L. (2011). Developmental control of oocyte maturation and egg activation in metazoan models. *Cold Spring Harbor Perspectives in Biology, 3*, a005553.

Wakai, T., & Fissore, R. A. (2013). Ca(2+) homeostasis and regulation of ER Ca(2+) in mammalian oocytes/eggs. *Cell Calcium, 53*, 63–67.

Watson, C. S., Jeng, Y. J., & Kochukov, M. Y. (2008). Nongenomic actions of estradiol compared with estrone and estriol in pituitary tumor cell signaling and proliferation. *The FASEB Journal, 22*(9), 3328–3336.

Webb, S. J., Geoghegan, T. E., Prough, R. A., & Michael Miller, K. K. (2006). The biological actions of dehydroepiandrosterone involves multiple receptor. *Drug Metabolism Reviews, 38*(1–2), 89–116.

White, B. A., & Porterfield, S. P. (2013). Endocrine and reproductive physiology: Mosby physiology monograph series (with student consult online access). In *Endocrine and reproductive physiology* (4th ed., p. 164). Elsevier Health Sciences. ISBN: 0-323-08704-3.

William, F., & Ganong, M. D. (2005). *Review of medical physiology* (22nd ed., p. 362). McGraw Hill.

Yanagida, K., Katayose, H., Yazawa, H., Kimura, Y., Sato, A., Yanagimachi, H., et al. (1999). Successful fertilization and pregnancy following ICSI and electrical oocyte activation. *Human Reproduction, 14*, 1307–1311.

Yding Andersen, C. (1993). Characteristics of human follicular fluid associated with successful conception after in vitro fertilization. *The Journal of Clinical Endocrinology and Metabolism*, 77, 1227–1234.

Yeste, M., Jones, C., Amdani, S. N., Patel, S., & Coward, K. (2016). Oocyte activation deficiency: A role for an oocyte contribution? *Human Reproduction Update, 22*, 23–47. Advance access published Sep 7. Pii: dmv040.

Yoon, S. Y., Eum, J. H., Lee, J. E., Lee, H. C., Kim, Y. S., Han, J. E., et al. (2012). Recombinant human phospholipase C zeta I induces intracellular calcium oscillations and oocyte activation in mouse and human oocytes. *Human Reproduction*, *27*, 1768–1780.
Young, J. M., & McNeilly, A. S. (2010). Theca: The forgotten cell of the ovarian follicle. *Reproduction*, *140*(4), 489–504.
Zuke, K. A., & Brackett, B. G. (1990). Luteinizing hormone enhanced in-vitro maturation of bovine oocyte with or without protein supplementation. *Biology of Reproduction*, *43*, 784–787.
Zylinska, L., Kowalska, I., & Ferene, B. (2009). Calmodulin effects on steroids-regulated plasma membrane calcium pump activity. *Cell Biochemistry and Function*, *27*, 111–117.
Zylinska, L., Kowalska, I., & Kozaczuk, A. (2008). Fast action of neuroactive steroids on plasma membrane calcium pump in PC12 cells. *Annals of the New York Academy of Sciences*, *1148*, 515–519. https://doi.org/10.1196/annals.1410.077.

FURTHER READING

Shulman, R. A., & Dean, C. (2007). *DHEA dehydroepiandrosterone is a common hormone produced in the adrenal glands, the gonads, and the brain* (p. 100). New York City: Rodale, Inc, ISBN: 978-1-57954-942-8. Solve it with supplements.

CHAPTER NINE

Dehydroepiandrosterone and Bone

Shuanhu Zhou*,†,1, Julie Glowacki*,‡
*Brigham and Women's Hospital, Harvard Medical School, Boston, MA, United States
†Harvard Stem Cell Institute, Harvard University, Cambridge, MA, United States
‡Harvard School of Dental Medicine, Boston, MA, United States
[1]Corresponding author: e-mail addresses: szhou@bwh.harvard.edu; shuanhuzhou@yahoo.com

Contents

Abstract

In humans, dehydroepiandrosterone (DHEA), secreted mainly from the adrenal cortex, and its sulfate ester, DHEAS, are the most abundant circulating steroids. DHEA/DHEAS possess pleiotropic effects in human aging, bone, metabolic diseases, neurologic function/neurodegenerative diseases, cancer, immune system and disorders, cardiovascular diseases, diabetes, muscle function, sexual dysfunction, and other health conditions. The age-related reduced levels of DHEA and DHEAS are associated with bone mineral density measures of osteopenia and osteoporosis. Clinical, epidemiological, and experimental studies indicate that DHEA replacement therapy may be beneficial for bone health through its inhibition of skeletal catabolic IL-6 and stimulation of osteoanabolic IGF-I-mediated mechanisms. Studies with primary cultures of human bone marrow-derived mesenchymal stem cells (hMSCs) were used to show that DHEA stimulates osteoblastogenesis. The in vitro stimulation of both osteoblastogenesis and *IGF-I* gene expression by DHEA in hMSCs requires IGF-I receptor, PI3K, p38 MAPK, or p42/44 MAPK signaling pathways. The in vitro inhibition of IL-6 secretion in hMSCs by DHEA was more consistent and extensive than by estradiol or dihydrotestosterone. In summary, evidence from us and others indicates that DHEA may be useful for treating bone diseases through its inhibition of skeletal catabolic IL-6 and stimulation of anabolic IGF-I-mediated mechanisms.

Vitamins and Hormones, Volume 108
ISSN 0083-6729
https://doi.org/10.1016/bs.vh.2018.01.005

1. INTRODUCTION

In humans, dehydroepiandrosterone (DHEA) and its sulfate ester, DHEAS, are the most abundant circulating steroids. DHEA/DHEAS possess pleiotropic effects in human aging, bone, metabolic diseases, neurologic function/neurodegenerative diseases, cancer, immunosenescence, and immune disorders, cardiovascular diseases, diabetes, muscle function, sexual dysfunction, sleep, and other health conditions (Bellino, Daynes, Hornsby, Lavrin, & Nestler, 1995; Binello & Gordon, 2003; Diamanti-Kandarakis et al., 2017; Gordon, Glowacki, & LeBoff, 1999; Nieuwenhu & Thijssen, 1999; Watson, Huls, Araghinikuam, & Chung, 1996). DHEA is produced from cholesterol mainly in the adrenals but also by the testis, the ovaries, skin, and brain; DHEAS is exclusively produced from DHEA in the adrenal zona reticularis, which is rich in sulfuryl transferase activity (Samaras, Samaras, Frangos, Forster, & Philippe, 2013). The secretion of DHEA and DHEAS from adrenals is regulated by the hypothalamo–pituitary–adrenal axis and GH–IGF-I axis (Belgorosky, Baquedano, Guercio, & Rivarola, 2008; Gupta & Morley, 2014; Kasayama et al., 2007; Lamberts, van den Beld, & van der Lely, 1997). DHEA can be transformed in peripheral tissues to more potent androgens and estrogens (Binello & Gordon, 2003; Genazzani, Lanzoni, & Genazzani, 2007) by aromatase activity (Gennari, Merlotti, & Nuti, 2011). Steroid production at a tissue level permits autoregulation of the local hormonal environment according to local needs, with less systemic effects, called intracrinology (Samaras et al., 2013). In premenopausal women, 50%–75% of estrogens and the majority of androgens are produced through intracrine mechanisms from DHEA; after the menopause, practically all androgens and estrogens are synthesized at a tissue level. In men, whose testes continue to secrete androgens throughout life, local hormone production also occurs but is more difficult to assess (Genazzani et al., 2007; Labrie, Martel, Bélanger, & Pelletier, 2017; Panjari & Davis, 2007; Samaras et al., 2013).

DHEA has direct and indirect mechanisms of action (Binello & Gordon, 2003). DHEA mediates its action via multiple signaling pathways involving specific membrane G protein-coupled receptors and via transformation into androgen and estrogen derivatives (androgens, estrogens, 7α and 7β DHEA, and 7α and 7β epiandrosterone derivatives) acting through their specific nuclear receptors (Traish, Kang, Saad, & Guay, 2011). Early studies on the biological action of DHEA mainly focused on its indirect action via the metabolism to testosterone and estradiol, and the subsequent effect on the activation of steroid receptors AR, ERα, and ERβ. Recent studies showed that DHEA acts with much lower binding affinities for nuclear receptors AR, ERα, and ERβ or an activator/agonist for nuclear receptors, such as constitutive-activated receptor β (CARβ), peroxisome proliferator-activated receptor α (PPARα), and pregnane X receptor (PXR), or membrane G protein-coupled receptors (Binello & Gordon, 2003; Prough, Clark, & Klinge, 2016).

DHEA and DHEAS serum levels are high in fetal life, decrease after birth, and show a marked pubertal increase to a maximal level during young adulthood; after young adulthood, levels of DHEA decline and are between 10% and 20% of young adult levels by age 70 (Gordon et al., 1999). Clinical and epidemiological studies suggest that low DHEA levels may be associated with ischemic heart disease, endothelial dysfunction, atherosclerosis, bone loss, inflammatory diseases, and sexual dysfunction (Traish et al., 2011). Evidence is accruing in support of DHEA supplementation in adrenal insufficiency, hypopituitarism, osteoporosis, systemic lupus erythematosus, depression, and schizophrenia (Binello & Gordon, 2003). This review summarizes recent information concerning the role of DHEA in skeletal physiology including its effects on human mesenchymal stem cells (MSCs).

2. DHEA IN AGE-RELATED BONE LOSS AND OSTEOPOROSIS

Osteoporosis affects millions of people worldwide and is the most common bone disease in humans, representing a major public health problem (Cosman et al., 2014). Osteoporosis and low bone mass affected an estimated 53.6 million or 54% of people 50 years and older in the United States in 2010 (Wright et al., 2014). Due in part to an increasing aged population, the cost of care of patients with osteoporotic fractures is expected to rise to $25.3 billion by 2025 (Cosman et al., 2014).

Bone mineral density (BMD) increases during adolescence accompanying the rise in adrenal androgens and gonadal steroids, and declines with menopause and aging (Bonjour, Theintz, Buchs, Slosman, & Rizzoli, 1991). In adults, serum concentrations of DHEA and DHEAS are associated with age-related loss of BMD and subsequent osteopenia, osteoporosis, and fracture risk (Cormier, Souberbielle, & Kahan, 2001; Gordon et al., 1999). Our study of a cohort of 102 healthy women between 29 and 95 years of age showed an age-related decrease of DHEAS levels, significantly associated with the decrease in BMD (Haden, Glowacki, Hurwitz, Rosen, & LeBoff, 2000).

The MrOS Sweden Study reported that low serum DHEAS is a risk marker of mainly nonvertebral fractures in older men ($n=2568$, aged 69–81 years) (Ohlsson et al., 2017).

The age-related reduced levels of DHEA and DHEAS in parallel with the appearance of age-related osteopenia and osteoporosis suggest that DHEA replacement therapy may be beneficial for bone health in older patients. Positive changes in BMD have been repeatedly observed in studies, but the changes have been small, specific to site and sex, and not always reproducible between studies (Stewart, 2006). In a 2-year, randomized, placebo-controlled, double-blind study involving 87 elderly men and 57 elderly women with low levels of sulfated DHEA, DHEA replacement (at a dose of 75 mg per day in men and 50 mg per day in women) had no physiologically relevant beneficial effects on body composition, physical performance, insulin sensitivity, or quality of life. That trial, however, showed that men in the DHEA treatment group had an increase in BMD at the femoral neck, and women who received DHEA had an increase in BMD at the ultradistal radius (Nair et al., 2006). Jankowski et al. reported similar skeletal results after 12 months of therapy in 69 elderly men and women, with a significant 1.2%–1.6% increase in BMD at the hip, as compared with that in 71 placebo-treated controls (Jankowski et al., 2006). The DHEAge study demonstrated improvements only in women; those BMD increases occurred at Ward's triangle in women under the age of 70 years and at the radius in those over 70 years (Baulieu et al., 2000). Daily oral doses that restore DHEAS blood levels to the values seen in young adults are well tolerated in the short-term (6 months to 1 year) and seem to have a positive although modest beneficial effect on bone in elderly women, particularly those with low pretreatment DHEAS levels (Cormier et al., 2001). Although recent evidence suggests that DHEA administration does not significantly impact metabolic markers in postmenopausal women with normal adrenal function (Elraiyah et al., 2014), DHEA supplementation may be of benefit in two specific groups of women: those with the lowest circulating levels of DHEA and those with established osteoporosis (Raven & Hinson, 2007). In elderly women with the lowest baseline DHEAS levels, the effects of DHEA administration on bone were the most marked (Marcelli, 2003).

The data regarding DHEA replacement therapy in older men are inconsistent and even contradictory, with some investigators concluding that DHEA therapy has positive effects on BMD in both women and men (Samaras et al., 2013), others that the effects of DHEA therapy on bone turnover and BMD can be greater in females than in males (Weiss et al., 2009), and others that there is no effect of DHEA supplementation (Corona et al., 2013). A large cross-sectional study with 1089 healthy Korean men supported previous evidence that DHEAS has favorable effects on bone mass in men and that a low serum DHEAS level may be a potential risk factor for male osteoporosis (Lee et al., 2015). Some studies showed that DHEA has a protective effect against osteoporosis in women after menopause through the intracrine mechanism in osteoblasts, where DHEA is converted to estrogen through aromatase activity (Nawata et al., 1995; Takayanagi et al., 2002). The effects of DHEA on bone may be related to conversion of DHEA into estradiol and testosterone. Some investigators propose that this may explain why there is no consistent effect on bone of men, whose testes may continue to produce enough testosterone for skeletal benefits (Cormier et al., 2001).

Although DHEA and sulfated DHEA are the most abundant steroids secreted from the adrenal cortex in humans, as compared with the corticosteroids, cortisol, and aldosterone, they remain something of an enigma. DHEA is not essential for life, rodent adrenal glands do not make DHEA, and the involvement of tissue-specific receptors or intracellular signaling pathways is unclear (Stewart, 2006). Although it is debatable whether the correction of age-related decreases in the secretion of DHEA and testosterone through "physiologic" replacement regimens should be attempted (Stewart, 2006), the studies of DHEA replacement therapy so far do not exclude the possibility that DHEA treatment may prove beneficial in specific subgroups of elderly subjects (Perrini, Laviola, Natalicchio, & Giorgino, 2005) or for specific diseases such as osteoporosis.

3. DHEA AND IGF-I IN BONE HEALTH

The GH/IGF-I axis is an important regulator of bone homeostasis and is central to the achievement of normal longitudinal bone growth and bone mass. Secretion of GH and IGF-I decreases with aging, and abnormalities in the GH/IGF-I axis play a role in the pathogenesis of the osteoporosis of anorexia nervosa and after glucocorticoid exposure (Giustina, Mazziotti, & Canalis, 2008). The GH/IGF-I axis is important for ACTH-dependent DHEA and DHEAS production in human adrenals. Aging directly influences adrenal DHEA production (adrenopause) and also has an indirect influence via the GH/IGF-I axis (somatopause) (Kasayama et al., 2007). Several studies examining the effect of DHEA administration on circulating IGF-I concentrations demonstrated a possibility that DHEA treatment influences serum IGF-I levels in some individuals (Arlt et al., 1999; Morales, Haubrich, Hwang, Asakura, & Yen, 1998; Morales, Nolan, Nelson, & Yen, 1994; Papierska, Rabijewski, Kasperlik-Zaluska, & Zgliczynski, 2012; van Thiel et al., 2005; Yen, Morales, & Khorram, 1995). Nevertheless, the mechanisms by which DHEA/DHEAS increase serum IGF-I are unresolved.

IGF-I is a hormone that functions in an endocrine, paracrine, and autocrine manner in most tissues. Although it was postulated initially that liver-derived IGF-I is the major source of IGF-I, it is also produced in a wide variety of tissues including bone, cartilage, and all cell lineages in the skeleton (Guntur & Rosen, 2013). The role of IGF-I in bone metabolism has been established from clinical studies of patients with growth hormone deficiency or *IGF-I* gene disruption, as well as from transgenic animal studies that disrupt or overexpress components of the IGF-I axis (Mohan & Kesavan, 2012). IGF-I is essential for the development and growth of the skeleton and maintenance of bone mass (Agnusdei & Gentilella, 2005).

The age-associated decline in circulating DHEAS is accompanied by a decline in IGF-I, with a strong correlation between them (Haden et al., 2000). The dramatic rise in DHEA and IGF-I levels during adolescence and bone accretion, the age-related decline in levels of DHEA and IGF-I, and evidence that DHEA and IGF-I levels are positively correlated with bone density suggest that DHEA and IGF-I may be critical for the development of peak bone mass (reviewed in Gordon et al., 1999). Adolescents and young adults with anorexia nervosa have reduced peak bone mass and an increased risk of osteoporotic fractures in conjunction with subnormal DHEA and IGF-I levels that may be linked to their reduced bone mass (Gordon et al., 2002; Soyka, Grinspoon, Levitsky, Herzog, & Klibanski, 1999). In addition, both IGF-I and DHEAS were significantly correlated with BMD and bone mineral content in young adults with cystic fibrosis (Gordon et al., 2006). Relative insufficiencies of IGF-I and DHEAS may be additional risk factors for low bone density in postmenopausal women with endogenous subclinical hyperthyroidism (Földes, Lakatos, Zsadányi, & Horváth, 1997). Clinical studies demonstrated that DHEA therapy increased bone formation and serum IGF-I in specific subgroups of elderly subjects (Morales et al., 1998, 1994; Papierska et al., 2012; Yen et al., 1995). In a clinical trial of 19 women, aged 50–78 years, with glucocorticoid-induced osteoporosis, a significant elevation of serum IGF-I and increase of BMD were observed after 6 and 12 months of

DHEA treatment (Papierska et al., 2012). In another study, daily oral DHEA administration for 6 months resulted in elevation of serum IGF-I levels in both men and women (Morales et al., 1998, 1994). In a randomized trial with older women and men, DHEA (50 mg/day) plus vitamin D (16 μg/day) and calcium (700 mg/day) supplements significantly improved spine BMD and increased serum IGF-I after 1 year in women, but not in men (Weiss et al., 2009). That trial included a second-year, open-label extension, in which all subjects received DHEA supplements; women on continuous DHEA showed a 3.6% increase in spine BMD over baseline, and men showed a small but significant 1.6% increase. In a 1-year randomized, placebo-controlled, double-blinded trial in 58 women and 61 men, aged 60–88 years, oral DHEA (50 mg/day) produced significant increases in hip BMD in older adults as well as serum IGF-I in women (Jankowski et al., 2008). The women in the DHEA group but not the men also had a significant increase in lumbar spine BMD. Thus, the gender differences in BMD responses to DHEA may be site-specific and mediated through estrogen and IGF-I.

4. DHEA AND IGF-I IN BONE CELLS

DHEA can be transformed in peripheral tissues to more potent androgens and estrogens (Binello & Gordon, 2003; Genazzani et al., 2007). Bone is one of the peripheral tissues that transform DHEA to androgens and estrogens. It was suggested that human MSCs contain an intracrine enzyme machinery potentially able to synthesize the cell's own supply of dihydrotestosterone (DHT), testosterone (T), and 17β-estradiol (E2) from DHEA (Pino et al., 2006; Sillat et al., 2009). DHEA significantly increased aromatase, ERα, ERβ, and AR expression in primary murine osteoblasts (Wang, Wang, Wang, & Li, 2009) and stimulated osteoblast differentiation in murine MSCs (Qiu, Gui, Xu, Li, & Wang, 2015).

There is increasing information about the effects of DHEA and IGF-I on bone cells. Several animal studies have reported that DHEA promotes osteoblast differentiation and inhibits osteoclast maturation (Qiu et al., 2015; Wang, Wang, Li, & Wang, 2006; Wang et al., 2009; Wang, Wang, Wang, Zhu, & Li, 2007). After adult OVX mice were treated with DHEA or 17-β-estradiol (E2) for 12 weeks, DHEA improved bone mass and displayed action like that of E2 with regard to decreasing osteoclast-related markers of bone resorption (Zhang et al., 2016). Primary human osteoblasts were used to show that both DHEA and DHT stimulated proliferation and differentiation and those effects required the androgen receptor, and that DHEA effects did not require conversion to androgens or estrogens (Kasperk, Wakley, Hierl, & Ziegler, 1997). Osteoblast differentiation of human MSCs in osteoblastogenic medium can be stimulated by E2, DHT, and DHEA in a dose-dependent manner (Fig. 1). It is notable that P450 aromatase was required for DHEA effects in osteoblasts isolated from women (Takayanagi et al., 2002). Osteoblasts from osteoporotic subjects show dysregulation of IGF-I receptor signaling (Perrini et al., 2008). Local IGF-I/IGF-IR signaling is critical for the anabolic bone actions of growth hormone, parathyroid hormone (PTH), and DHEA (Liang et al., 2016; Wang, Bikle, & Chang, 2013).

More information is accumulating about the regulation of in vitro differentiation of human osteoblast progenitors, most commonly obtained from bone marrow as MSCs. Those cells retain many of the properties of the individual from whom they were isolated, such as age and estrogen status (Zhou & Glowacki, 2017). There are numerous intrinsic age-related alterations in human MSCs that may contribute to the process of skeletal aging in humans (Zhou, 2015; Zhou et al., 2008). IGF-I and IGFBP-3 were shown to be secreted constitutively by human marrow cells in vitro, with an age-dependent increase in IGFBP-3 that was not evident in a subject receiving estrogen replacement therapy (Rosen, Verault, Steffens, Cheleuitte, & Glowacki, 1997). We and others showed that exogenous IGF-I stimulates osteoblast differentiation of MSCs (Granero-Moltó et al., 2011; Xian et al., 2012;

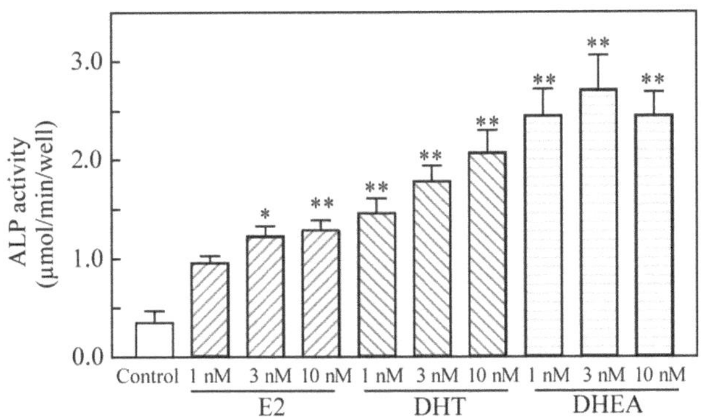

Fig. 1 The effects of steroids on alkaline phosphatase (ALP) activity in human bone marrow stromal KM101cells. The effects of 17β-estradiol (E2), dihydrotestosterone (DHT), and DHEA on osteoblast differentiation of KM101 cells in osteogenic medium were performed as described (Liang et al., 2016). E2, DHT, and DHEA stimulated osteoblast differentiation in a dose-dependent manner as shown with ALP activity (μmol/min/well), 4 days after treatments (*$P < 0.05$, **$P < 0.01$; steroid vs osteogenic control, $n = 4$–6; ANOVA).

Zhou, LeBoff, & Glowacki, 2010). In other studies with human MSCs, we demonstrated the critical role of IGF-I in mediating osteoanabolic effects of vitamin D metabolites and of PTH (Geng, Zhou, & Glowacki, 2011; Zhou et al., 2010). In a recent study with both primary cultures and a cell line of hMSCs, we showed that DHEA and other steroids stimulate osteoblastogenesis and *IGF-I* expression (Liang et al., 2016). In many systems, IGF-I regulates cell growth, survival, and differentiation by acting on the IGF-I receptor (IGF-IR), a tyrosine kinase receptor, which elicits diverse intracellular signaling responses by transducing signals to the nucleus and mitochondria primarily via the mitogen-activated protein kinase (MAPK) and PI3K/Akt pathways (Grimberg, 2003). Our studies showed that DHEA stimulates osteoblast differentiation and promotes *IGF-I* gene expression in human MSCs, and requires PI3K signaling (Liang et al., 2016). In mice, MSCs expressing IGF-I promote fracture healing and required PI3K signaling (Granero-Moltó et al., 2011). Our studies showed that DHEA also requires p42/44 MAPK but not JNK to upregulate *IGF-I* gene expression and requires p42/44 MAPK to stimulate alkaline phosphatase activity in hMSCs (Liang et al., 2016). Likewise, in osteoblast-like MC3T3-E1 cells, IGF-I-stimulated alkaline phosphatase activity by inducing the activation of p42/p44 MAPK but not JNK (Hanai et al., 2006). These data also indicate that DHEA does not require CREB to upregulate *IGF-I* gene. In other studies of the effects of PTH on hMSCs, there were both CREB-independent and CREB-dependent signaling of IGF-I (Geng et al., 2011). In summary, our data (Liang et al., 2016) reveal that DHEA stimulation of osteoblast differentiation is mediated by IGF-I signaling via IGF-I receptor, PI3K, P38 MAPK, and P42/44 MAPK pathways (Fig. 2). Together with clinical studies that show DHEA treatment increases serum IGF-I levels in some individuals (Arlt et al., 1999; Morales et al., 1998, 1994; Papierska et al., 2012; van Thiel et al., 2005; Yen et al., 1995), our study with MSCs (Liang et al., 2016) supports the view that DHEA may be useful for treating bone diseases through its stimulation of osteoanabolic IGF-I-mediated mechanisms.

5. DHEA AND IL-6 IN BONE HEALTH

It is clear that serum IL-6 is increased with subject age and is significantly higher in postmenopausal women compared with premenopausal subjects (Giuliani et al., 2001). In our analysis of 102 healthy women between 29 and 95 years of age, there was a significant age-related increase

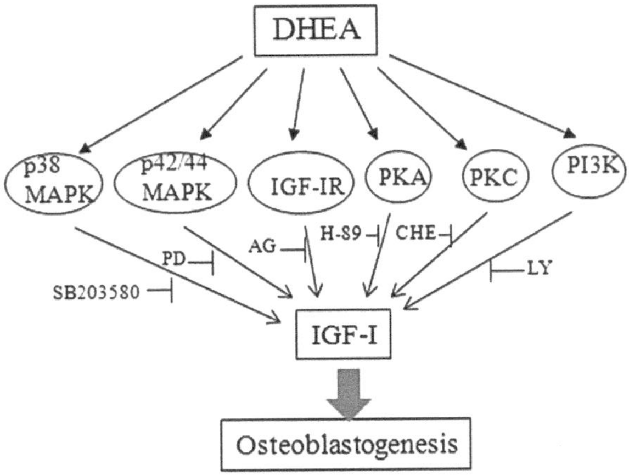

Fig. 2 A summary of mechanisms by which DHEA regulates IGF-I and osteoblastogenesis in human MSCs. DHEA requires p38 MAPK, p42/44 MAPK, IGF-I receptor (IGF-IR), PKA, PKC, and PI3K, but not JNK or CREB pathway to stimulate *IGF-I* gene expression, and needs p38 MAPK, p42/44 MAPK, IGF-IR, and PI3K to stimulate osteoblastogenesis in human bone marrow-derived mesenchymal stem cells. The small molecule inhibitors are p38 MAPK inhibitor SB203580, p42/44 MAPK inhibitor PD098059 (PD), IGF-R inhibitor AG1024 (AG), PKA inhibitor H-89, PKC inhibitor chelerythrine chloride (CHE), and PI3K inhibitor LY294002 (LY).

in serum IL-6 and an inverse correlation between serum IL-6 and DHEAS (Haden et al., 2000). Straub et al. also found an inverse correlation between serum IL-6 and DHEAS in men and women (Straub et al., 1998). In an attempt to find biomarkers of healthy aging rather than of age-associated diseases, the Long Life Family Study tested 4704 healthy subjects between 25 and 110 years of age upon enrollment, with eligibility criteria based upon evidence of familial longevity. Among the biomarkers with significant correlation with age were IL-6 and, with inverse correlations, DHEA and IGF-I (Sebastiani et al., 2016). A randomized, double-blind crossover study showed that DHEA supplementation reduced serum IL-6 (Casson et al., 1993). This effect is similar to the decrease in IL-6 produced by DHEA in aged female mice (Inserra et al., 1998). DHEA has also been shown to reduce IL-6 that is elevated in an acute mouse model of posttraumatic inflammation (Lichte et al., 2014).

6. DHEA AND IL-6 IN BONE CELLS

DHEA has been shown to inhibit bone resorption by mechanisms involving inflammatory mediators. Rat cell and mouse bone culture systems were used to first describe the role of IL-6 as mediator of effects of bone-resorbing agents, such as PTH (Löwik et al., 1989). There was dose–response stimulation of IL-6 production by bovine PTH(1–84) for primary rat osteoblast-like cells and the rat osteogenic sarcoma cell line UMR-106, concomitant with the expected rise in cyclic AMP production. In addition, PTH (1–84) increased IL-6 activity in cultures of metacarpal explants from 17-day-old fetal mice, and exogenous rhIL-6-stimulated osteoclastic resorption in those metacarpal cultures, as measured by ^{45}Ca release from prelabeled bones. Because 17-day-old metacarpal bones do not contain mature osteoclasts and do contain osteoclast precursors, the investigators concluded that the IL-6 effect on resorption entails osteoclast differentiation. As more osteolytic agents were discovered to act through IL-6, such as IL-1, TNFα, 1α,25 $(OH)_2D$, and endotoxin (Ishimi et al., 1990), the connection between bone resorption and inflammatory cytokines became clearer and led to the term osteoimmunology. It also became clear that the sources of proosteoclastogenic IL-6 and M-CSF were the osteoblast and osteoblast progenitor, which activate the respective receptors of those ligands, the IL-6 receptor and c-fms, on osteoclast progenitors in the hematopoietic lineage (Otsuka et al., 1994). Finding that 17β-estradiol inhibited IL-6 that was stimulated in osteoblasts and MSCs by such osteolytic agents indicates that IL-6 is under the inhibitory control of estrogens; hence, the bone loss associated with estrogen withdrawal might be due, at least in part, to the removal of those inhibitory effects (Girasole et al., 1992). Those observations were followed by similar studies with androgens and DHEAS showing that they inhibited IL-6 production and transcriptional activity of the IL-6 promoter (Bellido et al., 1995).

We first showed production of IL-6 by human MSCs and its stimulation by osteolytic agents in 1995 (Glowacki, 1995). A subsequent analysis of human MSCs from well-characterized subjects showed that secretion of IL-6 was dependent upon the age and estrogen status of the subject from whom the cells were obtained. There was a profound age-related increase in constitutive IL-6 secretion by MSCs for women and for men; IL-6 secretion by MSCs from postmenopausal women receiving estrogen replacement therapy was 8% of that from age-matched control women (Cheleuitte, Mizuno, & Glowacki, 1998).

In investigations to determine the mechanisms by which adrenal and gonadal steroids have protective effects on the skeleton, it was shown that IL-6 secretion by the adherent fraction of human marrow cells, which includes osteoblast progenitors (MSCs), was significantly inhibited to different magnitudes by DHEA, DHT, and E2 (Gordon, LeBoff, & Glowacki, 2001).

With cells from a cohort of 19 postmenopausal women and 7 who were receiving systemic estrogen replacement therapy (ERT) at the time of MSC isolation, constitutive secretion of IL-6 was more consistently and more dramatically inhibited by DHEA than by DHT or E2. Furthermore, with cells from ERT-supplemented women, constitutive IL-6 secretion tended to be lower than that from age-matched women who were not receiving ERT. Of note, cells from women receiving ERT showed further reduction of IL-6 secretion by in vitro DHEA and to a lesser degree by DHT but no effect at all by E2. That MSCs from subjects receiving ERT were refractory to inhibition by in vitro estrogen treatment indicates that estrogen status affects the cells and suggests that DHEA may act independent of aromatization. Likewise, DHEA suppressed IL-6 production in two human clonal marrow stromal lines, KM101 (Gordon et al., 2001) and KM-102 (Liu et al., 2005). There is some aging research with human peripheral blood mononuclear cells that shows the generalizability of the DHEA and IL-6 relationship (Straub et al., 1998). DHEA inhibited IL-6 production in peripheral blood mononuclear cells, with a U-shaped dose–response curve that suggested 10^{-8} mol/L as optimal for immunomodulation.

The role of inflammatory cytokines in osteoclast differentiation was well appreciated in the 1990s, as was the involvement of the osteoblast and stromal preosteoblast (MSCs) in providing mediators that regulate osteoclastogenesis. Other key mediators upregulated by PTH, $1\alpha,25(OH)_2D$, IL-1, and IL-6 have been identified, such as RANKL and osteoprotegerin (OPG) which are produced by osteoblasts and MSCs, the former being a ligand for the cell membrane receptor RANK on osteoclast progenitors in the hematopoietic lineage, and the latter being a decoy receptor that interferes with RANKL–RANK signaling (Suda et al., 1999). The ratio of RANKL/OPG has been shown to be an important index of bone mass (Tudpor et al., 2015), is increased in MSCs with age, and is correlated with the extent of in vitro osteoclast differentiation of human marrow cells (Chung et al., 2014). DHEA decreased the ratio of RANKL/OPG in murine calvarial osteoblasts and decreased bone resorption in cocultures of osteoblasts and osteoclast progenitors (Wang et al., 2006).

7. SUMMARY AND CONCLUSION

Four hormone systems show decreasing circulating hormone concentrations during normal aging: (i) DHEA and DHEAS in adrenopause, (ii) estrogen in menopause and (iii) testosterone in andropause, and

(iv) the GH/IGF-I axis in somatopause (LeBoff & Glowacki, 1999; Lamberts et al., 1997) (Fig. 3). Steroid production at a tissue level provides autoregulation of the local hormonal environment according to local needs, with less systemic effects, called intracrinology (Samaras et al., 2013). In premenopausal women, 50%–75% of estrogens and the majority of androgens are produced through intracrine mechanisms from DHEA; after menopause, nearly all androgens and estrogens are synthesized at a tissue level. In men, the testes continue to secrete androgens at a declining level, while local hormone production also occurs but is more difficult to assess (Genazzani et al., 2007; Panjari & Davis, 2007; Samaras et al., 2013). Because DHEA can be transformed in peripheral tissues to more potent androgens and estrogens (Binello & Gordon, 2003; Genazzani et al., 2007), the age-related decreased circulating DHEA and DHEAS in adrenopause are associated with declines of estrogen in menopause and testosterone in andropause. Aromatase is one of the enzymes responsible for the conversion of DHEA and testosterone into estrogens. Several lines of clinical and experimental evidence indicates that aromatase activity and estrogen production are necessary for longitudinal bone growth, attainment of peak bone mass, the pubertal growth spurt, epiphyseal closure, and normal bone remodeling in young individuals; moreover, with aging, individual differences in aromatase activity and thus in estrogen levels may significantly affect bone loss and fracture risk in both genders (Gennari et al., 2011). GH and IGF-I secretion are decreased with aging, and abnormalities in the GH/IGF-I

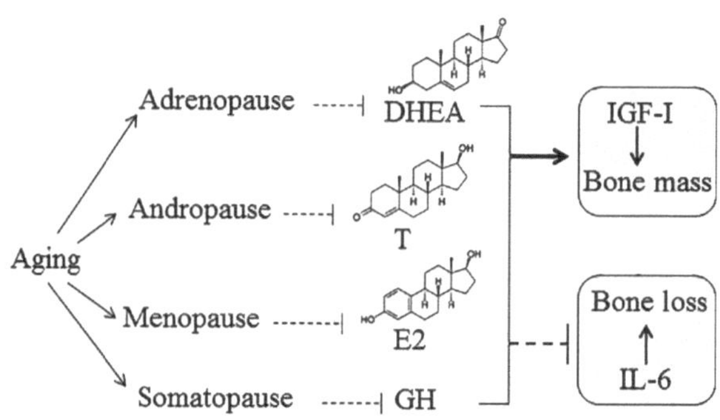

Fig. 3 Proposed mechanisms of bone loss associated with age-related declines in hormones. The age-related decreasing circulating levels of DHEA/DHEAS in adrenopause, estrogen (E2) in menopause, testosterone (T) in andropause, and GH in somatopause are associated with bone mass loss via the age-related decline of IGF-I and increase of IL-6. *Solid line arrow* represents stimulation or result; *dashed line with bar* represents inhibition or reduction.

axis play a role in the pathogenesis of the osteoporosis of anorexia nervosa and after glucocorticoid exposure (Giustina et al., 2008; Perrini et al., 2010). The GH/IGF-I axis is important for ACTH-dependent DHEA and DHEAS production in human adrenals and aging directly influences adrenal DHEA production (adrenopause) and also has an indirect influence via the GH/IGF-I axis (somatopause) (Kasayama et al., 2007) (Fig. 4). In addition, DHEA/DHEAS also regulates the GH/IGF-I axis as demonstrated by data that DHEA increases the levels of GH and IGF-I in postmenopausal women (Genazzani et al., 2001).

In summary, evidence indicates that the age-related decreases in circulating DHEA/DHEAS, estrogen, testosterone, and growth hormone are associated with bone mass loss via the age-related decline of IGF-I and increase of IL-6 (Fig. 3). The age-related reduced levels of DHEA and DHEAS in parallel with age-related osteopenia and osteoporosis lead to the suggestion that DHEA replacement therapy may be beneficial for bone

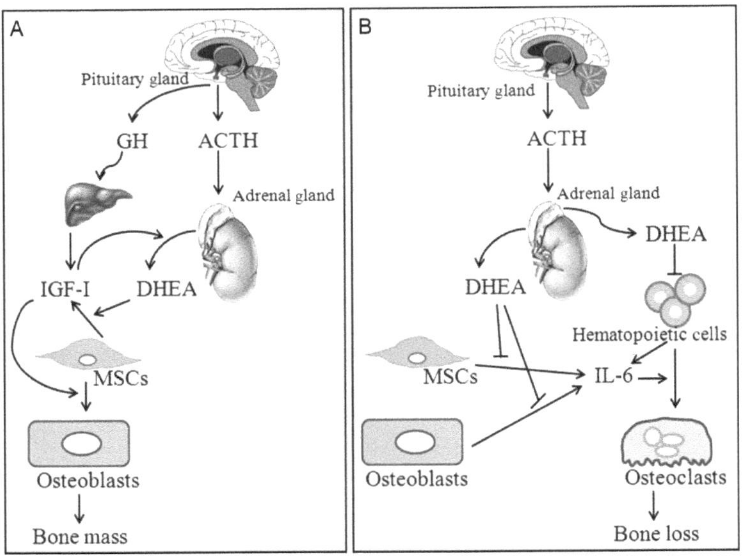

Fig. 4 The roles of DHEA, IGF-I, and IL-6 in bone. (A) The production of DHEA in adrenal glands is regulated by ACTH from pituitary gland. The GH–IGF-I axis is important for ACTH-dependent DHEA and DHEAS production in human adrenals; DHEA stimulates IGF-I in bone cells including MSCs and osteoblasts and regulates osteoblast differentiation in MSCs and bone formation by osteoblasts. (B) DHEA inhibits IL-6 in MSCs, osteoblasts, and hematopoietic mononuclear cells; IL-6 stimulates osteoclast differentiation, maturation, and bone resorption. *Solid line arrow* represents stimulation/production or result; *solid line with bar* represents inhibition or reduction.

health in older patients. Clinical, epidemiological, and experimental evidence indicates that DHEA replacement therapy may be useful for managing osteopenia and osteoporosis. The stimulation by DHEA of both IGF-I and osteoblastogenesis in hMSCs requires IGF-I receptor, PI3K, p38 MAPK, or p42/44 MAPK signaling pathways (Fig. 2). The production of DHEA in adrenal glands is regulated by ACTH from pituitary gland and GH/IGF-I axis (Fig. 4). Although it is unclear whether DHEA modulates IGF-I production in liver, DHEA increases the plasma levels of GH and IGF-I in postmenopausal women (Genazzani et al., 2001); DHEA stimulates IGF-I in MSCs and their differentiation to osteoblasts that produce bone matrix (Fig. 4). On the osteolytic aspect of bone metabolism, DHEA inhibits IL-6 in bone cells including MSCs (Cheleuitte et al., 1998; Glowacki, 1995; Gordon et al., 2001; Liu et al., 2005) and osteoblasts (Harding et al., 2006), and in hematopoietic mononuclear cells (Straub et al., 1998) (Fig. 4). DHEA also reduces the stimulation of IL-6 by LPS in monocytes (Delpedro, Barjavel, Mamdouh, & Bakouche, 1998) and macrophages (Kim et al., 2006; Padgett & Loria, 1998). IL-6 stimulates osteoclast differentiation of mononuclear precursors, osteoclast maturation, and bone resorption (Fig. 4). Thus, taken together, animal and human studies indicate that DHEA acts on preosteoblast MSCs or osteoblasts to promote osteoblast differentiation and bone formation activity while it also acts on those same cells to reduce proosteoclastogenic factors, thus having two ways to enhance bone mass (Fig. 4). In summary, evidence from us and others indicates that DHEA may be useful for treating bone diseases through its inhibition of skeletal catabolic IL-6 and stimulation of anabolic IGF-I-mediated mechanisms.

ACKNOWLEDGMENTS

This work was supported by grants from Department of Orthopedic Surgery, Brigham and Women's Hospital, The Gillian Reny Stepping Strong Fund, NIH (AG025015, AG028114, and U19A1068021), the Department of Defense, and Brigham and Women's Hospital BRI Fund. The authors greatly appreciate Drs. Josef Simon and Xiaonan Liang for contributions made in the early stage of this project and also thank colleagues, especially Professor Meryl S. LeBoff, MD, of Brigham and Women's Hospital and Professor Catherine M. Gordon, MD, of Cincinnati Children's Hospital Medical Center, who have collaborated in investigation of DHEA. The authors thank the surgeons and nurses of Department of Orthopedic Surgery, Brigham and Women's Hospital for their help with discarded human specimens.

DISCLOSURE STATEMENT

The authors have nothing to disclose.

REFERENCES

Agnusdei, D., & Gentilella, R. (2005). GH and IGF-I as therapeutic agents for osteoporosis. *Journal of Endocrinological Investigation, 28*, 32–36.

Arlt, W., Callies, F., van Vlijmen, J. C., Koehler, I., Reincke, M., Bidlingmaier, M., et al. (1999). Dehydroepiandrosterone replacement in women with adrenal insufficiency. *New England Journal of Medicine, 341*, 1013–1020.

Baulieu, E. E., Thomas, G., Legrain, S., Lahlou, N., Roger, M., Debuire, B., et al. (2000). Dehydroepiandrosterone (DHEA), DHEA sulfate, and aging: Contribution of the DHEAge study to a sociobiomedical issue. *Proceedings of the National Academy of Sciences of the United States of America, 97*, 4279–4284.

Belgorosky, A., Baquedano, M. S., Guercio, G., & Rivarola, M. A. (2008). Adrenarche: Postnatal adrenal zonation and hormonal and metabolic regulation. *Hormone Research, 70*, 257–267.

Bellido, T., Jilka, R. L., Boyce, B. F., Girasole, G., Broxmeyer, H., Dalrymple, S. A., et al. (1995). Regulation of interleukin-6, osteoclastogenesis, and bone mass by androgens. The role of the androgen receptor. *Journal of Clinical Investigation, 95*, 2886–2895.

Bellino, F. L., Daynes, R. A., Hornsby, P. J., Lavrin, D. H., & Nestler, J. E. (1995). Dehydroepiandrosterone (DHEA) and Aging. *Annals of the New York Academy of Sciences: Vol. 774*. New York, NY: The New York Academy of Sciences.

Binello, E., & Gordon, C. M. (2003). Clinical uses and misuses of dehydroepiandrosterone. *Current Opinion in Pharmacology, 3*, 635–641.

Bonjour, J. P., Theintz, G., Buchs, B., Slosman, D., & Rizzoli, R. (1991). Critical years and stages of puberty for spinal and femoral bone mass accumulation during adolescence. *Journal of Clinical Endocrinology and Metabolism, 73*, 555–563.

Casson, P. R., Andersen, R. N., Herrod, H. G., Stentz, F. B., Straughn, A. B., Abraham, G. E., et al. (1993). Oral dehydroepiandrosterone in physiologic doses modulates immune function in postmenopausal women. *American Journal of Obstetrics and Gynecology, 169*, 1536–1539.

Cheleuitte, D., Mizuno, S., & Glowacki, J. (1998). In vitro secretion of cytokines by human bone marrow: Effects of age and estrogen status. *Journal of Clinical Endocrinology and Metabolism, 83*, 2043–2051.

Chung, P.-L., Zhou, S., Eslami, B., Shen, L., LeBoff, M. S., & Glowacki, J. (2014). Effect of age on regulation of human osteoclast differentiation. *Journal of Cellular Biochemistry, 115*, 1412–1419.

Cormier, C., Souberbielle, J. C., & Kahan, A. (2001). DHEA in bone and joint diseases. *Joint, Bone, Spine, 68*, 588–594.

Corona, G., Rastrelli, G., Giagulli, V. A., Sila, A., Sforza, A., Forti, G., et al. (2013). Dehydroepiandrosterone supplementation in elderly men: A meta-analysis study of placebo-controlled trials. *Journal of Clinical Endocrinology and Metabolism, 98*, 3615–3626.

Cosman, F., de Beur, S. J., LeBoff, M. S., Lewiecki, E. M., Tanner, B., Randall, S., et al. (2014). Clinician's guide to prevention and treatment of osteoporosis. *Osteoporos International, 25*, 2359–2381.

Delpedro, A. D., Barjavel, M. J., Mamdouh, Z., & Bakouche, O. (1998). Activation of human monocytes by LPS and DHEA. *Journal of Interferon & Cytokine Research, 18*, 125–135.

Diamanti-Kandarakis, E., Dattilo, M., Macut, D., Duntas, L., Gonos, E. S., Goulis, D. G., et al. (2017). Mechanisms in endocrinology: Aging and anti-aging: A combo-endocrinology overview. *European Journal of Endocrinology, 176*, R283–R308.

Elraiyah, T., Sonbol, M. B., Wang, Z., Khairalseed, T., Asi, N., Undavalli, C., et al. (2014). Clinical review: The benefits and harms of systemic dehydroepiandrosterone (DHEA) in postmenopausal women with normal adrenal function: A systematic review and meta-analysis. *Journal of Clinical Endocrinology and Metabolism, 99*, 3536–3542.

Földes, J., Lakatos, P., Zsadányi, J., & Horváth, C. (1997). Decreased serum IGF-I and dehydroepiandrosterone sulphate may be risk factors for the development of reduced bone mass in postmenopausal women with endogenous subclinical hyperthyroidism. *European Journal of Endocrinology*, *136*, 277–281.

Genazzani, A. D., Lanzoni, C., & Genazzani, A. R. (2007). Might DHEA be considered a beneficial replacement therapy in the elderly? *Drugs & Aging*, *24*, 173–185.

Genazzani, A. D., Stomati, M., Strucchi, C., Puccetti, S., Luisi, S., & Genazzani, A. R. (2001). Oral dehydroepiandrosterone supplementation modulates spontaneous and growth hormone-releasing hormone-induced growth hormone and insulin-like growth factor-1secretion in early and late postmenopausal women. *Fertility and Sterility*, *76*, 241–248.

Geng, S., Zhou, S., & Glowacki, J. (2011). Age-related decline in osteoblastogenesis and 1α-hydroxylase/CYP27B1 in human mesenchymal stem cells: Stimulation by parathyroid hormone. *Aging Cell*, *10*, 962–971.

Gennari, L., Merlotti, D., & Nuti, R. (2011). Aromatase activity and bone loss. *Advances in Clinical Chemistry*, *54*, 129–164.

Girasole, G., Jilka, R. L., Passeri, G., Boswell, S., Boder, G., Williams, D. C., et al. (1992). 17 beta-estradiol inhibits interleukin-6 production by bone marrow-derived stromal cells and osteoblasts in vitro: A potential mechanism for the antiosteoporotic effect of estrogens. *Journal of Clinical Investigation*, *89*, 883–891.

Giuliani, N., Sansoni, P., Girasole, G., Vescovini, R., Passeri, G., Passeri, M., et al. (2001). Serum interleukin-6, soluble interleukin-6 receptor and soluble gp130 exhibit different patterns of age- and menopause-related changes. *Experimental Gerontology*, *36*, 547–557.

Giustina, A., Mazziotti, G., & Canalis, E. (2008). Growth hormone, insulin-like growth factors, and the skeleton. *Endocrine Reviews*, *29*, 535–559.

Glowacki, J. (1995). Influence of age on human marrow. *Calcified Tissue International*, *56*, S50–S51.

Gordon, C. M., Binello, E., LeBoff, M. S., Wohl, M. E., Rosen, C. J., & Colin, A. A. (2006). Relationship between insulin-like growth factor I, dehydroepiandrosterone sulfate and proresorptive cytokines and bone density in cystic fibrosis. *Osteoporos International*, *17*, 783–790.

Gordon, C. M., Glowacki, J., & LeBoff, M. S. (1999). DHEA and the skeleton (through the ages). *Endocrine*, *11*(1), 11.

Gordon, C. M., Grace, E., Emans, S. J., Feldman, H. A., Goodman, E., Becker, K. A., et al. (2002). Effects of oral dehydroepiandrosterone on bone density in young women with anorexia nervosa: A randomized trial. *Journal of Clinical Endocrinology and Metabolism*, *87*, 4935–4941.

Gordon, C. M., LeBoff, M. S., & Glowacki, J. (2001). Adrenal and gonadal steroids inhibit IL-6 secretion by human marrow cells. *Cytokine*, *16*, 178–186.

Granero-Moltó, F., Myers, T. J., Weis, J. A., Longobardi, L., Li, T., Yan, Y., et al. (2011). Mesenchymal stem cells expressing insulin-like growth factor-I (MSCIGF) promote fracture healing and restore new bone formation in Irs1 knockout mice: Analyses of MSCIGF autocrine and paracrine regenerative effects. *Stem Cells*, *29*, 1537–1548.

Grimberg, A. (2003). Mechanisms by which IGF-I may promote cancer. *Cancer Biology & Therapy*, *2*, 630–635.

Guntur, A. R., & Rosen, C. J. (2013). IGF-1 regulation of key signaling pathways in bone. *BoneKEy Reports*, *2*, 437.

Gupta, D., & Morley, J. E. (2014). Hypothalamic-pituitary-adrenal (HPA) axis and aging. *Comprehensive Physiology*, *4*, 1495–1510.

Haden, S. T., Glowacki, J., Hurwitz, S., Rosen, C., & LeBoff, M. S. (2000). Effects of age on serum dehydroepiandrosterone, IGF-I, and IL-6 in women. *Calcified Tissue International*, *66*, 414–418.

Hanai, Y., Tokuda, H., Ishisaki, A., Matsushima-Nishiwaki, R., Nakamura, N., Yoshida, M., et al. (2006). Involvement of p44/p42 MAP kinase in insulin-like growth factor-I-induced alkaline phosphatase activity in osteoblast-like-MC3T3-E1 cells. *Molecular and Cellular Endocrinology*, *251*, 42–48.

Harding, G., Mak, Y. T., Evans, B., Cheung, J., MacDonald, D., & Hampson, G. (2006). The effects of dexamethasone and dehydroepiandrosterone (DHEA) on cytokines and receptor expression in a human osteoblastic cell line: potential steroid-sparing role for DHEA. *Cytokine*, *36*, 57–68.

Inserra, P., Zhang, Z., Ardestani, S. K., Araghi-Niknam, M., Liang, B., Jiang, S., et al. (1998). Modulation of cytokine production by dehydroepiandrosterone (DHEA) plus melatonin (MLT) supplementation of old mice. *Proceedings of the Society for Experimental Biology and Medicine*, *218*, 76–82.

Ishimi, Y., Miyaura, C., Jin, C. H., Akatsu, T., Abe, E., Nakamura, Y., et al. (1990). IL-6 is produced by osteoblasts and induces bone resorption. *Journal of Immunology*, *145*, 3297–3303.

Jankowski, C. M., Gozansky, W. S., Kittelson, J. M., Van Pelt, R. E., Schwartz, R. S., & Kohrt, W. M. (2008). Increases in bone mineral density in response to oral dehydroepiandrosterone replacement in older adults appear to be mediated by serum estrogens. *Journal of Clinical Endocrinology and Metabolism*, *93*, 4767–4773.

Jankowski, C. M., Gozansky, W. S., Schwartz, R. S., Dahl, D. J., Kittelson, J. M., Scott, S. M., et al. (2006). Effects of dehydroepiandrosterone replacement therapy on bone mineral density in older adults: A randomized, controlled trial. *Journal of Clinical Endocrinology and Metabolism*, *91*, 2986–2993.

Kasayama, S., Morita, S., Otsuki, M., Asanuma, N., Saito, H., Mukai, M., et al. (2007). Independent association between insulin-like growth factor-I and dehydroepiandrosterone sulphate in women in middle adulthood. *Clinical Endocrinology (Oxf)*, *66*, 797–802.

Kasperk, C. H., Wakley, G. K., Hierl, T., & Ziegler, R. (1997). Gonadal and adrenal androgens are potent regulators of human bone cell metabolism in vitro. *Journal of Bone and Mineral Research*, *12*, 464–471.

Kim, S. K., Shin, M. S., Jung, B. K., Shim, J. Y., Won, H. S., Lee, P. R., et al. (2006). Effect of dehydroepiandrosterone on lipopolysaccharide-induced interleukin-6 production in DH82 cultured canine macrophage cells. *Journal of Reproductive Immunology*, *70*, 71–81.

Labrie, F., Martel, C., Bélanger, A., & Pelletier, G. (2017). Androgens in women are essentially made from DHEA in each peripheral tissue according to intracrinology. *Journal of Steroid Biochemistry and Molecular Biology*, *168*, 9–18.

Lamberts, S. W., van den Beld, A. W., & van der Lely, A. J. (1997). The endocrinology of aging. *Science*, *278*, 419–424.

LeBoff, M., & Glowacki, J. (1999). Sex steroids, aging, and bone. In: C. Rosen, J. Glowacki, & J. P. Bilzekian (Eds.), *The aging skeleton* (pp. 159–174). San Diego, CA: Academic Press.

Lee, D., Kim, H., Ahn, S. H., Lee, S. H., Bae, S. J., Kim, E. H., et al. (2015). The association between serum dehydroepiandrosterone Sulphate (DHEA-S) level and bone mineral density in Korean men. *Clinical Endocrinology (Oxf)*, *83*, 173–179.

Liang, X., Glowacki, J., Hahne, J., Xie, L., LeBoff, M. S., & Zhou, S. (2016). Dehydroepiandrosterone stimulation of Osteoblastogenesis in human MSCs requires IGF-I signaling. *Journal of Cellular Biochemistry*, *117*, 1769–1774.

Lichte, P., Pfeifer, R., Werner, B. E., Ewers, P., Tohidnezhad, M., Pufe, T., et al. (2014). Dehydroepiandrosterone modulates the inflammatory response in a bilateral femoral shaft fracture model. *European Journal of Medical Research*, *19*, 27.

Liu, S., Ishikawa, H., Li, F. J., Ma, Z., Otsuyama, K., Asaoku, H., et al. (2005). Dehydroepiandrosterone can inhibit the proliferation of myeloma cells and the interleukin-6 production of bone marrow mononuclear cells from patients with myeloma. *Cancer Research*, *65*, 2269–2276.

Löwik, C. W., van der Pluijm, G., Bloys, H., Hoekman, K., Bijvoet, O. L., Aarden, L. A., et al. (1989). Parathyroid hormone (PTH) and PTH-like protein (PLP) stimulate interleukin-6 production by osteogenic cells: A possible role of interleukin-6 in osteoclastogenesis. *Biochemical and Biophysical Research Communications*, *162*, 1546–1552.

Marcelli, C. (2003). Can DHEA be used to prevent bone loss and osteoporosis? *Joint, Bone, Spine*, *70*, 1–2.

Mohan, S., & Kesavan, C. (2012). Role of insulin-like growth factor-1 in the regulation of skeletal growth. *Current Osteoporosis Reports*, *10*, 178–186.

Morales, A. J., Haubrich, R. H., Hwang, J. Y., Asakura, H., & Yen, S. S. (1998). The effect of six months treatment with a 100 mg daily dose of dehydroepiandrosterone (DHEA) on circulating sex steroids, body composition and muscle strength in age-advanced men and women. *Clinical Endocrinology (Oxf)*, *49*, 421–432.

Morales, A. J., Nolan, J. J., Nelson, J. C., & Yen, S. S. (1994). Effects of replacement dose of dehydroepiandrosterone in men and women of advancing age. *Journal of Clinical Endocrinology and Metabolism*, *78*, 1360–1367.

Nair, K. S., Rizza, R. A., O'Brien, P., Dhatariya, K., Short, K. R., Nehra, A., et al. (2006). DHEA in elderly women and DHEA or testosterone in elderly men. *New England Journal of Medicine*, *355*, 1647–1659.

Nawata, H., Tanaka, S., Tanaka, S., Takayanagi, R., Sakai, Y., Yanase, T., et al. (1995). Aromatase in bone cell: Association with osteoporosis in postmenopausal women. *Journal of Steroid Biochemistry and Molecular Biology*, *53*, 165–174.

Nieuwenhu, H., & Thijssen, J. H. H. (1999). *DHEA: A comprehensive review*. New York: The Parthenon Publishing Group.

Ohlsson, C., Nethander, M., Kindmark, A., Ljunggren, Ö., Lorentzon, M., Rosengren, B. E., et al. (2017). Low serum DHEAS predicts increased fracture risk in older Men: The MrOS Sweden study. *Journal of Bone and Mineral Research*, *32*, 1607–1614.

Otsuka, T., Ogo, T., Nakano, T., Niiro, H., Kuga, S., Satoh, H., et al. (1994). Expression of the c-kit ligand and interleukin 6 genes in mouse bone marrow stromal cell lines. *Stem Cells*, *12*, 409–415.

Padgett, D. A., & Loria, R. M. (1998). Endocrine regulation of murine macrophage function: Effects of dehydroepiandrosterone, androstenediol, and androstenetriol. *Journal of Neuroimmunology*, *84*, 61–68.

Panjari, M., & Davis, S. R. (2007). DHEA therapy for women: Effect on sexual function and wellbeing. *Human Reproduction*, *13*, 239–248.

Papierska, L., Rabijewski, M., Kasperlik-Zaluska, A., & Zgliczynski, W. (2012). Effect of DHEA supplementation on serum IGF-1, osteocalcin, and bone mineral density in postmenopausal, glucocorticoid-treated women. *Advances in Medical Sciences*, *57*, 51–57.

Perrini, S., Laviola, L., Carreira, M. C., Cignarelli, A., Natalicchio, A., & Giorgino, F. (2010). The GH/IGF1 axis and signaling pathways in the muscle and bone: Mechanisms underlying age-related skeletal muscle wasting and osteoporosis. *Journal of Endocrinology*, *205*, 201–210.

Perrini, S., Laviola, L., Natalicchio, A., & Giorgino, F. (2005). Associated hormonal declines in aging: DHEAS. *Journal of Endocrinological Investigation*, *28*, 85–93.

Perrini, S., Natalicchio, A., Laviola, L., Cignarelli, A., Melchiorre, M., De Stefano, F., et al. (2008). Abnormalities of insulin-like growth factor-I signaling and impaired cell proliferation in osteoblasts from subjects with osteoporosis. *Endocrinology*, *149*, 1302–1313.

Pino, A. M., Rodríguez, J. M., Ríos, S., Astudillo, P., Leiva, L., Seitz, G., et al. (2006). Aromatase activity of human mesenchymal stem cells is stimulated by early differentiation, vitamin D and leptin. *Journal of Endocrinology*, *191*, 715–725.

Prough, R. A., Clark, B. J., & Klinge, C. M. (2016). Novel mechanisms for DHEA action. *Journal of Molecular Endocrinology*, *56*, 139–155.

Qiu, X., Gui, Y., Xu, Y., Li, D., & Wang, L. (2015). DHEA promotes osteoblast differentiation by regulating the expression of osteoblast-related genes and Foxp3(+) regulatory T cells. *Bioscience Trends*, *9*, 307–314.

Raven, P. W., & Hinson, J. P. (2007). Dehydroepiandrosterone (DHEA) and the menopause: An update. *Menopause International*, *13*, 75–78.

Rosen, C. J., Verault, D., Steffens, C., Cheleuitte, D., & Glowacki, J. (1997). Effects of age and estrogen status on the skeletal IGF regulatory system. Studies with human marrow. *Endocrine*, *7*, 77–80.

Samaras, N., Samaras, D., Frangos, E., Forster, A., & Philippe, J. (2013). A review of age-related dehydroepiandrosterone decline and its association with well-known geriatric syndromes: Is treatment beneficial? *Rejuvenation Research*, *16*, 285–294.

Sebastiani, P., Thyagarajan, B., Sun, F., Honig, L. S., Schupf, N., Cosentino, S., et al. (2016). Age and sex distributions of age-related biomarker values in healthy older adults from the long life family study. *Journal of the American Geriatrics Society*, *64*, e189–e194.

Sillat, T., Pöllänen, R., Lopes, J. R., Porola, P., Ma, G., Korhonen, M., et al. (2009). Intracrine androgenic apparatus in human bone marrow stromal cells. *Journal of Cellular and Molecular Medicine*, *13*, 3296–3302.

Soyka, L. A., Grinspoon, S., Levitsky, L. L., Herzog, D. B., & Klibanski, A. (1999). The effects of anorexia nervosa on bone metabolism in female adolescents. *Journal of Clinical Endocrinology and Metabolism*, *84*, 4489–4496.

Stewart, P. M. (2006). Aging and fountain-of-youth hormones. *New England Journal of Medicine*, *355*, 1724–1726.

Straub, R. H., Konecna, L., Hrach, S., Rothe, G., Kreutz, M., Schölmerich, J., et al. (1998). Serum dehydroepiandrosterone (DHEA) and DHEA sulfate are negatively correlated with serum interleukin-6 (IL-6), and DHEA inhibits IL-6 secretion from mononuclear cells in man in vitro: Possible link between endocrinosenescence and immunosenescence. *Journal of Clinical Endocrinology and Metabolism*, *83*, 2012–2017.

Suda, T., Takahashi, N., Udagawa, N., Jimi, E., Gillespie, M. T., & Martin, T. J. (1999). Modulation of osteoclast differentiation and function by the new members of the tumor necrosis factor receptor and ligand families. *Endocrine Reviews*, *20*, 345–357.

Takayanagi, R., Goto, K., Suzuki, S., Tanaka, S., Shimoda, S., & Nawata, H. (2002). Dehydroepiandrosterone (DHEA) as a possible source for estrogen formation in bone cells: Correlation between bone mineral density and serum DHEA-sulfate concentration in postmenopausal women, and the presence of aromatase to be enhanced by 1,25-dihydroxyvitamin D3 in human osteoblasts. *Mechanisms of Ageing and Development*, *123*, 1107–1114.

Traish, A. M., Kang, H. P., Saad, F., & Guay, A. T. (2011). Dehydroepiandrosterone (DHEA)—A precursor steroid or an active hormone in human physiology. *Journal of Sexual Medicine*, *8*, 2960–2982.

Tudpor, K., van der Eerden, B. C., Jongwattanapisan, P., Roelofs, J. J., van Leeuwen, J. P., Bindels, R. J., et al. (2015). Thrombin receptor deficiency leads to a high bone mass phenotype by decreasing the RANKL/OPG ratio. *Bone*, *72*, 14–22.

van Thiel, S. W., Romijn, J. A., Pereira, A. M., Biermasz, N. R., Roelfsema, F., van Hemert, A., et al. (2005). Effects of dehydroepiandrostenedione, superimposed on growth hormone substitution, on quality of life and insulin-like growth factor I in patients with secondary adrenal insufficiency: A randomized, placebo-controlled, cross-over trial. *Journal of Clinical Endocrinology and Metabolism*, *90*, 3295–3303.

Wang, Y., Bikle, D. D., & Chang, W. (2013). Autocrine and paracrine actions of IGF-I signaling in skeletal development. *Bone Research*, *1*, 249–259.

Wang, Y. D., Wang, L., Li, D. J., & Wang, W. J. (2006). Dehydroepiandrosterone inhibited the bone resorption through the upregulation of OPG/RANKL. *Cellular & Molecular Immunology*, *3*, 41–45.

Wang, L., Wang, Y. D., Wang, W. J., & Li, D. J. (2009). Differential regulation of dehydroepiandrosterone and estrogen on bone and uterus in ovariectomized mice. *Osteoporosis International*, *20*, 79–92.

Wang, L., Wang, Y. D., Wang, W. J., Zhu, Y., & Li, D. J. (2007). Dehydroepiandrosterone improves murine osteoblast growth and bone tissue morphometry via mitogenactivated protein kinase signaling pathway independent of either androgen receptor or estrogen receptor. *Journal of Molecular Endocrinology*, *38*, 467–479.

Watson, R. R., Huls, A., Araghinikuam, M., & Chung, S. (1996). Dehydroepiandrosterone and diseases of aging. *Drugs & Aging*, *9*, 274–291.

Weiss, E. P., Shah, K., Fontana, L., Lambert, C. P., Holloszy, J. O., & Villareal, D. T. (2009). Dehydroepiandrosterone replacement therapy in older adults: 1- and 2-y effects on bone. *American Journal of Clinical Nutrition*, *89*, 1459–1467.

Wright, N. C., Looker, A. C., Saag, K. G., Curtis, J. R., Delzell, E. S., Randall, S., et al. (2014). The recent prevalence of osteoporosis and low bone mass in the United States based on bone mineral density at the femoral neck or lumbar spine. *Journal of Bone and Mineral Research*, *29*, 2520–2526.

Xian, L., Wu, X., Pang, L., Lou, M., Rosen, C. J., Qiu, T., et al. (2012). Matrix IGF-1 maintains bone mass by activation of mTOR in mesenchymal stem cells. *Nature Medicine*, *18*, 1095–1101.

Yen, S. S., Morales, A. J., & Khorram, O. (1995). Replacement of DHEA in aging men and women. Potential remedial effects. *Annals of the New York Academy of Sciences*, *774*, 128–142.

Zhang, N., Gui, Y., Qiu, X., Tang, W., Li, L., Gober, H. J., et al. (2016). DHEA prevents bone loss by suppressing the expansion of CD4(+) T cells and TNFa production in the OVX-mouse model for postmenopausal osteoporosis. *Bioscience Trends*, *10*, 277–287.

Zhou, S. (2015). Paracrine effects of haematopoietic cells on human mesenchymal stem cells. *Scientific Reports*, *5*, 10573.

Zhou, S., & Glowacki, J. (2017). Chronic kidney disease and vitamin D metabolism in human bone marrow-derived MSCs. *Annals of the New York Academy of Sciences-Marrow*, *1402*, 43–55.

Zhou, S., Greenberger, J. S., Epperly, M. W., Goff, J. P., Adler, C., Leboff, M. S., et al. (2008). Age-related intrinsic changes in human bone-marrow-derived mesenchymal stem cells and their differentiation to osteoblasts. *Aging Cell*, *7*, 335–343.

Zhou, S., LeBoff, M. S., & Glowacki, J. (2010). Vitamin D metabolism and action in human bone marrow stromal cells. *Endocrinology*, *151*, 14–22.

FURTHER READING

Baskaran, C., Misra, M., & Klibanski, A. (2017). Effects of anorexia nervosa on the endocrine system. *Pediatric Endocrinology Reviews*, *14*, 302–311.

Labrie, F., Bélanger, A., Pelletier, G., Martel, C., Archer, D. F., & Utian, W. H. (2017). Science of intracrinology in postmenopausal women. *Menopause*, *24*, 702–712.

Samaras, N., Papadopoulou, M. A., Samaras, D., & Ongaro, F. (2014). Off-label use of hormones as an antiaging strategy: A review. *Clinical Interventions in Aging*, *9*, 1175–1186.

Zhou, S., Glowacki, J., Kim, S. W., Hahne, J., Geng, S., Mueller, S. M., et al. (2012). Clinical characteristics influence *in vitro* action of 1,25-dihydroxyvitamin D3 in human marrow stromal cells. *Journal of Bone and Mineral Research*, *27*, 1992–2000.

CHAPTER TEN

Autophagy and Autophagic Cell Death: Uncovering New Mechanisms Whereby Dehydroepiandrosterone Promotes Beneficial Effects on Human Health

Rolando Vegliante*, Maria R. Ciriolo†,‡,1
*MN3T, The Microenvironmental Niche in Tumorigenesis and Targeted Therapy, Hopital Civil—Institut d'Hématologie et Immunologie, Strasbourg, France
†University of Rome 'Tor Vergata', Rome, Italy
‡IRCCS San Raffaele 'La Pisana', Rome, Italy
[1]Corresponding author: e-mail address: ciriolo@bio.uniroma2.it

Contents

Vitamins and Hormones, Volume 108
ISSN 0083-6729
https://doi.org/10.1016/bs.vh.2018.01.006

Abstract

Dehydroepiandrosterone (DHEA) is the most abundant steroid hormone in human serum and a precursor of sexual hormones. Its levels, which are maximum between the age of 20 and 30, dramatically decline with aging thus raising the question that many pathological conditions typical of the elderly might be associated with the decrement of circulating DHEA. Moreover, since its very early discovery, DHEA and its metabolites have been shown to be active in many pathophysiological contexts, including cardiovascular disease, brain disorders, and cancer. Indeed, treatment with DHEA has beneficial effects for the cure of these and many other pathologies in vitro, in vivo, and in patient studies. However, the molecular mechanisms underlying DHEA effects have been only partially elucidated.

Autophagy is a self-digestive process, by which cell homeostasis is maintained, damaged organelles removed, and cell survival assured upon stress stimuli. However, high rate of autophagy is detrimental and leads to a form of programmed cell death known as autophagic cell death (ACD).

In this chapter, we describe the process of autophagy and the morphological and biochemical features of ACD. Moreover, we analyze the beneficial effects of DHEA in several pathologies and the molecular mechanisms with particular emphasis on its regulation of cell death processes. Finally, we review data indicating DHEA and structurally related steroid hormones as modulators of both autophagy and ACD, a research field that opens new avenues in the therapeutic use of these compounds.

1. INTRODUCTION

Cell death is a physiological process controlling homeostasis in the whole organism or a response to different stress stimuli and, importantly, its modulation (activation or suppression) can be exploited for clinical purposes in different pathological conditions. Autophagic cell death (ACD) is a particular form of caspase-independent programmed cell death, whose accomplishment relies on the autophagic machinery (Liu & Levine, 2015). Autophagy is a catabolic self-digesting cell process that, although usually exerting a prosurvival role, can turn detrimental following proper stimulation (Nakatogawa, Suzuki, Kamada, & Ohsumi, 2009). The recent discovery of ACD makes its definition controversial; therefore in this chapter we review part of the knowledge about its morphological and biochemical features, the molecular mechanisms and its involvement in pathophysiological conditions.

Dehydroepiandrosterone (DHEA) is the most abundant circulating steroid hormone in humans. Although initially considered a mere precursor of sexual hormones, several bodies of evidence have demonstrated that it has biological activities in many tissues and activates several signaling pathways through binding to membrane and nuclear receptors (Webb, Geoghegan, Prough, & Michael Miller, 2006). DHEA is known as "the fountain of youth hormone" due to its antiaging properties, but even more it exerts beneficial effects in several pathologies. Here, we analyzed its effects on some pathological conditions, including cardiovascular diseases, brain disorders, and cancer, highlighting possible correlations with serum levels of DHEA and its metabolites, the outcomes of exogenous DHEA administration and emphasizing the DHEA capacity to regulate cell death. In fact, recent breakthroughs have demonstrated that DHEA, its active metabolites, and structurally related steroids are involved in the regulation of autophagy and ACD. This chapter will provide insights into these previously unrevealed effects to demonstrate that some potentially beneficial properties of DHEA are still poorly characterized and in need of further investigation to broadly exploit this hormone as treatment for several human disorders.

2. ACD: TYPE II PROGRAMMED CELL DEATH

Programmed cell death is a fundamental cellular process in metazoans occurring both in physiological conditions (e.g., during development) and in response to different types of stimuli and/or stress (Tait, Ichim, & Green, 2014). Since its first discovery, which can be traced back to the mid-1960s (Kerr, 1965; Lockshin & Williams, 1965), many efforts have been made by the scientific community to systematically classify cell death types. The first effort, by Schweichel and Merker (1973), shed light on a morphology-based classification of three cell death types, which was considered insufficient due to lack of attention to biochemical features. In order to fill this gap, the Nomenclature Committee on Cell Death (NCCD) established unequivocal criteria and terminology to define cell death (Galluzzi et al., 2012; Kroemer et al., 2005, 2009). In particular, the principal goal of the NCCD is "to provide a forum in which names describing distinct modalities of cell death are critically evaluated and recommendations on their definition and use are formulated, hoping that a nonrigid, yet uniform nomenclature will facilitate the communication among scientists and ultimately accelerate the pace of discovery" (Kroemer et al., 2009). Based on the most recent classification, four main cell death types have been described, namely, apoptosis, ACD, programmed necrosis (best known as necroptosis), and mitotic catastrophe (Galluzzi et al., 2012). Albeit the characterization of each type goes largely beyond the purpose of this chapter, we summarized the main features of the three most characterized forms of programmed cell death (Table 1) (Bursch,

Table 1 Hallmarks of the Main Types of Programmed Cell Death

	Apoptosis	Autophagic Cell Death	Necroptosis
Morphology	Nuclear fragmentation, plasma membrane blebbing, chromatin condensation, cell shrinkage	Massive presence of autophagosomes and autophagolysosomes	Organelle swollen, early plasma membrane rupture
Biochemical features	Caspase activation, MMP, PS exposure	mTOR repression, ATGs activation	TNFR1 activation
Effectors	Bcl-2 family members, caspases, mitochondria	ATGs, mTOR, JNK, Class III PI3PK, beclin-1	RIP1, RIP3, MLKL
Chemical inhibitors	zVAD-FMK (caspases)	Wortmannin, 3-methyladenine (Class III PI3PK); choloroquine, bafilomycin A1 (autophagolysosome formation)	Necrostatin-1 (RIP1); GSK843, GSK872 (RIP3); Necrosulfonamide (MLKL)

The targets of each chemical inhibitor are indicated in parentheses. *Abbreviations*: ATG, autophagy-related protein; Bcl-2, B-cell lymphoma 2; JNK, c-Jun N-terminal kinase; MLKL, mixed lineage kinase domain-like protein; MMP, mitochondrial membrane permeabilization; mTOR, mammalian target of rapamycin; PI3PK, phosphatidylinositol-3-kinase; PS, phosphatidylserine; RIP1/3, receptor-interacting serine/threonine-protein kinase 1/3; TNFR1, tumor necrosis factor receptor 1; zVAD-FMK, z-Val-Ala-DL-Asp-fluoromethylketone.

Ellinger, Gerner, Frohwein, & Schulte-Hermann, 2000; Galluzzi et al., 2012). Indeed, our major focus will be ACD, historically known as programmed cell death Type II.

2.1 Overview on Autophagy

Before proceeding with a detailed description of ACD, we believe it is unavoidable to provide a concise snapshot of the autophagic process and its regulation at the molecular level. Indeed, ACD occurs when autophagy is uncontrolled or overstimulated. Macroautophagy (hereafter referred to as autophagy) is a conserved catabolic process aimed at the control of cellular homeostasis by degrading damaged organelles and aggregated or misfolded proteins (Nakatogawa et al., 2009). Controlled autophagic flux is usually regarded as a survival promoting mechanism since it is activated by a variety of stress stimuli such as nutrient starvation and hypoxia. In particular, upon energetic demand, simple molecules such as amino acids and fatty acids are derived from autophagy-driven recycling of complex macromolecules (Choi, Ryter, & Levine, 2013a). Other autophagy effects encompass control of cellular senescence, elimination of intracellular pathogens, and muscle differentiation (Gamberi et al., 2016; Wileman, 2013). Such a wide range of effects has been confirmed by the observation that autophagy is involved in many diseases including neurodegenerative disorders and cancer (Choi, Ryter, & Levine, 2013b).

The autophagic process is accomplished through the coordinated activity of a series of evolutionary conserved proteins, the autophagy-related (Atg) proteins, which cooperate to form a double-lipid bilayer vesicle called the autophagosome. The degradation of the *cargo* that is engulfed by the autophagosome is achieved within the autophagolysosome deriving from the fusion of autophagosomes and lysosomes (Janousek, 1990). In mammals, the induction of autophagy is triggered by a cytosolic complex consisting of unc-51-like kinase 1 or 2 (ULK1 or 2), Atg13, Atg101, and focal adhesion kinase family interacting protein of 200 kDa (FIP200), which participates in the formation of a preautophagic structure known as phagophore or isolation membrane. It is still largely debated whether the phagophore is built from organelle preexisting membranes (best candidates are ER, Golgi-apparatus, and the outer mitochondria membrane) or ex novo lipid biosynthesis (Abounit, Scarabelli, & McCauley, 2012). Further maturation of the phagophore requires the recruitment of the class III phosphatidylinositol-3-phosphate kinase (PI3PK)/vesicular sorting protein 34 (vps34) and beclin-1 (mammalian orthologue of yeast Atg6) complex, which is stabilized by beclin-1-associated autophagy-related key regulator (Barkor) and UV radiation-associated resistance gene (UVRAG). PI3PK activity plays a crucial role in the induction of autophagy to such an extent that its inhibitors (e.g., wortmannin and 3-methyladenine) are among the most commonly used drugs to arrest autophagy. It is not completely clarified how the ULK1 complex activates its downstream targets, but one study demonstrated that it phosphorylates the activating molecule in beclin-1-regulated autophagy protein 1 (AMBRA1), a key component of the class III PI3PK complex (Di Bartolomeo et al., 2010). The elongation of the phagophore into a complete autophagosome, which usually has an average diameter of 700 nm, involves two ubiquitin-like conjugation systems, namely, Atg5/Atg12 and microtubule-associated protein 1A/1B-light chain 3 (LC3) (the mammalian orthologue of Atg8), which share the protein

Atg7. Particular attention has to be paid to LC3 processing, which starts with its C-terminus cleavage by Atg4 and culminates with conjugation to the amino group of phosphatidylethanolamine that is required for association with the autophagosome. Besides a functional role, lipidated-LC3 is recognized as the most widely used marker to monitor the autophagic flux (Nakatogawa, 2013).

Autophagy is regulated by multiple signaling cascades, which mainly converge on the mammalian target of rapamycin (mTOR) that can bind to and phosphorylate ULK1, thus inhibiting autophagy. Conversely, one of the main inducers of autophagy is the energy-sensing AMP-activated protein kinase (AMPK), which upon proper stimulation (e.g., glucose deprivation or hypoxia) activates Rheb and Raptor leading to dissociation of mTOR form the ULK1 complex. Moreover, it has been demonstrated that, subsequent to mTOR dissociation, ULK1 can be directly phospho-activated by AMPK (Kim, Kundu, Viollet, & Guan, 2011). Further mechanisms have been shown to contribute to autophagy regulation, including oxidative stress (Desideri, Filomeni, & Ciriolo, 2012; Filomeni, Desideri, Cardaci, Rotilio, & Ciriolo, 2010). Indeed, the cysteine protease Atg4 is targeted by reactive oxygen species (ROS), specifically hydrogen peroxide (H_2O_2), on a cysteine located near the catalytic site and it is well established that several autophagy inducers, such as nutrient starvation, produce massive ROS bursts (Scherz-Shouval et al., 2007). In this context, our group demonstrated that the mitogen-activated protein kinase (MAPK) p38 promotes ROS buffering and hinders potentially detrimental excessive autophagy thus exerting a prosurvival role in different cancer cell lines challenged by nutrient starvation (Desideri et al., 2014).

2.2 Definition and Assessment of ACD

Historically, the term "autophagic cell death" was attributed to all those instances in which cell death was accompanied by autophagy that is to say, from a morphological point of view, in presence of large vacuolization (Liu & Levine, 2015). There is solid evidence supporting autophagy as directly committed in cell death processes, as demonstrated for some steps of *Drosophila melanogaster* development (Mohseni, McMillan, Chaudhary, Mok, & Reed, 2009). Indeed, it has been demonstrated that autophagy-mediated cell death, not apoptosis, is required for the elimination of midgut during *Drosophila* metamorphosis. In fact, midgut cell death is significantly repressed by autophagy inhibitors but not apoptotic ones, despite the presence of caspase activity. This is, to our best knowledge, the first in vivo evidence for the existence of ACD. Autophagy was shown to play a mechanistic role in cell death also in mammalian cell lines challenged with different cytotoxic drugs, especially if pivotal apoptotic effectors, such as BCL2-associated X (Bax), BCL2-antagonist/killer (Bak), and caspases, were lacking (Lamy et al., 2013; Shimizu et al., 2004, 2010). Although autophagosomes can be easily recognized by transmission electron microscopy as double-membrane vacuoles engulfing either cytosol or intact organelles, cytoplasmic vacuolization is not necessarily linked to autophagy and, more importantly, it should be demonstrated that autophagy is causally associated with occurring cell death. Indeed, it has been shown that in most instances autophagy is activated as a drastic attempt to cope with death-inducing stress and its inhibition accelerates, rather than blocks, cell death (Boya et al., 2005). This is the case of mammalian cells subjected to starvation, growth factor withdrawal, infectious agents, and cytotoxic chemotherapeutics drugs. In order to avoid misinterpretation, the NCCD established new systematic criteria to define ACD based on functional and biochemical features. In particular, "ACD" describes all those cell death instances mediated by autophagy, implying that they can be reverted by the use of the autophagic pathway inhibitors, including both drugs (e.g., class III PI3PK inhibitors: 3-methyladenine and wortmannin) and genetic silencing (e.g., targeting *ATG5, ATG12, AMBRA1*). In this context, large priority must be given to genetic silencing experiments over pharmacologic inhibitors, since the latter may have other targets than autophagic proteins. For example, 3-methyladenine is usually used at high (millimolar) concentration and might also affect class I PI3PK and the mitochondrial permeability transition pore (a death-triggering feature shared by apoptosis and necrosis) (Chen, Feng, Nie, & Zheng, 2012). A more stringent definition imposes that at least two different autophagic proteins must be silenced to ascertain the presence of ACD, since many Atg proteins are involved in non-autophagic functions (Galluzzi et al., 2012). This particularly applies to beclin-1 that is involved in the control of vesicular trafficking (Zeng, Overmeyer, & Maltese, 2006) and is known to interact with proapoptotic proteins such as B-cell CLL/lymphoma 2 (Bcl-2) and B-cell lymphoma-extra-large (Bcl-XL) (Pattingre et al., 2005). Another important issue that most probably contributed to slow down the unique identification of ACD in the last two decades of studies is that evidence for increased autophagic flux in cell dying by ACD was often missing. Autophagic flux is defined as the amount of degradable material that is carried from the autophagosome to the lysosome (Meijer, 2009). Indeed, autophagy is a dynamic process and the increase of autophagic markers (e.g., LC3 lipidation) might be the consequence of arrested autophagic flux due to the impairment of autophagosome maturation/degradation, rather than its increase. Autophagic flux can be easily monitored through the use of late stage-autophagy inhibitors such as bafilomycin A_1 (lysosome fusion inhibition) or chloroquine (lysosomal degradation inhibition) and fortunately, guidelines providing the tools for the interpretation of autophagy assays are being constantly updated (Klionsky et al., 2016). Overall, only cell death instances that are positive for active autophagic flux markers and can be restored by inhibition of at least two components of the autophagic pathway can be considered as ACD.

2.3 Molecular Mechanisms Underlying ACD

Despite full elucidation of the criteria to properly define and identify ACD, clues about the molecular effectors triggering and participating in the execution of cell death have been missing for long time. One of the first mechanisms was proposed by the studies of Shimizu and colleagues using etoposide- and staurosporine-challenged Bax/*Bak* double-knockout mouse embryonic fibroblasts (MEFs), a widespread model system to study ACD (Shimizu et al., 2010). The authors demonstrated that the activation of the MAPK c-Jun N-terminal kinase (JNK) is required for triggering ACD in these cells. Indeed, pharmacological inhibition of JNK by SP600125 impaired autophagy-dependent cell death. Remarkably, JNK activation was dependent on autophagy, since it was abolished by the use of 3-methyladenine or siRNA against *Atg5* and conversely JNK did not affect autophagy. Surprisingly, the authors also showed that these findings did not apply to starvation-induced ACD, suggesting that JNK-mediated control of ACD might be strictly stimulus dependent. JNK activity, alone or in combination with other effectors, is required in other ACD in in vitro models: (i) mouse fibroblast L929 cell line treated with zVAD, a pan-caspase inhibitor (Yu et al., 2004). Here, the authors showed that ACD requires Src-mediated activation of JNK and extracellular signal-regulated kinase (ERK) (Chen et al., 2011); (ii) chronic myelogenous leukemia cells upon treatment with the phytoalexin resveratrol. In this case, ACD was activated by JNK-dependent accumulation of p62/SQSTM1 (an autophagy substrate) in cooperation with AMPK (Puissant et al., 2010). In the context of natural compounds, the triterpenoid avicin D decreases ATP levels, thus triggering AMPK-mediated ACD in various cell models (Xu et al., 2007). Azad and colleagues provided further insight in ACD induced by hypoxia in several cancer cell lines (Azad et al., 2008). Although it is well known that oxygen deprivation induces autophagy as a defense mechanism, the authors proved that prolonged hypoxia induced ACD regardless of an intact apoptotic machinery and this process involved the Bcl-2/E1B 19 kDa-interacting protein (BNIP3), a death-inducing Bcl-2 family member.

2.4 Story of a Threatened Identity

Since the first observations of autophagy-associated cell death, many questions have arisen about the existence of ACD. In particular, whether ACD constitutes a death mechanism or acts as a primer igniting other cell death pathways (e.g., apoptosis or necrosis) has been largely discussed. In 2006, a first evidence demonstrated that autophagy is necessary to trigger apoptosis: the authors showed that *BECN1* (encoding for beclin-1) and *ATG7* knockdown completely abolished HIV envelope glycoprotein-induced apoptosis in $CD4^+$ T lymphocytes (Espert et al., 2006). Similar results were obtained by studies on oxidative stress-induced apoptosis in photoreceptors, in which pharmacologic and genetic ablation of autophagic proteins significantly reduced cell death and caspase-3 activity both in in vivo and in vitro (Kunchithapautham & Rohrer, 2007). Moreover, it has been shown that p53-mediated apoptosis requires the activity of its target gene product damage-regulated autophagy modulator (DRAM), which is known to induce autophagy (Crighton et al., 2006). These studies argue for a concurring function of autophagy in triggering the execution of apoptotic cell death rather than being a death mechanism by itself. In 2008, Kroemer and Levine openly claimed that no evidence had convincingly demonstrated that cells die by autophagy in a physiopathological in vivo context and proposed as a solution a general effort to use the term ACD exclusively in those instances in which autophagy mediates cell demise (Kroemer & Levine, 2008). The reason of all these inconsistencies needs further elucidations but it is likely to depend on the cell type and on the nature of the stimuli.

2.5 Autosis: A Particular Subtype of ACD?

In an effort to dissect the mechanism of autophagy-induced cell death, Liu et al. identified a peptide, termed Tat-beclin-1, generated by the fusion of the evolutionary conserved domain of beclin-1 and the HIV Tat protein transduction domain (Shoji-Kawata et al., 2013). Tat-beclin-1 induced massive protective autophagy upon a variety of stimuli both in vitro and in mouse models, but also time- and dose-dependent cell death in vitro, which was dramatically suppressed by autophagy inhibition and was then termed "autosis" (Liu et al., 2013). Autosis, which is not triggered by ROS, has been also demonstrated in vivo in hippocampal neurons of rat subjected to hypoxia-ischemia. Morphologically, cells dying by autosis displayed augmented cell-substrate adherence, late ER disappearance, and unique nuclear membrane changes (i.e., swelling of the perinuclear space, separation of the inner and outer membranes). The repression by pharmacological and genetic inhibition of Na^+/K^+ ATPase is another typical autosis signature, which is not shared by other programmed cell death types (Liu et al., 2013). The strong reliance on the membrane pump was demonstrated by the use of cardiac glycosides, known antagonists of the Na^+/K^+ ATPase, which repressed death triggered by autophagy-inducing peptides or starvation to a larger extent than autophagy inhibitors. Autosis dependence on Na^+/K^+ ATPase may partially explain some morphological features, such as the enhanced substrate adhesion of dying cells. In fact, beyond the well-known role of generating Na^+ and K^+ gradient across plasma membrane, the Na^+/K^+ ATPase participated in many cellular pathways including regulation of cell-substrate adhesion (Contreras, Shoshani, Flores-Maldonado, Lazaro, & Cereijido, 1999). Moreover, the Na^+/K^+ ATPase also localizes to the nuclear inner membrane and thus it may mediate the autotic typical swelling of perinuclear space through disrupted ionic transport (Galva, Artigas, & Gatto, 2012). The detailed molecular mechanisms of autosis need further investigation and may offer novel insight in autophagy-mediated cell death. Whether autosis can be considered as a subtype of ACD or an independent cell death mechanism remains a question mark.

3. BIOLOGY OF DHEA

DHEA (3b-hydroxyandrost-5-en-17-one) is, along with its sulfated form dehydroepiandrosterone sulfate (DHEA-S), the most abundant plasma steroid hormone in humans and a precursor of sexual hormones (i.e., testosterone and estrogen). Adrenal cortex is the main site of DHEA synthesis, while testes, brain, and ovaries account for a minor extent. DHEA is synthesized through a series of reactions involving P450-dependent monooxygenases and hydroxysteroid dehydrogenases. Although 90% of produced DHEA is rapidly metabolized into DHEA-S, the nonsulfated form is mainly used for hormone synthesis, whereas DHEA-S can be regarded as a reservoir. The two forms can be interconverted by DHEA sulfotransferases and hydroxysteroid sulfatases and their relative abundance is determined by the tissue-specific expression of these enzymes (Allolio & Arlt, 2002). DHEA active metabolites are released by several tissues including muscle, skin, and brain, but liver is the main source. Only 5% of DHEA circulates free in the bloodstream since it is usually conjugated to albumin. Although DHEA was isolated in 1934 from male urine (its name was proposed in 1949), the first human studies came only since the 1990s, when its beneficial effects in a plethora of human physiopathological conditions, were discovered (Rutkowski, Sowa, Rutkowska-Talipska, Kuryliszyn-Moskal, & Rutkowski, 2014). These studies shed light on the concept that DHEA is a biologically active molecule rather than a mere hormone precursor. In particular, DHEA is popular for its antiaging properties to such an extent that it is known all over the world as the "fountain of youth hormone." Indeed, DHEA plasma levels reach their maximum between the age of 20 and 30 years and dramatically decrease with aging, both in men and women, due to reduced secretion by adrenal cortex, although other adrenal hormones are not affected. This progressive decline has been linked to age-related pathologies such as osteopenia, atherosclerosis, cognitive impairment, and cancer (Maninger, Wolkowitz, Reus, Epel, & Mellon, 2009).

3.1 Controversies of DHEA and Identification of Its Receptors

A number of arguments have been raised against a physiological role of DHEA. Some have argued that studies in animal models are poorly reliable because DHEA is essentially a human-produced molecule and animal tissues may respond differently from human ones. Others claim that the rapid conversion into its metabolites weakens the concept that DHEA is mechanistically involved in the observed effects. Investigation on DHEA, which is commercially available as a dietary supplement, is sometimes neglected and is not sustained by financial support by pharmaceutical companies since an endogenous molecule cannot be patented. Initially, one of the major concerns about DHEA action in human physiology was the lack of identification of a DHEA receptor. In fact, it is well known that steroid hormones, including glucocorticoids, androgens, and estrogens, elicit their functions via binding to nuclear receptors (Traish, Kang, Saad, & Guay, 2011). However, a growing body of evidence has identified nuclear receptors either bound to or activated by DHEA. Among these, it is worth mentioning that DHEA can promote the expression of several genes including cytochrome

P450 genes, through the activation of the peroxisome proliferator-activated receptor α (PPARα). Surprisingly, DHEA is not a ligand of PPARα, but rather influences its phosphorylation status (Webb et al., 2006; Webb, Xiao, Geoghegan, & Prough, 1996). Other nuclear receptors mediating DHEA effects are the pregnane X receptor, the estrogen receptor, and the constitutive androstane receptor (Kohalmy et al., 2007; Webb et al., 2006). In 2002, Liu and Dillon provided the first evidence of a membrane receptor for DHEA. In fact, they identified a high-affinity putative plasma membrane receptor and its associated G-proteins, through [3*H*]DHEA experiments in endothelial cells. Remarkably, structurally related steroids did not compete for DHEA binding to isolated plasma membrane (Liu & Dillon, 2002). It was further clarified that DHEA triggers the activation of endothelial nitric oxide synthase (eNOS) and production of nitric oxide, which culminated in increased cyclic guanosine monophosphate (cGMP). DHEA-induced nitric oxide release was shown to be dependent on the activity of the MAPK ERK1/2. Importantly, DHEA effects are not repressed by the use of androgen, estrogen, and other steroid hormone receptors, suggesting that it acts through a specific receptor (Simoncini et al., 2003). The presence of a G protein-coupled membrane receptor was confirmed in neuronal studies on PC12 cell line and rat hippocampus, in which DHEA promoted an antiapoptotic effect via the activation of anti-apoptotic Bcl-2 and Src kinase, globally exerting neuroprotective functions (Charalampopoulos et al., 2006). The biochemical nature of these receptors remains obscure to date and requires further investigation, in order to provide better understanding of the mechanism underlying DHEA pleiotropic action.

3.2 DHEA Regulation of Cell Death Mechanisms: A Jack of All Trades in Human Physiopathology

DHEA function has been implicated in a plethora of human pathological conditions, including obesity, diabetes, atherosclerosis, autoimmune and cardiovascular disease, systemic lupus erythematosus, osteoporosis, depression, sexual dysfunction, impaired memory, and cancer, as demonstrated by many prospective and retrospective studies on human cohorts that have identified correlations between DHEA serum levels and pathophysiological conditions. Moreover, treatment with exogenous DHEA in preclinical and clinical studies has provided exciting results, although sometimes contradictory. An exhaustive analysis of all these studies goes largely beyond the scope of this chapter. However, hereafter we provide a snapshot of three of the most characterized and relevant effects of DHEAs (DHEA + DHEA-S) on human physiology with a focus on molecular mechanisms, particularly those related to cell death.

3.2.1 Cardiovascular Disease

The first evidence of a correlation between DHEA and DHEA-S plasma levels was postulated at the end of the 20th century. First, it was observed that DHEA levels were lower in patients with a history of cardiac disease compared to those who had never been affected. More importantly, mortality from any heart disease inversely correlated with DHEA concentration in men over age 50 (Barrett-Connor, Khaw, & Yen, 1986). Since then, unfortunately, studies comparing DHEAs serum levels in cardiovascular disease patients vs controls were often contradictory and inconsistent, and seemed highly sex dependent. For instance, one study on women with ischemic heart disease showed an inverse correlation between DHEA-S concentration and the occurrence of pathology, which was not highlighted by another study (Golden et al., 2002; Haffner, Moss, Klein, & Klein, 1996). Results in men are slightly more consistent, and extensively reviewed by Tchernof and Labrie (2004). Although endogenous DHEAs levels cannot be considered reliable markers for cardiovascular disease outcome, a large body of evidence showed that exogenous DHEA supplementation protects the cardiovascular system against different stimuli-mediated damage. The presence of proliferating and apoptosis-resistant vascular smooth muscle cells is a feature of injury-induced conditions such as restenosis. It was demonstrated both in cell models (human carotid vascular smooth muscle cell and saphenous vein grafts in tissue culture) and in vivo (rat carotid injury model) that DHEA reduced cell proliferation with a parallel increase of apoptosis (Bonnet et al., 2009). Notably, these effects were mediated by the inhibition of Akt (also known as protein kinase B) and subsequent activation of glycogen synthase kinase-3β (GSK-3β), thorough a G-coupled membrane receptor. Moreover, DHEA has been showed to beneficially impact on pulmonary arterial hypertension, a pathology characterized by pulmonary artery smooth muscle (PASM) cell overproliferation and death-resistance, mainly sustained by the activation of the Src/signal transducer and activator of transcription 3 (STAT3) axis (Paulin et al., 2011). Pulmonary arterial cells derived from hypertension patient were challenged by DHEA and displayed decreased activation of Src/STAT3 accompanied by increased cell death and reduced proliferation. Similar results were obtained in rats where DHEA reversed pulmonary hypertension.

Conversely, DHEA can also exert a prosurvival function via inhibition, rather than activation, of apoptosis. In fact, DHEA was shown to protect human and bovine vascular endothelial cells against serum withdrawal-mediated apoptosis, through activation of PI3PK/Akt pathway and the anti-apoptotic protein Bcl-2 (Liu et al., 2007). These effects were maximal at physiological concentrations (0.1–10 nM) but were not confirmed in vivo. Overall, the discovery of these mechanisms of action of DHEA could pave the way to the identification of new candidates to treat cardiovascular diseases.

3.2.2 Brain Disorders

The concept of the presence of DHEAs in the brain was supported by rat models, in which DHEA was found in the anterior and posterior parts of brain independently of adrenal secretion (Corpechot, Robel, Axelson, Sjovall, & Baulieu, 1981). Intriguingly, DHEAs levels in central nervous system are considerably higher than circulating ones both in rodents and humans (Corpechot et al., 1981; Weill-Engerer et al., 2003). Due to its synthesis in brain, DHEA has also been designated as a neurosteroid and many effects on brain functionality have been attributed to this hormone (Robel & Baulieu, 1995). Although studies correlating DHEAs serum levels of patients with dementia, including Alzheimer's disease, failed to produce unequivocal results, attention paid to psychological and neuropsychological disorders has revealed that DHEA's concentration correlates with indicators of mood and memory (Starka, Duskova, & Hill, 2015). Moreover, despite sometimes disputable, in most studies it has been proved that DHEA serum levels are lower in schizophrenic patients compared to healthy individuals and affect responsiveness to antipsychotic treatment (Ritsner, 2011). Neurosteroids are known to modulate neuron activity via interactions with neurotransmitter receptors impacting neuron excitability and neurotransmitter secretion (Schumacher et al., 2000). One of the first documented actions of DHEA on a neurotransmitter receptor was provided by studies on the gamma-aminobutyric $acid_A$ ($GABA_A$) receptor (Imamura & Prasad, 1998), where DHEAs were shown to noncompetitively inhibit $GABA_A$ receptor-mediated chloride influx in rat neurons. Most evidence about the effects of exogenous DHEAs on brain derives from animal experimentation, particularly on rodents. The notion that DHEAs can improve memory has been confirmed by different approaches (e.g., avoidance-based paradigms, tests of spatial memory involving different types of mazes) in aged rodents (Mayo et al., 1993; Melchior & Ritzmann, 1996). With regard to DHEA administration to patients, promising results came from the first single case study on memory, conducted in a 47-year-old woman who displayed learning disabilities and poor levels of serum DHEA. Extended administration of DHEA resulted in recovery of DHEAs serum concentration and some memory functions (Racchi, Balduzzi, & Corsini, 2003). While other clinical trials had no relevant effects on cognitive and memory parameters (Wolf, Koster, et al., 1997; Wolf, Neumann, et al., 1997), encouraging results on DHEAs plasma levels restoring and improvement of depression and some memory aspects were obtained by administration of DHEA (30–90 mg/day for 4 weeks) (Wolkowitz et al., 1997). Studies aimed at dissecting the molecular mechanisms of DHEA in neurons demonstrated that this hormone exerts a neuroprotective function in different conditions, and this may have beneficial consequence for all those pathologies linked to neuronal cell death, including neurodegenerative disorders (e.g., Alzheimer's and Parkinson's diseases) (Yuan & Yankner, 2000). A feature of many neurological disorders is persistent deprivation of glucose and hypometabolism, which causes neuron damage and death. DHEA was able to reverse hypometabolism through increased oxidative metabolism thus preserving neuron viability in different models of neurodegenerative disorders (Vieira-Marques et al., 2016). Moreover, glucose deprivation-induced damage was hampered by treatment with DHEA in hippocampal neurons and SH-SY5Y neuroblastoma cells. In a rat model of ischemia/reperfusion, treatment with DHEA (2–4 mg/day per rat) protected cell integrity and viability, measured as membrane Na^+/K^+ ATPase activity and lactate dehydrogenase release, in isolated synaptosomes through reduction of ROS production (Aragno et al., 2000). Indeed, it is well accepted that exogenous DHEA can exert dual effects, either antioxidant or prooxidant, depending on doses and on tissue specificity in most experimental systems (Celebi et al., 2004; Kim & Novak, 2007). For instance, DHEA protects cell viability and prevents oxidative stress-induced lipid peroxidation in a study conducted in human hippocampal tissue from Alzheimer's disease patients vs age-matched controls (Bastianetto, Ramassamy, Poirier, & Quirion, 1999). Several lines of evidence revealed that DHEA action is involved in neurogenesis and a study on neural precursors from the embryonic forebrain of rat demonstrated that DHEA promoted Akt activation thus decreasing apoptosis (Zhang et al., 2002). Surprisingly, DHEA-S played opposite role in the same system via impairing Akt activation. Nevertheless, other data speak in favor of a proapoptotic effect of DHEA in GT1–7 hypothalamic neurons deprived of trophic support (Lin, Cui, Yusta, & Belsham, 2004). The cytotoxic action of DHEA is not due to conversion into other steroids and is associated with decreased activity of the Akt/GSK-3β axis. Moreover, DHEA reduced cell viability via induction of apoptosis also in BV-2 murine microglia cells, and this negatively correlated with glucose concentration (Yang, Jeng, Ho, Chou, & Hu, 2000). Thus, it appears that both the role of DHEA in brain (beneficial or detrimental) and the underlying mechanisms are still elusive and require further investigation.

3.2.3 Cancer

The anticancer properties of DHEA are known since a long time and hints were provided by the first data demonstrating that this hormone has anti-mutagenic effects against different mutagens in the Ames test (Wilpart, Speder, Ninane, & Roberfroid, 1986). Some attempts to find correlations between DHEAs serum concentration and cancer incidence have been made by case–control studies, but results are often inconsistent. Indeed, data suggest a significant risk of developing some cancers, including breast (in premenopausal women), ovarian (in postmenopausal women), gastric, and bladder, in patients with low serum levels of these steroids, whereas other studies failed to find any trends for breast and ovarian cancers, melanoma, and squamous cell carcinoma of the skin (Matsuzaki & Honda, 2006).

Many reports have demonstrated the role of exogenous DHEA (used at pharmacological doses, i.e., 20–1200 mg/kg/day) in inhibiting tumor growth in rats and mice (Svec & Porter, 1998). An interesting study was carried on C3(1)/SV40 large T-antigen mice, which spontaneously develop mammary lesions/invasive carcinomas (females) and prostate cancer (males) (Green et al., 2001). No differences in the number of preinvasive neoplasia lesions were observed in mice receiving DHEA, whereas mammary cancer growth was significantly inhibited. No relevant results were obtained in male mice concerning prostate cancer. These results strongly suggested that DHEA is involved in tumor progression/repression, not initiation. Regarding humans, to our best knowledge, only a single case report revealed some beneficial, although doubtful, effects of DHEA treatment in a hepatocellular carcinoma female patient (Jozuka, Jozuka, Suzuki, Takeuchi, & Takatsu, 2003). Indeed, the patient was treated against major depression with a combination of the antidepressant fluvoxamine, glycyrrhizic acid (the main component of licorice), and DHEA and medical examinations revealed normal serum levels of DHEA, improvement of liver functionality, and alpha-fetoprotein (a marker of hepatoma) decrease. However, scarcity of data does not permit to establish whether the patient's recovery was spontaneous or mediated by the combinational therapy. Although DHEA anticancer properties have been proved in many tissues, including liver in rats, it has been speculated that DHEA may have also some hepatocarcinogenic functions in rats (Mayer, Forstner, & Kopplow, 2003). In particular, DHEA promoted hepatomegaly, peroxisome proliferation, and, upon prolonged treatment, development of hepatocellular carcinoma (Mayer et al., 2003). Indeed, only 3 days of DHEA administration (100 mg/kg body weight/day) led to increased levels of cyclin E and Ki-67 staining, indicative of increased proliferation in rat livers (Kopplow, Wayss, Enzmann, & Mayer, 2005). This was not associated with enhanced serum concentration of either transaminases or bilirubin, suggesting that livers were not experiencing damage-triggered proliferation. Albeit the mechanisms have not been fully clarified, one possible explanation may be the peroxisome proliferator role of DHEA, which results in enhanced peroxisomal β-oxidation and production of hydrogen peroxide (H_2O_2) and consequent potential DNA damage. To what extent these data obtained from rats are relevant in humans is not clear, but other molecules, such as clofibrate, are peroxisome-proliferators and hepatocarcinogenic in rats, whereas they induce no peroxisome proliferation in hepatic human cells and are used in patients for hyperlipidemia treatment without any side effects on liver (Blaauboer et al., 1990).

The antiproliferative effects of DHEA have extensively been proved in several types of human cancer cell lines, including hepatoma, prostate, myeloma, leukemia, human colon adenocarcinoma, and cervical cancer (Arnold et al., 2007; Catalina, Milewich, Kumar, & Bennett, 2003; Giron, Montano, Escobar, & Lopez-Marure, 2009; Jiang et al., 2005; Liu et al., 2005). Potential mechanisms have been proposed. One explanation is the strong uncompetitive inhibitory activity of DHEA on glucose-6-phosphate dehydrogenase (G6PD), the first and rate-limiting enzyme of the pentose phosphate pathway, whose major products are ribose-5-phosphate (necessary for the synthesis of nucleic acids) and NADPH, the main intracellular electron donor for biosynthetic reactions (e.g., fatty acids) (Schwartz & Pashko, 2004). The utilization of the pentose phosphate pathway confers several advantages to malignant cells, whose main feature is rapid proliferation rate and thus high nucleic acid and lipid demand for DNA and membrane synthesis, respectively. Several human tumors upregulate some enzymes of the pentose phosphate pathway, such as G6PD and transketolase-like 1 (Diaz-Moralli et al., 2016). Accordingly, DHEA induced cell growth inhibition in hepatoma cell lines, due to ribose-5-phosphate depletion, because this effect was abrogated by the use of ribonucleosides (Yoshida et al., 2003). G6PD inhibition by

DHEA may also, at least in part, account for the in vitro antimutagenic properties of the hormone, since most chemical carcinogens require activation by NADPH-dependent cytochrome P450 prior to conversion into mutagenic molecules (Miller & Miller, 1981). Notably, it was demonstrated in two breast cancer cell lines (MCF-7 and MDA-MB-231) that the inhibitory effect of DHEA on G6PD and proliferation was observed only at pharmacological concentrations (above 10 μM), whereas it was not evident at physiological (up to 10 nM) concentrations (Di Monaco et al., 1997). Moreover, DHEA repressed cell growth also in G6PD-null cells, indicating that some mechanisms other than G6PD inhibition are involved in DHEA antiproliferative function (Biaglow et al., 2000). Another appealing hypothesis is that DHEA reduces the activity and the expression of 3-hydroxy-3-methylglutaryl-CoA reductase (HMGR), the rate-limiting enzyme of cholesterol biosynthesis, whose immediate product is mevalonate (Pascale et al., 1995). Indeed, mevalonate is pivotal for cholesterol and thus biological membranes biosynthesis and its abundance are related to the progression of different kinds of human cancers (Mullen, Yu, Longo, Archer, & Penn, 2016). Mevalonate is also a precursor for isoprenoids, including farnesyl pyrophosphate, which are important mediators of protein posttranslational modifications. In fact, mevalonate is essential for the farnesylation of Ras oncogene by farnesyl transferase (Murphy, Moore, & Nesnow, 2012). Farnesylated-Ras in turn is involved in signaling promoting cell proliferation via activation of MAPK ERK1/2 pathway. Consistently, DHEA delayed cell growth and induced cell cycle arrest, due to depletion of intracellular mevalonate and other isoprenoids, in human colonic adenocarcinoma cells (Schulz, Klann, Schonfeld, & Nyce, 1992). Ho and colleagues' work revealed that DHEA delayed hepatoma cell growth through a mechanism different from G6PD or HMGR repression. They found that DHEA treatment was accompanied by impaired expression of nuclear-encoded mitochondrial genes and mitochondria functionality, resulting in decreased ATP levels (Ho, Cheng, Chiu, Weng, & Chiu, 2008). Indeed, pyruvate supplementation restored mitochondrial activity and reversed the antiproliferative effects of DHEA.

Much attention has been paid to DHEA active metabolites (e.g., DHEA-S, androsterone, epiandrosterone, etiocholanolone) that were tested for antiproliferative properties in different human cancer cell lines, including HepG2 (hepatocarcinoma), Caco-2 (colon carcinoma), and HT-29 cells (colon adenocarcinoma) (Yoshida et al., 2003). Results obtained from 3-[4,5-dimethylthiazol]-2yl-2,5-diphenyltetrazolium bromide (MTT) and bromodeoxyuridine incorporation assays demonstrated a time- and dose-dependent inhibition of cell proliferation by all tested compounds. Moreover, these effects were not dependent on inhibition of either G6PD or HMGR, suggesting that this mechanism is peculiar of DHEA. Another important mechanism whereby DHEA may promote its anticancer functions is the induction of cell death, either programmed or non-programmed. Results in different breast cancer cell lines, namely, MCF-7 (estrogen receptor positive) and MDA-MB-231and Hs578T (estrogen receptor negative), showed that DHEA reduced cell proliferation in all these cell lines, whereas, used at a concentration of 100 μM for hours, it induced necrosis only in MCF-7 (Lopez-Marure, Contreras, & Dillon, 2011). A previous work has instead demonstrated a proproliferation role of DHEA in MCF-7 cell lines, rather than an inhibition (Schmitt, Klinga, Schnarr, Morfin, & Mayer, 2001). However, differently from the other study, here cells were treated with 100 nM DHEA (i.e., physiological concentration) for 4 days and demonstrated that the conversion of DHEA into estrogens (particularly estradiol) was the real inducer of cell proliferation, since these effects were reduced by the estrogen receptor down regulator ICI 182,780 and tamoxifen. Thus, it appears that, at least in breast cancer, DHEA exerts very different effects depending on the dose, the presence of estrogen receptors, the duration of the treatment, and the possibility of conversion into its active metabolites. Another study on a panel of three cell lines from human uterine cervical cancer confirmed that the anticancer properties of DHEA are partially ascribable to cell death induction. Indeed, DHEA inhibited cell proliferation in both human papilloma virus (HPV)-positive CASKI and HeLa and HPV-negative C33A cells through an androgen- and estrogen-receptor independent mechanism (Giron et al., 2009). Cell death was evaluated by flow cytometry and TUNEL assay and revealed induction of cell death by two different mechanisms: C33A cell lines died by necrosis, while CASKI and HeLa elimination was mediated by apoptosis. Induction of apoptosis in HeLa cells by DHEA treatment was confirmed by our results obtained from caspase-9 Western blots and recovery of cell death upon use of pan-caspase inhibitor zVAD (Vegliante, Desideri, Di Leo, & Ciriolo, 2016). On the contrary, DHEA exerted a protective function against serum deprivation-induced apoptosis in prostate cancer (DU145 and LNCaP) and colon cancer (Caco2) cell lines and this effect, which was antagonized by testosterone, was mediated by the tropomyosin-related kinase TrkA receptor, a membrane receptor of neurotrophin nerve growth factor (NGF) (Anagnostopoulou et al., 2013). Indeed, this receptor is known for its prosurvival role and to be expressed in DU145 and LNCaP cell lines.

Remarkably, DHEA can bind TrkA thus confirming the neurotrophic function of this steroid. Exogenous expression of TrkA receptor restored DHEA prosurvival effects also in pheochromocytoma PC12 cell clone nnr5, which is TrkA-null (Lazaridis et al., 2011). Although promotion of cancer cell death may not be a general function of DHEA, we speculate that the described antiapoptotic effects of DHEA may be circumscribed to few cell lines, such as those expressing particular receptors. In vivo experiments are needed to clarify this and many other aspects of DHEA properties to induce/repress cell death in order to discriminate the effects of DHEA metabolites and the role of hormone receptors in tumors.

4. DHEA AND ACD

Occurrence of programmed cell death types other than apoptosis is sometimes neglected during in vitro experiments. In particular, many authors overlook the importance of using the appropriate inhibitors (described in Section 1) to infer the cell death type. As a consequence, the possibility that DHEA might induce nonapoptotic programmed cell death cannot be excluded. To our best knowledge, no studies reported the occurrence of necroptosis upon treatment with DHEA in any cell system, and this could also raise from shallow investigation of cell death mechanisms in works focusing more on biological outcomes than molecular signaling. Recently, we demonstrated for the first time that DHEA can trigger in HepG2 hepatoma cell line a cell death type that has all the features of ACD (Vegliante et al., 2016). Providing further insight into the morphological and biochemical mechanism of action of DHEA, we highlighted a strong increase of the lipidated form of LC3, autophagic flux, and number of autophagolysosomes in dying cells. Importantly, the use of zVAD not only failed to prevent DHEA-induced cell death, but even exacerbated the percentage of dead cells. Conversely, autophagy inhibition by both 3-methyladenine and siRNA against *ATG5*, significantly reduced the extent of cell death upon treatment with DHEA. One of the molecular effectors of ACD is the proautophagic protein beclin-1. Indeed, it has been proved that beclin-1 knockdown is able to reverse ceramide-induced ACD in CNE2 and Hep3B cell lines (Li et al., 2009). In this experimental system, JNK acted upstream of beclin-1 and its inhibition by both SP600125 and genetic silencing inhibited ACD. Consistently, in our cell model we also showed an increase of the levels of beclin-1 and JNK, but we did not investigate for a causal relation between the two proteins. We also demonstrated that JNK activation was not triggered by ROS as confirmed by the use of the antioxidant vitamin E analog Trolox, albeit DHEA induced an early ROS burst (Vegliante et al., 2016). This is not the unique evidence of JNK involvement in DHEA mechanisms. Indeed, DHEA-activated JNK in HAEC endothelial cell line (Lee et al., 2015). These cells were challenged by different physiological concentrations of DHEA (0.5–50 nM) for different times (1–4 h) and an overall increase of the phosphorylated forms of all MAPKs (i.e., p38, ERK, and JNK) was observed, indicative of their activation, but JNK, which can also be activated by the two other MAPKs, had a prominent role in the effects of DHEA. Other structurally related molecules, such as 1α,25(OH)2-vitamin D3, the hormonally active form of vitamin D, have been shown to induce MAPK phosphorylation in C2C12 muscle cells (Buitrago, Ronda, de Boland, & Boland, 2006). Therefore, it is tempting to speculate that cholesterol-derived molecules might share the ability of activating MAPK and, depending on target cells, used concentration and other factors, autophagy/ACD. DHEA-activated JNK-triggered ACD was accomplished through an accumulation of the autophagic substrate p62 in HepG2 cells, since p62 silencing decreased the number of cells dying by autophagy (Vegliante et al., 2016). Despite p62 levels accumulation is usually indicative of disrupted autophagic flux, we observed an augmented expression of the protein, since DHEA promoted increased nuclear migration of the transcription factor Nuclear factor (erythroid-derived 2)-like 2 (Nrf2), a target gene of which is p62. Puissant and collaborators also demonstrated that JNK–p62 axis can mediate ACD (Puissant et al., 2010). They showed that the phytoalexin resveratrol eliminated chronic myelogenous leukemia cells via JNK-mediated accumulation of p62. Remarkably, the same result was observed in CD34+ progenitors from patients with chronic myelogenous leukemia challenged with resveratrol.

Although we observed that DHEA-induced ACD in HepG2, another report demonstrated that it instead induces apoptosis in the same cell line (Jiang et al., 2005). They also showed that this effect was due to reduced Akt phosphorylation on threonine 308 and serine 473 and consequently its antiapoptotic role. This is a common mechanism by which DHEA regulates, either positively or negatively, programmed cell death in other cell systems (see Section 3). We motivate the discrepancy between the two studies with the different concentrations of DHEA that were used on HepG2: we challenged cells with high pharmacological doses (300–400 μM) (Vegliante et al., 2016), whereas they used lower doses (100 μM) (Jiang et al., 2005). However, Jiang et al. measured apoptotic cell death only by flow cytometry analysis of phosphatidylserine exposure (Annexin V) and did not confirm the inhibition of cell death by zVAD use (Jiang et al., 2005). Since apoptosis and ACD are intertwined and share some biochemical features, we cannot exclude the occurrence of ACD alone or in combination with apoptosis in this cell model. Indeed, a series of novel peptidomimetics bearing DHEA moiety has been recently designed and their anticancer properties tested on different cancer cell lines including HepG2, human lung cancer (A549), and melanoma (A875) (Wang, Su, Wang, Chen, & Cao, 2016). In particular, this study highlighted the capacity of some of these peptides to kill HepG2 cells at least initially via apoptosis, but the authors did not confirm the real mechanisms of cell death. Therefore, it seems that DHEA has different or multistage effects to induce cell death in HepG2. Nevertheless, these results also offer an encouraging perspective for tumor treatment since they show that modulating DHEA concentration two different cell death pathways can be activated, and this might be particularly relevant for those cancers that are drug-induced apoptosis resistant. The induction of cell death mediated by DHEA on another hepatoma cell line, Hep3B, seems more defined. In fact, these cells die by apoptosis when challenged by DHEA as demonstrated by the use of zVAD that significantly reduces dead cell percentage (Fig. 1). One of the possible explanation of the different effects of DHEA on the two hepatoma cell lines, HepG2 and Hep3B, is the status of the tumor suppressor p53. Indeed, it is well known that Hep3B cells

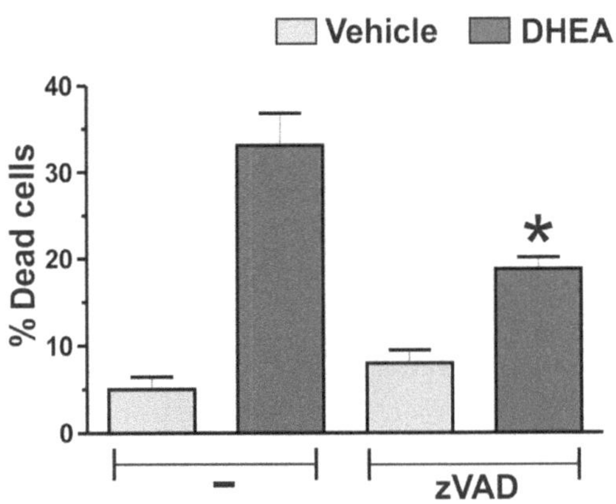

Fig. 1 DHEA induces apoptosis in Hep3B cell line. Hep3B cells were pretreated for 1 h with either vehicle or 10 μM of the caspase inhibitor zVAD and then treated for 24 h with 400 μM DHEA. The percentage of dead cells was assayed by Trypan blue staining procedure. Data are presented as mean ± SD of $n = 3$ independent experiments. $^{*}P < 0.05$ against DHEA alone.

are p53-null and p53 is involved in a plethora of cell processes including apoptosis and autophagy (Green & Kroemer, 2009). This could be an interesting aspect of hepatoma cell treatment with DHEA and requires further investigation. Other hints of DHEA (or DHEA metabolites) inducing ACD exist in literature. In particular, the neurosteroids 3β-androstene-17α-diol (17α-AED), a metabolite of DHEA produced in neuroectodermal tissue, which differs from the α epimeric form for different biological effects, was shown to induce a larger amount of acidic vacuoles during death induction of T98G and U251 glioma compared to U937 lymphoma (Graf, Jia, & Loria, 2007). These vacuoles were demonstrated to be autophagolysosomes since bafilomycin A1 inhibited their formation. Although the authors failed to provide more biochemical and genetic details to identify the molecular process inducing cell death, they speculate that 17α-AED *bona fide* induces ACD in glioma cells but not in lymphoma cells, since the cytotoxic effect of the neurosteroid in the former cells was not accompanied by DNA fragmentation, caspase-3 activation, and its substrate poly [ADP-ribose] polymerase 1 (PARP-1) cleavage. Importantly, ACD is at least in part responsible for tamoxifen-treated breast cancers cell death. Indeed, zVAD only weakly reduces tamoxifen-induced cell death, indicating that apoptosis marginally contributes to the anticancer properties of this drug. Although tamoxifen, which competes with estrogens for the estrogen receptor binding, is at the forefront for the cure of estrogen receptor-positive tumors, drug resistance often impairs the positive outcome of the cure (Yu et al., 2016). It was demonstrated that miR-21 is partially responsible for tamoxifen-resistance in MCF-7 cell lines, and this effect is abrogated by its silencing, which results in increased autophagy markers and ACD induction through the inhibition of the PI3PK/Akt/mTOR pathway. Albeit tamoxifen has a nonsteroid structure, it shares the same receptors (i.e., estrogen receptor) and signaling pathways (PI3PK/Akt/mTOR) as other steroid hormones, indicating that induction of ACD might be a common effect of virtually all steroid hormones.

4.1 DHEA and Autophagy

A *sine qua non* hallmark of ACD is the induction of autophagy as causal stimuli triggering cell death (see Section 2). Therefore, to strengthen the concept that DHEA can induce ACD we will review the main data showing that the hormone can at least induce autophagy. We are aware that the triggering of autophagy is necessary but not sufficient to ignite ACD, but a better comprehension of these effects paves the way to the possibility of modulating DHEA doses and times in order to tune the balance between prosurvival autophagy and detrimental ACD depending on the pathophysiological context.

DHEA capacity to promote autophagy contributes to the beneficial effects of this hormone in cardiovascular disease. In fact, one of main cause of aging-associated vascular pathology is the occurrence of senescence, which impairs angiogenesis and endothelial vessels repair, and promotes expression of proinflammatory cytokines such as interleukin-1 (IL-1) (Erusalimsky & Kurz, 2005). Remarkably, DHEA ameliorated the features of linoleic-induced senescence (e.g., acidic β-galactosidase staining and hyperphosphorylation of retinoblastoma protein—pRB) in endothelial HAEC cell lines through the induction of autophagy (Lee et al., 2015). In particular, DHEA restored the autophagic flux, as highlighted by the analysis of LC3 and p62, and autophagosome fusion to lysosomes. In the context of cardiovascular disease, a study showed an opposite function of DHEA in the regulation of autophagy. Pulmonary arterial hypertension is a pathological condition characterized by right ventricular hypertrophy leading to left ventricular compression and overall diastolic heart failure, and is often a consequence of ROS burst and autophagy induction (Rawat et al., 2014). DHEA administration in pulmonary arterial hypertension rat models, improved the left ventricular diastolic activity through reduction of oxidative stress and, remarkably, inhibition of the autophagic flux (Rawat et al., 2014). These effects were mediated by the inhibitory activity of DHEA on G6PD. Conversely, autophagy can be triggered by DHEA and serves as a prosurvival mechanism in macrophages challenged by *Mycobacterium tuberculosis*, the major pathological agent causing tuberculosis (Bongiovanni et al., 2015). Indeed, the role of autophagy in removing intracellular pathogens is well known and DHEA promoted autophagy-mediated clearance of *M. tuberculosis* in human macrophage-like THP-1 cells. However, the mechanisms by which DHEA facilitates the entry of the bacterium into macrophages and triggers the induction of autophagy remain to be elucidated.

In the context of cancer, besides the aforementioned role of DHEA in inducing autophagy and ACD in hepatoma HepG2 cancer cells (see Section 4), we also observed induction of autophagic flux measured by LC3-I conversion to LC3-II in presence of chloroquine, in two breast cancer cell lines, namely, MCF-7 and SK-BR-3 (Fig. 2). It is noteworthy how this hormone can induce similar effects in two cell lines with different features: indeed, MCF-7 cells are p53 wild type, estrogen receptor-positive,

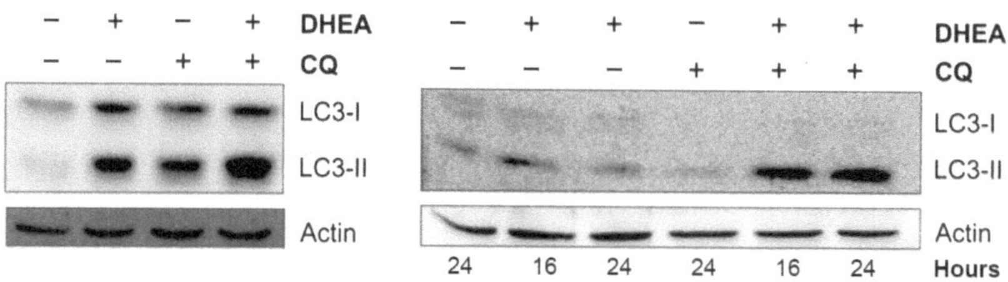

Fig. 2 DHEA promotes the autophagic flux in breast cancer cell lines. (*Left*) MCF-7 cells were treated for 24h with 300μM DHEA. (*Right*) SK-BR-3 cells were treated for the indicated times with 300μM DHEA. Where indicated, 50μM chloroquine (CQ) was added 4h before the end of the experiment. Western blot analyses of LC3 was performed. Actin was used as loading control. Western blots are from one experiment representative of three that gave similar results.

low human epidermal growth factor-2 (HER2) expressing, whereas SK-BR-3 are p53 mutated, estrogen receptor-negative, and HER2 positive (Papanikolaou et al., 2009), suggesting that DHEA may act through mechanisms other than p53 signaling, and estrogen/epidermal growth factor receptors. Further experiments in our laboratory are currently ongoing to investigate whether DHEA treatment might induce ACD in these breast cancer cell lines in a dose- or time-dependent manner, thus identifying previously uncovered death mechanisms.

Given the causal role of uncontrolled autophagy in several diseases, the opposite functions of DHEA in regulating the autophagic process in the context of diverse pathologies highlight the importance of new investigations in both in vitro and in vivo systems in order to find new possibilities of exploiting this hormone for clinical purposes.

4.2 Steroid-Regulated Autophagy

Induction of autophagy by exogenous administration of cholesterol-derived steroids has been largely validated in literature in different contexts. For instance, it was demonstrated that 6h treatment with 5mM cholesterol was able to induce apoptosis by lipotoxicity in pancreatic β-cells (Kong, Wu, Sun, & Zhou, 2017). On the other hand, cholesterol also induced autophagy as a prosurvival mechanism to an extent that was not able to counteract the detrimental induction of apoptosis. Indeed, inhibition of autophagy by E64d/pepstatin A worsened the effects of cholesterol-induced damage, thus suggesting that moderate levels of autophagy may help controlling cholesterol lipotoxicity. Induction of autophagy by cholesterol treatment was identified also in senescent bone marrow mesenchymal stem cells (BMSCs) (Zhang et al., 2016). Impairment of autophagy contributed to the senescent phenotype of these cells, as demonstrated by increased positivity of acidic β-galactosidase staining in presence of 3-methyladenine, which was reversed by cholesterol-restoration of the autophagic flux. Similarly, 17β-estradiol plays a dual role in the modulation of autophagy in a cell type-specific manner. In fact, the hormone exerts its growth effects in BT474 breast cancer cells through induction of autophagy, since treatment with 17β-estradiol increases the levels of LC3-II, beclin-1, Atg5/12, and autophagosome formation in a time-dependent manner (Wang et al., 2014). On the contrary, 17β-estradiol promotes protective functions in a rat model of spinal cord injury via suppression of autophagy, which seems to be crucial in the course of nerve injury (Lin, Chen, Huang, Dai, & Teng, 2016). 17β-Estradiol decreased the levels of LC3-II in tissues and similar results were obtained in PC12 cell lines, where the hormone blunted the extent of rapamycin-induced autophagy.

Testosterone seems to have a negative role in the modulation of autophagy. Indeed, it was shown that 2-month-old male castrated mice undergo profound muscle loss (similar to patients with prostate cancer treated with androgen deprivation) due to exacerbated autophagy-mediated protein degradation in some muscles, including the levator ani muscle and triceps (Serra et al., 2013). Testosterone administration rescued muscle atrophy, by reducing the autophagic flux via inhibition of AMPK and activation of mTOR. Testosterone negative modulation of autophagy impacts also testis homeostasis. In fact, androgen-binding protein, an essential protein for spermatogenesis in the seminiferous tubules and sperm maturation in the epididymis, was shown to be protected by autophagy-mediated degradation through testosterone-suppression of the autophagic flux in rat Sertoli cells (Ma et al., 2015).

Collectively, these data suggest that DHEA along with its precursors (e.g., cholesterol) and metabolites (e.g., estrogen, testosterone) are implicated in the modulation of autophagy and their effects are highly dependent on the pathophysiological context in which they are used.

5. CONCLUSIONS AND FUTURE DIRECTIONS

Evidence of DHEA beneficial effects on human health has considerably increased during the last decades. Epidemiological studies revealing correlations between DHEA serum levels and the occurrence of some pathologies, as well as experimental results obtained by administration of exogenous DHEA(s) in several cell types are incalculable, and an extensive description was largely beyond the scope of this chapter. Here, we preferred to focus on the molecular mechanisms whereby DHEA is able to regulate cell death, either as an inducer or a repressor, and other cellular processes (e.g., metabolism, oxidative stress) to ameliorate pathological features of many disorders, including cardiovascular disease, brain disorders, and cancer (Fig. 3). In particular, we focused on the recently unraveled role of DHEA in controlling autophagy and ACD. Data unequivocally show that DHEA can modulate such processes in a cell type- and dose-dependent manner. For instance, we reported that the hormone can induce apoptosis or ACD in the same hepatoma cell line depending on the used concentration and, conversely, the same dose has divergent death outcomes in two different hepatoma cell lines. Therefore, deep investigation is required for determining optimal dosage and duration of DHEA treatment. However, as an endogenous molecule, DHEA cannot be patented and for this reason interest and economic support by pharmaceutical companies have been missing. Nevertheless, DHEA is currently available over-the-counter, marketed as a dietary supplement and is therefore not regulated by the Food and Drug Administration (FDA) and not subject to specific regulations. Incautious extrapolations from animal studies to humans have been overstated and more large-scale clinical trials, involving both the medical and scientific communities, are needed to (i) determine potential adverse effects before DHEA enters routine clinical practice and (ii) confirm the exciting results delivered by basic research, which turned DHEA into a panacea for the cure of several pathologies.

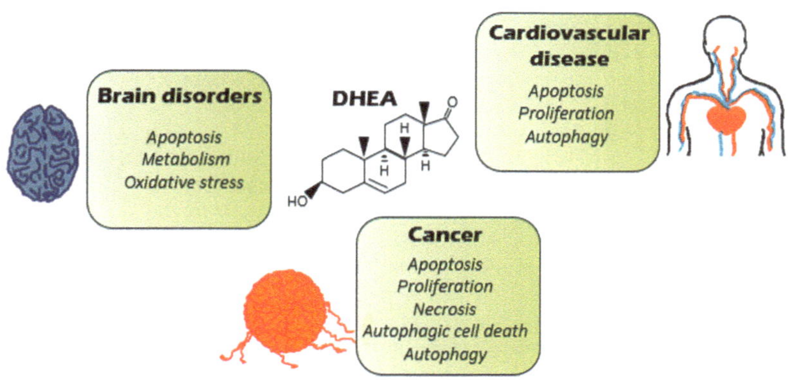

Fig. 3 Schematic representation of the cellular processes (indicated in *italics* in the *green boxes*) positively/negatively regulated by DHEA, whereby this hormone exerts beneficial effects on cardiovascular disease, brain disorders, and cancer.

ACKNOWLEDGMENT

Associazione Italiana per la Ricerca sul Cancro (AIRC) (IG 10636 and IG 15403).

REFERENCES

Abounit, K., Scarabelli, T. M., & McCauley, R. B. (2012). Autophagy in mammalian cells. *World Journal of Biological Chemistry*, *3*(1), 1–6.

Allolio, B., & Arlt, W. (2002). DHEA treatment: Myth or reality? *Trends in Endocrinology and Metabolism*, *13*(7), 288–294.

Anagnostopoulou, V., Pediaditakis, I., Alkahtani, S., Alarifi, S. A., Schmidt, E. M., Lang, F., et al. (2013). Differential effects of dehydroepiandrosterone and testosterone in prostate and colon cancer cell apoptosis: The role of nerve growth factor (NGF) receptors. *Endocrinology*, *154*(7), 2446–2456.

Aragno, M., Parola, S., Brignardello, E., Mauro, A., Tamagno, E., Manti, R., et al. (2000). Dehydroepiandrosterone prevents oxidative injury induced by transient ischemia/reperfusion in the brain of diabetic rats. *Diabetes*, *49*(11), 1924–1931.

Arnold, J. T., Liu, X., Allen, J. D., Le, H., McFann, K. K., & Blackman, M. R. (2007). Androgen receptor or estrogen receptor-beta blockade alters DHEA-, DHT-, and E(2)-induced proliferation and PSA production in human prostate cancer cells. *Prostate*, *67*(11), 1152–1162.

Azad, M. B., Chen, Y., Henson, E. S., Cizeau, J., McMillan-Ward, E., Israels, S. J., et al. (2008). Hypoxia induces autophagic cell death in apoptosis-competent cells through a mechanism involving BNIP3. *Autophagy*, *4*(2), 195–204.

Barrett-Connor, E., Khaw, K. T., & Yen, S. S. (1986). A prospective study of dehydroepiandrosterone sulfate, mortality, and cardiovascular disease. *The New England Journal of Medicine*, *315*(24), 1519–1524.

Bastianetto, S., Ramassamy, C., Poirier, J., & Quirion, R. (1999). Dehydroepiandrosterone (DHEA) protects hippocampal cells from oxidative stress-induced damage. *Brain Research. Molecular Brain Research*, *66*(1–2), 35–41.

Biaglow, J. E., Ayene, I. S., Koch, C. J., Donahue, J., Stamato, T. D., & Tuttle, S. W. (2000). G6PD deficient cells and the bioreduction of disulfides: Effects of DHEA, GSH depletion and phenylarsine oxide. *Biochemical and Biophysical Research Communications*, *273*(3), 846–852.

Blaauboer, B. J., van Holsteijn, C. W., Bleumink, R., Mennes, W. C., van Pelt, F. N., Yap, S. H., et al. (1990). The effect of beclobric acid and clofibric acid on peroxisomal beta-oxidation and peroxisome proliferation in primary cultures of rat, monkey and human hepatocytes. *Biochemical Pharmacology*, *40*(3), 521–528.

Bongiovanni, B., Mata-Espinosa, D., D'Attilio, L., Leon-Contreras, J. C., Marquez-Velasco, R., Bottasso, O., et al. (2015). Effect of cortisol and/or DHEA on THP1-derived macrophages infected with *Mycobacterium tuberculosis*. *Tuberculosis (Edinburgh, Scotland)*, *95*(5), 562–569.

Bonnet, S., Paulin, R., Sutendra, G., Dromparis, P., Roy, M., Watson, K. O., et al. (2009). Dehydroepiandrosterone reverses systemic vascular remodeling through the inhibition of the Akt/GSK3-{beta}/NFAT axis. *Circulation*, *120*(13), 1231–1240.

Boya, P., Gonzalez-Polo, R. A., Casares, N., Perfettini, J. L., Dessen, P., Larochette, N., et al. (2005). Inhibition of macroautophagy triggers apoptosis. *Molecular and Cellular Biology*, *25*(3), 1025–1040.

Buitrago, C. G., Ronda, A. C., de Boland, A. R., & Boland, R. (2006). MAP kinases p38 and JNK are activated by the steroid hormone 1alpha,25(OH)2-vitamin D3 in the C2C12 muscle cell line. *Journal of Cellular Biochemistry*, *97*(4), 698–708.

Bursch, W., Ellinger, A., Gerner, C., Frohwein, U., & Schulte-Hermann, R. (2000). Programmed cell death (PCD). Apoptosis, autophagic PCD, or others? *Annals of the New York Academy of Sciences*, *926*, 1–12.

Catalina, F., Milewich, L., Kumar, V., & Bennett, M. (2003). Dietary dehydroepiandrosterone inhibits bone marrow and leukemia cell transplants: Role of food restriction. *Experimental Biology and Medicine (Maywood, NJ)*, *228*(11), 1303–1320.

Celebi, F., Yilmaz, I., Aksoy, H., Gumus, M., Taysi, S., & Oren, D. (2004). Dehydroepiandrosterone prevents oxidative injury in obstructive jaundice in rats. *The Journal of International Medical Research*, *32*(4), 400–405.

Charalampopoulos, I., Alexaki, V. I., Lazaridis, I., Dermitzaki, E., Avlonitis, N., Tsatsanis, C., et al. (2006). G protein-associated, specific membrane binding sites mediate the neuroprotective effect of dehydroepiandrosterone. *The FASEB Journal*, *20*(3), 577–579.

Chen, S. Y., Chiu, L. Y., Maa, M. C., Wang, J. S., Chien, C. L., & Lin, W. W. (2011). zVAD-induced autophagic cell death requires c-Src-dependent ERK and JNK activation and reactive oxygen species generation. *Autophagy*, 7(2), 217–228.

Chen, W., Feng, L., Nie, H., & Zheng, X. (2012). Andrographolide induces autophagic cell death in human liver cancer cells through cyclophilin D-mediated mitochondrial permeability transition pore. *Carcinogenesis*, *33*(11), 2190–2198.

Choi, A. M., Ryter, S. W., & Levine, B. (2013a). Autophagy in human health and disease. *The New England Journal of Medicine*, *368*(19), 1845–1846.

Choi, A. M., Ryter, S. W., & Levine, B. (2013b). Autophagy in human health and disease. *The New England Journal of Medicine*, *368*(7), 651–662.

Contreras, R. G., Shoshani, L., Flores-Maldonado, C., Lazaro, A., & Cereijido, M. (1999). Relationship between Na(+),K(+)-ATPase and cell attachment. *Journal of Cell Science*, *112*(Pt. 23), 4223–4232.

Corpechot, C., Robel, P., Axelson, M., Sjovall, J., & Baulieu, E. E. (1981). Characterization and measurement of dehydroepiandrosterone sulfate in rat brain. *Proceedings of the National Academy of Sciences of the United States of America*, *78*(8), 4704–4707.

Crighton, D., Wilkinson, S., O'Prey, J., Syed, N., Smith, P., Harrison, P. R., et al. (2006). DRAM, a p53-induced modulator of autophagy, is critical for apoptosis. *Cell*, *126*(1), 121–134.

Desideri, E., Filomeni, G., & Ciriolo, M. R. (2012). Glutathione participates in the modulation of starvation-induced autophagy in carcinoma cells. *Autophagy*, *8*(12), 1769–1781.

Desideri, E., Vegliante, R., Cardaci, S., Nepravishta, R., Paci, M., & Ciriolo, M. R. (2014). MAPK14/p38alpha-dependent modulation of glucose metabolism affects ROS levels and autophagy during starvation. *Autophagy*, *10*(9), 1652–1665.

Di Bartolomeo, S., Corazzari, M., Nazio, F., Oliverio, S., Lisi, G., Antonioli, M., et al. (2010). The dynamic interaction of AMBRA1 with the dynein motor complex regulates mammalian autophagy. *The Journal of Cell Biology*, *191*(1), 155–168.

Di Monaco, M., Pizzini, A., Gatto, V., Leonardi, L., Gallo, M., Brignardello, E., et al. (1997). Role of glucose-6-phosphate dehydrogenase inhibition in the antiproliferative effects of dehydroepiandrosterone on human breast cancer cells. *British Journal of Cancer*, *75*(4), 589–592.

Diaz-Moralli, S., Aguilar, E., Marin, S., Coy, J. F., Dewerchin, M., Antoniewicz, M. R., et al. (2016). A key role for transketolase-like 1 in tumor metabolic reprogramming. *Oncotarget*, 7(32), 51875–51897.

Erusalimsky, J. D., & Kurz, D. J. (2005). Cellular senescence in vivo: Its relevance in ageing and cardiovascular disease. *Experimental Gerontology*, *40*(8–9), 634–642.

Espert, L., Denizot, M., Grimaldi, M., Robert-Hebmann, V., Gay, B., Varbanov, M., et al. (2006). Autophagy is involved in T cell death after binding of HIV-1 envelope proteins to CXCR4. *The Journal of Clinical Investigation*, *116*(8), 2161–2172.

Filomeni, G., Desideri, E., Cardaci, S., Rotilio, G., & Ciriolo, M. R. (2010). Under the ROS...thiol network is the principal suspect for autophagy commitment. *Autophagy*, *6*(7), 999–1005.
Galluzzi, L., Vitale, I., Abrams, J. M., Alnemri, E. S., Baehrecke, E. H., Blagosklonny, M. V., et al. (2012). Molecular definitions of cell death subroutines: Recommendations of the nomenclature committee on cell death 2012. *Cell Death and Differentiation*, *19*(1), 107–120.
Galva, C., Artigas, P., & Gatto, C. (2012). Nuclear Na+/K+-ATPase plays an active role in nucleoplasmic Ca2+ homeostasis. *Journal of Cell Science*, *125*(Pt. 24), 6137–6147.
Gamberi, T., Modesti, A., Magherini, F., D'Souza, D. M., Hawke, T., & Fiaschi, T. (2016). Activation of autophagy by globular adiponectin is required for muscle differentiation. *Biochimica et Biophysica Acta*, *1863*(4), 694–702.
Giron, R. A., Montano, L. F., Escobar, M. L., & Lopez-Marure, R. (2009). Dehydroepiandrosterone inhibits the proliferation and induces the death of HPV-positive and HPV-negative cervical cancer cells through an androgen- and estrogen-receptor independent mechanism. *The FEBS Journal*, *276*(19), 5598–5609.
Golden, S. H., Maguire, A., Ding, J., Crouse, J. R., Cauley, J. A., Zacur, H., et al. (2002). Endogenous postmenopausal hormones and carotid atherosclerosis: A case-control study of the atherosclerosis risk in communities cohort. *American Journal of Epidemiology*, *155*(5), 437–445.
Graf, M. R., Jia, W., & Loria, R. M. (2007). The neuro-steroid, 3beta androstene 17alpha diol exhibits potent cytotoxic effects on human malignant glioma and lymphoma cells through different programmed cell death pathways. *British Journal of Cancer*, *97*(5), 619–627.
Green, D. R., & Kroemer, G. (2009). Cytoplasmic functions of the tumour suppressor p53. *Nature*, *458*(7242), 1127–1130.
Green, J. E., Shibata, M. A., Shibata, E., Moon, R. C., Anver, M. R., Kelloff, G., et al. (2001). 2-difluoromethylornithine and dehydroepiandrosterone inhibit mammary tumor progression but not mammary or prostate tumor initiation in C3(1)/SV40 T/t-antigen transgenic mice. *Cancer Research*, *61*(20), 7449–7455.
Haffner, S. M., Moss, S. E., Klein, B. E., & Klein, R. (1996). Sex hormones and DHEA-SO4 in relation to ischemic heart disease mortality in diabetic subjects. The Wisconsin epidemiologic study of diabetic retinopathy. *Diabetes Care*, *19*(10), 1045–1050.
Ho, H. Y., Cheng, M. L., Chiu, H. Y., Weng, S. F., & Chiu, D. T. (2008). Dehydroepiandrosterone induces growth arrest of hepatoma cells via alteration of mitochondrial gene expression and function. *International Journal of Oncology*, *33*(5), 969–977.
Imamura, M., & Prasad, C. (1998). Modulation of GABA-gated chloride ion influx in the brain by dehydroepiandrosterone and its metabolites. *Biochemical and Biophysical Research Communications*, *243*(3), 771–775.
Janousek, J. (1990). Diagnosis and therapy of heart rhythm disorders after surgery of congenital heart defects. *Ceskoslovenská Pediatrie*, *45*(8), 468–472.
Jiang, Y., Miyazaki, T., Honda, A., Hirayama, T., Yoshida, S., Tanaka, N., et al. (2005). Apoptosis and inhibition of the phosphatidylinositol 3-kinase/Akt signaling pathway in the antiproliferative actions of dehydroepiandrosterone. *Journal of Gastroenterology*, *40*(5), 490–497.
Jozuka, H., Jozuka, E., Suzuki, M., Takeuchi, S., & Takatsu, Y. (2003). Psycho-neuro-immunological treatment of hepatocellular carcinoma with major depression—A single case report. *Current Medical Research and Opinion*, *19*(1), 59–63.
Kerr, J. F. (1965). A histochemical study of hypertrophy and ischaemic injury of rat liver with special reference to changes in lysosomes. *The Journal of Pathology and Bacteriology*, *90*(2), 419–435.
Kim, J., Kundu, M., Viollet, B., & Guan, K. L. (2011). AMPK and mTOR regulate autophagy through direct phosphorylation of Ulk1. *Nature Cell Biology*, *13*(2), 132–141.

Kim, S. K., & Novak, R. F. (2007). The role of intracellular signaling in insulin-mediated regulation of drug metabolizing enzyme gene and protein expression. *Pharmacology & Therapeutics*, *113*(1), 88–120.

Klionsky, D. J., Abdelmohsen, K., Abe, A., Abedin, M. J., Abeliovich, H., Acevedo Arozena, A., et al. (2016). Guidelines for the use and interpretation of assays for monitoring autophagy (3rd edition). *Autophagy*, *12*(1), 1–222.

Kohalmy, K., Tamasi, V., Kobori, L., Sarvary, E., Pascussi, J. M., Porrogi, P., et al. (2007). Dehydroepiandrosterone induces human CYP2B6 through the constitutive androstane receptor. *Drug Metabolism and Disposition*, *35*(9), 1495–1501.

Kong, F. J., Wu, J. H., Sun, S. Y., & Zhou, J. Q. (2017). The endoplasmic reticulum stress/autophagy pathway is involved in cholesterol-induced pancreatic beta-cell injury. *Scientific Reports*, 7, 44746.

Kopplow, K., Wayss, K., Enzmann, H., & Mayer, D. (2005). Dehydroepiandrosterone causes hyperplasia and impairs regeneration in rat liver. *International Journal of Oncology*, *27*(6), 1551–1558.

Kroemer, G., El-Deiry, W. S., Golstein, P., Peter, M. E., Vaux, D., Vandenabeele, P., et al. (2005). Classification of cell death: Recommendations of the nomenclature committee on cell death. *Cell Death and Differentiation*, *12*(Suppl. 2), 1463–1467.

Kroemer, G., Galluzzi, L., Vandenabeele, P., Abrams, J., Alnemri, E. S., Baehrecke, E. H., et al. (2009). Classification of cell death: Recommendations of the nomenclature committee on cell death 2009. *Cell Death and Differentiation*, *16*(1), 3–11.

Kroemer, G., & Levine, B. (2008). Autophagic cell death: The story of a misnomer. *Nature Reviews. Molecular Cell Biology*, *9*(12), 1004–1010.

Kunchithapautham, K., & Rohrer, B. (2007). Apoptosis and autophagy in photoreceptors exposed to oxidative stress. *Autophagy*, *3*(5), 433–441.

Lamy, L., Ngo, V. N., Emre, N. C., Shaffer, A. L., 3rd, Yang, Y., Tian, E., et al. (2013). Control of autophagic cell death by caspase-10 in multiple myeloma. *Cancer Cell*, *23*(4), 435–449.

Lazaridis, I., Charalampopoulos, I., Alexaki, V. I., Avlonitis, N., Pediaditakis, I., Efstathopoulos, P., et al. (2011). Neurosteroid dehydroepiandrosterone interacts with nerve growth factor (NGF) receptors, preventing neuronal apoptosis. *PLoS Biology*, *9*(4), e1001051.

Lee, M. J., Kim, E. H., Lee, S. A., Kang, Y. M., Jung, C. H., Yoon, H. K., et al. (2015). Dehydroepiandrosterone prevents linoleic acid-induced endothelial cell senescence by increasing autophagy. *Metabolism*, *64*(9), 1134–1145.

Li, D. D., Wang, L. L., Deng, R., Tang, J., Shen, Y., Guo, J. F., et al. (2009). The pivotal role of c-Jun NH2-terminal kinase-mediated Beclin 1 expression during anticancer agents-induced autophagy in cancer cells. *Oncogene*, *28*(6), 886–898.

Lin, C. W., Chen, B., Huang, K. L., Dai, Y. S., & Teng, H. L. (2016). Inhibition of autophagy by estradiol promotes locomotor recovery after spinal cord injury in rats. *Neuroscience Bulletin*, *32*(2), 137–144.

Lin, S. Y., Cui, H., Yusta, B., & Belsham, D. D. (2004). IGF-I signaling prevents dehydroepiandrosterone (DHEA)-induced apoptosis in hypothalamic neurons. *Molecular and Cellular Endocrinology*, *214*(1–2), 127–135.

Liu, D., & Dillon, J. S. (2002). Dehydroepiandrosterone activates endothelial cell nitric-oxide synthase by a specific plasma membrane receptor coupled to Galpha(i2,3). *The Journal of Biological Chemistry*, *277*(24), 21379–21388.

Liu, S., Ishikawa, H., Li, F. J., Ma, Z., Otsuyama, K., Asaoku, H., et al. (2005). Dehydroepiandrosterone can inhibit the proliferation of myeloma cells and the interleukin-6 production of bone marrow mononuclear cells from patients with myeloma. *Cancer Research*, *65*(6), 2269–2276.

Liu, Y., & Levine, B. (2015). Autosis and autophagic cell death: The dark side of autophagy. *Cell Death and Differentiation*, *22*(3), 367–376.

Liu, Y., Shoji-Kawata, S., Sumpter, R. M., Jr., Wei, Y., Ginet, V., Zhang, L., et al. (2013). Autosis is a Na+,K+-ATPase-regulated form of cell death triggered by autophagy-inducing peptides, starvation, and hypoxia-ischemia. *Proceedings of the National Academy of Sciences of the United States of America*, *110*(51), 20364–20371.
Liu, D., Si, H., Reynolds, K. A., Zhen, W., Jia, Z., & Dillon, J. S. (2007). Dehydroepiandrosterone protects vascular endothelial cells against apoptosis through a Galphai protein-dependent activation of phosphatidylinositol 3-kinase/Akt and regulation of antiapoptotic Bcl-2 expression. *Endocrinology*, *148*(7), 3068–3076.
Lockshin, R. A., & Williams, C. M. (1965). Programmed cell death—I. Cytology of degeneration in the intersegmental muscles of the Pernyi Silkmoth. *Journal of Insect Physiology*, *11*, 123–133.
Lopez-Marure, R., Contreras, P. G., & Dillon, J. S. (2011). Effects of dehydroepiandrosterone on proliferation, migration, and death of breast cancer cells. *European Journal of Pharmacology*, *660*(2–3), 268–274.
Ma, Y., Yang, H. Z., Xu, L. M., Huang, Y. R., Dai, H. L., & Kang, X. N. (2015). Testosterone regulates the autophagic clearance of androgen binding protein in rat Sertoli cells. *Scientific Reports*, *5*, 8894.
Maninger, N., Wolkowitz, O. M., Reus, V. I., Epel, E. S., & Mellon, S. H. (2009). Neurobiological and neuropsychiatric effects of dehydroepiandrosterone (DHEA) and DHEA sulfate (DHEAS). *Frontiers in Neuroendocrinology*, *30*(1), 65–91.
Matsuzaki, Y., & Honda, A. (2006). Dehydroepiandrosterone and its derivatives: Potentially novel anti-proliferative and chemopreventive agents. *Current Pharmaceutical Design*, *12*(26), 3411–3421.
Mayer, D., Forstner, K., & Kopplow, K. (2003). Induction and modulation of hepatic preneoplasia and neoplasia in the rat by dehydroepiandrosterone. *Toxicologic Pathology*, *31*(1), 103–112.
Mayo, W., Dellu, F., Robel, P., Cherkaoui, J., Le Moal, M., Baulieu, E. E., et al. (1993). Infusion of neurosteroids into the nucleus basalis magnocellularis affects cognitive processes in the rat. *Brain Research*, *607*(1–2), 324–328.
Meijer, A. J. (2009). Autophagy research: Lessons from metabolism. *Autophagy*, *5*(1), 3–5.
Melchior, C. L., & Ritzmann, R. F. (1996). Neurosteroids block the memory-impairing effects of ethanol in mice. *Pharmacology, Biochemistry, and Behavior*, *53*(1), 51–56.
Miller, E. C., & Miller, J. A. (1981). Searches for ultimate chemical carcinogens and their reactions with cellular macromolecules. *Cancer*, *47*(10), 2327–2345.
Mohseni, N., McMillan, S. C., Chaudhary, R., Mok, J., & Reed, B. H. (2009). Autophagy promotes caspase-dependent cell death during Drosophila development. *Autophagy*, *5*(3), 329–338.
Mullen, P. J., Yu, R., Longo, J., Archer, M. C., & Penn, L. Z. (2016). The interplay between cell signalling and the mevalonate pathway in cancer. *Nature Reviews. Cancer*, *16*(11), 718–731.
Murphy, L. A., Moore, T., & Nesnow, S. (2012). Propiconazole-enhanced hepatic cell proliferation is associated with dysregulation of the cholesterol biosynthesis pathway leading to activation of Erk1/2 through Ras farnesylation. *Toxicology and Applied Pharmacology*, *260*(2), 146–154.
Nakatogawa, H. (2013). Two ubiquitin-like conjugation systems that mediate membrane formation during autophagy. *Essays in Biochemistry*, *55*, 39–50.
Nakatogawa, H., Suzuki, K., Kamada, Y., & Ohsumi, Y. (2009). Dynamics and diversity in autophagy mechanisms: Lessons from yeast. *Nature Reviews. Molecular Cell Biology*, *10*(7), 458–467.
Papanikolaou, V., Iliopoulos, D., Dimou, I., Dubos, S., Tsougos, I., Theodorou, K., et al. (2009). The involvement of HER2 and p53 status in the regulation of telomerase in irradiated breast cancer cells. *International Journal of Oncology*, *35*(5), 1141–1149.

Pascale, R. M., Simile, M. M., De Miglio, M. R., Nufris, A., Seddaiu, M. A., Muroni, M. R., et al. (1995). Inhibition of 3-hydroxy-3-methylglutaryl-CoA reductase activity and gene expression by dehydroepiandrosterone in preneoplastic liver nodules. *Carcinogenesis*, *16*(7), 1537–1542.

Pattingre, S., Tassa, A., Qu, X., Garuti, R., Liang, X. H., Mizushima, N., et al. (2005). Bcl-2 antiapoptotic proteins inhibit Beclin 1-dependent autophagy. *Cell*, *122*(6), 927–939.

Paulin, R., Meloche, J., Jacob, M. H., Bisserier, M., Courboulin, A., & Bonnet, S. (2011). Dehydroepiandrosterone inhibits the Src/STAT3 constitutive activation in pulmonary arterial hypertension. *American Journal of Physiology. Heart and Circulatory Physiology*, *301*(5), H1798–1809.

Puissant, A., Robert, G., Fenouille, N., Luciano, F., Cassuto, J. P., Raynaud, S., et al. (2010). Resveratrol promotes autophagic cell death in chronic myelogenous leukemia cells via JNK-mediated p62/SQSTM1 expression and AMPK activation. *Cancer Research*, *70*(3), 1042–1052.

Racchi, M., Balduzzi, C., & Corsini, E. (2003). Dehydroepiandrosterone (DHEA) and the aging brain: Flipping a coin in the "fountain of youth". *CNS Drug Reviews*, *9*(1), 21–40.

Rawat, D. K., Alzoubi, A., Gupte, R., Chettimada, S., Watanabe, M., Kahn, A. G., et al. (2014). Increased reactive oxygen species, metabolic maladaptation, and autophagy contribute to pulmonary arterial hypertension-induced ventricular hypertrophy and diastolic heart failure. *Hypertension*, *64*(6), 1266–1274.

Ritsner, M. S. (2011). The clinical and therapeutic potentials of dehydroepiandrosterone and pregnenolone in schizophrenia. *Neuroscience*, *191*, 91–100.

Robel, P., & Baulieu, E. E. (1995). Dehydroepiandrosterone (DHEA) is a neuroactive neurosteroid. *Annals of the New York Academy of Sciences*, *774*, 82–110.

Rutkowski, K., Sowa, P., Rutkowska-Talipska, J., Kuryliszyn-Moskal, A., & Rutkowski, R. (2014). Dehydroepiandrosterone (DHEA): Hypes and hopes. *Drugs*, *74*(11), 1195–1207.

Scherz-Shouval, R., Shvets, E., Fass, E., Shorer, H., Gil, L., & Elazar, Z. (2007). Reactive oxygen species are essential for autophagy and specifically regulate the activity of Atg4. *The EMBO Journal*, *26*(7), 1749–1760.

Schmitt, M., Klinga, K., Schnarr, B., Morfin, R., & Mayer, D. (2001). Dehydroepiandrosterone stimulates proliferation and gene expression in MCF-7 cells after conversion to estradiol. *Molecular and Cellular Endocrinology*, *173*(1–2), 1–13.

Schulz, S., Klann, R. C., Schonfeld, S., & Nyce, J. W. (1992). Mechanisms of cell growth inhibition and cell cycle arrest in human colonic adenocarcinoma cells by dehydroepiandrosterone: Role of isoprenoid biosynthesis. *Cancer Research*, *52*(5), 1372–1376.

Schumacher, M., Akwa, Y., Guennoun, R., Robert, F., Labombarda, F., Desarnaud, F., et al. (2000). Steroid synthesis and metabolism in the nervous system: Trophic and protective effects. *Journal of Neurocytology*, *29*(5–6), 307–326.

Schwartz, A. G., & Pashko, L. L. (2004). Dehydroepiandrosterone, glucose-6-phosphate dehydrogenase, and longevity. *Ageing Research Reviews*, *3*(2), 171–187.

Schweichel, J. U., & Merker, H. J. (1973). The morphology of various types of cell death in prenatal tissues. *Teratology*, *7*(3), 253–266.

Serra, C., Sandor, N. L., Jang, H., Lee, D., Toraldo, G., Guarneri, T., et al. (2013). The effects of testosterone deprivation and supplementation on proteasomal and autophagy activity in the skeletal muscle of the male mouse: Differential effects on high-androgen responder and low-androgen responder muscle groups. *Endocrinology*, *154*(12), 4594–4606.

Shimizu, S., Kanaseki, T., Mizushima, N., Mizuta, T., Arakawa-Kobayashi, S., Thompson, C. B., et al. (2004). Role of Bcl-2 family proteins in a non-apoptotic programmed cell death dependent on autophagy genes. *Nature Cell Biology*, *6*(12), 1221–1228.

Shimizu, S., Konishi, A., Nishida, Y., Mizuta, T., Nishina, H., Yamamoto, A., et al. (2010). Involvement of JNK in the regulation of autophagic cell death. *Oncogene*, *29*(14), 2070–2082.

Shoji-Kawata, S., Sumpter, R., Leveno, M., Campbell, G. R., Zou, Z., Kinch, L., et al. (2013). Identification of a candidate therapeutic autophagy-inducing peptide. *Nature, 494*(7436), 201–206.

Simoncini, T., Mannella, P., Fornari, L., Varone, G., Caruso, A., & Genazzani, A. R. (2003). Dehydroepiandrosterone modulates endothelial nitric oxide synthesis via direct genomic and nongenomic mechanisms. *Endocrinology, 144*(8), 3449–3455.

Starka, L., Duskova, M., & Hill, M. (2015). Dehydroepiandrosterone: A neuroactive steroid. *The Journal of Steroid Biochemistry and Molecular Biology, 145*, 254–260.

Svec, F., & Porter, J. R. (1998). The actions of exogenous dehydroepiandrosterone in experimental animals and humans. *Proceedings of the Society for Experimental Biology and Medicine, 218*(3), 174–191.

Tait, S. W., Ichim, G., & Green, D. R. (2014). Die another way—Non-apoptotic mechanisms of cell death. *Journal of Cell Science, 127*(Pt. 10), 2135–2144.

Tchernof, A., & Labrie, F. (2004). Dehydroepiandrosterone, obesity and cardiovascular disease risk: A review of human studies. *European Journal of Endocrinology, 151*(1), 1–14.

Traish, A. M., Kang, H. P., Saad, F., & Guay, A. T. (2011). Dehydroepiandrosterone (DHEA)—A precursor steroid or an active hormone in human physiology. *The Journal of Sexual Medicine, 8*(11), 2960–2982 [quiz 2983].

Vegliante, R., Desideri, E., Di Leo, L., & Ciriolo, M. R. (2016). Dehydroepiandrosterone triggers autophagic cell death in human hepatoma cell line HepG2 via JNK-mediated p62/SQSTM1 expression. *Carcinogenesis, 37*(3), 233–244.

Vieira-Marques, C., Arbo, B. D., Ruiz-Palmero, I., Ortiz-Rodriguez, A., Ghorbanpoor, S., Kucharski, L. C., et al. (2016). Dehydroepiandrosterone protects male and female hippocampal neurons and neuroblastoma cells from glucose deprivation. *Brain Research, 1644*, 176–182.

Wang, R., He, W., Li, Z., Chang, W., Xin, Y., & Huang, T. (2014). Caveolin-1 functions as a key regulator of 17beta-estradiol-mediated autophagy and apoptosis in BT474 breast cancer cells. *International Journal of Molecular Medicine, 34*(3), 822–827.

Wang, X., Su, H., Wang, W., Chen, C., & Cao, X. (2016). Peptidomimetics based on dehydroepiandrosterone scaffold: Synthesis, antiproliferation activity, structure-activity relationship, and mechanisms. *Scientific Reports, 6*, 32654.

Webb, S. J., Geoghegan, T. E., Prough, R. A., & Michael Miller, K. K. (2006). The biological actions of dehydroepiandrosterone involves multiple receptors. *Drug Metabolism Reviews, 38*(1–2), 89–116.

Webb, S. J., Xiao, G. H., Geoghegan, T. E., & Prough, R. A. (1996). Regulation of CYP4A expression in rat by dehydroepiandrosterone and thyroid hormone. *Molecular Pharmacology, 49*(2), 276–287.

Weill-Engerer, S., David, J. P., Sazdovitch, V., Liere, P., Schumacher, M., Delacourte, A., et al. (2003). In vitro metabolism of dehydroepiandrosterone (DHEA) to 7alpha-hydroxy-DHEA and Delta5-androstene-3beta,17beta-diol in specific regions of the aging brain from Alzheimer's and non-demented patients. *Brain Research, 969*(1–2), 117–125.

Wileman, T. (2013). Autophagy as a defence against intracellular pathogens. *Essays in Biochemistry, 55*, 153–163.

Wilpart, M., Speder, A., Ninane, P., & Roberfroid, M. (1986). Antimutagenic effects of natural and synthetic hormonal steroids. *Teratogenesis, Carcinogenesis, and Mutagenesis, 6*(4), 265–273.

Wolf, O. T., Koster, B., Kirschbaum, C., Pietrowsky, R., Kern, W., Hellhammer, D. H., et al. (1997). A single administration of dehydroepiandrosterone does not enhance memory performance in young healthy adults, but immediately reduces cortisol levels. *Biological Psychiatry, 42*(9), 845–848.

Wolf, O. T., Neumann, O., Hellhammer, D. H., Geiben, A. C., Strasburger, C. J., Dressendorfer, R. A., et al. (1997). Effects of a two-week physiological dehydroepiandrosterone substitution on cognitive performance and well-being in healthy elderly women and men. *The Journal of Clinical Endocrinology and Metabolism*, *82*(7), 2363–2367.

Wolkowitz, O. M., Reus, V. I., Roberts, E., Manfredi, F., Chan, T., Raum, W. J., et al. (1997). Dehydroepiandrosterone (DHEA) treatment of depression. *Biological Psychiatry*, *41*(3), 311–318.

Xu, Z. X., Liang, J., Haridas, V., Gaikwad, A., Connolly, F. P., Mills, G. B., et al. (2007). A plant triterpenoid, avicin D, induces autophagy by activation of AMP-activated protein kinase. *Cell Death and Differentiation*, *14*(11), 1948–1957.

Yang, N. C., Jeng, K. C., Ho, W. M., Chou, S. J., & Hu, M. L. (2000). DHEA inhibits cell growth and induces apoptosis in BV-2 cells and the effects are inversely associated with glucose concentration in the medium. *The Journal of Steroid Biochemistry and Molecular Biology*, *75*(2–3), 159–166.

Yoshida, S., Honda, A., Matsuzaki, Y., Fukushima, S., Tanaka, N., Takagiwa, A., et al. (2003). Anti-proliferative action of endogenous dehydroepiandrosterone metabolites on human cancer cell lines. *Steroids*, *68*(1), 73–83.

Yu, L., Alva, A., Su, H., Dutt, P., Freundt, E., Welsh, S., et al. (2004). Regulation of an ATG7-beclin 1 program of autophagic cell death by caspase-8. *Science*, *304*(5676), 1500–1502.

Yu, X., Li, R., Shi, W., Jiang, T., Wang, Y., Li, C., et al. (2016). Silencing of MicroRNA-21 confers the sensitivity to tamoxifen and fulvestrant by enhancing autophagic cell death through inhibition of the PI3K-AKT-mTOR pathway in breast cancer cells. *Biomedicine & Pharmacotherapy*, 77, 37–44.

Yuan, J., & Yankner, B. A. (2000). Apoptosis in the nervous system. *Nature*, *407*(6805), 802–809.

Zeng, X., Overmeyer, J. H., & Maltese, W. A. (2006). Functional specificity of the mammalian Beclin-Vps34 PI 3-kinase complex in macroautophagy versus endocytosis and lysosomal enzyme trafficking. *Journal of Cell Science*, *119*(Pt. 2), 259–270.

Zhang, M., Du, Y., Lu, R., Shu, Y., Zhao, W., Li, Z., et al. (2016). Cholesterol retards senescence in bone marrow mesenchymal stem cells by modulating autophagy and ROS/p53/p21Cip1/Waf1 pathway. *Oxidative Medicine and Cellular Longevity*, *2016*, 7524308.

Zhang, L., Li, B., Ma, W., Barker, J. L., Chang, Y. H., Zhao, W., et al. (2002). Dehydroepiandrosterone (DHEA) and its sulfated derivative (DHEAS) regulate apoptosis during neurogenesis by triggering the Akt signaling pathway in opposing ways. *Brain Research. Molecular Brain Research*, *98*(1–2), 58–66.

FURTHER READING

Schulz, S., & Nyce, J. W. (1991). Inhibition of protein isoprenylation and p21ras membrane association by dehydroepiandrosterone in human colonic adenocarcinoma cells in vitro. *Cancer Research*, *51*(24), 6563–6567.

CHAPTER ELEVEN

Dehydroepiandrosterone (DHEA)-SO_4 Depot and Castration-Resistant Prostate Cancer

Trevor M. Penning[1]
Center of Excellence in Environmental Toxicology, Perelman School of Medicine, University of Pennsylvania, Philadelphia, PA, United States
[1]Corresponding author: e-mail address: penning@upenn.edu

Contents

Abstract

Dehydroepiandrosterone (DHEA)-SO_4 of adrenal origin is the major C19 steroid in the serum. It is a precursor of intratumoral androgen biosynthesis in patients with advanced prostate cancer following chemical or surgical castration. DHEA is a product of the P450c17 (17α-hydroxylase-17,20-lyase) enzyme. Despite inhibition of P450c17 with

Vitamins and Hormones, Volume 108
ISSN 0083-6729
https://doi.org/10.1016/bs.vh.2018.01.007

new agents, e.g., Abiraterone acetate, Orterenol, and Galeterone, the level of enzyme inhibition rarely exceeds 90% leaving behind a significant depot for androgen biosynthesis within the tumor. For DHEA-SO_4 to be utilized there is uptake by organic anion transporter polypeptides, deconjugation catalyzed by steroid sulfatase, and adaptive upregulation of prostate steroidogenic enzymes that will convert DHEA into either testosterone or dihydrotestosterone. The depot of DHEA-SO_4 that remains after P450c17 inhibition and the adaptive responses that occur within the tumor to promote DHEA utilization contribute to mechanisms of drug resistance observed with P450c17 inhibitors. Knowledge of these mechanisms identify new targets for therapeutics that could be used to surmount drug resistance in prostate cancer.

1. INTRODUCTION

Prostate cancer is the leading cause of cancer in males in the United States and has the second highest mortality. Approximately 161,000 new cases are diagnosed annually resulting in 27,000 deaths. It is estimated that more than 3.0 million men in the United States live with prostate cancer (SEER, 2017). Advanced localized prostate cancer is androgen dependent and Huggins and coworkers showed that surgical castration, followed by adrenalectomy and replacement glucocorticoids led to remission (Huggins, 1965; Huggins, Stevens, & Hodges, 1941). This surgery has been supplanted by chemical castration which consists of administration of a LH-RH agonist, e.g., Leuprolide (Sharifi et al., 1996) which is often coadministered with an androgen receptor (AR) antagonist (*R*-bicalutamide also known as Casodex) (Furr et al., 1987). This first-line androgen deprivation therapy (ADT) leads to remission of the cancer but invariably there is relapse and "castration-resistant prostate cancer (CRPC)," the lethal form of the disease emerges.

CRPC is accompanied by a rising serum prostatic-specific antigen (PSA). PSA levels increase as a result of androgen induction of the *KLK3* gene which occurs in a background of castrate levels of circulating testosterone (T). This phenomenon occurs since the tumor adapts to ADT to reactivate androgen signaling. Androgen signaling is reactivated by intracrine androgen biosynthesis and by changes in the AR itself (Knudsen & Penning, 2010; Knudsen & Scher, 2009). Intratumoral androgen biosynthesis can occur by either de novo steroidogenesis starting from cholesterol, or by conversion of adrenal androgens, e.g., dehydroepiandrosterone (DHEA)-SO_4 and DHEA into T and 5α-dihydrotestosterone (DHT). This chapter will examine the role of DHEA-SO_4 and DHEA as a major source of tumor androgens where the former is the most abundant steroid in the male circulation and is present in a 400- to 1000-fold excess over DHEA (Damgaard-Olesen et al., 2016; Shackleton, Kletke, Wudy, & Pratt, 1990; Tamae et al., 2013).

2. INTRACRINE FORMATION OF ANDROGENS USING DHEA AS A PRECURSOR

The concept that DHEA-SO_4 and DHEA could be important precursors of prostatic testosterone T and DHT was originally proposed by Labrie and coworkers (Labrie et al., 1998; Labrie, Bélanger, Simard, Luu-The, & Labrie, 1995). DHEA is synthesized in the zona recticularis of the adrenal cortex from pregnenolone which is converted to 17α-hydroxypregnenolone via the 17α-hydroxylase activity catalyzed by P450c17. The resultant 17α-hydroxypregnenolone is converted to DHEA via the 17,20-lyase activity of the same enzyme (Chung et al., 1987; Nakajin, Shinoda, Haniu,

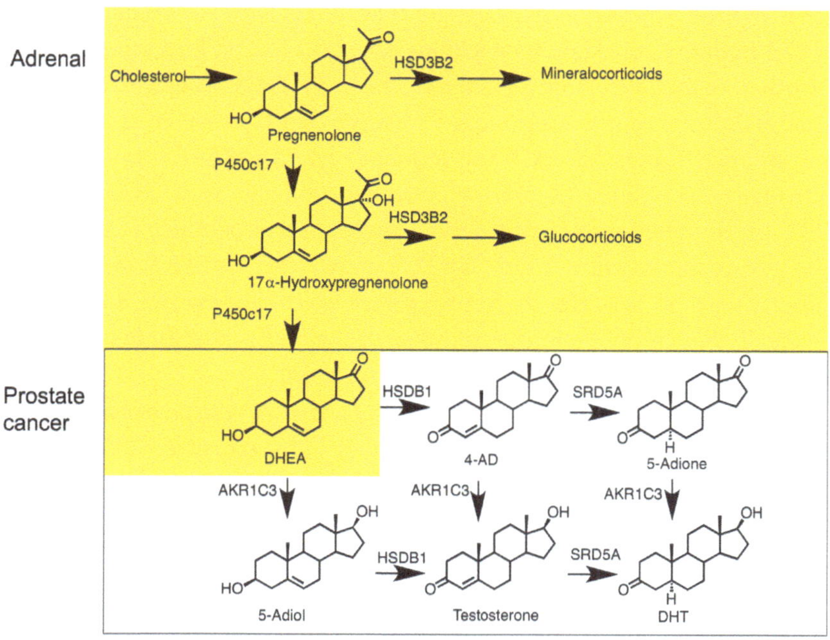

Fig. 1 Adrenal and intraprostatic androgen biosynthesis. Conversion of cholesterol to DHEA is shown in the adrenal (*yellow box*). The conversion of DHEA to DHEA-SO_4 and is subsequent desulfonation by STS is not shown for simplicity. *4-AD*, Δ^4-androstene-3,17-dione; *5-Adiol*, Δ^5-androstene-3β,17β-diol; *5-Adione*, 5α-androstane-3,17-dione; *DHEA*, dehydroepiandrosterone; *HSD3B*, 3β-hydroxysteroid dehydrogenase; *SRD5A*, steroid 5α-reductase.

Shively, & Hall, 1984), Fig. 1. The electron donor for P450c17 is NADPH cytochrome P450 oxidoreductase (Estabrook, Shet, Fisher, Jenkins, & Waterman, 1996), even though the activity of the lyase is determined by the availability of cytochrome *b*5 (Auchus, Lee, & Miller, 1998; Peng et al., 2016). In the absence of cytochrome *b*5, 17α-hydroxypregnenolone is converted to glucocorticoids in the zona fasciculata. Thus, cytochrome *b*5 is the molecular switch that determines whether the 17,20-lyase activity will proceed. Once formed a significant portion of DHEA is sulfonated by SULT2A1 (hydroxysteroid sulfotransferase) to form DHEA-SO_4 (Saner et al., 2005). DHEA-SO_4 is then transported into the prostate by organic anion transporter polypeptides (OATPs), e.g., encoded by *SLCO1A2*, *SLCO1B1*, *SLCO1B3*, and *SLCO2B1*, (Arakawa et al., 2012; Cho, Montgomery, & Mostaghel, 2014), deconjugated by steroid sulfatase (STS) and then converted to T and DHT.

The conversion of DHEA to T and DHT can occur by three routes. In the first route, DHEA is converted to Δ^4-androstene-3,17-dione (Δ^4-AD) by 3β-hydroxysteroid dehydrogenase type 1 (HSD3B1), which is reduced by 17β-hydroxysteroid dehydrogenase type 5 (AKR1C3) to yield T, which is further reduced by steroid 5α-reductase (SRD5A) to yield DHT (this route is known as the canonical pathway) (Fankhauser et al., 2014); in the second route, Δ^4-AD is reduced by SRDA5A to yield 5α-androstane-3,17-dione (5-Adione) and then reduced by AKR1C3 to yield DHT (in this alternative pathway, T is by-passed altogether) (Chang et al., 2011); and in the third route, DHEA is reduced by AKR1C3 to yield 5-androstene-3β,17β-diol (5-Adiol), which is then converted by HSD3B1 to yield T, which can be further reduced to DHT. Each of these steps will be discussed in greater detail in subsequent sections.

DHEA can also be by-passed in the backdoor pathway to prostatic androgens (Andriole et al., 2004; Auchus, 2004; Mohler, Titus, Bai, et al., 2011; Mohler, Titus, & Wilson, 2011). In this pathway, pregnenolone of adrenal origin is converted to progesterone which is reduced to 5α-dihydroprogesterone by SRD5A, and subsequently reduced to allopregnanolone and then converted to androsterone by P450c17. Androsterone is then reduced to 5α-androstane-3α,17β-diol (3α-diol) by AKR1C3, which is oxidized by HSD17B6 to DHT.

3. ADT FOR ADVANCED PROSTATE CANCER

Standard treatment of care for prostate cancer using first-line ADT has evolved to aggressively suppress all sources of androgen by which the tumor can synthesize its own T and DHT. Starting with Leuprolide additional agents have been added to the regimen, including *R*-bicalutamide and dutasteride (a dual inhibitor of steroid 5α-reductase type 1 and type 2) (Andriole et al., 2004), and ketoconazole (a P450c17 inhibitor) (Kühn-Velten & Lessmann, 1992; Trachtenberg, Halpern, & Pont, 1983).

3.1 Targeted Androgen Pathway Suppression

The effect of combined first-line ADT was assessed in a targeted androgen pathway suppression (TAPS) clinical trial (Mostaghel et al., 2014). Patients with localized prostate cancer were randomized into four groups: group 1 patients received goserelin (LH-RH agonist) plus bicalutamide; group 2 patients received goserelin plus dutasteride; group 3 patients received goserelin, bicalutamide, and dutasteride; and group 4 patients received goserelin, bicalutamide, dutasteride, and ketoconozale. All treatment arms showed a statistically significant reduction in prostate tissue DHT, but the largest change was observed in group 4, which also showed the largest change in serum T, androsterone, and DHEA-SO_4, suggesting that adrenal P450c17 was an important source of prostatic androgen precursors.

3.2 P450c17 Inhibitors, Abiraterone, Orterenol, and Galeterone

Inhibitors of P450c17 have been developed to block the production of DHEA in the adrenal to prevent intracrine synthesis of T and DHT in prostate cancer. Abiraterone acetate (AA) is the lead agent that is FDA approved for this indication, Fig. 2 (Attard et al., 2008). Since AA blocks the formation of 17α-hydroxypregnenolone it also inhibits the production of glucocorticoids. AA must therefore be coadministered with prednisone to prevent overstimulation of the adrenal by ACTH from the anterior pituitary. In the absence of adrenal–pituitary axis blockade, pregnenolone will be converted to progesterone and diverted to desoxycorticosterone, a potent mineralocorticoid, leading to mineralocorticoid excess and hypertension (Attard et al., 2012). Clinical trials in which CRPC patients that receive AA plus prednisone show an increase in progression free survival. This outcome is seen in CRPC patients, (Attard et al., 2009) and in metastatic CRPC patients that are either chemotherapy naïve (Fizazi et al., 2012) or have been treated with docetaxel (Danila et al., 2010; Reid et al., 2010).

AA is metabolized to Δ^4-abiraterone by HSD3B1 where it represents (4% of the total metabolites). Δ^4-Abiraterone has almost the same IC_{50} value as abiraterone for P450c17, but also acts as an inhibitor of HSD3B1 and as an inhibitor of SRD5A (Li et al., 2015). Δ^4-Abiraterone also acts as an

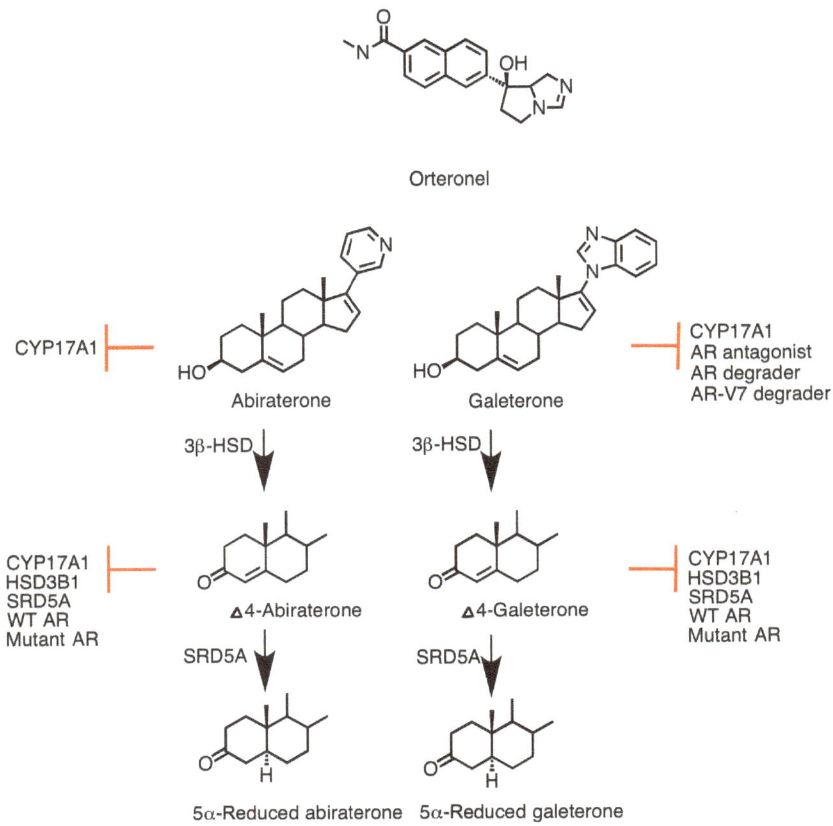

Fig. 2 P450c17 inhibitors and their metabolites.

antagonist of AR and the mutated AR, suggesting that Δ^4-abiraterone may have a superior drug profile to AA (Li et al., 2015). However, Δ^4-abiraterone is reduced by SRD5A to the corresponding 5α-reduced metabolite which acts as an AR agonist (Li et al., 2016).

Second-generation inhibitors have been developed for P450c17 with the goal of preferentially inhibiting the 17,20-lyase activity to block DHEA production while sparing the 17α-hydroxylase activity. These inhibitors are desirable since they would permit the synthesis of cortisol in the adrenal which would negate the need to coadminister prednisone. For example, Orteronel (a nonsteroidal inhibitor of P450c17) shows preferential inhibition of the 17,20-lyase activity ($IC_{50}=140$ nM) over the 17α-hydroxylase activity ($IC_{50}=760$ nM) (Yamaoka et al., 2012), Fig. 2. However, clinical trials showed that Orteronel had no clinical benefit in terms of survival when the treatment arms of Orteronel plus prednisone vs prednisone alone were compared (Fizazi et al., 2015). As a result the further development of Orteronel has been abandoned.

Galeterone (TOK-001 or VN/124-1) was also developed to act as a P450c17 inhibitor, (Njar & Brodie, 2015) and showed a superior multifunctional activity profile to AA (Alyamani et al., 2017). Galeterone acts as a direct AR-antagonist on the wild-type AR and mutated AR and prevents the binding of the liganded receptor to chromatin in a manner similar to that reported for enzalutamide (ENZ). The drug was also found to be a selective AR degrader of the mutant AR (T878A) but not of the wild-type AR (Yu et al., 2014). Interestingly, Galeterone was also found to enhance degradation of the AR-V7 splice variant (Kwegyir-Afful, Ramalingam, Purushottamachar, Ramamurthy, & Njar, 2015). AR-V7 lacks the ligand-binding domain of the AR and is constitutively active in CRPC patients that become resistant to AA (Cao et al., 2014; Mostaghel et al., 2011). However, in the ARMOR3-SV trial this multifunctional activity was not found to be superior to the AR antagonist ENZ alone and the clinical development of Galeterone has been halted (Highlights, 2016).

Galeterone is also metabolized to Δ^4-galeterone and suppresses steroidogenesis, AR stability, and AR target gene expression with similar potency to the parent compound. But further metabolism by SRD5A limits these desirable properties (Alyamani et al., 2017). Thus, both Abiraterone and Galeterone are metabolized to Δ^4-3-ketosteroids that can enhance the properties of the parent drug. This has led to the concept that the superior profile of the Δ^4-3-ketosteroid metabolites could be exploited if either AA or Galeterone were coadministered with dutasteride to prevent metabolism to their 5α-reduced metabolites but this benefit has not yet been realized in clinical trials (McKay et al., 2017).

Crystal structures of P450c17 with both Abiraterone and Galeterone show that the ring nitrogen of the pyridine or benzimidazole ring at C17 in these respective drugs coordinates with the heme iron (DeVore & Scott, 2012). While the α-face of the steroid is enclosed by the active site, the active site is not complimentary to the β-face of the steroid. The largest extension of the active site exists where the 3β-hydroxy group binds. The overall binding mode for both inhibitors suggests similarity in either the binding of the 3β-hydroxyl group or the binding of the 3-keto group present in Δ^4-abiraterone and Δ^4-galeterone similar to that observed in binding steroids to the AR, and predicts that these metabolites will act as ligands for AR.

3.3 Prednisone Treatment

Prednisone is coadministered with P450c17 inhibitors to block the adrenal–pituitary axis to prevent the overstimulation of the adrenal and prevent mineralocorticoid excess. As prednisone by itself would block the production of DHEA, it is reasonable to ask whether there is any benefit to adding a P450c17 inhibitor to a regimen of prednisone alone, and whether a P450c17 inhibitor can be given in the absence of prednisone clinically. In clinical trials with AA and prednisone vs prednisone alone, there was a benefit from the drug combination (Ryan et al., 2015) but this was not seen when Orterenol when used in place of AA (see earlier) (Fizazi et al., 2015). Administration of AA alone resulted in an increase in mineralocorticoid production upstream from P450c17 as expected leading to hypertension, hypokalemia, and fluid overload. However, these symptoms could be reversed by the administration of either an antimineralocorticoid or dexamethasone (Attard et al., 2012). The requirement for coadministration of glucocorticoids could have a dark side in that as prostate cancer progresses mutations in the AR occur to make it ligand promiscuous so that the AR will use glucocorticoids as substitute ligands (Zhao et al., 2000). In addition, AR and GR share the same response element in the promoter regions of responsive genes so that the liganded-GR may substitute for the liganded-AR under conditions of androgen deprivation (Arora et al., 2013).

3.4 Drug Resistance Mechanisms

CRPC patients on P450c17 inhibitors invariably develop drug resistance (Buttigliero et al., 2015). Resistance mechanisms can be multifaceted. These include: (i) incomplete inhibition of P450c17 (Tamae et al., 2015; Taplin et al., 2014); (ii) increased expression of steroidogenic enzymes in the tumor to exploit residual DHEA (Liu et al., 2017; Mostaghel et al., 2011); (iii) the presence of allelic variants in *HSD3BI* that increase enzyme stability (Chang et al., 2013); (iv) use of the backdoor pathway to DHT to by-pass DHEA (Auchus, 2004; Mohler, Titus, Bai, et al., 2011; Mohler, Titus, & Wilson, 2011): (v) AR gene amplification (Visakorpi et al., 1995); (vi) AR mutations (Taplin et al., 1995); and (vii) emergence of AR-SV's (Cao et al., 2014; Mostaghel et al., 2011). These mechanisms are not mutually exclusive. Some of these mechanisms could be surmounted by the use of an AR receptor antagonist, e.g., ENZ (Tran et al., 2009). Use of combination therapy of AA and ENZ has met with mixed success. Some clinical benefit is associated with the sequential use of AA followed by ENZ but the reverse does not hold (Miyake et al., 2017). Interrogation of the mechanisms of drug resistance can be improved by the sensitive and accurate quantitation of serum and prostatic androgens.

4. ANALYTICAL CHEMISTRY TO MEASURE KETO- AND HYDROXYANDROGENS

Traditional methods of measuring androgens have used radioimmunoassays which are unreliable for the following reasons: (i) they can only measure one analyte at a time and multiple different antibodies have to be used of different specificity and sensitivity; (ii) there is no structural identity of the analyte being measured; and (iii) it is not possible to control for unknowns in the biospecimens that may react with the antibody leading to false positives (Penning, Lee, Jin, Gutierrez, & Blair, 2010).

The gold-standard is to use stable-isotope dilution liquid chromatography tandem mass spectrometry (SID-LC-MS/MS). This method separates out the analytes by HPLC and identifies the analyte based on three criteria: retention time, exact mass, and precursor-product ion transitions. The method uses either ^{13}C or ^{2}H internal standards that only differ from the analyte being measured in mass due to the presence of the heavy isotope. Thus, exact quantitation can then be interpolated by the ratio method (Penning et al., 2010).

Androgens do pose a challenge to be measured by this technology. First, androgens are present as a number of regio- or stereoisomers that will have the same exact mass, e.g., DHEA, T, and epi-T are all isomers; and similarly, DHT, androsterone, and epi-androsterone are all isomers. Thus, chromatographic separation of these analytes is essential. Second, androgens have poor ionization properties and thus derivatization with preionization derivatives is essential to attain the desired sensitivity (Penning et al., 2010). Third, androgens may contain only keto- or hydroxy groups and therefore two derivatization procedures are required to measure a complete androgen profile. We have used Girard-T oximes to derivatize ketoandrogens, where the charged trimethylamine group gives a strong positive ion and mass transition to follow (Tamae et al., 2013). We have also used picolinic acid to derivatize hydroxyandrogens, where the charged picolinate ester gives a strong positive ion and mass transition (Zang, Tamae, et al., 2017). Using these methods, nine androgens plus their glucuronide and sulfate conjugates can be quantified with precision and accuracy in the low or sub-pg range. Detection of the glucuronide or sulfate conjugates requires enzymatic hydrolysis with β-glucuronidase and aryl sulfatase, respectively. These methods have been used in a number of clinical trials: (i) the TAPS trial (Mostaghel et al., 2014) and (ii) the neoadjuvant Leuprolide plus AA clinical trial (Taplin et al., 2014).

5. EVIDENCE FOR A DEPOT OF DHEA-SO_4

5.1 DHEA-SO_4

Using SID-LC-MS/MS methods, we examined the efficacy of a neoadjuvant clinical trial using Leuoprolide and AA to reduce serum androgens (Taplin et al., 2014). This trial had two groups. In group 1, patients received Leuprolide for 12 weeks and then for a subsequent 12 weeks they received Leuprolide, AA, and prednisone. In group 2, patients received Leuprolide, AA, and prednisone for all 24 weeks. Serum androgen levels were measured at base-line, at 12 and 24 weeks for all patients. Group 1 allowed us to compare the efficacy of Leuprolide alone vs the combined drug treatment. Data showed that after 12 weeks of Leuprolide alone testicular androgens, e.g., T and DHT fell to castrate levels, whereas DHEA-SO_4 and DHEA were unaffected. Subsequent administration of AA caused no further change in T and DHT but now DHEA-SO_4 and DHEA were reduced by 90%. However, since DHEA-SO_4 is present in μM concentrations, the amount that remains after Leuprolide, AA plus prednisone is 4000 times higher than the castrate levels of circulating T achieved. This suggests that despite best efforts to inhibit P450c17 a substantial reservoir of DHEA-SO_4 remains to fuel intraprostatic and androgen biosynthesis. It was predicted that all inhibitors that target P450c17 would have the same property and that this will lead to their therapeutic failure (Tamae et al., 2015). In several clinical trials with AA, the lower limit of quantitation for DHEA-SO_4 was 15 μg/dL which is 2000 times higher than castrate levels of T and the presence of this reservoir may not have been appreciated, (Attard et al., 2008; Ryan et al., 2010).

5.2 5-Androstenediol-SO_4 and Epi-Androsterone-SO_4 as a Source of Intratumoral Androgens

We subsequently applied our SID-LC-MS/MS method for hydroxyandrogen measurement to patient samples from the neoadjuvant Leuprolide/AA clinical trial (Zang, Taplin, et al., 2017). This enabled us to measure diols not previously measured, e.g., 5-androstene-3β,17β-diol (5-Adiol), 5α-androstane-3α,17β-diol (3α-diol), and 5α-androstane-3β,17-β-diol (3β-diol) in the serum samples. We reasoned that the design of the trial would permit us for the first time to discriminate between testicular and adrenal sources of androgens in prostate cancer patients. Those androgens that were reduced significantly after Leuprolide treatment would be of testicular origin, while those androgens that were only reduced after the addition of AA plus prednisone would be of adrenal origin. We find that T, DHT, 3α-adiol, and 3β-adiol were predominately of testicular origin. By contrast, DHEA, epi-androsterone and their conjugates, 5-Adiol-SO_4, and 5-Adiol glucuronide were predominately of adrenal origin. Our findings also show that ADT failed to completely suppress DHEA-SO_4 levels and that two unappreciated sources of intratumoral androgens that were not suppressed by Leuprolide acetate alone were 5-Adiol-SO_4 and epi-androsterone-SO_4. Pathways by which 5-Adiol-SO_4 and epi-androsterone-SO_4 could be converted to T and DHT are shown in Fig. 3.

5.3 Steroid Sulfate Transport

Steroid sulfates that could act as depots for intratumoral androgen biosynthesis in prostate cancer include DHEA-SO_4, 5-Adiol-SO_4, and epi-androsterone-SO_4. *SLCO*-encoded organic anion transporting polypeptides (OATP) will dictate whether these steroids can be taken up by prostate tumors (Arakawa et al., 2012; Green et al., 2017). Microarray measurements of *SLCO* transcripts in LNCaP and 22Rv1 cells grown under conditions of

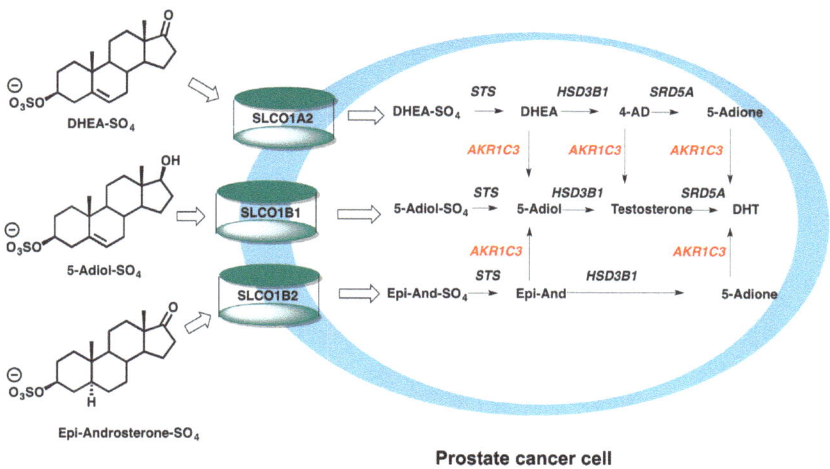

Fig. 3 Utilization of steroid sulfates in CRPC. *SLCO*, genes for organic anion transporter polypeptides; and *STS*, steroid sulfatase. Otherwise abbreviations are the same as in Fig. 1.

androgen depletion to mimic ADT demonstrated that *SLCO1A2* mRNA expression was markedly upregulated (4-fold), and to a lesser extent *SLCO1B1* (2.5-fold) and *SLCO1B2* (1.74-fold) were increased in expression as well. These changes were confirmed by qRT-PCR by monitoring increases in expression at 3 and 7 days postandrogen depletion. Increased expression was accompanied by increased cellular uptake of [^{3}H]-DHEA-SO_4 and increased cellular proliferation stimulated by DHEA-SO_4. Cellular proliferation in response to DHEA-SO_4 was attenuated by *SLCO1A2* knockdown. *SLCO1A2* was also downregulated by DHT, suggesting that its expression is repressed by androgens (Arakawa et al., 2012).

SLCO1B1- and *SLCO2B1*-expressing prostate cancer xenografts showed a 3.9- and 1.9-fold increase in tumor accumulation of DHEA-SO_4, respectively, and a 1.6- and 2.7-fold increase in DHEA, respectively (Green et al., 2017). The same genes were found to be overexpressed in CRPC metastases and were associated with prostate cancer mortality when 1309 Caucasian prostate cancer patients were genotyped (Wright et al., 2011). Importantly, SNPs in these transporters can affect patient outcome: Three SNPs in *SLCO2B1* rs1242149 [935G > A; Arg312Gln], rs1789693, and rs1077858 were associated with decreased time to progression (TTP) on ADT ($P < 0.05$). The differences in median TTP for each of these polymorphisms were about 10 months (Yang et al., 2012). In LNCaP cells transfected with the *SLCO2B1* SNP variant rs1242149 (935G > A; Arg312Gln), the variant exhibited higher maximal DHEA-SO_4 uptake when compared with the wild-type allele (Yang et al., 2012). This *SLCO2B1* genotype, which allows more efficient import of androgen and enhances cell growth (Yang et al., 2012). Collectively, these findings indicate that changes in expression of *SLCO* transporters and genotype can determine response to ADT, and that they increase the availability of DHEA-SO_4 for intracrine androgen biosynthesis.

5.4 Steroid Sulfatase

DHEA-SO_4 and other sulfates that are transported into prostate tumors must be hydrolyzed by STS to liberate the free steroid for downstream steroidogenesis, (Klein, Molwitz, & Bartsch, 1989), Fig. 3. Only one STS is known to be expressed in humans (Reed, Purohit, Woo, Newman, & Potter, 2005; Yen et al., 1987). The crystal structure of human STS identified a formylglycine residue at the active site that can be hydrated to form a hydroxyl-formylglycine (gem-diol) which is sulfonated during the hydrolysis of sulfonated steroids (Gosh, 2007). The sulfonated enzyme intermediate then hydrolyzes to liberate HSO_4^- and regenerate the formylglycine residue.

STS is regulated by six different tissue-specific promoters, however, there have been few studies on STS expression by tissue-specific promoters in human prostate. Nevertheless, STS has been detected in prostate tissue where the highest activity is in the prostatic epithelium (Farnsworth, 1973; Klein et al., 1989); and it is also expressed in LNCaP cells (Dalla Valle et al., 2007; Selcer, Kabler, Sarap, Xiao, & Li, 2002). The pivotal role of STS in the liberation of free DHEA for intracrine androgen synthesis has prompted the development of both irreversible steroidal- and nonsteroidal-based STS inhibitors, STX213 (Fischer, Chander, et al., 2003; Fischer, Woo, et al., 2003) and STX64, respectively (Woo, Purohit, Malini, Reed, & Potter, 2000). These compounds each contain a sulfamate group which can inactivate STS by one of two mechanisms. The sulfamate can either attack the formylgycine residue or a nucleophile on the enzyme can attack the sulfamate to produce a sulfamoylated enzyme. In either case, irreversible enzyme inhibition is achieved, Fig. 4. These compounds have

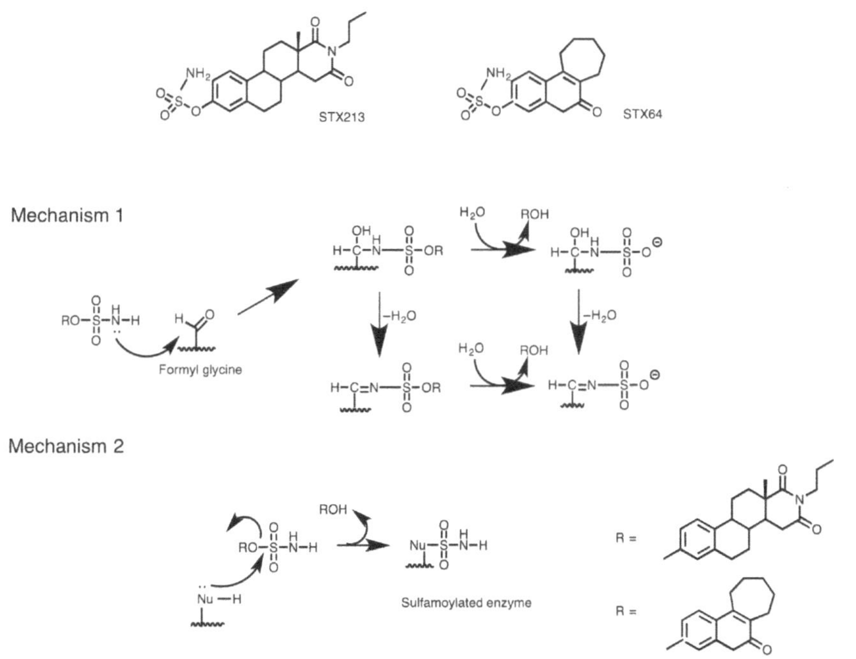

Fig. 4 Irreversible inhibition of steroid sulfatase by sulfamate-based inhibitors. In mechanism 1 the sulfamate attacks the formylglycine residue at the active site. In mechanism 2, attack by an enzyme nucleophile gives rise to sulfamoylated enzyme.

yet to be used in clinical trials of CRPC. It is noteworthy that X-linked ichthyosis is characterized by a loss of STS, raising concerns about the possible long-term adverse side effects of using sulfatase inhibitors for prostate cancer (Alperin & Shapiro, 1997; Webster, France, Shapiro, & Weiss, 1978).

6. EVIDENCE FOR DE NOVO SYNTHESIS OF T AND DHT FROM CHOLESTEROL OR PREGNENOLONE IN CRPC TUMORS

The large amount of DHEA-SO_4 that remains in the serum even after P450c17 inhibition raises the issue as to whether prostate cancer cells need to synthesize androgens from cholesterol when there is such a significant amount of downstream precursor available. Patient-derived xenografts that are AA resistant show a 1.72-fold increase in *CYP17A1* transcript and still synthesize a basal level of T and DHT (Mostaghel et al., 2011). Measurement of [^{14}C]-acetate conversion into androgens in LNCaP cells showed that whereas there was conversion into a lipid fraction containing cholesterol, there was only minor conversion into progesterone, 17α-hydroxyprogesterone, and DHT (Locke et al., 2008). When CRPC tumors isolated from castrate mice were incubated ex vivo with [^{14}C]-acetic acid, formation of progesterone (1.5%); 17α-hydroxyprogesterone (4.0%); and DHT (8.3%) based on radiochromatography was observed (Locke et al., 2008). Because other analytes might coelute with DHT and no MS data were presented to support analyte identity these findings are equivocal at best. MS data were provided to support the formation of 5α-dihydroprogesterone, but without evidence for a functional P450c17 reaction it is uncertain whether these tumor specimens can really convert acetic acid to DHT. In a follow-up study, LNCaP cells were incubated with [^{3}H]-progesterone as precursor. While significant amounts were converted into 5α-pregnane-3,17-diol-20-one (a precursor in the backdoor pathway to DHT) less than 0.2% and 0.04% was converted to T and DHT, respectively (Locke et al., 2009). Collectively, these data do not strongly support de novo steroidogenesis nor do they support the conversion of C21 steroids into C19 steroids in CRPC.

7. ROLE OF HSD3B1 IN INTRATUMORAL ANDROGEN BIOSYNTHESIS

HSD3B1 is the isoform that converts DHEA to Δ^4-AD, and 5-Adiol to T in the prostate. The enzyme is bifunctional in that it catalyzes the

NAD^+-dependent dehydrogenation of the 3β-hydroxy group of a Δ^5-ene and then catalyzes the isomerization of the resultant Δ^5-3-ketosteroid to yield the Δ^4-3-ketosteroid. In the case of epi-androsterone only the dehydrogenation step is required to produce 5-Adione. A common allelic *HSD3B1*(1245C) exists in approximately 22% of the Caucasian population but varies widely by ethnicity. The mutation N367T does not affect the steady-state kinetic constants for the enzyme but increases enzyme stability by preventing its proteasomal degradation (Chang et al., 2013). This stability mutant will increase the flux to DHEA down the steroidogenic pathways to DHT and could contribute to AA drug resistance. The same mutant would, however, increase formation of Δ^4-abiraterone and Δ^4-galeterone which would be advantageous based on their superior properties. The differential effects of this mutation on promoting steroidogenesis and improving the P450c17 inhibitor profile remain to be resolved.

8. ROLE OF AKR1C3 IN INTRATUMORAL ANDROGEN BIOSYNTHESIS

One of the most overexpressed genes in CRPC is *AKR1C3*, type 5 17β-hydroxysteroid dehydrogenase. This is seen in prostate cancer cells grown in androgen-depleted media (Hofland et al., 2010), in xenografts in castrate mice (Hofland et al., 2010; Liu et al., 2017, 2015), in patient tumor samples, and in soft-tissue metastases (Stanbrough et al., 2006). Nine clinical studies have reported similar findings for AKR1C3 expression in biospecimens from human patients (Fankhauser et al., 2014; Hamid et al., 2012; Hofland et al., 2010; Jernberg et al., 2013; Mitsiades et al., 2012; Pfeiffer, Smit, Sedelaar, & Schalken, 2011; Powell et al., 2015; Stanbrough et al., 2006; Tian et al., 2014). AKR1C3 plays a pivotal role in all pathways to DHT. It catalyzes the conversion of Δ^4-AD to T, the conversion of 5-Adione to DHT, and the conversion of DHEA to 5-Adiol. It is also involved in the backdoor pathway where it converts androsterone to 3α-diol (Byrns, Mindnich, Duan, & Penning, 2012; Lin et al., 1997; Penning et al., 2000). AKR1C3 expression is upregulated by ADT and repressed by androgens and its overexpression is part of the resistance mechanism to AA and ENZ (Liu et al., 2017, 2015). AKR1C3 expression is driven by the TMPRSS2-ERG fusion protein by binding to the *AKR1C3* promoter (Powell et al., 2015). A feed-forward mechanism has been proposed in which TMPRSS2-ERG expression increases as the tumor begins to synthesize more T and DHT and this displaces the AR from the *AKR1C3* promoter to enhance intratumoral androgen biosynthesis (Powell et al., 2015).

We have proposed that the residual DHEA-SO_4 that exists after P450c17 inhibition coupled with AKR1C3 overexpression creates a "perfect-storm" for AA drug resistance (Tamae et al., 2015). In mouse xenografts of CWR22Rv1 prostate tumors that are resistant to AA and ENZ, drug resistance can be surmounted by indomethacin (Liu et al., 2017, 2015), a selective and potent inhibitor of AKR1C3 identified by us (Byrns, Steckelbroeck & Penning, 2008). AKR1C3 inhibitors thus have a place in CRPC treatment especially in patients who progress on second-line ADT.

There are a number of nonsteroidal- (Adeniji et al., 2012; Kikuchi et al., 2014; Liedtke et al., 2013; Yepuru et al., 2013; Yin et al., 2014) and steroidal-based inhibitors of AKR1C3 (Bothe, Cancho-Grande, Irlbach, & Ray, 2015) in preclinical development. Several are deserving of mention: SN33638, a sulfonyl piperidine was found to be only effective in high-AKR1C3 expressing prostate cancer cells, e.g., LAPC4 or 22Rv1 (Yin et al., 2014). ASP9521 developed by Astellas with desirable preclinical properties for the inhibition of AKR1C3 was taken into a phase1/11b clinical trial where it was well tolerated but without efficacy in 7/13 patients (Kikuchi et al., 2014; Loriot et al., 2014). The authors concluded that the lack of drug efficacy may have been due to the fact that individuals were excluded who had been recently on ADT and that therefore AKR1C3 was not expressed in sufficient amounts to be targeted (Loriot et al., 2014). Collectively, these data suggest that when AKR1C3 is overexpressed as part of the adaptive response to ADT a therapeutic benefit of AKR1C3 inhibition is anticipated.

9. CONCLUSIONS AND FUTURE DIRECTIONS

DHEA-SO_4 is present at 400–1000 times the concentration of DHEA and is of adrenal origin. Inhibition of P450c17, the primary source of DHEA, leaves a sizable pool of DHEA-SO_4 available for intratumoral biosynthesis of T and DHT in CRPC. As CRPC progresses the tumor can stay dependent upon androgens due to adaptive responses that can exploit this residual pool of DHEA-SO_4. These adaptive responses include increased expression of *SLCO* transporters and *AKR1C3*. Whether changes in *STS* expression coincide with these adaptive responses remains to be determined. STS inhibitors and AKR1C3 inhibitors could be used to block steroidogenesis downstream from P450c17, they could be used to surmount P450c17 inhibitor drug resistance and could be used as monotherapies.

ACKNOWLEDGMENTS

This work is supported by Grants P01-CA163227 and P30-ES013508 from the National Cancer Institute and the National Institute of Environmental Health Sciences, respectively.

REFERENCES

Adeniji, A. O., Twenter, B. M., Byrns, M. C., Jin, Y., Chen, M., Winkler, J. D., et al. (2012). Development of potent and selective inhibitors of aldo-keto reductase 1C3 (type 5 17β-hydroxysteroid dehydrogenase) based on N-phenyl-aminobenzoates and their structure-activity relationships. *Journal of Medicinal Chemistry*, *55*, 2311–2323.

Alperin, E. S., & Shapiro, L. J. (1997). Characterization of point mutations in patients with X-linked ichthyosis. Effects on the structure and function of the steroid sulfatase protein. *Journal of Biological Chemistry*, *272*, 20746–20763.

Alyamani, M., Li, Z., Berk, M., Li, J., Tang, J., Upadhyay, S., et al. (2017). Steroidogenic metabolism of galeterone reveals a diversity of biochemical activities. *Cell Chemical Biology*, *24*, 825–832.

Andriole, G. L., Humphrey, P., Ray, P., Gleave, M. E., Trachtenberg, J., Thomas, L. N., et al. (2004). Effect of the dual 5alpha-reductase inhibitor dutasteride on markers of tumor regression in prostate cancer. *Journal of Urology*, *172*, 915–919.

Arakawa, H., Nakanishi, T., Yanagihara, C., Nishimoto, T., Wakayama, T., Mizokami, A., et al. (2012). Enhanced expression of organic anion transporting polypeptides (OATPs) in androgen receptor-positive prostate cancer cells: Possible role of OATP1A2 in adaptive cell growth under androgen-depleted conditions. *Biochemical Pharmacology*, *84*, 1070–1077.

Arora, V. K., Schenkein, E., Murali, R., Subudhi, S. K., Wongvipat, J., Balbas, M. D., et al. (2013). Glucocorticoid receptor confers resistance to antiandrogens by bypassing androgen receptor blockade. *Cell*, *155*, 1309–1322.

Attard, G., Reid, A. H., A'Hern, R., Parker, C., Oommen, N. B., Folkerd, E., et al. (2009). Selective inhibition of CYP17 with abiraterone acetate is highly active in the treatment of castration-resistant prostate cancer. *Journal of Clinical Oncology*, *27*, 3742–37482.

Attard, G., Reid, A. H., Auchus, R. J., Hughes, B. A., Cassidy, A. M., Thompson, E., et al. (2012). Clinical and biochemical consequences of CYP17A1 inhibition with abiraterone given with and without exogenous glucocorticoids in castrate men with advanced prostate cancer. *The Journal of Clinical Endocrinology and Metabolism*, *97*, 507–516.

Attard, G., Reid, A. H., Yap, T. A., Raynaud, F., Dowsett, M., Settatree, S., et al. (2008). Phase 1 clinical trial of a selective inhibitor of CYP17, abiraterone acetate, confirms that castration-resistant prostate cancer commonly remains hormone driven. *Journal of Clinical Oncology*, *28*, 4563–4571.

Auchus, R. J. (2004). The backdoor pathway to dihydrotestosterone. *Trends in Endocrinology and Metabolism*, *15*, 432–438.

Auchus, R. J., Lee, T. C., & Miller, W. L. (1998). Cytochrome b5 augments the 17,20-lyase activity of human P450c17 without direct electron transfer. *Journal of Biological Chemistry*, *273*, 3158–3165.

Bothe, U., Cancho-Grande, Y., Irlbach, E. R., & Ray, N. C. (2015). 3-Nitrogen or sulphur substitutd oestra-1,3,5(10),16-tetraene AKR1C3 inhibitors world intellectual property organization WO 2016037956.

Buttigliero, C., Tucci, M., Bertaglia, V., Vignani, F., Bironzo, P., Di Maio, M., et al. (2015). Understanding and overcoming the mechanisms of primary and acquired resistance to abiraterone and enzalutamide in castration resistant prostate cancer. *Cancer Treatment Reviews*, *41*, 884–892.

Byrns, M. C., Mindnich, R., Duan, L., & Penning, T. M. (2012). Overexpression of aldo-keto reductase 1C3 (AKR1C3) in LNCaP cells diverts androgen metabolism towards testosterone resulting in resistance to the 5α-reductase inhibitor finasteride. *The Journal of Steroid Biochemistry and Molecular Biology, 130*, 7–15.

Byrns, M. C., Steckelbroeck, S., & Penning, T. M. (2008). An indomethacin analogue, N-(4-chlorobenzoyl)-melatonin, is a selective inhibitor of aldo-keto reductase 1C3 (type 2 3alpha-HSD, type 5 17beta-HSD, and prostaglandin F synthase), a potential target for the treatment of hormone dependent and hormone independent malignancies. *Biochemical Pharmacology, 75*, 484–493.

Cao, B., Qi, Y., Zhang, G., Xu, D., Zhan, Y., Alvarez, X., et al. (2014). Androgen receptor splice variants activating the full-length receptor in mediating resistance to androgen-directed therapy. *Oncotarget, 5*, 1646–1656.

Chang, K.-H., Li, R., Kuri, B., Lotan, Y., Roehrborn, C. G., Liu, J., et al. (2013). A gain-of-function mutation in DHT synthesis in castration-resistant prostate cancer. *Cell, 154*, 1074–1084.

Chang, K. H., Li, R., Papari-Zareei, M., Watumull, L., Zhao, Y. D., Auchus, R. J., et al. (2011). Dihydrotestosterone synthesis bypasses testosterone to drive castration-resistant prostate cancer. *Proceedings of the National Academy of Sciences of the United States of America, 108*, 13728–13733.

Cho, E., Montgomery, R. B., & Mostaghel, E. A. (2014). Minireview: SLCO and ABC transporters: A role for steroid transport in prostate cancer progression. *Endocrinology, 155*, 4124–4132.

Chung, B. C., Picado-Leonard, J., Haniu, M., Bienkowski, M., Hall, P. F., Shively, J. E., et al. (1987). Cytochrome P450c17 (steroid 17 alpha-hydroxylase/17,20 lyase): Cloning of human adrenal and testis cDNAs indicates the same gene is expressed in both tissues. *Proceedings of the National Academy of Sciences of the United States of America, 84*, 407–411.

Dalla Valle, L., Toffolo, V., Nardi, A., Fiore, C., Armanini, D., Belvedere, P., et al. (2007). The expression of the human steroid sulfatase-encoding gene is driven by alternative first exons. *The Journal of Steroid Biochemistry and Molecular Biology, 107*, 22–29.

Damgaard-Olesen, A., Johannsen, T. H., Holmboe, S. A., Søeborg, T., Petersen, J. H., Andersson, A., et al. (2016). Reference ranges of 17-hydroxyprogesterone, DHEA, DHEAS, androstenedione, total and free testosterone determined by TurboFlow-LC-MS/MS and associations to health markers in 304 men. *Clinical Chimica Acta, 454*, 82–88.

Danila, D. C., Morris, M. J., de Bono, J. S., Ryan, C. J., Denmeade, S. R., Smith, M. R., et al. (2010). Phase II multicenter study of abiraterone acetate plus prednisone therapy in patients with docetaxel-treated castration-resistant prostate cancer. *Journal of Clinical Oncology, 28*, 1496–1501.

DeVore, N. M., & Scott, E. E. (2012). Structures of cytochrome P450 17A1 with prostate cancer drugs abiraterone and TOK-001. *Nature, 482*, 116–119.

Estabrook, R. W., Shet, M. S., Fisher, C. W., Jenkins, C. M., & Waterman, M. R. (1996). The interaction of NADPH-P450 reductase with P450: An electrochemical study of the role of the flavin mononucleotide-binding domain. *Archives of Biochemistry and Biophysics, 333*, 308–315.

Fankhauser, M., Tan, Y., Macintyre, G., Haviv, I., Hong, M. K. H., Nguyen, A., et al. (2014). Canoncial androstenedione reduction is the predominat source of signaling androgens in hormone-refractory prostate cancer. *Clinical Cancer Research, 20*, 5547–5557.

Farnsworth, W. E. (1973). Human prostatic dehydroepiandrosterone sulfate sulfatase. *Steroids, 21*, 647–664.

Fischer, D. S., Chander, S. K., Woo, L. W., Fenton, J. C., Purohit, A., Reed, M. J., et al. (2003). Novel D-ring modified steroid derivatives as potent, non-estrogenic, steroid sulfatase inhibitors with in vivo activity. *The Journal of Steroid Biochemistry and Molecular Biology, 84*, 343–349.

Fischer, D. S., Woo, L. W. L., Mahon, M. F., Purohit, A., Reed, M. J., & Potter, B. V. L. (2003). D-ring modified estrone derivatives as novel potent inhibitors of steroid sulfatase. *Bioorganic & Medicinal Chemistry, 11*, 1685–1700.

Fizazi, K., Jones, R., Oudard, S., Efstathiou, E., Saad, F., de Wit, R., et al. (2015). Phase III, randomized, double-blind, multicenter trial comparing orteronel (TAK-700) plus prednisone with placebo plus prednisone in patients with metastatic castration-resistant prostate cancer that has progressed during or after docetaxel-based therapy: ELM-PC 5. *Journal of Clinical Oncology, 33*, 723–731.

Fizazi, K., Scher, H. I., Molina, A., Logothetis, C. J., Chi, K. N., Jones, R. J., et al. (2012). Abiraterone acetate for treatment of metastatic castration-resistant prostate cancer: Final overall survival analysis of the COU-AA-301 randomised, double-blind, placebo-controlled phase 3 study. *Lancet Oncology*, 983–992.

Furr, B. J., Valcaccia, B., Curry, B., Woodburn, J. R., Chesterson, G., & Tucker, H. (1987). ICI 176,334: A novel non-steroidal, peripherally selective antiandrogen. *Journal of Endocrinology, 113*, R7–R9.

Gosh, D. (2007). Human sulfatases: A structural perspective to catalysis. *Cellular and Molecular Life Sciences, 64*, 2013–2022.

Green, S. M., Kaipainen, A., Bullock, K., Zhang, A., Lucas, J. M., Matson, C., et al. (2017). Role of OATP transporters in steroid uptake by prostate cancer cells in vivo. *Prostate Cancer and Prostatic Diseases, 20*, 20–27.

Hamid, A. R., Pfeiffer, M. J., Verhaegh, G. W., Schaafsma, E., Brandt, A., Sweep, F. C., et al. (2012). Aldo-keto reductase family 1 member C3 (AKR1C3) is a biomarker and therapeutic target for castration-resistant prostate cancer. *Molecular Medicine, 18*, 1449–1455.

Highlights, G. N. (2016). *Tokai ends phase III trial of lead candidate galeterone.* Genetic Engineering & Biotechnology News. https://www.genengnews.com/gen-news-highlights/tokai-ends-phase-iii-trial-of-lead-candidate-galeterone/81253001 [Press release; July 26, 2016].

Hofland, J., van Weerden, W. M., Dits, N. F. J., Steenbergen, J., van Leenders, G. H., Jenster, G., et al. (2010). Evidence of limited contributions for intratumoral steroidogenesis in prostate cancer. *Cancer Research, 70*, 1256–1264.

Huggins, C. B. (1965). Two principles in endocrine therapy of cancers: Hormone deprival and hormone interference. *Cancer Research, 25*, 1163–1167.

Huggins, C., Stevens, R., & Hodges, C. V. (1941). Studies on prostatic cancer: II. The effects of castration on advanced carcinoma of the prostate gland. *Archives of Surgery, 43*, 209–223.

Jernberg, E., Thysell, E., Ylitalo, E. B., Rudolfsson, S., Crnalic, S., Widmark, A., et al. (2013). Characterization of the prostate cancer bone metastases according to expression levels of steroidogenic enzymes and androgen receptor splice variants. *PLoS One, 8*, e77407.

Kikuchi, A., Furutani, T., Azami, H., Watanbe, K., Nimi, T., Kamiyama, Y., et al. (2014). In vitro and in vivo characterization of ASP9521: A novel selective, orally bioavailable inhibitor of 17beta-hydroxysteroid dehydrogenase type 5 (17beta-HSD5; AKR1C3). *Investigational New Drugs, 32*, 860–870.

Klein, H., Molwitz, T., & Bartsch, W. (1989). Steroid sulfate sulfatase in human benign prostatic hyperplasia: Characterization and quantification of the enzyme in epithelium and stroma. *Journal of Steroid Biochemistry, 33*, 195–200.

Knudsen, K., & Penning, T. M. (2010). Partners in crime: Deregulation of AR activity and androgen synthesis in prostate cancer. *Trends in Endocrinology and Metabolism, 21*, 315–324.

Knudsen, K., & Scher, H. I. (2009). Starving the addiction: New opportunities for durable suppression of AR signaling in prostate cancer. *Clinical Cancer Research, 15*, 4792–4798.

Kühn-Velten, W. N., & Lessmann, M. (1992). Ketoconazole inhibition of the bifunctional cytochrome P450c17 does not affect androgen formation from the endogenous lyase substrate. The catalytic site remains refractory in the course of intermediary hydroxyprogesterone processing. *Biochemical Pharmacology, 44*, 2371–2378.

Kwegyir-Afful, A. K., Ramalingam, S., Purushottamachar, P., Ramamurthy, V. P., & Njar, V. C. (2015). Galeterone and VNPT55 induce proteasomal degradation of AR/AR-V7, induce significant apoptosis via cytochrome *c* release and suppress growth of castration resistant prostate cancer xenografts in vivo. *Oncotarget, 6*, 27440–27460.

Labrie, F., Bélanger, A., Luu-The, V., Labrie, C., Simard, J., Cusan, L., et al. (1998). DHEA and the intracrine formation of androgens and estrogens in peripheral target tissues: Its role during aging. *Steroids, 63*, 322–328.

Labrie, F., Bélanger, A., Simard, J., Luu-The, V., & Labrie, C. (1995). DHEA and peripheral androgen and estrogen formation: Intracinology. *Annals of the New York Academy of Sciences, 774*, 16–28.

Li, Z., Alyamani, M., Li, J., Rogacki, K., Abazeed, M., Upadhyay, S. K., et al. (2016). Redirecting abiraterone metabolism to fine-tune prostate cancer anti-androgen therapy. *Nature, 533*, 547–551.

Li, Z., Bishop, A. C., Alyamani, M., Garcia, J. A., Dreicer, R., Bunch, D., et al. (2015). Conversion of abiraterone to D4A drives anti-tumour activity in prostate cancer. *Nature, 523*, 347–351.

Liedtke, A. J., Adeniji, A. O., Chen, M., Byrns, M. C., Jin, Y., Christianson, D. W., et al. (2013). Development of potent and selective indomethacin analogues for the inhibition of AKR1C3 (type 5 17beta-hydroxysteroid dehydrogenase/prostaglandin F synthase) in castrate-resistant prostate cancer. *Journal of Medicinal Chemistry, 56*, 2429–2446.

Lin, H.-K., Jez, J. M., Schlegel, B. P., Peehl, D. M., Pachter, J. A., & Penning, T. M. (1997). Expression and characterization of recombinant type 2 3alpha-hydroxysteroid dehydrogenase (HSD) from human prostate: Demonstration of bifunctional 3alpha/17beta-HSD activity and cellular distribution. *Molecular Endocrinology, 11*, 1971–1984.

Liu, C., Armstrong, C. M., Lou, W., Lombard, A., Evans, C. P., & Gao, A. C. (2017). Inhibition of AKR1C3 activation overcomes resistance to abiraterone in advanced prostate cancer. *Molecular Cancer Therapeutics, 16*, 35–44.

Liu, C., Lou, W., Zhu, Y., Yang, J. C., Nadiminty, N., Gaikwad, N. W., et al. (2015). Intracrine androgens and AKR1C3 activation confer resistance to enzalutamide in prostate cancer. *Cancer Research, 75*, 1413–1422.

Locke, J. A., Guns, E. S., Lubik, A. A., Adomat, H. H., Hendy, S. C., Wood, C. A., et al. (2008). Androgen levels increase by intratumoral de novo steroidogenesis during progression of castration-resistant prostate cancer. *Cancer Research, 68*, 6407–6415.

Locke, J. A., Nelson, C. C., Adomat, H. H., Hendy, S. C., Gleave, M. E., & Guns, E. S. (2009). Steroidogenesis inhibitors alter but do not eliminate androgen synthesis mechanisms during progression to castration-resistance in LNCaP prostate xenografts. *The Journal of Steroid Biochemistry and Molecular Biology, 115*, 126–136.

Loriot, Y., Fizazi, K., Jones, R. J., Brand, V. D. J., Molife, R. L., Omlin, A., et al. (2014). Safety, tolerability and anti-tumor activity of the androgen biosynthesis inhibitor ASP9521 in patients with metastatic castration-resistant prostate cancer: Multi-centre phase I/II study. *Investigational New Drugs, 32*, 995–1004.

McKay, R. R., Werner, L., Mostaghel, E. A., Lis, R., Voznesensky, O., Zhang, Z., et al. (2017). A phase II trial of abiraterone combined with dutasteride for men with metastatic castration-resistant prostate cancer. *Clinical Cancer Research, 23*, 935–945.

Mitsiades, N., Sung, C. C., Schultz, N., Danila, D. C., He, B., Eedunuri, V. K., et al. (2012). Distinct patterns of dysregulated expression of enzymes involved in androgen synthesis and metabolism in metastatic prostate cancer tumors. *Cancer Research, 72*, 6142–6152.

Miyake, H., Hara, T., Tamura, K., Sugiyama, T., Furuse, H., Ozono, S., et al. (2017). Comparative assessment of efficacies between 2 alternative therapeutic sequences with novel androgen receptor-axis-targeted agents in patients with chemotherapy-naïve metastatic castration-resistant prostate cancer. *Clinical Genitourinary Cancer, 15*, e591–e597.

Mohler, J. L., Titus, M. A., Bai, S., Kennerley, B. J., Lih, F. B., Tomer, K. B., et al. (2011). Activation of the androgen receptor by intratumoral bioconversion of androstanediol to dihydrotestosterone in prostate cancer. *Cancer Research, 71*, 1486–1496.

Mohler, J. L., Titus, M. A., & Wilson, E. M. (2011). Potential prostate cancer drug target: Bioactivation of androstanediol by conversion to dihydrotestosterone. *Clinical Cancer Research, 17*, 5844–5849.

Mostaghel, E. A., Marck, B. T., Plymate, S. R., Vessella, R. L., Balk, S., Matsumoto, A. M., et al. (2011). Resistance to CYP17A1 inhibition with abiraterone in castrate-resistance prostate cancer: Induction of steroidogenesis and androgen receptor splice variants. *Clinical Cancer Research, 17*, 5913–5925.

Mostaghel, E. A., Nelson, P. S., Lange, P., Lin, D. W., Taplin, M. E., Balk, S., et al. (2014). Targeted androgen pathway suppression in localized prostate cancer: A pilot study. *Journal of Clinical Oncology, 32*, 229–237.

Nakajin, S., Shinoda, M., Haniu, M., Shively, J. E., & Hall, P. F. (1984). C21 steroid side chain cleavage enzyme from porcine adrenal microsomes. Purification and characterization of the 17 alpha-hydroxylase/C17,20-lyase cytochrome P-450. *Journal of Biological Chemistry, 259*, 3971–3976.

Njar, V. C., & Brodie, A. M. (2015). Discovery and development of galeterone (TOK-001 or VN/124-1) for the treatment of all stages of prostate cancer. *Journal of Medicinal Chemistry, 58*, 2077–2087.

Peng, H. M., Im, S. C., Pearl, N. M., Turcu, A. F., Rege, J., Waskell, L., et al. (2016). Cytochrome b5 activates the 17,20-lyase activity of human cytochrome P450 17A1 by increasing the coupling of NADPH consumption to androgen production. *Biochemistry, 55*, 4356–4365.

Penning, T. M., Burczynski, M. E., Jez, J. M., Hung, C.-F., Lin, H.-K., Ma, H., et al. (2000). Human 3alpha-hydroxysteroid dehydrogenase isoforms (AKR1C1-AKR1C4) of the aldo keto reductase superfamily: Functional plasticity and tissue distribution reveals roles in the inactivation and formation of male and female sex hormones. *Biochemical Journal, 351*, 67–77.

Penning, T. M., Lee, S. H., Jin, Y., Gutierrez, A., & Blair, I. A. (2010). Liquid chromatography-mass spectrometry (LC-MS) of steroid hormone metabolites and its applications. *The Journal of Steroid Biochemistry and Molecular Biology, 118*, 177–187.

Pfeiffer, M. J., Smit, F. P., Sedelaar, J. P., & Schalken, J. A. (2011). Steroidogenic enzymes and stem cell markers are upregulated during androgen deprivation in prostate cancer. *Molecular Medicine, 17*, 657–664.

Powell, K., Semaan, L., Conley-LaComb, M. K., Asangani, I., Wu, Y. M., Ginsburg, K. B., et al. (2015). ERG/AKR1C3/AR constitutes a feed-forward loop for AR signaling in prostate cancer cells. *Clinical Cancer Research, 21*, 2569–2579.

Reed, M. J., Purohit, A., Woo, L. W., Newman, S. P., & Potter, B. V. (2005). Steroid sulfatase: Molecular biology, regulation, and inhibition. *Endocrine Reviews, 26*, 171–202.

Reid, A. H., Attard, G., Danila, D. C., Oommen, N. B., Olmos, D., Fong, P. C., et al. (2010). Significant and sustained antitumor activity in post-docetaxel, castration-resistant prostate cancer with the CYP17 inhibitor abiraterone acetate. *Journal of Clinical Oncology, 28*, 1489–1495.

Ryan, C. J., Smith, M. R., Fizazi, K., Saad, F., Mulders, P. F., Sternberg, C. N., et al. (2015). Abiraterone acetate plus prednisone versus placebo plus prednisone in chemotherapy-naive men with metastatic castration-resistant prostate cancer (COU-AA-302): Final overall survival analysis of a randomised, double-blind, placebo-controlled phase 3 study. *Lancet Oncology, 16*, 152–160.

Ryan, C. J., Smith, M. R., Fong, L., Rosenberg, J. E., Kantoff, P., Raynaud, F., et al. (2010). Phase I clinical trial of the CYP17 inhibitor abiraterone acetate demonstrating clinical activity in patients with castration-resistant prostate cancer who received prior ketoconazole therapy. *Journal of Clinical Oncology*, *28*, 1481–1488.

Saner, K. J., Suzuki, T., Sasano, H., Pizzey, J., Ho, C., Strauss, J. F., 3rd., et al. (2005). Steroid sulfotransferase 2A1 gene transcription is regulated by steroidogenic factor 1 and GATA-6 in the human adrenal. *Molecular Endocrinology*, *19*, 184–197.

SEER. (2017). *Surveilance, epidemiology, and end results program*. Retrieved from http://seer.cancer.gov/statfacts/html/prost.html.

Selcer, K. W., Kabler, H., Sarap, J., Xiao, Z., & Li, P.-K. (2002). Inhibition of steryl sulfatase activity in LNCaP human prostate cancer cells. *Steroids*, *67*, 821–826.

Shackleton, C. H., Kletke, C., Wudy, S., & Pratt, J. H. (1990). Dehydroepiandrosterone sulfate quantification in serum using high-performance liquid chromatography/mass spectrometry and a deuterated internal standard: A technique suitable for routine use or as a reference method. *Steroids*, *55*, 472–478.

Sharifi, R., Bruskewitz, R. C., Gittleman, M. C., Graham, S. D., Jr., Hudson, P. B., & Stein, B. (1996). Leuprolide acetate 22.5 mg 12-week depot formulation in the treatment of patients with advanced prostate cancer. *Clinical Therapeutics*, *18*, 647–657.

Stanbrough, M., Bubley, G. J., Ross, K., Golub, T. R., Rubin, M. A., Penning, T. M., et al. (2006). Increased expression of genes converting adrenal androgens to testosterone in androgen-independent prostate cancer. *Cancer Research*, *66*, 2815–2825.

Tamae, D., Byrns, M., Marck, B., Mostaghel, E. A., Nelson, P. S., Lange, P., et al. (2013). Development, validation and application of a stable isotope dilution liquid chromatography electrospray ionization/selected reaction monitoring/mass spectrometry (SID-LC/ESI/SRM/MS) method for quantification of keto-androgens in human serum. *The Journal of Steroid Biochemistry and Molecular Biology*, *138C*, 281–289.

Tamae, D., Mostaghel, E., Montgomery, B., Nelson, P. S., Balk, S. P., Kantoff, P. W., et al. (2015). The DHEA-sulfate depot following P450c17 inhibition supports the case for AKR1C3 inhibition in high risk localized and advanced castration resistant prostate cancer. *Chemico-Biological Interactions*, *234*, 332–338.

Taplin, M. E., Bubley, G. J., Shuster, T. D., Frantz, M. E., Spooner, A. E., Ogata, G. K., et al. (1995). Mutation of the androgen-receptor gene in metastatic androgen-independent prostate cancer. *The New England Journal of Medicine*, *332*, 1393–1398.

Taplin, M. E., Montgomery, B., Logothetis, C. J., Bubley, G. J., Richie, J. P., Dalkin, B. L., et al. (2014). Intense androgen-deprivation therapy with abiraterone acetate plus leuprolide acetate in patients with localized high-risk prostate cancer: Results of a randomized phase II neoadjuvant study. *Journal of Clinical Oncology*, *32*, 3705–3715.

Tian, Y., Zhao, L., Zhang, H., Liu, X., Zhao, L., Zhao, X., et al. (2014). AKR1C3 overexpression may serve as a promising biomarker for prostate cancer progression. *Diagnostic Pathology*, *9*, 42–48.

Trachtenberg, J., Halpern, N., & Pont, A. (1983). Ketoconazole: A novel and rapid treatment for advanced prostatic cancer. *The Journal of Urology*, *130*, 152–153.

Tran, C., Ouk, S., Clegg, N. J., Chen, Y., Watson, P. A., Arora, V., et al. (2009). Development of a second-generation antiandrogen for treatment of advanced prostate cancer. *Science*, *324*, 787–790.

Visakorpi, T., Hyytinen, E., Koivisto, P., Tanner, M., Keinänen, R., Palmberg, C., et al. (1995). In vivo amplification of the androgen receptor gene and progression of human prostate cancer. *Nature Genetics*, *9*, 401–406.

Webster, D., France, J. T., Shapiro, L. J., & Weiss, R. (1978). X-linked ichthyosis due to steroid-sulphatase deficiency. *Lancet*, *1*, 70–72.

Woo, L. W. L., Purohit, A., Malini, B., Reed, M. J., & Potter, B. V. L. (2000). Potent active site-directed inhibition of steroid sulphatase by tricyclic coumarin-based sulphamates. *Chemistry & Biology*, *7*, 773–791.

Wright, J. L., Kwon, E. M., Ostrander, E. A., Montgomery, R. B., Lin, D. W., Vessella, R., et al. (2011). Expression of SLCO transport genes in castration-resistant prostate cancer and impact of genetic variation in SLCO1B3 and SLCO2B1 on prostate cancer outcomes. *Cancer Epidemiology, Biomarkers & Prevention, 20*, 619–627.

Yamaoka, M., Hara, T., Hitaka, T., Kaku, T., Takeuchi, T., Takahashi, J., et al. (2012). Orteronel (TAK-700), a novel non-steroidal 17,20-lyase inhibitor: Effects on steroid synthesis in human and monkey adrenal cells and serum steroid levels in cynomolgus monkeys. *The Journal of Steroid Biochemistry and Molecular Biology, 129*, 115–128.

Yang, M., Xie, W., Mostaghel, E., Nakabayashi, M., Werner, L., Sun, T., et al. (2012). SLCO2B1 and SLCO1B3 may determine time to progression for patients receiving androgen deprivation therapy for prostate cancer. *Journal of Clinical Oncology, 29*, 2565–2573.

Yen, P. H., Allen, E., Marsh, B., Mohandas, T., Wang, N., Taggart, R. T., et al. (1987). Cloning and expression of steroid sulfatase cDNA and the frequent occurrence of deletions in STS deficiency: Implications for X-Y interchange. *Cell, 49*, 443–454.

Yepuru, M., Wu, Z., Kyulkarni, A., Yin, F., Barrett, C. M., Kim, J., et al. (2013). Steroidogenic enzyme AKR1C3 is a novel androgen receptor-selective coactivator that promotes prostate cancer growth. *Clinical Cancer Research, 19*, 5613–5625.

Yin, Y. D., Fu, M., Bokke, D. G., Heinrich, D. M., Denny, W. A., & Jamieson, S. M. F. (2014). The activity of SN33638, an inhibitor of AKR1C3, on testosterone and 17beta-estradiol production and function in castration resistant prostate cancer and ER-positive breast cancer. *Frontiers in Oncology, 4*, 159, 1-12.

Yu, Z., Cai, C., Gao, S., Simon, N. I., Shen, H. C., & Balk, S. P. (2014). Galeterone prevents androgen receptor binding to chromatin and enhances degradation of mutant androgen receptor. *Clinical Cancer Research, 20*, 4075–4085.

Zang, T., Tamae, D., Mesaros, C., Wang, Q., Huang, M., Blair, I. A., et al. (2017). Simultaneous quantitation of nine hydroxy-androgens and their conjugates in human serum by stable isotope dilution liquid chromatography electrospray ionization tandem mass spectrometry. *The Journal of Steroid Biochemistry and Molecular Biology, 165*, 342–355.

Zang, T., Taplin, M. E., Tamae, D., Xie, W., Mesaros, C., Zhang, Z., et al. (2017). Testicular vs adrenal sources of hydroxy-androgens in prostate cancer. *Endocrine-Related Cancer, 24*, 393–404.

Zhao, X. Y., Malloy, P. J., Krishnan, A. V., Swami, S., Navone, N. M., Peehl, D. M., et al. (2000). Glucocorticoids can promote androgen-independent growth of prostate cancer cells through a mutated androgen receptor. *Nature Medicine, 6*, 703–706.

CHAPTER TWELVE

Dehydroepiandrosterone and Cardiovascular Disease

P. Mannella, T. Simoncini, M. Caretto, A.R. Genazzani[1]
Department of Clinical and Experimental Medicine, University of Pisa, Pisa, Italy
[1]Corresponding author: e-mail address: agenazzi@gmail.com

Contents

Abstract

The dehydroepiandrosterone and its metabolite, dehydroepiandrosterone sulfate, have been for a long while at the center of interest for endocrinologists and cardiologists. Consolidated data show that the dehydroepiandrosterone and the dehydroepiandrosterone sulfate present protective actions on the cardiovascular system. These actions are accomplished directly through target tissues such as endothelial cells, smooth muscle cells, and cardiomyocytes. At this level, they are able to activate a complex group of receptor, not completely identified, which modulate important functions such as vasodilation, antiinflammation, and antithrombosis. These data support the hypothesis that dehydroepiandrosterone could be used as drug for primary prevention of cardiovascular disease especially during aging and potentially also in addition of the common therapeutic strategy for the treatment and prevention of cardiovascular disease recurrence. In this publication, the effects of dehydroepiandrosterone and dehydroepiandrosterone sulfate on the cardiovascular system have been elucidated, starting with an analysis of the molecular action at target organ levels. In the second part, we evaluated the clinical effects of this administration, considering ultimately possible implications in introducing this hormone into clinical practice.

Vitamins and Hormones, Volume 108
ISSN 0083-6729
https://doi.org/10.1016/bs.vh.2018.05.001

1. INTRODUCTION

Cardiovascular diseases (CVD) represent a serious public health matter. CVD, mainly corresponding to coronary heart disease, cardiomyopathy, congenital heart disease, heart failure, stroke, atherosclerosis, and hypertension, are among the first causes of death in the developed countries nowadays. The World Health Organization (WHO) estimates that the total number of CVD deaths is increased globally to 17.7 million, representing 31% of all global deaths. Of these deaths, an estimated 7.4 million is due to coronary heart disease and 6.7 million is due to stroke.

Most cardiovascular diseases can be prevented by addressing behavioral risk factors such as tobacco use, unhealthy diet and obesity, physical inactivity, and harmful use of alcohol using population-wide strategies (WHO: World Health Organization Report, 2017).

However, people with cardiovascular disease or who are at high cardiovascular risk (due to the presence of one or more risk factors such as hypertension, diabetes, hyperlipidemia, or already established disease) need early detection and management using counseling and medicines, as appropriate.

An excess of steroid hormones such as mineralocorticoids has been linked to heart hypertrophy, arrhythmias, inflammation, fibrosis, and apoptosis as well as to endothelial and smooth muscle vascular dysfunctions. Benefits of antagonizing aldosterone action in patients with CVD have demonstrated a causal relationship between steroid hormones and CVD (Pitt et al., 2003).

On the other hand, protective steroid hormones, such as dehydroepiandrosterone (DHEA), decrease with aging. DHEA, and particularly its sulfated derivative, DHEA sulfate (DHEAS), are the most abundant circulating steroid hormones (Table 1). However, these concentrations present wide variations, which can depend on different factors: stress, central nervous system function, and psychiatric disorders as depression, insulin sensitivity, immunological function, osteoporosis, and cardiovascular disorders (Kedziora-Kornatowska, Beszczynska-Oles, Kornatowski, & Szadujkis-Szadurski, 2007; Kroboth, Salek, Pittenger, Fabian, & Frye, 1999). Certainly, plasma concentrations of DHEA and DHEAS progressively decline with age either in women or men (Belanger et al., 1994; Lasley et al., 2002), suggesting that they may be implicated in the aging process (Table 2).

At the same time, reduced concentration of these molecules has been associated to the risk to develop cardiovascular diseases (Haffner, Newcomb,

Table 1 Chemistry of DHEA and DHEAS

	DHEA (Dehydroepiandrosterone; Prasterone)	**DHEAS (Dehydroepiandrosterone Sulfate; Prasterone Sulfate)**
	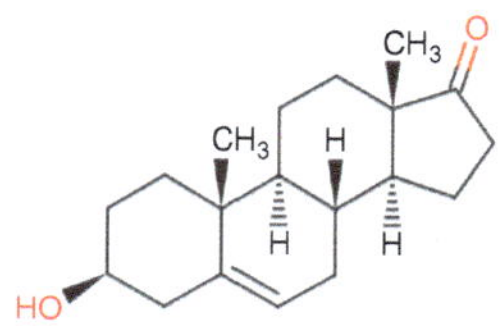	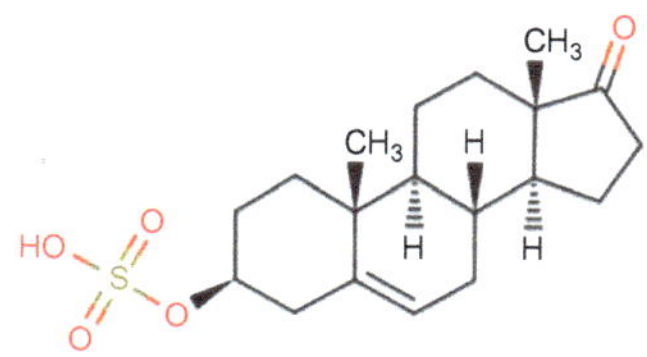
Chemical name	3β-hydroxyandrost-5-en-17-one	3β-hydroxyandrost-5-en-17-one 3-sulfate
Molecular formula	$C_{19}H_{28}O_2$ hydrophobic and lipophilic	$C_{19}H_{28}O_5S$ hydrophilic
Molecular weight (g/mol)	288.42442	368.49
Daily production rate	6–8 mg/day of active hormone	3.5–20 mg/day of inactivated, sulfated form
Serum concentration in adults	Males: 10.6–28.9 nmol/L	Males: 5.41–9.09 lmol/L
	Females: 9.77–26.7 nmol/L	Females: 2.23–9.20 lmol/L
Metabolic clearance rate (L/day)	2000	5–20
Hepatic clearance	Rapid	Slow
Half-life time (h)	1–3	10–20

Marcus, Klein, & Klein, 1995). DHEAS levels are also significantly lower in men with a history of heart disease and other risk factors as obesity or smoking than in those without such a history (Barrett-Connor, Khaw, & Yen, 1986; Khaw, 1996), even if this is not completely clear in women (Barrett-Connor & Ferrara, 1996; Thijs, Fagard, Forette, Nawrot, & Staessen, 2003). DHEAS levels seem to be inversely correlated with male mortality, whereas there was no association with female mortality (Mazat et al., 2001).

Many publications have hypothesized that the development of DHEA deficiencies with age might play a key role in the degradation of many functions and contribute to the genesis of several disorders, including cardiovascular defects as arterial stiffness (Weiss, Villareal, Ehsani, Fontana, & Holloszy, 2012) and development of carotid atherosclerosis (Yoshida et al., 2010).

Table 2 DHEA and DHEAS Concentration During the Life

DHEAS	**Sex**	**Age Range (Years)**	**DHEAS Range (ng/mL)**
	Men	15–39	1500–5000
		40–49	1000–4000
		50–59	600–3000
		60	300–2000
	Women	15–29	1000–5000
		30–39	600–3500
		40–49	400–2500
		50	200–1500
DHEA	**Sex**	**Age Range (Years)**	**DHEA Range (ng/L)**
	Men	6–24 months	2500
		2–3	630
		4–5	930
		6–7	60–1930
		7–9	100–2080
		10–11	320–3080
		14–15	930–6040
		16–17	1170–6520
		18–40	1330–7780
		40–67	630–4700
	Women	6–24 months	1990
		2–3	630
		4–5	1030
		6–7	120–1520
		7–9	140–2350
		10–11	430–3780
		12–13	890–6210
		14–15	1,220–7010
		16–17	1,420–9000
		18–40	1,330–7780
		40–67	630–4700

Nevertheless, in postmenopausal women as in the WISE (Women's Ischemia Syndrome Evaluation) study, a positive correlation between DHEAS levels and CVD outcome was reported. This was potentially explained by a better atherosclerosis state of patients with higher DHEAS levels (Shufelt et al., 2010).

Other studies have reported a beneficial role for DHEAS in various physiologic or pathophysiologic conditions such as brain development, aging, osteoporosis, immune system-mediated rheumatologic diseases, diabetes mellitus obesity, chronic heart failure (particularly when linked with oxidative stress), and recently, vascular remodeling as occurring in pulmonary arterial hypertension (Savineau, Marthan, & Dumas de la Roque, 2013).

Over the past 10 years, preparations of DHEA have been available and sold as the "fountain of youth." However, DHEA replacement may seem an attractive theoretical concept, there are very few definitive reports on the biological functions of this steroid, and it is still under debate. A better understanding of the role of DHEA in CVD thus appears relevant for public health purpose.

2. BASIC PROPERTIES OF DHEA: EXPRESSION, STRUCTURE, BIOLOGICAL ACTIVITIES

In 1934 Butenandt and Dannenbaum isolated DHEA (3β-hydroxyandrost-5-en-17-oneprasterone) from male urine, and in 1954 Migeon and Plager isolated it from human plasma. Dehydroepiandrosterone sulfate (DHEAS; dehydroepiandrosterone-3-sulfate), a sulfated form of DHEA was obtained by Munson, Gallagher, and Koch in 1944. Ten years later, Baulieu suggested that DHEAS was the most abundant steroid hormone in human plasma. The name dehydroepiandrosterone was proposed by Fieser in 1949.

These two hormones, DHEA and DHEAS, are derived from cholesterol and are principally produced by the adrenal cortex; however, small amounts of them have been proposed to be produced by other organs such as the heart.

For almost 50 years after discovery, DHEA was considered an inert compound involved mainly in the bioconversion of cholesterol to androgens and estrogens and its real role was unknown. There had been no large human studies until the 1990s. During the first half of the 20th century, many publications have hypothesized that the development of DHEA deficiencies with age might play a key role in the degradation of many functions and contribute to the genesis of several disorders, including cardiovascular defects

(Weiss et al., 2012). For these reasons, in the last years, DHEA and DHEAS have been rediscovered for their biological activities on different tissues such as cardiovascular system and brain and for their potential clinical role (Table 3) (Rutkowski, Sowa, Rutkowska-Talipska, Kuryliszyn-Moskal, & Rutkowski, 2014).

Table 3 Biology of DHEA and DHEAS

	DHEA (Dehydroepiandosterone; Prasterone)	DHEAS (Dehydroepiandosterone Sulfate; Prasterone Sulfate)
Site of synthesis	75%–90% of DHEA synthesized in zona reticularis, remainder produced by testes, ovaries, and brain	Almost exclusively in zona reticularis
Biological activity	A large reservoir of precursors for intracellular production of androgens and estrogens in nonreproductive tissues; around 30% of total androgens in men and 75% of estrogens in premenopausal and close to 100% in postmenopausal women produced by intracrine conversion of DHEA/DHEAS steroids in peripheral tissue	A large and stable plasma reservoir of DHEA
	Acts indirectly on peripheral sex steroid target tissues by conversion to androgens, estrogens, or both, or directly as a neurosteroid via interaction with neurotransmitter receptors in the brain. Exerts its effect via multiple signalling pathways involving nitric oxide synthase activation, modulation of c-amino acid receptors, *N*-methyl D-aspartate, receptor sigma receptors	
Clinical role	Supplementation in the elderly, menopause, ovarian dysfunction, Addison's disease, Crohn's disease, ulcerative colitis, systemic lupus erythematosus, asthma, allergic rhinitis, urticaria, chronic obstructive pulmonary disease	Marker in Cushing's disease; screening test in asthma treated with inhaled glucorticosteroids

3. CELL AND TISSUE EFFECTS RELEVANT TO CARDIOPROTECTION

3.1 Molecular Action of DHEA on Cardiovascular System: Tissue Targets

Several molecular mechanisms by which DHEA and DHEAS exert their action on CVS have been proposed. These steroids serve as precursors for estrogens and androgens, and DHEA could be merely an inactive precursor pool for the formation of bioactive steroid hormones. As precursor, it is converted in different peripheral tissues or even within cells by the steroidogenic enzymes specifically in each cell type in individual target tissues (Labrie, Luu-The, Labrie, & Simard, 2001). In fact, oral supplementation of DHEA in postmenopausal women increases the concentrations of 17β estradiol, estrone, androstenedione, testosterone, and dihydrotestosterone (Stomati et al., 2000).

However, different in vivo studies demonstrate beneficial effects of DHEA on different tissues directly, on blood vessels (Hayashi et al., 2000; Simoncini et al., 2003).

This supports that DHEA could act through a dedicated, not yet identified, receptor. A steroid-specific receptor could be a G-protein-coupled receptor which is expressed on endothelial cell plasma membranes and coupled to eNOS activity through Galpha(i2) and Galpha(i3) protein (Liu & Dillon, 2004).

The binding of DHEA to the cell membrane induces a rapid activation of intracellular signaling cascades, which leads to the activation of endothelial nitric oxide enzyme and the release of nitric oxide (Fig. 1) which modulate vessels dilatation.

This action requires also a later recruitment of ERK 1/2 MAPK. Even if these pathways are activated even by the recruitment of estrogen or androgen receptor, in endothelial cells this action is maintained even when other steroid receptors are selectively blocked, supporting again the hypothesis that DHEA uses a specific receptor in these cells (Simoncini et al., 2003).

Beyond this rapid, nongenomic mechanism, DHEA induces eNOS expression and function in long term (Simoncini et al., 2003). These effects are present even in endothelial cells of ovariectomized rats where DHEA acts as estrogen agonist activating the same molecular pathway used by estrogen (Simoncini et al., 2003).

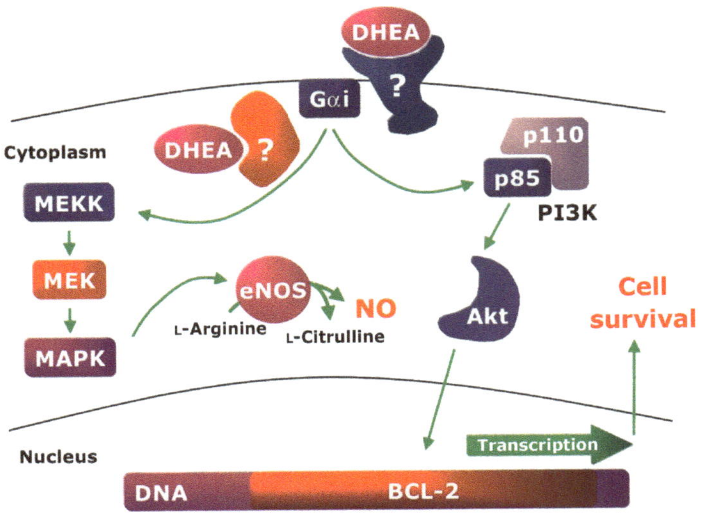

Fig. 1 Intracellular signaling of DHEA through G proteins in endothelial cells. *eNOS*, endothelial nitric oxide synthase; *MEK*, MAPK kinase; *MEKK*, MEK kinase; *PI3K*, phosphatidylinositol 3OH kinase. *Courtesy provided by Simoncini, T. (2007). Dehydroepiandrosterone, the endothelium, and cardiovascular protection.* Endocrinology, 148*(7), 3065–3067.*

Still by the recruitment of specific receptor, not related to ER or AR, DHEA regulates proliferation and apoptosis of smooth muscle cells through the inhibition of the Akt pathway and subsequent activation of GSK-3beta (Bonnet et al., 2009) preventing and reversing the remodeling which determinates restenosis after cardiovascular injuries. DHEA inhibits voltage-gated T-type Ca^{2+} channels. This action leads to the inhibition of pulmonary artery contraction and relates to the therapeutic action and/or physiological effects in cardiovascular diseases (Savineau et al., 2013).

Some authors propose the human sigma-1 receptor (Sig-1R) as potential receptor target of DHEA. By binding with this receptor, DHEA is able to increase phosphorylation of eNOS and Akt activity with protective effects in rats (Bhuiyan & Fukunaga, 2010). This occurs only with DHEA as agonist of Sigma-1R but not with 17β estradiol improving cardiac function via stimulation of the Sigma-1R–Akt–eNOs signaling pathway in pressure overload-induced hypertrophy in ovariectomized rats. These studies suggest that Sigma-1R could be an important target of DHEA in CVD and overall this could justify a different action of DHEA and estrogen on cardiovascular system (Tagashira, Bhuiyan, Shioda, & Fukunaga, 2011).

However, much remains to be clarified about the role of Sig-1R. It is expressed in various tissues such as brain, retina, lens, liver, lung, heart, but also in many tumor lines but in fact the real role is not known. Located at both plasma and endoplasmic reticulum levels, its activation may modulate different processes such as inflammation, tumor cell proliferation, and cytoskeletal remodeling (Su, Su, Nakamura, & Tsai, 2016). Such functions would seem to be disconnected from the biological role of DHEA. On the other hand, these actions seem to be closely related to a cross talk with ER and AR.

DHEA and its metabolites are able to activate several receptors as peroxisome proliferators activated receptor alpha, pregnane X receptor, and even estrogen receptor directly which could suggest an inverse association with cardiovascular risk, morbidity, and mortality (Mannic, Viguie, & Rossier, 2015; Webb, Geoghegan, Prough, & Michael Miller, 2006). However, there is still going to be a way of isolating a single receptor for DHEA.

Indeed, beyond these direct effects on vascular cells mediated by a hypothetic-specific receptor, DHEA could mimic vascular actions of insulin to acutely activate signaling pathways in endothelium-mediating production of nitric oxide and endothelin 1 (Formoso et al., 2006) (Table 4).

3.2 Cardiovascular Tissue-Specific Effects of DHEA

Sex hormones play a major role in determining the risk of cardiovascular disease and DHEA and DHEAS are highly concentrated in human blood. Although a specific receptor has not been yet isolated, DHEA is highly active on the vascular system, acting directly on various targets: endothelial cells, smooth muscular cellular, and cardiomyocytes (Table 5).

The final biological action is globally protective, and this is accomplished through various mechanisms of action that determine a final antiatherosclerotic effect.

DHEA inhibits monocyte adhesion and inflammatory process generally preventing the formation and progression of the atherosclerotic plaque (Mannic et al., 2015). Indeed DHEA is able to dilate vessels by stimulating the synthesis and release of nitric oxide (Simoncini et al., 2003).

Also at the pulmonary artery level, inducing vasodilation in the pulmonary circulation promotes the reduction of downstream heart resistance by reducing the risk of developing pulmonary hypertension (Dumas de la Roque, Savineau, & Bonnet, 2010).

This action indirectly promotes a protective effect on the heart and on cardiomyocytes reducing the risk of ventricular hypertrophy induced by high levels of endothelin and improving cardiac function (Alzoubi et al., 2013).

Table 4 Different Mechanisms of Action of DHEA in Cardiovascular System

Signaling Pathways	Tissue/Cell Type	Pathophysiological Role of DHEA
eNos phosphorylation/NO production (+)	EC	Prevention of endothelial dysfunction
Oxidative stress (−)	EC and cardiac cells	Prevention of oxidative injury in cardiovascular system
Akt/GSK-3β/NFAT (−)	EC and VSMC	Antivascular remodeling effect
Src/STAT3 (—)	PASMC	Antivascular remodeling effect
Ionic channels		
BK_{Ca} (+)	PASMC	Membrane hyperpolarization, vasorelaxation
T-type Ca channels (−)		Prevention of PH
RhoA/RhoA kinase (−)	PASMC	Prevention of PH
(HIF-1*a*) (−)	PASMC	Prevention of PH
s/cGMP (+)	PASMC	Prevention of PH
DHEA receptors		
Plasma membrane associated (+)	EC and PASMC	NO production, vasorelaxation
Cytosolic/nuclear hormone receptor (+)	EC and VSMC	Antivascular remodeling effect
Endoplasmic reticulum membrane-associated receptor: Sigma-1 receptor (+)	Cardiac cell	Protection of cardiac injury and hypertension

(−), (+) indicate an activating or inhibitory action of DHEA, respectively. EC, VSM, and PASMC are endothelial cells, vascular smooth muscle cells, and pulmonary artery smooth muscle cells, respectively. *PH*, pulmonary hypertension.

3.3 Tissue-Specific Effects of DHEA Relevant to Cardioprotection: In Vivo and In Vitro Evidence

When analyzing the cellular and molecular mechanism of DHEA action, it is important to realize that DHEA and DHEAS may act differently on their targets cells in terms of both efficiency and specificity of the involved signaling pathways because of their quite different chemical properties. Moreover, both genomic and nongenomic mechanisms can be recruited by these hormones with divergent consequences for the cell function (Table 5).

Table 5 Effects of DHEA on Endothelial cells, Smooth Muscular Cellular, and Cardiomyocytes and Possible Mechanisms

Target Tissue/Cell (Species)	Macroscopic Biologic Response to DHEA(S)	Receptor Involved (or Messenger)	Cellular Signaling Pathway or Molecular Mechanism Suggested
Endothelial cells			
Coronary artery (human)	Inhibition of monocyte adhesion	AR and ER (via testosterone or estradiol)	Increased expression of polysialylated NCAM
Umbilical vein HUVEC (human)	Inhibition of monocyte adhesion and of TNFα-induced inflammation	N/D	Inhibition of NFκB translocation, reduction of ICAM-1 and E-selectin expression, diminution of ROS production
Umbilical vein HUVEC (human)	Prevention of monocyte adhesion induced by high glucose	N/D	Inhibition of NFκB translocation and adhesion molecule expression, antioxidant properties of DHEAS
Aorta (human)	Inhibition of TNFα-induced inflammation	PPARα	Attenuation of IL-8, ICAM-1 and VCAM-1 expression, inhibition of NFκB translocation
Umbilical vein HUVEC (human)	Inhibition of cell proliferation	NOT through AR or ER (not DHEAS)	N/D
Aorta (bovine) and HUVEC (human)	Increase of cell proliferation and protection against superoxide injury	NOT through AR or ER	Antioxidant properties of DHEA(S), increase of ERK 1, 2, and eNOS expression
Aorta (bovine) and HUVEC (human)	Increase of NO release	NOT through ER cell surface GPCR	NOS activation through acute (nongenomic) and Pertussis toxin-sensitive pathway
Aorta (rats)	Preservation of endothelial function upon aortic stenosis	Sigma-1R	Akt and eNOS phosphorylation and expression, inhibition of Sig-1R downregulation

Continued

Table 5 Effects of DHEA on Endothelial cells, Smooth Muscular Cellular, and Cardiomyocytes and Possible Mechanisms—cont'd

Target Tissue/Cell (Species)	Macroscopic Biologic Response to DHEA(S)	Receptor Involved (or Messenger)	Cellular Signaling Pathway or Molecular Mechanism Suggested
Vascular smooth muscle cells			
Pulmonary and aortic artery (rats)	Relaxation of contracted (KCl) rings	N/D	Opening of potassium channels (cell hyperpolarization)
Pulmonary artery (rats)	Reduction of hypotoxic pulmonary hypertension, relaxation of artery rings	N/D	Upregulation of the soluble guanylyl cyclase (increased sensitivity to NO)
Pulmonary artery (rats)	Relaxation of recontracted (KCl) rings	NOT through AR or ER	Inhibition of T-type calcium channels through a Pertussis toxin-sensitive G protein
Pulmonary artery (human)	Inhibition of cell proliferation and sensitization to apoptosis (PAH patients)	N/D	Inhibition of STAT3 and NFAT, induction of mitochondria depolarization, and reduction of cell calcium concentration
Cardiomyocytes			
Ventricle (rat)	Reduction of hypertrophy induced by endothelin	N/D	Inhibition of BNP expression
Ventricle (rat)	Prevention of corticosteroid-induced hypertrophy and chronotropic response	N/D	Decrease of T-type calcium channel expression and activity, inhibition of ANP and BNP expression, antioxidant properties of DHEA
Ventricle (rat)	Improvement of cardiac function and remodeling after chronic hypoxia	N/D	Activation of CREB and PGC1α, increase of myocyte proliferation and respiratory chain activity, reduction of superoxide anions

3.3.1 *Effects on Vascular Wall*

The initiating event of atherosclerosis is endothelial dysfunction followed by smooth muscle proliferation and architectural disruption. Monocytes–macrophages are then recruited by the endothelium and invade the subintimal space by a complex mechanism characterized by the enhanced expression of chemoattractant interleukins (such as interleukin-8 and MCP-1), adhesion molecule proteins, and various cell-membrane cytokine-receptors. This altered pattern of expression is further destabilized by the secondary appearance of low-density lipoproteins (LDL) which are mildly oxidized by free radicals derived from activated macrophages, endothelial and smooth muscle cells. The mechanisms by which oxidized LDL induces the expression of adhesion molecules on endothelial cells are still unknown; nevertheless, the signal transduction pathways that lead to the enhanced gene expression of these binding molecules require the translocation of the transcription factor NFκB and the degradation of IκB protein, as well as the intracellular production of reactive oxygen species.

The formation of the atheroma plaque requires the adhesion of monocytes to vascular endothelium through binding of endothelial neural cell adhesion molecule to monocyte. In human coronary artery endothelial cells, DHEA inhibited monocyte adhesion (Curatola, Huang, & Naftolin, 2012) via different mechanisms such as inhibiting in vitro TNFα-induced inflammatory response including NFκB translocation, ICAM-1, and E-selectin expression (Gutierrez et al., 2007).

Indeed, DHEA is able to increase synthesis and release of nitric oxide, a powerful vasodilator and antiinflammatory agent. The induction of nitric oxide synthesis by DHEA is relevant for vascular function and may provide an explanation for the antiatherogenic effects of this steroid in animals and in humans (Hayashi et al., 2000). Beyond that, the antiaging actions of DHEA in vessels may reflect a reduction of programmed cell death in the endothelium, via a rapid recruitment of the phosphatidylinositol 3-OH kinase/Akt pathway, which mediates the antiapoptosis action of DHEA (Liu et al., 2007).

The association of low DHEAS levels with carotid artery atherosclerosis (Zubairi, Idrees, Jabeen, Kamal, & Zafar, 2017), obesity (Savineau et al., 2013), decreased diurnal blood pressure variability (Rutkowski et al., 2014), and ischemic heart disease mortality in diabetic women (Labrie et al., 2001) also suggests a vasoprotective role of DHEAS in women (Stomati et al., 2000). Human intervention study data also suggest a beneficial effect of DHEA on angiographic evidence of atherosclerosis

(Hayashi et al., 2000), markers of vascular risk (Liu & Dillon, 2004; Simoncini et al., 2003), and vascular function (Bonnet et al., 2009). The low-plasma DHEAS levels were also associated with endothelial dysfunction in postmenopausal women (Stomati et al., 2000). Moreover, in postmenopausal women, DHEAS was inversely associated with left ventricular mass index.

Also different in vivo studies have demonstrated the antiatherogenic and vasculotonic effects of DHEAS. Oral administration of DHEA to cholesterol-fed rabbits results in a 50% reduction in plaque size (Gordon, Bush, & Weisman, 1988). Reduced lipid-lowering effects in both rodents and rabbits and relax vascular tissue through a calcium-related mechanism, inhibit smooth muscle cell proliferation, and inhibit the accumulation of cholesterol ester in macrophage cells (Chen, Xu, & Huang, 2014).

As potent regulator of nitric oxide synthesis and release, DHEA could present an other important role on regulation of platelet aggregation and thrombus formation. DHEA could play a relevant role in the molecular mechanisms that physiologically maintain platelets in a resting state. DHEAS may also contribute to the protection from thrombosis in circulation. It has been reported that DHEAS is able to inhibit human platelet aggregation promoted by arachidonic acid in plasma. DHEAS, but not DHEA, inhibits in vitro thrombin-dependent platelet aggregation in a dose-dependent manner (Bertoni et al., 2012). DHEAS at physiological concentration inhibited platelet aggregation induced only by low doses of thrombin. DHEAS seems to inhibit the aggregation of platelets stimulated with various agonists including thrombin, collagen, and the thromboxane A2 analogue (Bertoni et al., 2012).

3.3.2 Effects on Vascular Smooth Muscle Cells

Pulmonary arterial hypertension is a severe disease of the pulmonary vasculature, which is defined by an increased pulmonary arterial pressure and is characterized by proliferation and resistance to apoptosis of arterial smooth muscle cells. DHEAS has been reported to be a pulmonary vasodilator that prevents acute hypoxic pulmonary vasoconstriction by opening potassium channels (Williams et al., 2002).

In addition, DHEA reverses human vascular remodeling by inhibiting Src/STAT3 constitutive activation reversing pulmonary hypertension (Paulin et al., 2011).

The beneficial effect of DHEA on the increase in pulmonary blood pressure is also connected at a prevention of right cardiac hypertrophy and pulmonary arterial remodeling, especially pulmonary artery smooth muscle. Several animal models demonstrated both the protective and the reversal effects of DHEA against pulmonary hypertension (Homma et al., 2008). Chronic DHEA treatment nearly normalized the increased RhoA/RhoA kinase signaling pathway activity which contributes to both vasoconstriction and vascular remodeling associated with pulmonary hypertension. The use of DHEA could have beneficial effects in treating pulmonary hypertension (Patel, Kandhi, Kelly, Neo, & Wolin, 2014).

3.3.3 Effects on Cardiomyocytes

The cardioprotective action of DHEA is exerted even on cardiomyocites. It was found that the presence of DHEA significantly inhibited the increase in myocyte cell size and BNP (B-type natriuretic peptide) mRNA levels upregulated by endothelin-1 (Nakamura et al., 2004). Interestingly, in the same study, the authors reported that DHEA levels in humans, measured at the coronary sinuses and aortic roots during cardiac catheterization, were suppressed in patients with heart failure in contrast to controls.

Recently, another study investigated the ability of DHEA to counteract the chronotropic and hypertrophic actions of aldosterone and glucocorticoids (Mannic et al., 2013). DHEA was found to reduce this chronotropic response induced by mineralocorticoid or the glucocorticoid receptor, decreasing the T-type calcium channel expression and activity within cardiomyocytes.

4. DHEA AND CARDIOVASCULAR DISEASE

DHEA/DHEAS production in humans changes with age (Kushnir et al., 2010). At birth, circulatory levels are high and reach maximal values between the age of 20 and 30 years. After reaching the intraindividual maximum levels during the third decade of life, DHEAS steadily decline down to 10%–20% of its maximum level by around the age of 70 years (Table 2).

With the aging, endothelial cells and smooth muscle cells, which are key cellular components of the blood vessel and play important roles in cardiovascular health and disease, deeply change. Indeed with the development and progression of CVD, changes occur in both structure and function of these cells resulting in a wide range of abnormalities.

4.1 Role of DHEA on Compromised Endothelium

DHEA and DHEAS have been studied in the context of endothelial function in postmenopausal women with coronary risks factors. Results have shown that lower levels of DHEA are associated with endothelial dysfunction and there is a positive correlation between DHEA and percentage of flow-mediated dilatation, a marker of endothelial function (Akishita et al., 2008). This correlation seems to be independent of other coronary risk factors such as age, body mass index, blood pressure, total and high-density lipoprotein cholesterol, fasting glucose levels, or smoking habits.

Endothelium dysfunction is a landmark of some CVD such as hypertension and atherosclerosis: correction of this dysfunction is one of the goals of the therapeutic strategy for these diseases. Plasma DHEA levels is significantly related to endothelial dysfunction in postmenopausal women and it is not associated with other coronary risk factors.

4.2 Role of DHEA on Compromised Vascular Smooth Muscle Cells

Pulmonary hypertension depends on pulmonary vascular resistance, which requires higher effort by the right ventricle to maintain cardiac output. This causes the failure of the after load-intolerant right ventricle and ultimately it leads to death.

DHEAS is reported to be a pulmonary vasodilator that prevent acute hypoxic pulmonary vasoconstriction by opening potassium channels and upregulating the NO-modulated soluble guanylyl cyclase, both mechanisms leading to relaxation of vascular smooth muscle cells (Mannic et al., 2015).

4.3 Role of DHEA on Cardiovascular Function

A variety of research strategies suggest a beneficial effect of DHEA on cardiovascular functions. Several epidemiological studies have indicated that low plasma levels of DHEA and DHEAS are associated with elevated cardiovascular risk, cardiovascular morbidity, coronary artery disease, or atherosclerotic vascular diseases (Savineau et al., 2013).

In the WISE study, including 270 postmenopausal women undergoing coronary angiography for suspected ischemia, was reported a positive correlation between DHEAS levels and CVD outcome, which was potentially explained by a better atherosclerosis state of patients with higher DHEAS levels (Shufelt et al., 2010).

In order to examine the association between DHEA and atherosclerosis, recent studies evaluated whether DHEA is associated with development of carotid atherosclerosis, a situation predicting cardiovascular complications (Yoshida et al., 2010).

Then, diabetes mellitus is frequently associated with CVD, and significantly lower DHEAS concentrations were observed in 51 diabetic patients with a specific subtype of type 2 DM in comparison with 49 healthy controls (Liu et al., 2013). The putative cardioprotective effect of DHEA has been investigated in 62 male patients with type 2 DM by determining the association between the hormone levels and the activated protein C, C-reactive protein, and the IMT of the carotid artery. Activated protein C is an anticoagulant factor generated by the thrombin–thrombomodulin complex on vascular endothelial cells and imbalance in the coagulation–fibrinolysis system is an important risk factor of vascular events (for example, myocardial infarction). The study concluded that lower circulating levels of DHEA are associated with decreased activated protein C generation and higher IMT in patients type 2 DM (Suzuki et al., 2012).

In conclusion, DHEA levels are a good markers for predicting CVD in the general population but still some concerns need to be raised in the presence of other risk factors. Gender, age, smoking, diabetes could interfere with the plasma concentration of DHEA and there is not a linear association with the risk to develop CVD.

5. CONCLUSION AND FUTURE DIRECTIONS

Over the past 15 years, hormone preparations of DHEA have been available over-the-counter and have been sold as the "fountain of youth." DHEA serves as a precursor for estrogens and androgens from fetal life to postmenopause, and many people believe that DHEA is merely an inactive precursor pool for the formation of bioactive steroid hormones. On the contrary, DHEA acts by itself through dedicated receptors.

Several studies have shown that DHEA exerts pleiotropic effects and reduced levels of DHEA and DHEAS may be associated with different pathologies. A possible role of DHEA in the pathophysiology of diseases such as insulin resistance, cancer, reduction of the immune defense, Alzheimer's disease, depression, or deterioration in the sensation of well-being as well as cardiovascular diseases has been hypothesized. In human vessels, DHEA binds with high affinity to the endothelial cell membrane driving several functions which demonstrated the high activity of DHEA. The large difference between low and high serum DHEA levels has a major clinical impact. Among postmenopausal women with coronary risk factors, lower DHEAS levels were linked with higher mortality from cardiovascular disease and all-cause mortality.

Several studies had previously demonstrated that 1-year treatment (Genazzani et al., 2011; Pluchino et al., 2008), using administration of 10 mg DHEA daily in symptomatic postmenopausal women with lower (5th percentile) baseline DHEAS levels, improved climacteric and sexual symptoms and directly reversed some age-related changes in adrenal enzymatic pathways, including adrenal DHEA and progesterone synthesis.

In addition, before drawing definitive conclusions on DHEA replacement therapy, further aspects need to be better investigated, such as the genetics of DHEA intracrinology and adrenal aging as well their relation with climacteric symptoms (Genazzani & Pluchino, 2015). In human, DHEA supplementation studies suggest a beneficial effect of DHEA on angiographic evidence of atherosclerosis, markers of vascular risk, and vascular function. DHEA exhibits well-documented antiatherogenic and vasculotonic effects in various animal models of atherosclerosis or vascular injury. Indeed, oral administration of DHEA inhibited atherosclerosis and plaque progression in an experimental model. DHEAS also has a beneficial effect on ischemia–reperfusion injury in the microcirculation.

DHEA is a well-tolerated molecule as well as an inexpensive drug. DHEA supplementation might be useful for the clinical management of some types of CVD. Additional large multicentric clinical studies are required to understand the role and clinical efficacy of DHEA in CVD.

REFERENCES

Akishita, M., Hashimoto, M., Ohike, Y., Ogawa, S., Iijima, K., Eto, M., et al. (2008). Association of plasma dehydroepiandrosterone-sulfate levels with endothelial function in postmenopausal women with coronary risk factors. *Hypertension Research*, *31*(1), 69–74.

Alzoubi, A., Toba, M., Abe, K., O'Neill, K. D., Rocic, P., Fagan, K. A., et al. (2013). Dehydroepiandrosterone restores right ventricular structure and function in rats with severe pulmonary arterial hypertension. *American Journal of Physiology. Heart and Circulatory Physiology*, *304*(12), H1708–H1718.

Barrett-Connor, E., & Ferrara, A. (1996). Dehydroepiandrosterone, dehydroepiandrosterone sulfate, obesity, waist-hip ratio, and noninsulin-dependent diabetes in postmenopausal women: The Rancho Bernardo Study. *The Journal of Clinical Endocrinology and Metabolism*, *81*(1), 59–64.

Barrett-Connor, E., Khaw, K. T., & Yen, S. S. (1986). A prospective study of dehydroepiandrosterone sulfate, mortality, and cardiovascular disease. *The New England Journal of Medicine*, *315*(24), 1519–1524.

Belanger, A., Candas, B., Dupont, A., Cusan, L., Diamond, P., Gomez, J. L., et al. (1994). Changes in serum concentrations of conjugated and unconjugated steroids in 40- to 80-year-old men. *The Journal of Clinical Endocrinology and Metabolism*, *79*(4), 1086–1090.

Bertoni, A., Rastoldo, A., Sarasso, C., Di Vito, C., Sampietro, S., Nalin, M., et al. (2012). Dehydroepiandrosterone-sulfate inhibits thrombin-induced platelet aggregation. *Steroids*, *77*(3), 260–268.

Bhuiyan, S., & Fukunaga, K. (2010). Stimulation of Sigma-1 receptor by dehydroepiandrosterone ameliorates hypertension-induced kidney hypertrophy in ovariectomized rats. *Experimental Biology and Medicine (Maywood, N.J.)*, *235*(3), 356–364.

Bonnet, S., Paulin, R., Sutendra, G., Dromparis, P., Roy, M., Watson, K. O., et al. (2009). Dehydroepiandrosterone reverses systemic vascular remodeling through the inhibition of the Akt/GSK3-{beta}/NFAT axis. *Circulation*, *120*(13), 1231–1240.

Chen, J., Xu, L., & Huang, C. (2014). DHEA inhibits vascular remodeling following arterial injury: A possible role in suppression of inflammation and oxidative stress derived from vascular smooth muscle cells. *Molecular and Cellular Biochemistry*, *388*(1–2), 75–84.

Curatola, A. M., Huang, K., & Naftolin, F. (2012). Dehydroepiandrosterone (DHEA) inhibition of monocyte binding by vascular endothelium is associated with sialylation of neural cell adhesion molecule. *Reproductive Sciences*, *19*(1), 86–91.

Dumas de la Roque, E., Savineau, J. P., & Bonnet, S. (2010). Dehydroepiandrosterone: A new treatment for vascular remodeling diseases including pulmonary arterial hypertension. *Pharmacology & Therapeutics*, *126*(2), 186–199.

Formoso, G., Chen, H., Kim, J. A., Montagnani, M., Consoli, A., & Quon, M. J. (2006). Dehydroepiandrosterone mimics acute actions of insulin to stimulate production of both nitric oxide and endothelin 1 via distinct phosphatidylinositol 3-kinase- and mitogen-activated protein kinase-dependent pathways in vascular endothelium. *Molecular Endocrinology*, *20*(5), 1153–1163.

Genazzani, A. R., & Pluchino, N. (2015). DHEA replacement for postmenopausal women: Have we been looking in the right direction? *Climacteric: The Journal of the International Menopause Society*, *18*(5), 669–671.

Genazzani, A. R., Stomati, M., Valentino, V., Pluchino, N., Pot, E., Casarosa, E., et al. (2011). Effect of 1-year, low-dose DHEA therapy on climacteric symptoms and female sexuality. *Climacteric: The Journal of the International Menopause Society*, *14*(6), 661–668.

Gordon, G. B., Bush, D. E., & Weisman, H. F. (1988). Reduction of atherosclerosis by administration of dehydroepiandrosterone. A study in the hypercholesterolemic New Zealand white rabbit with aortic intimal injury. *The Journal of Clinical Investigation*, *82*(2), 712–720.

Gutierrez, G., Mendoza, C., Zapata, E., Montiel, A., Reyes, E., Montano, L. F., et al. (2007). Dehydroepiandrosterone inhibits the TNF-alpha-induced inflammatory response in human umbilical vein endothelial cells. *Atherosclerosis*, *190*(1), 90–99.

Haffner, S. M., Newcomb, P. A., Marcus, P. M., Klein, B. E., & Klein, R. (1995). Relation of sex hormones and dehydroepiandrosterone sulfate (DHEA-SO4) to cardiovascular risk factors in postmenopausal women. *American Journal of Epidemiology*, *142*(9), 925–934.

Hayashi, T., Esaki, T., Muto, E., Kano, H., Asai, Y., Thakur, N. K., et al. (2000). Dehydroepiandrosterone retards atherosclerosis formation through its conversion to estrogen: The possible role of nitric oxide. *Arteriosclerosis, Thrombosis, and Vascular Biology*, *20*(3), 782–792.

Homma, N., Nagaoka, T., Karoor, V., Imamura, M., Taraseviciene-Stewart, L., Walker, L. A., et al. (2008). Involvement of RhoA/Rho kinase signaling in protection against monocrotaline-induced pulmonary hypertension in pneumonectomized rats by dehydroepiandrosterone. *American Journal of Physiology. Lung Cellular and Molecular Physiology*, *295*(1), L71–L78.

Kedziora-Kornatowska, K., Beszczynska-Oles, R., Kornatowski, T., & Szadujkis-Szadurski, L. (2007). The analysis of dehydroepiandrosterone sulphate concentration in elderly age women depending on coexisting disease states. *Advances in Medical Sciences*, *52*(Suppl. 1), 126–130.

Khaw, K. T. (1996). Dehydroepiandrosterone, dehydroepiandrosterone sulphate and cardiovascular disease. *The Journal of Endocrinology*, *150*(Suppl), S149–S153.

Kroboth, P. D., Salek, F. S., Pittenger, A. L., Fabian, T. J., & Frye, R. F. (1999). DHEA and DHEA-S: A review. *Journal of Clinical Pharmacology*, *39*(4), 327–348.

Kushnir, M. M., Blamires, T., Rockwood, A. L., Roberts, W. L., Yue, B., Erdogan, E., et al. (2010). Liquid chromatography-tandem mass spectrometry assay for androstenedione, dehydroepiandrosterone, and testosterone with pediatric and adult reference intervals. *Clinical Chemistry*, *56*(7), 1138–1147.

Labrie, F., Luu-The, V., Labrie, C., & Simard, J. (2001). DHEA and its transformation into androgens and estrogens in peripheral target tissues: Intracrinology. *Frontiers in Neuroendocrinology*, *22*(3), 185–212.

Lasley, B. L., Santoro, N., Randolf, J. F., Gold, E. B., Crawford, S., Weiss, G., et al. (2002). The relationship of circulating dehydroepiandrosterone, testosterone, and estradiol to stages of the menopausal transition and ethnicity. *The Journal of Clinical Endocrinology and Metabolism*, *87*(8), 3760–3767.

Liu, D., & Dillon, J. S. (2004). Dehydroepiandrosterone stimulates nitric oxide release in vascular endothelial cells: Evidence for a cell surface receptor. *Steroids*, *69*(4), 279–289.

Liu, D., Si, H., Reynolds, K. A., Zhen, W., Jia, Z., & Dillon, J. S. (2007). Dehydroepiandrosterone protects vascular endothelial cells against apoptosis through a Galphai protein-dependent activation of phosphatidylinositol 3-kinase/Akt and regulation of antiapoptotic Bcl-2 expression. *Endocrinology*, *148*(7), 3068–3076.

Liu, L., Wang, M., Yang, X., Bi, M., Na, L., Niu, Y., et al. (2013). Fasting serum lipid and dehydroepiandrosterone sulfate as important metabolites for detecting isolated postchallenge diabetes: Serum metabolomics via ultra-high-performance LC-MS. *Clinical Chemistry*, *59*(9), 1338–1348.

Mannic, T., Mouffok, M., Python, M., Yoshida, T., Maturana, A. D., Vuilleumier, N., et al. (2013). DHEA prevents mineralo- and glucocorticoid receptor-induced chronotropic and hypertrophic actions in isolated rat cardiomyocytes. *Endocrinology*, *154*(3), 1271–1281.

Mannic, T., Viguie, J., & Rossier, M. F. (2015). In vivo and in vitro evidences of dehydroepiandrosterone protective role on the cardiovascular system. *International Journal of Endocrinology and Metabolism*, *13*(2), e24660.

Mazat, L., Lafont, S., Berr, C., Debuire, B., Tessier, J. F., Dartigues, J. F., et al. (2001). Prospective measurements of dehydroepiandrosterone sulfate in a cohort of elderly subjects: Relationship to gender, subjective health, smoking habits, and 10-year mortality. *Proceedings of the National Academy of Sciences of the United States of America*, *98*(14), 8145–8150.

Nakamura, S., Yoshimura, M., Nakayama, M., Ito, T., Mizuno, Y., Harada, E., et al. (2004). Possible association of heart failure status with synthetic balance between aldosterone and dehydroepiandrosterone in human heart. *Circulation*, *110*(13), 1787–1793.

Patel, D., Kandhi, S., Kelly, M., Neo, B. H., & Wolin, M. S. (2014). Dehydroepiandrosterone promotes pulmonary artery relaxation by NADPH oxidation-elicited subunit dimerization of protein kinase G 1alpha. *American Journal of Physiology. Lung Cellular and Molecular Physiology*, *306*(4), L383–L391.

Paulin, R., Meloche, J., Jacob, M. H., Bisserier, M., Courboulin, A., & Bonnet, S. (2011). Dehydroepiandrosterone inhibits the Src/STAT3 constitutive activation in pulmonary arterial hypertension. *American Journal of Physiology. Heart and Circulatory Physiology*, *301*(5), H1798–H1809.

Pitt, B., Remme, W., Zannad, F., Neaton, J., Martinez, F., Roniker, B., et al. (2003). Eplerenone, a selective aldosterone blocker, in patients with left ventricular dysfunction after myocardial infarction. *The New England Journal of Medicine*, *348*(14), 1309–1321.

Pluchino, N., Ninni, F., Stomati, M., Freschi, L., Casarosa, E., Valentino, V., et al. (2008). One-year therapy with 10 mg/day DHEA alone or in combination with HRT in postmenopausal women: Effects on hormonal milieu. *Maturitas*, *59*(4), 293–303.

Rutkowski, K., Sowa, P., Rutkowska-Talipska, J., Kuryliszyn-Moskal, A., & Rutkowski, R. (2014). Dehydroepiandrosterone (DHEA): Hypes and hopes. *Drugs*, *74*(11), 1195–1207.

Savineau, J. P., Marthan, R., & Dumas de la Roque, E. (2013). Role of DHEA in cardiovascular diseases. *Biochemical Pharmacology*, *85*(6), 718–726.

Shufelt, C., Bretsky, P., Almeida, C. M., Johnson, B. D., Shaw, L. J., Azziz, R., et al. (2010). DHEA-S levels and cardiovascular disease mortality in postmenopausal women: Results from the National Institutes of Health—National Heart, Lung, and Blood Institute (NHLBI)-sponsored Women's ischemia syndrome evaluation (WISE). *The Journal of Clinical Endocrinology and Metabolism*, *95*(11), 4985–4992.

Simoncini, T., Mannella, P., Fornari, L., Varone, G., Caruso, A., & Genazzani, A. R. (2003). Dehydroepiandrosterone modulates endothelial nitric oxide synthesis via direct genomic and nongenomic mechanisms. *Endocrinology*, *144*(8), 3449–3455.

Stomati, M., Monteleone, P., Casarosa, E., Quirici, B., Puccetti, S., Bernardi, F., et al. (2000). Six-month oral dehydroepiandrosterone supplementation in early and late postmenopause. *Gynecological Endocrinology: The Official Journal of the International Society of Gynecological Endocrinology*, *14*(5), 342–363.

Su, T. P., Su, T. C., Nakamura, Y., & Tsai, S. Y. (2016). The Sigma-1 receptor as a pluripotent modulator in living systems. *Trends in Pharmacological Sciences*, *37*(4), 262–278.

Suzuki, T., Yano, Y., Sakamoto, M., Uemura, M., Yasuma, T., Onishi, Y., et al. (2012). Correlation of circulating dehydroepiandrosterone with activated protein C generation and carotid intima-media thickness in male patients with type 2 diabetes. *Diabetic Medicine*, *29*(7), e41–e46.

Tagashira, H., Bhuiyan, S., Shioda, N., & Fukunaga, K. (2011). Distinct cardioprotective effects of 17beta-estradiol and dehydroepiandrosterone on pressure overload-induced hypertrophy in ovariectomized female rats. *Menopause*, *18*(12), 1317–1326.

Thijs, L., Fagard, R., Forette, F., Nawrot, T., & Staessen, J. A. (2003). Are low dehydroepiandrosterone sulphate levels predictive for cardiovascular diseases? A review of prospective and retrospective studies. *Acta Cardiologica*, *58*(5), 403–410.

Webb, S. J., Geoghegan, T. E., Prough, R. A., & Michael Miller, K. K. (2006). The biological actions of dehydroepiandrosterone involves multiple receptors. *Drug Metabolism Reviews*, *38*(1–2), 89–116.

Weiss, E. P., Villareal, D. T., Ehsani, A. A., Fontana, L., & Holloszy, J. O. (2012). Dehydroepiandrosterone replacement therapy in older adults improves indices of arterial stiffness. *Aging Cell*, *11*(5), 876–884.

WHO: World Health Organization Report. (2017). 2017. http://www.who.int/news-room/fact-sheets/detail/cardiovascular-diseases-(cvds).

Williams, M. R., Ling, S., Dawood, T., Hashimura, K., Dai, A., Li, H., et al. (2002). Dehydroepiandrosterone inhibits human vascular smooth muscle cell proliferation independent of ARs and ERs. *The Journal of Clinical Endocrinology and Metabolism*, *87*(1), 176–181.

Yoshida, S., Aihara, K., Azuma, H., Uemoto, R., Sumitomo-Ueda, Y., Yagi, S., et al. (2010). Dehydroepiandrosterone sulfate is inversely associated with sex-dependent diverse carotid atherosclerosis regardless of endothelial function. *Atherosclerosis*, *212*(1), 310–315.

Zubairi, A. B., Idrees, F., Jabeen, K., Kamal, S., & Zafar, A. (2017). Coinfection with Lichtheimia corymbifera and Aspergillus flavus in an immune-competent patient mimicking as pulmonary-renal syndrome. *Mycopathologia*, *182*, 727–731.

CHAPTER THIRTEEN

Effect of Dehydroepiandrosterone (DHEA) on Diabetes Mellitus and Obesity

Kazutaka Aoki*,†,1, Yasuo Terauchi†,1
*Internal Medicine, Kanagawa Dental University, Yokosuka, Japan
†Yokohama City University Graduate School of Medicine, Yokohama, Japan
[1]Corresponding authors: e-mail address: k.aoki@kdu.ac.jp; terauchi@yokohama-cu.ac.jp

Contents

Abstract

Type 2 diabetes is a metabolic disorder that is characterized by an impaired capacity to secrete insulin, insulin resistance, or both. Dehydroepiandrosterone (DHEA), a steroid hormone produced by the adrenal cortex, has been reported to have beneficial effects on diabetes mellitus and obesity in animal models. DHEA and DHEA-sulfate (DHEA-S) have been reported to increase not only insulin secretion of the pancreas but also insulin sensitivity of the liver, adipose tissue, and muscle. We investigated the effects of DHEA on glucose metabolism in animal models and reported decrease of liver gluconeogenesis. Recently, we reported the effect of DHEA on the liver and muscle by using insulin-stimulated insulin receptor substrate 1 and 2 (IRS1 and IRS2)-deficient mice. DHEA increased Akt phosphorylation in the liver of C57BL6 IRS1- and IRS2-deficient mice fed with a high-fat diet (HFD), which suggests that the increase in DHEA-induced Akt signaling is sufficient in the presence of IRS1 or IRS2. In addition, other studies have also reported the effect of DHEA on diabetes mellitus in the liver, muscle, adipose tissue, and pancreatic β-cell and its effect on obesity in animal models.

Vitamins and Hormones, Volume 108
ISSN 0083-6729
https://doi.org/10.1016/bs.vh.2018.01.008

A meta-analysis in elderly men and women has found that DHEA supplementation has no effects on blood glucose levels. However, DHEA supplementation to patients with type 2 diabetes has not been fully elucidated. Therefore, further studies are needed to provide greater insight into the effect of DHEA on diabetes and obesity in animal and human models.

1. INTRODUCTION

Dehydroepiandrosterone (DHEA), a steroid hormone produced by the adrenal cortex, is synthesized from pregnenolone and is further metabolized to androstenedione, testosterone, and estrogens. DHEA and its sulfate (DHEA-S) are secreted mainly from the adrenal cortex and are the most abundant adrenal steroids in humans. DHEA and DHEA-S levels peak at one's twenties and gradually decrease with age, indicating a possible association with "antiaging." In the Baltimore Longitudinal Study of Aging, longevity was associated with high DHEA-S concentration (Roth et al., 2002). DHEA has also been reported to have beneficial effects against diseases such as diabetes mellitus, obesity, atherosclerosis, osteoporosis, and collagen disease in animal models (Coleman, Leiter, & Schwizer, 1982; Coleman, Schwizer, & Leiter, 1984; Wang, Wang, Li, & Wang, 2006; Wang, Wang, Wang, Zhu, & Li, 2007; Yamakawa et al., 2009). However, the physiological role of DHEA and its mechanism has not been clearly defined to date.

According to the International Diabetes Federation (IDF) Diabetes Atlas (7th edition), there are 415 million patients with diabetes in the world, which shows that 1 in 11 adults have diabetes. Moreover, by 2040, 1 adult in 10 (642 million) will have diabetes (http://www.diabetesatlas.org/key-messages.html#sthash.JrAYwyI3.dpuf). Since the number of patients with diabetes has increased worldwide, investigating the mechanism of the onset of diabetes and its novel therapeutic strategies is very important.

In this chapter, we introduce the effect of DHEA in animal and human models on diabetes and obesity. We also discuss whether the effect of DHEA is exerted by DHEA itself or its metabolites.

2. EFFECTS OF DHEA ON DIABETES MELLITUS

Type 2 diabetes is characterized by an impaired capacity to secrete insulin, insulin resistance, or both. We analyzed the results of studies on the effects of DHEA on diabetes mellitus in animal and human studies.

2.1 Animal Studies

Obese, hyperglycemic, and hyperinsulinemic (db/db) mice were used as an animal model for type 2 diabetes. In 1982, Coleman et al. reported that administration of DHEA to db/db mice improved hyperglycemia, increased insulin sensitivity by oral glucose tolerance test (OGTT), and preserved function and β-cell structure (Coleman et al., 1982). As shown in Fig. 1, other reports also indicated that DHEA and DHEA-S increase not only insulin secretion but also insulin sensitivity in animal models (Dillon et al., 2000; Han, Hansen, Chen, & Holloszy, 1998; Ishizuka et al., 1999; Kimura et al., 1998; Yue et al., 2013). Here, the effects of DHEA on the liver, muscle, adipose tissue, and insulin secretion have been described.

2.1.1 Effect of DHEA on the Liver

McIntosh et al. reported that DHEA administration decreases hepatic glucose production in isolated hepatocytes from prediabetic male BHE/cdb rats (McIntosh & Berdanier, 1991). As no reports have investigated the antidiabetic effect of DHEA in db/db mice since that of Coleman et al.'s described above, in 1999, we investigated the effect of DHEA on glucose metabolism in the liver and muscle of db/db mice. The activity of gluconeogenic enzyme glucose-6-phosphatase (G6Pase) of db/db mice was elevated compared to its heterogeneous db/+m mice. The administration of DHEA suppresses the increased activity of G6Pase in db/db mice (Aoki et al., 1999) and the increased mRNA expression of the hepatic G6Pase and hepatic glucose production in db/db mice, compared to control db/+m mice (Aoki et al., 2000, 2004). The addition of DHEA or DHEA-S to the primary hepatocytes in db/db mice resulted in

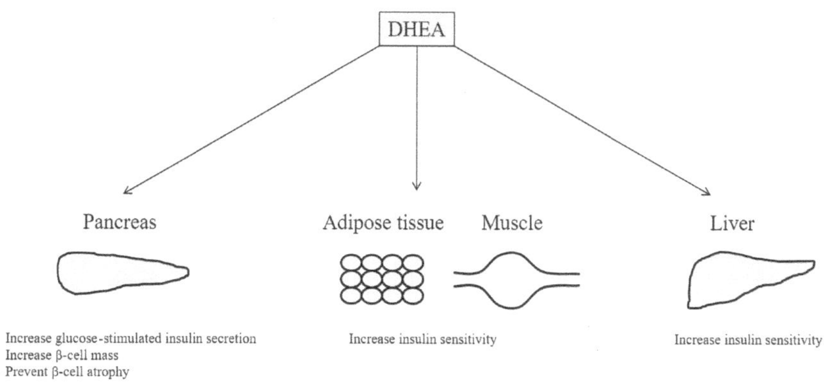

Fig. 1 Effect of DHEA on diabetes mellitus in animal models.

decreased gluconeogenesis. Additionally, DHEA suppresses the activity, protein expression, and gene expression of G6Pase and enhances 2-deoxyglucose uptake in HepG2 cells (Yamashita et al., 2005). Therefore, administration of DHEA is considered to decrease gluconeogenesis in the liver.

We investigated insulin signaling and glucose metabolism in the liver in the phosphoinositide 3 kinase (PI3K) p85 $\alpha^{-/-}$ mice (Aoki et al., 2009). Jacob et al. reported that DHEA administration increases pAkt/Akt in the liver of rats (Jacob et al., 2011). According to Campbell et al., the PI3K/Akt pathway is activated in the liver, and DHEA increases the tyrosine phosphorylation of insulin-induced insulin receptor substrates 1 and 2 (IRS1 and IRS2) in the liver of Wistar rats (Campbell et al., 2004). We also investigated the changes in Akt phosphorylation in the liver using C57BL6, IRS1$^{-/-}$, and IRS2$^{-/-}$ mice fed with a high-fat diet (HFD) (Aoki, Tajima, Taguri, & Terauchi, 2016). C57BL6, IRS1$^{-/-}$, and IRS2$^{-/-}$ mice fed with a HFD containing DHEA for 4 weeks showed increased Akt phosphorylation. This result suggested that the increase in Akt signaling induced by DHEA is sufficient in the presence of IRS1 or IRS2. We previously reported that DHEA suppressed the activity and mRNA expression of G6Pase in the liver (Aoki et al., 2000, 2004; Yamashita et al., 2005). As the deletion of Akt2 increases gluconeogenesis (Cho et al., 2001), the administration of DHEA may suppress gluconeogenesis by increasing Akt phosphorylation. Similarly, Kang et al. reported the effect of DHEA on the liver in rats treated with HFD (Kang, Ge, Yu, Li, & Ma, 2016). DHEA supplementation for 8 weeks increased hepatic PI3K and Akt mRNA expressions, hepatic glycolytic enzyme PFK-2 activity, and hepatic glycogen content. DHEA may activate the PI3K/Akt-PFK-2 signal pathway.

2.1.2 Effect of DHEA on Muscle and Adipose Tissue

In our laboratory, Mukasa et al. observed that DHEA ameliorates insulin sensitivity in older rats by using the glucose clamp test (Mukasa et al., 1998). Kimura et al. also showed that DHEA treatment for 10 days decreases serum tumor necrosis factor-alpha and restores insulin sensitivity in genetically obese Zucker rats, using glucose clamp test (Kimura et al., 1998). This effect was independent of secondary weight reduction by DHEA. Administration of DHEA increased the glycolytic enzyme hexokinase (HK) + glucokinase (GK) and phosphofructokinase (PFK) activities in the muscle of db/db mice (Aoki et al., 1999).

Several reports existed to study insulin signaling in animal models. A single bout of DHEA injection relieved the impaired activation of Akt and the protein kinase C ζ/λ (PKCζ/λ)-GLUT4 pathway in the skeletal muscle of streptozotocin-induced diabetic rats as a model of type 1 diabetes (Sato et al., 2012). Jahn et al. reported that DHEA administration for 5 weeks decreases Akt levels in the muscle of Wistar rats, but has no effect on Akt phosphorylation in the muscle of streptozotocin-induced diabetic rats (Sato, Iemitsu, Aizawa, & Ajisaka, 2008). A study by Campbell et al. reported that administering DHEA for a week increases PKCζ/λ phosphorylation in the muscle, though it does not increase the activity of the Akt phosphorylation, and increases the insulin-induced tyrosine phosphorylation of only IRS1 in the muscle of rats (Aoki et al., 2016). DHEA activated HK and PFK activity, phosphorylation of Akt and PKC ζ/λ, and GLUT4 protein expression in cultured skeletal muscle cells from SD rats (Jahn et al., 2010). In our study, the administration of DHEA for 4 weeks did not increase Akt and PKCζ phosphorylation in the muscle of C57BL6 $IRS1^{-/-}$ and $IRS2^{-/-}$ mice fed with HFD. The results also showed that PKCζ phosphorylation was reduced in $IRS1^{-/-}$ mice as compared to $IRS2^{-/-}$ mice. Therefore, long-term administration of DHEA alone may not increase Akt signaling in the muscle. The role of IRS2 in the muscle might be less important for the effect of DHEA on glucose metabolism. However, more investigations would be required to confirm this finding.

Concerning adipose tissue, DHEA has been shown to modulate insulin signaling pathways by increasing tyrosine phosphorylation of IRS1 and IRS2 and stimulate IRS1- and IRS2-associated PI3K activity in 3T3-L1 adipocytes (Perrini et al., 2004). However, no effect on either insulin receptor or Akt phosphorylation was found. Activation of PI3K and PKCζ in adipocytes was significantly elevated in DHEA-treated Otsuka Long-Evans Tokushima Fatty (OLETF) rats (Ishizuka et al., 2007).

2.1.3 Effect of DHEA on Insulin Secretion

DHEA is reported to increase glucose-stimulated insulin secretion when administered in vivo to rats or in vitro to pancreatic β-cell lines (Dillon et al., 2000). Pancreatic islets from DHEA-treated rats showed an increased β-cell mass accompanied by increased Akt1 protein level but reduced insulin receptor (IR), IRS1, IRS2 levels and enhanced glucose-stimulated insulin secretion (Medina et al., 2006). Coleman et al. reported that the administration of DHEA to db/db mice preserved the β-cell function and structure

(Coleman et al., 1982). In our previous study, the plasma insulin levels have been observed to be increased in DHEA-treated db/db mice at 15 weeks of age (DHEA was administered from 13 weeks) compared to control db/db mice (Aoki et al., 1999). Thus, DHEA has beneficial effects on the β-cell function in animal models.

3. HUMAN STUDIES

Regarding the relationship between endogenous DHEA and diabetes, one study reported that urine DHEA levels were low or not detected in obese patients with diabetes in 1964 (Sonka, Gregorva, Pav, & Skrha, 1964). In our laboratory, Yamaguchi et al. also reported that patients with type 2 diabetes whose serum insulin levels were high had significantly lower DHEA and DHEA-S serum levels than normal subjects and patients with type 2 diabetes whose serum insulin levels were not high (Yamaguchi et al., 1998).

Regarding the administration of DHEA in human studies, there were two results of meta-analysis of DHEA supplementation to elderly men and postmenopausal women. In elderly men, DHEA supplementation reduced fat mass (Corona et al., 2013). However, DHEA had no effects on glycaemia, insulin, total cholesterol (TC), and bone mineral density (BMD) compared to a placebo group. This effect is dependent on DHEA conversion into its metabolites such as androgens or estrogens. Postmenopausal women supplemented with DHEA showed no effects on TC, high-density lipoprotein cholesterol (HDL-C), low-density lipoprotein cholesterol (LDL), triglycerides, serum glucose, weight, body mass index, or BMD (Elraiyah et al., 2014). From these results, it could be concluded that DHEA supplementation has no effect on glucose metabolism in elderly men and postmenopausal women. However, DHEA supplementation to patients with diabetes and its effects to insulin sensitivity were not fully evaluated.

The change of insulin sensitivity by DHEA supplementation varies between reports. In the patients without diabetes, the administration of DHEA (25 mg/day for 12 weeks) did not decrease plasma glucose; however, it decreases steady-state plasma glucose using octreotide acetate (Kawano et al., 2003). In another study, 50 mg/day of DHEA administered to elderly subjects did not change the area under the curve (AUC) of glucose, as measured using an oral glucose tolerance test, and decreased the AUC of insulin using OGTT after 6 months (Villareal & Holloszy, 2004). These results indicate that DHEA supplementation improved insulin sensitivity. However, in a different study, 50 mg/day of DHEA supplementation for elderly subjects with low serum DHEA-S concentrations did not change AUC of glucose and insulin using OGTT after 12 months (Jankowski et al., 2011). In our laboratory, Yamada et al. reported the effect of 25 mg of DHEA intake for 2 weeks in 22 subjects without diabetes (Yamada et al., 2007). According to this study, homeostatic model assessment-IR (HOMA-IR) of all three subjects with >2.0 decreased after DHEA treatment.

We found a report on DHEA administration to patients with type 2 diabetes. In the randomized controlled trial (RCT) of patients with type 2 diabetes, whose mean BMI was 25, 20 patients were divided into DHEA ($n=10$) and placebo group ($n=10$). DHEA supplementation (50 mg/day for 12 weeks) did not change plasma glucose, HbA1c, HOMA index, or BMI (Brignardello et al., 2007). DHEA supplementation ameliorated oxidative imbalance and prevented advanced glycation end product formation. The effect of DHEA on insulin sensitivity on 30 patients with impaired glucose tolerance (IGT) was reported (Talaei, Amini, Siavash, & Zare, 2010). DHEA supplementation of 50 mg administered to 30 IGT patients for 6 months (crossover study) did not reduced, but they tend to improve, the HOMA-IR.

Thus, DHEA supplementation for patients with type 2 diabetes has not been completely elucidated. Thereby, further studies of patients with type 2 diabetes are necessary.

3.1 Effects of DHEA on Obesity

As described earlier, DHEA supplementation decreased the fat mass in elderly men in a meta-analysis (Corona et al., 2013). DHEA helps to reduce the body weight gain in pair-fed obese rats (Kimura et al., 1998). We also found that DHEA administration decreased the body weight gain in C57BL6, IRS1$^{-/-}$, and IRS2$^{-/-}$ mice. The antiobesity effects of DHEA may be mediated by futile substrate cycling in hepatocytes, as reported previously (McIntosh & Berdanier, 1991).

Ashida et al. identified that DHEA-induced dual specificity protein (DDSP) and that this protein regulated 38 mitogen-activated protein kinases (MAPKs) (Ashida et al., 2005). Indeed, using DDSP-Tg mice showed that

DDSP prevents diet-induced obesity and genetic obesity (db/db mice); the antiobesity effects of DHEA might be mediated through DDSP (Watanabe et al., 2015).

4. IS DHEA ITSELF OR ITS METABOLITES EFFECTIVE?

DHEA is synthesized from pregnenolone and is further metabolized to androstenedione, testosterone, and estrogens. Therefore, the action of DHEA is considered an effect of DHEA itself or of its metabolites. The receptor of DHEA itself is not known. As described earlier, the effect of DHEA on reduction of fat mass is considered to be due to its metabolites (Corona et al., 2013). Dihydrotestosterone (DHT) converted from DHEA may increase phosphorylation of Akt and PKC ζ/λ in cultured skeletal muscle cells from Sprague–Dawley (SD) rats, using DHT inhibitor (Sato et al., 2008).

To determine whether DHEA or its metabolites are responsible for improving hyperglycemia, we administered androstenedione to C57BL6 mice fed with HFD, which is considered to be metabolized from DHEA in vivo. Androstenedione did not increase Akt phosphorylation in the liver of C57BL6 mice fed with HFD, which suggests that the increase in Akt signaling in the liver was due to DHEA or DHEA-S (Aoki et al., 2016). The administration of androstenedione to db/db mice slightly lowered blood glucose levels as compared to the administration of DHEA to db/db mice. This also suggests that the hypoglycemic effect of DHEA is due to DHEA or DHEA-S (Aoki et al., 1999). Therefore, DHEA may exert its effect through both DHEA/DHEA-S itself and its metabolites.

5. CONCLUSION AND FUTURE DIRECTIONS

We have described the effects of DHEA on diabetes and obesity in animal and human studies. A meta-analysis of DHEA supplementation in elderly men showed an effect of fat mass reduction, while in postmenopausal women, DHEA showed no effect on metabolic markers. However, few reports have investigated DHEA supplementation in type 2 diabetes. Thereby, further studies are necessary to investigate the effects of DHEA on diabetes and obesity in human and animal models and whether DHEA itself or its metabolites produce these effects.

ACKNOWLEDGMENTS

This work was supported in part by Grants-in-Aid for Scientific Research (C) 16K09806, (B) 21390282, and (B) 24390235 from the Ministry of Education, Culture, Sports, Science and Technology (MEXT) of Japan, and a Medical Award from the Japan Medical Association.

REFERENCES

Aoki, K., Kikuchi, T., Mukasa, K., Ito, S., Nakajima, A., Satoh, S., et al. (2000). Dehydroepiandrosterone suppresses elevated hepatic glucose-6-phosphatase mRNA level in C57BL/KsJ-db/db mice: Comparison with troglitazone. *Endocrine Journal*, *47*, 799–804.

Aoki, K., Matsui, J., Kubota, N., Nakajima, H., Iwamoto, K., Takamoto, I., et al. (2009). Role of the liver in glucose homeostasis in PI 3-kinase p85alpha-deficient mice. *American Journal of Physiology. Endocrinology and Metabolism*, *296*, E842–853.

Aoki, K., Saito, T., Satoh, S., Mukasa, K., Kaneshiro, M., Kawasaki, S., et al. (1999). Dehydroepiandrosterone suppresses the elevated hepatic glucose-6-phosphatase and fructose-1,6-bisphosphatase activities in C57BL/Ksj-db/db mice: Comparison with troglitazone. *Diabetes*, *48*, 1579–1585.

Aoki, K., Tajima, K., Taguri, M., & Terauchi, Y. (2016). Effect of dehydroepiandrosterone (DHEA) on Akt and protein kinase C zeta (PKCζ) phosphorylation in different tissues of C57BL6, insulin receptor substrate (IRS)$1^{-/-}$, and $IRS2^{-/-}$ male mice fed a high-fat diet. *The Journal of Steroid Biochemistry and Molecular Biology*, *159*, 110–120.

Aoki, K., Taniguchi, H., Ito, Y., Satoh, S., Nakamura, S., Muramatsu, K., et al. (2004). Dehydroepiandrosterone decreases elevated hepatic glucose production in C57BL/KsJ-db/db mice. *Life Sciences*, *74*, 3075–3084.

Ashida, K., Goto, K., Zhao, Y., Okabe, T., Yanase, T., Takayanagi, R., et al. (2005). Dehydroepiandrosterone negatively regulates the p38 mitogen-activated protein kinase pathway by a novel mitogen-activated protein kinase phosphatase. *Biochimica et Biophysica Acta*, *1728*, 84–94.

Brignardello, E., Runzo, C., Aragno, M., Catalano, M. G., Cassader, M., Perin, P. C., et al. (2007). Dehydroepiandrosterone administration counteracts oxidative imbalance and advanced glycation end product formation in type 2 diabetic patients. *Diabetes Care*, *30*, 2922–2927.

Campbell, C. S., Caperuto, L. C., Hirata, A. E., Araujo, E. P., Velloso, L. A., Saad, M. J., et al. (2004). The phosphatidylinositol/AKT/atypical PKC pathway is involved in the improved insulin sensitivity by DHEA in muscle and liver of rats in vivo. *Life Sciences*, *76*, 57–70.

Cho, H., Mu, J., Kim, J. K., Thorvaldsen, J. L., Chu, Q., Crenshaw, E. B., 3rd, et al. (2001). Insulin resistance and a diabetes mellitus-like syndrome in mice lacking the protein kinase Akt2 (PKB beta). *Science*, *292*, 1728–1731.

Coleman, D. L., Leiter, E. H., & Schwizer, R. W. (1982). Therapeutic effects of dehydroepiandrosterone (DHEA) in diabetic mice. *Diabetes*, *31*, 830–833.

Coleman, D. L., Schwizer, R. W., & Leiter, E. H. (1984). Effect of genetic background on the therapeutic effects of dehydroepiandrosterone (DHEA) in diabetes-obesity mutants and in aged normal mice. *Diabetes*, *33*, 26–32.

Corona, G., Rastrelli, G., Giagulli, V. A., Sila, A., Sforza, A., Forti, G., et al. (2013). Dehydroepiandrosterone supplementation in elderly men: A meta-analysis study of placebo-controlled trials. *The Journal of Clinical Endocrinology and Metabolism*, *98*, 3615–3626.

Dillon, J. S., Yaney, G. C., Zhou, Y., Voilley, N., Bowen, S., Chipkin, S., et al. (2000). Dehydroepiandrosterone sulfate and beta-cell function: Enhanced glucose-induced insulin secretion and altered gene expression in rodent pancreatic beta-cells. *Diabetes*, *49*, 2012–2020.

Han, D. H., Hansen, P. A., Chen, M. M., & Holloszy, J. O. (1998). DHEA treatment reduces fat accumulation and protects against insulin resistance in male rats. *The Journals of Gerontology. Series A, Biological Sciences and Medical Sciences, 53*, B19–24.

Ishizuka, T., Kajita, K., Miura, A., Ishizawa, M., Kanoh, Y., Itaya, S., et al. (1999). DHEA improves glucose uptake via activations of protein kinase C and phosphatidylinositol 3-kinase. *The American Journal of Physiology, 276*, E196–204.

Ishizuka, T., Miura, A., Kajita, K., Matsumoto, M., Sugiyama, C., Matsubara, K., et al. (2007). Effect of dehydroepiandrosterone on insulin sensitivity in Otsuka Long-Evans Tokushima-fatty rats. *Acta Diabetologica, 44*, 219–226.

Jacob, M. H., Janner Dda, R., Araújo, A. S., Jahn, M. P., Kucharski, L. C., Moraes, T. B., et al. (2011). Dehydroepiandrosterone improves hepatic antioxidant reserve and stimulates Akt signaling in young and old rats. *The Journal of Steroid Biochemistry and Molecular Biology, 127*, 331–336.

Jahn, M. P., Jacob, M. H., Gomes, L. F., Duarte, R., Araújo, A. S., Belló-Klein, A., et al. (2010). The effect of long-term DHEA treatment on glucose metabolism, hydrogen peroxide and thioredoxin levels in the skeletal muscle of diabetic rats. *The Journal of Steroid Biochemistry and Molecular Biology, 120*, 38–44.

Jankowski, C. M., Gozansky, W. S., Van Pelt, R. E., Wolfe, P., Schwartz, R. S., & Kohrt, W. M. (2011). Oral dehydroepiandrosterone replacement in older adults: Effects on central adiposity, glucose metabolism and blood lipids. *Clinical Endocrinology, 75*, 456–463.

Kang, J., Ge, C., Yu, L., Li, L., & Ma, H. (2016). Long-Term Administration of dehydroepiandrosterone accelerates glucose catabolism via activation of PI3K/Akt-PFK-2 signaling pathway in rats fed a high-fat diet. *PLoS One, 11*(7), e0159077.

Kawano, H., Yasue, H., Kitagawa, A., Hirai, N., Yoshida, T., Soejima, H., et al. (2003). Dehydroepiandrosterone supplementation improves endothelial function and insulin sensitivity in men. *The Journal of Clinical Endocrinology and Metabolism, 88*, 3190–3195.

Kimura, M., Tanaka, S., Yamada, Y., Kiuchi, Y., Yamakawa, T., & Sekihara, H. (1998). Dehydroepiandrosterone decreases serum tumor necrosis factor-alpha and restores insulin sensitivity: Independent effect from secondary weight reduction in genetically obese Zucker fatty rats. *Endocrinology, 139*, 3249–3253.

McIntosh, M. K., & Berdanier, C. D. (1991). Antiobesity effects of dehydroepiandrosterone are mediated by futile substrate cycling in hepatocytes of BHE/cdb rats. *The Journal of Nutrition, 121*, 2037–2043.

Medina, M. C., Souza, L. C., Caperuto, L. C., Anhê, G. F., Amanso, A. M., Teixeira, V. P., et al. (2006). Dehydroepiandrosterone increases beta-cell mass and improves the glucose-induced insulin secretion by pancreatic islets from aged rats. *FEBS Letters, 580*(1), 285–290.

Mukasa, K., Kanesiro, M., Aoki, K., Okamura, J., Saito, T., Satoh, S., et al. (1998). Dehydroepiandrosterone (DHEA) ameliorates the insulin sensitivity in older rats. *The Journal of Steroid Biochemistry and Molecular Biology, 67*, 355–358.

Perrini, S., Natalicchio, A., Laviola, L., Belsanti, G., Montrone, C., Cignarelli, A., et al. (2004). Dehydroepiandrosterone stimulates glucose uptake in human and murine adipocytes by inducing GLUT1 and GLUT4 translocation to the plasma membrane. *Diabetes, 53*, 41–52.

Roth, G. S., Lane, M. A., Ingram, D. K., Mattison, J. A., Elahi, D., Tobin, J. D., et al. (2002). Biomarkers of caloric restriction may predict longevity in humans. *Science, 297*, 811.

Sato, K., Iemitsu, M., Aizawa, K., & Ajisaka, R. (2008). Testosterone and DHEA activate the glucose metabolism-related signaling pathway in skeletal muscle. *American Journal of Physiology. Endocrinology and Metabolism, 294*, E961–968.

Sato, K., Iemitsu, M., Aizawa, K., Mesaki, N., Ajisaka, R., & Fujita, S. (2012). DHEA administration and exercise training improves insulin resistance in obese rats. *Nutrition & Metabolism (London), 9*, 47.

Sonka, J., Gregorva, I., Pav, J., & Skrha, F. (1964). Dehydroepiandrosterone in obese diabetics. *Lancet*, *2*, 44–45.
Elraiyah, T., Sonbol, M. B., Wang, Z., Khairalseed, T., Asi, N., Undavalli, C., et al. (2014). The benefits and harms of systemic dehydroepiandrosterone (DHEA) in postmenopausal women with normal adrenal function: A systematic review and meta-analysis. *The Journal of Clinical Endocrinology and Metabolism*, *99*(10), 3536–3542.
Talaei, A., Amini, M., Siavash, M., & Zare, M. (2010). The effect of dehydroepiandrosterone on insulin resistance in patients with impaired glucose tolerance. *Hormones (Athens, Greece)*, *9*, 326–331.
Villareal, D. T., & Holloszy, J. O. (2004). Effect of DHEA on abdominal fat and insulin action in elderly women and men: A randomized controlled trial. *Journal of the American Medical Association*, *292*, 2243–2248.
Wang, Y. D., Wang, L., Li, D. J., & Wang, W. J. (2006). Dehydroepiandrosterone inhibited the bone resorption through the upregulation of OPG/RANKL. *Cellular & Molecular Immunology*, *3*, 41–45.
Wang, L., Wang, Y. D., Wang, W. J., Zhu, Y., & Li, D. J. (2007). Dehydroepiandrosterone improves murine osteoblast growth and bone tissue morphometry via mitogen-activated protein kinase signaling pathway independent of either androgen receptor or estrogen receptor. *Journal of Molecular Endocrinology*, *38*, 467–479.
Watanabe, T., Ashida, K., Goto, K., Nawata, H., Takayanagi, R., Yanase, T., et al. (2015). Dehydroepiandrosterone-enhanced dual specificity protein phosphatase (DDSP) prevents diet-induced and genetic obesity. *Biochemical and Biophysical Research Communications*, *468*, 196–201.
Yamada, Y., Sekihara, H., Omura, M., Yanase, T., Takayanagi, R., Mune, T., et al. (2007). Changes in serum sex hormone profiles after short-term low-dose administration of dehydroepiandrosterone (DHEA) to young and elderly persons. *Endocrine Journal*, *54*, 153–162.
Yamaguchi, Y., Tanaka, S., Yamakawa, T., Kimura, M., Ukawa, K., Yamada, Y., et al. (1998). Reduced serum dehydroepiandrosterone levels in diabetic patients with hyperinsulinaemia. *Clinical Endocrinology*, *49*, 377–383.
Yamakawa, T., Ogihara, K., Nakamura, M., Utsunomiya, H., Kadonosono, K., Kishikawa, S., et al. (2009). Effect of dehydroepiandrosterone on atherosclerosis in apolipoprotein E-deficient mice. *Journal of Atherosclerosis and Thrombosis*, *16*, 501–508.
Yamashita, R., Saito, T., Satoh, S., Aoki, K., Kaburagi, Y., & Sekihara, H. (2005). Effects of dehydroepiandrosterone on gluconeogenic enzymes and glucose uptake in human hepatoma cell line, HepG2. *Endocrine Journal*, *52*, 727–733.
Yue, J., Wang, L., Huang, R., Li, S., Ma, J., Teng, X., et al. (2013). Dehydroepiandrosterone-sulfate (DHEAS) promotes MIN6 cells insulin secretion via inhibition of AMP-activated protein kinase. *Biochemical and Biophysical Research Communications*, *440*, 756–761.

FURTHER READING

Weiss, E. P., Villareal, D. T., Ehsani, A. A., Fontana, L., & Holloszy, J. O. (2012). Dehydroepiandrosterone replacement therapy in older adults improves indices of arterial stiffness. *Aging Cell*, *11*(5), 876–884.

CHAPTER FOURTEEN

Dehydroepiandrosterone and Experimental Osteoarthritis

Kai Huang*[,1], Li-dong Wu[†], Jia-peng Bao[†]
*Tongde Hospital of Zhejiang Province, Health Bureau of Zhejiang Province, Hangzhou, PR China
[†]The Second Hospital of Medical College, Zhejiang University, Hangzhou, PR China
[1]Corresponding author: e-mail address: hzhuangk@163.com

Contents

Abstract

Despite an increased understanding of the pathogenesis of osteoarthritis (OA) and the availability of a number of drugs designed to ameliorate its symptoms, a successful disease-modifying therapy remains elusive. Recent lines of evidence suggest that dehydroepiandrosterone (DHEA), a 19-carbon steroid hormone classified as an adrenal androgen, exerts a chondroprotective effect in OA patients, and it has been proven to be an effective DMOAD candidate that slows OA progression. However, the exact mechanisms underlying its anti-OA effect is largely unknown. This review summarizes emerging observations from studies of cell biology, preclinical animal studies, and preliminary clinical trials and describes the findings of investigations on this topic to develop an initial blueprint of the mechanisms by which DHEA slows OA progression. Presently, studies on DMOADs are increasing in importance but have met limited success. Encouragingly, the current data on DHEA are promising and may prove that DHEA-based treatment is efficacious for preventing and slowing human OA progression.

Vitamins and Hormones, Volume 108
ISSN 0083-6729
https://doi.org/10.1016/bs.vh.2018.01.021

1. INTRODUCTION

Osteoarthritis (OA), a syndrome involving joint pain and dysfunction, affects more people than any other joint disease. It is characterized by degradation and loss of articular cartilage, subchondral bone remodeling, and at the clinical stage of the disease, inflammation of the synovial membrane. Currently the management of OA is symptomatic and targets the alleviation of pain and joint function but fails to interfere with the underlying biochemical processes involved in disease progression (Roos & Arden, 2016). Recent studies have begun to elucidate the potential of disease-modifying biological agents to slow or reverse cartilage degradation (Han, Fan, Bai, & Ding, 2017; Mantovani, Maccari, & Volpi, 2016; Sun et al., 2015).

Dehydroepiandrosterone (DHEA) is an endogenous steroid with a broad range of biological effects that is synthesized in the adrenal cortex, gonads, brain, and gastrointestinal tract. Its sulfated derivative, DHEA-S, is the most abundant circulating steroid in young adult humans, and its levels decline dramatically with age and have been correlated with degenerative changes related to aging (Rutkowski, Sowa, Rutkowska-Talipska, Kuryliszyn-Moskal, & Rutkowski, 2014). Thus, DHEA is known as an "antidote for aging," and a number of studies have examined its role in cardiovascular diseases (Hirokawa et al., 2016; Savineau, Marthan, & Dumas, 2013), cancer (Cheng, Chi, Wu, & Ho, 2016; Lopez-Marure et al., 2016), diabetes (Brahimaj et al., 2017; Kvasnickova, Hampl, & Vondra, 2015), obesity (Perić et al., 2016; Watanabe et al., 2015), senescence (Daniell, 2017; Lee et al., 2015), and inflammatory arthritis, such as rheumatoid arthritis (RA) (Mravcova et al., 2015). Although RA shares some clinical aspects with OA (Withrow et al., 2016), there is limited information about the effects of DHEA on OA. This chapter reviews the potential physiological and pharmaceutical role of DHEA in modifying the structure of cartilage and discusses the fundamental mechanisms by which DHEA protects cartilage in the progression of OA.

2. DHEA SYNTHESIS AND METABOLISM OVERVIEW

DHEA and its active metabolite, DHEA-S, are steroid hormones synthesized and excreted primarily by the zona reticularis of the adrenal cortex and are controlled by the hormonal signaling cascade of the hypothalamic–pituitary–adrenal axis (Gupta & Morley, 2014). These hormones exert weak androgenic effects and are therefore considered precursor hormones that need to be transformed into potent androgens or estrogens. They act on peripheral sex steroid target tissues and exert their biological effects via multiple signaling pathways (Lois, Kassi, Prokopiou, & Chrousos, 2000–2014). Prepubertal elevations in serum DHEA/DHEA-S are coincident with the differentiation of the adrenal zona reticularis in humans, and peak DHEA/DHEA-S levels are observed around age 20, followed by an age-dependent decline to prepubertal levels. Metabolism of DHEA to active androgens, including testosterone and 5-dihydrotestosterone (DHT), occurs in the gonads, liver, adrenal glands, and peripheral tissues. DHT and testosterone then can be metabolized to estradiol (E2) or estrone, respectively, by aromatase (Prough, Clark, & Klinge, 2016). Declining DHEA and DHEA-S levels are associated with age-related disorders, such as dementia (Maggio et al., 2015), osteoporosis (Park et al., 2017), and OA, and the underlying mechanism through which DHEA exerts its effects as a panacea to senescence still need to be explored.

3. ANTIINFLAMMATORY PROPERTIES OF DHEA

An increasing amount of evidence shows that DHEA plays an important role in the modulation of inflammatory responses in various diseases. For instance, DHEA reduces intestinal inflammation by rendering leukocytes hyporesponsive and balancing colitogenic inflammatory responses (Alves et al., 2016). DHEA also decreases the systemic inflammatory response induced by bilateral femoral fracture by downregulating serum levels of IL-6 and IL-1β in mice (Lichte et al., 2014) and inhibits vascular remodeling following arterial injury by suppressing inflammation and oxidative stress caused by signals from vascular smooth muscle cells (Chen, Xu, & Huang, 2014). In addition, there is evidence demonstrating that DHEA treatment provides diverse antiinflammatory benefits in rodent models of RA; benefits included reductions in joint inflammation, erosion, and synovial proliferation as measured via histological analysis, and reductions in levels of mRNA of proinflammatory cytokines, including tumor necrosis factor-alpha (TNF-α), IL-6, IL-1β, and IL-23 (Offner et al., 2009). Consistent with these findings, Auci et al. reported that an orally bioavailable synthetic analogue of DHEA can ameliorate ongoing disease in a collagen-induced arthritis mouse model with relevance to RA and presented correlated findings showing decreases in proinflammatory cytokine levels and increases in the number of regulatory T cells

(Auci et al., 2007). Although articular cartilage degeneration is a hallmark of OA, low-grade chronic inflammation in the joint can promote OA progression (Liu-Bryan, 2015). Thus, the protective effects of DHEA against OA might be partly attributable to its antiinflammatory properties. Indeed, our early study in a rabbit model of OA observed that the intraarticular injection of DHEA inhibited IL-1β expression by synovial tissues. HE staining of DHEA-treated synovia displayed decreased levels of tissue hyperplasia, inflammatory cell infiltration, and synovial fluid secretion compared with levels in the placebo group (Wu, Yu, Xiong, & Feng, 2006). Similarly, Okamoto et al. reported that DHEA may have a protective effect on osteoarthritic synovial tissue in temporomandibular joint disorders by enhancing fibromodulin formation and suppressing fibrous tissue hyperplasia in vitro and in vivo (Okamoto, Kiga, Shinohara, Tojyo, & Fujita, 2015). These data suggest that low-grade synovial inflammation is part of the pathogenesis of OA and that DHEA does reduce synovial inflammation during OA progression.

4. DHEA BALANCES ANABOLIC AND CATABOLIC PROCESSES IN ARTICULAR CARTILAGE

A proper balance of anabolic and catabolic activities is crucial for the maintenance of cartilage integrity and for the repair of molecular damage sustained during daily usage. During degenerative joint disease, a change occurs in metabolic activities, and tissue homeostasis is disturbed, resulting in a mismatch of anabolic and catabolic processes. If catabolism exceeds the regenerative capacity of the chondrocytes, articular cartilage degeneration occurs, which is the hallmark of OA (Thysen, Luyten, & Lories, 2015). Therefore, achieving anabolic/catabolic balance as a therapeutic target for OA has significant potential as a means to slow OA progression. Recent data demonstrate that DHEA has properties that can improve the balance of critical enzyme systems involved in OA pathogenesis.

4.1 The Matrix Metalloproteases/Tissue Inhibitor of Metalloproteinase System

The key enzymes of cartilage turnover and erosion include matrix metalloproteases (MMPs) and the closely related ADAMTS (a disintegrin and metalloproteinase with thrombospondin motifs). Aggrecan is degraded by both MMPs and ADAMTS, whereas type II collagen is primarily degraded by various MMPs (Troeberg & Nagase, 2012). MMP-3 is capable of cleaving aggrecan's core protein and type II collagen (Burrage, Mix, & Brinckerhoff, 2006; Haller et al., 2015). MMP-13 is the primary MMP involved in cartilage degradation through its particular ability to cleave type II collagen (Howes et al., 2014; Lee, Lamplugh, Kluczyk, Leblond, & Mort, 2009). Similarly, MMP-13-knockout mice have been shown to be resistant to cartilage erosion compared with wild-type animals in a surgical OA model (Little et al., 2009), and selective MMP-13 inhibition reduced glycosaminoglycan (GAG) release from IL-1α-treated joint cartilage in rabbits (Lim et al., 2015).

TIMPs are natural endogenous inhibitors of MMPs, a family of endopeptidases that regulate the turnover of the extracellular matrix (ECM) and cleave various bioactive molecules, including cytokines, chemokines, and growth factors. The activities of MMPs, which include collagenases, stromelysins, gelatinases, and matrilysins, are under tight control through inhibition by TIMPs. A finely tuned balance between the levels of MMPs and TIMPs controls the extent of local ECM degradation in the periphery of cells and thereby influences cellular processes, including migration, proliferation, differentiation, apoptosis, and survival (Ries, 2014). Tissue inhibitor of metalloproteinase (TIMP)-1 is a glycoprotein and inhibits all MMPs on a 1:1 basis by forming high-affinity complexes (Ries, 2014; Wang et al., 2011). A balance between MMPs and TIMP-1 regulates cartilage ECM remodeling and degradation in normal joints. However, a deregulation of this balance is found in pathological conditions, which is thought to be important in the progression of OA (Burrage & Brinckerhoff, 2007; Qu, Li, Wu, & Chen, 2016).

Jo et al. first reported that DHEA is capable of upregulating MMP-1 and MMP-3 gene expression and downregulating that of TIMP-1, an antagonist of MMPs, in both an osteoarthritic chondrocyte model induced by IL-1β (Jo et al., 2003) and a rabbit OA model induced by anterior cruciate ligament transection (Jo et al., 2004), implying that DHEA has an ability to modulate an imbalance between MMPs and TIMP-1, which could explain its protective role against articular cartilage loss during OA progression. In addition, they also found that DHEA treatment did not affect chondrocyte proliferation or GAG synthesis at concentrations up to 100 μM, suggesting that DHEA is safe for treating OA. Based on these two studies, our early work also found a similar effect of DHEA on osteoarthritic cartilage and further demonstrated that DHEA suppressed the mRNA expression of MMP-3 and enhanced that of TIMP-1 in the synovium. Our results suggest that DHEA protects against cartilage destruction and inhibits the progression of OA by modulating the imbalance between the MMPs and TIMP-1, in both the cartilage and synovium, which are two critical target tissues in treating OA progression (Wu et al., 2006). All these observations substantiate the idea that DHEA contributes to cartilage protection by modulating MMPs/TIMP-1 equilibrium in the progression of OA.

4.2 ADAMTS/TIMP-3 System

ADAMTS-4 and ADAMTS-5, also referred to as aggrecanase-1 and aggrecanase-2, are principally responsible for aggrecan degradation in degenerative cartilage diseases (Dancevic & McCulloch, 2014). Knockdown of the gene expression of both ADAMTS-4 and ADAMTS-5 in human cartilage explants results in reduced aggrecan depletion (Song et al., 2007). Knockout of ADAMTS-5 essentially eliminated cartilage erosion and fibrous overgrowth in an enforced uphill treadmill running model of murine arthritis, though it does not eliminate cartilage aggrecanase activity (Li et al., 2011). Recent data demonstrate the efficacy of ADAMTS inhibition in vivo using monoclonal antibodies to ameliorate disease progression in an animal model of OA (Larkin et al., 2015). Similarly, there is also evidence showing that synthetic inhibitors directed toward ADAMTS-4 and ADAMTS-5 might be a strategy to directly inhibit cartilage destruction and restore joint function in OA patients (Dancevic & McCulloch, 2014; Verma, Dalal, & Chopra, 2016; Wiley et al., 2016; Yang, Chanalaris, & Troeberg, 2017). All these observations suggest that ADAMTS-4 and ADAMTS-5 could be effective therapeutic targets in halting cartilage ECM loss during OA progression.

All ECM-degrading proteinases have a corresponding inhibitory cytokine, and the balance of these molecules is critical to the homeostasis of the organism. As mentioned earlier, MMPs bind to TIMP-1 to form a complex, and a balanced interaction between them allows MMPs to exert normal physiological functions in healthy cartilage, while an imbalance between them might lead to excess MMP production, thus shifting cartilage metabolism from anabolic to catabolic and accelerating OA progression. Similarly, for ADAMTS-4/5, TIMP-3 functions as an endogenous inhibitor, and an imbalance in the ADAMTS/TIMP-3 system contributes to the degradation of cartilage aggrecan (Li et al., 2016). With this knowledge, it seems logical to target the actions of these key players to stop the progression of cartilage loss in OA.

Although the suppressive effect of DHEA on some members of the MMP family in OA has been demonstrated, the effect of DHEA on

ADAMTS, which plays a more important role in aggrecan depletion in osteoarthritic cartilage, remains largely unclear. In our early paper, we established the hypothesis that DHEA might play a favorable role in osteoarthritic cartilage by inhibiting the expression of ADAMTS-4/5 (Huang & Wu, 2010). We established this hypothesis based on several pieces of evidence. First, now that DHEA had been proven to attenuate cartilage damage through MMP inhibition, it is not surprising that DHEA may also have an inhibitory effect on ADAMTS-4/5, which also belong to the same broader family of metalloproteinases as MMPs and have the same substrate, aggrecan, as MMPs. Second, DHEA has been found to counteract the proinflammatory effects of catabolic cytokines such as IL-1β, tumor necrosis factor, and IL-6 (Lichte et al., 2014). Additionally, considerable studies have demonstrated that ADAMTS-4/5 activity is induced in the presence of these proinflammatory factors (Kato et al., 2014; Sheu, Ho, Sun, Chen, & Ke, 2015). Therefore, it is reasonable to predict that DHEA has a favorable effect on cartilage via blocking proinflammatory pathways, which in turn suppresses the expression of ADAMTS-4/5. Third, as TIMP-3 is the most efficient endogenous inhibitor of ADAMTS-4/5, DHEA probably exerts its inhibitory role on these catabolic enzymes by upregulating TIMP-3 gene expression, that is, the steroid may have the ability to modulate the balance of ADAMTS/TIMP-3 gene expression. To test this hypothesized mechanism, we later carried out research on rabbit OA model with interventions of intraarticular DHEA administration, administering dimethyl sulfoxide to the contralateral joint as a control. Consistent with our expectations, the data showed that DHEA downregulated ADAMTS-4/5 and upregulated their endogenous inhibitor, TIMP-3, in articular cartilage. In addition, we assessed the levels of aggrecan, the main substrate of ADAMTS-4/5. Consistent with the data for ADAMTS-4/5 expression, the mRNA levels of aggrecan in the experimental group were significantly higher than in the control group. However, aggrecan is also a substrate of MMPs, as described earlier, and DHEA's upregulation of aggrecan might be due to MMP inhibition. Thus, we further investigated the protein levels of Aggrecan ARGxx, a downstream metabolite derived specifically from ADAMTS-4/5 cleavage of aggrecan. The result showed that DHEA treatment also decreased the level of Aggrecan ARGxx in cartilage, suggesting that DHEA does influence ADAMTS-4/5 activity and that positively modulating the ADAMTS/TIMP-3 balance might be a newly discovered mechanism of DHEA against OA (Huang & Wu, 2010; Huang, Zhang, Zhang, Bao, & Wu, 2011).

4.3 Cysteine Proteinases/Cystatin C System

Cysteine proteases are lysosomal enzymes of the papain family, among which cathepsin K, B, L, and S are regarded as the most relevant to the development of OA. The role of cathepsin K, which degrades both type II collagen and aggrecan in articular cartilage in the pathophysiological progression of OA, has been well documented previously (Miller, Lu, Tortorella, & Malfait, 2013; Mort et al., 2016). Most recent evidence demonstrates an increase in cathepsin K expression and activation in human osteoarthritic cartilage, suggesting possible crucial roles for this enzyme in the pathogenesis of osteoarthritic changes in articular cartilage (Kozawa et al., 2016). Studies using cathepsin K inhibitors in OA models in other species have shown beneficial effects on the structural progression of the disease (Connor et al., 2009; Hayami, Zhuo, Wesolowski, Pickarski, & Duong, 2012) and on joint pain in long-term spontaneous OA in Dunkin-Hartley guinea pigs (McDougall, Schuelert, & Bowyer, 2010). The presence and particular distribution of active cathepsin B in osteoarthritic cartilage at severely damaged sites suggest a pathological role for this enzyme in sustaining and perpetuating cartilage degradation (Baici, Horler, Lang, Merlin, & Kissling, 1995; Baici, Lang, Horler, Kissling, & Merlin, 1995). Cathepsin L also has collagenolytic activity and degrades basement membrane components (Maciewicz, Wotton, Etherington, & Duance, 1990), thereby facilitating cellular infiltration. In addition, cathepsin S activity correlates well with OA severity and joint inflammation and is a potent etiological therapeutic target (Ben-Aderet et al., 2015). Cystatin C, which is a small protein (13.3 kDa), is known as a specific endogenous inhibitor of cysteine proteinase (Lafarge, Naour, Clement, & Guerre-Millo, 2010). An imbalance of the cysteine proteinases/cystatin C system could be an important contributing factor in development of OA (Backlund et al., 2011; Chu et al., 2010; Kozawa et al., 2016; Morko, Soderstrom, Saamanen, Salminen, & Vuorio, 2004). In alignment with these observations, our previous studies on animal models revealed that cysteine proteinase levels, including levels of cathepsin K, B, L, and S, increased with the downregulation of cystatin C during the early to middle stages of OA, and the ratio of cystatin C to cysteine proteinases declined as OA progressed. We also found that the intraarticular administration of DHEA significantly suppressed the activity of all cysteine proteinases, especially at the middle stage of OA, and that the ratio of cystatin C to cysteine proteinases in the DHEA group was much higher than that in the OA group. These data imply that DHEA has a role in protecting articular cartilage by influencing the balance between cysteine proteinases and cystatin C enzymes (Bao et al., 2009).

4.4 The Urokinase Plasminogen Activator/Plasminogen Activator Inhibitor-1 Enzyme System

Urokinase plasminogen activator (uPA) has a direct role in degrading proteoglycan monomers and aggregates isolated from articular cartilage, and proteoglycan degradation is proportional to its concentration (Mochan & Keler, 1984). Plasminogen activator inhibitor-1 (PAI-1), the major circulating PAI, binds to the ECM, where it may regulate matrix breakdown and control the rate of plasmin generation by forming irreversible inhibitory complexes with uPA (Duffy, McGowan, Harbeck, Thomssen, & Schmitt, 2014; Knudsen, Harpel, & Nachman, 1987; Kwaan, Mazar, & McMahon, 2013). Clinical and experimental evidence suggests that the uPA/PAI-1 system contributes to joint pathogenesis in OA and RA (Bayram et al., 2012; Kim, Lee, Choi, Yoo, & Yang, 2012) and that a disturbed balance between uPA and PAI-1 ultimately determines the extent of degradation of the cartilage ECM (Zhan, Wu, Gu, Zhang, & Hu, 2006). Moreover, enhanced activity of the uPA/PAI-1 system has been associated with the clinical severity of OA (Brommer, Dooijewaard, Dijkmans, & Breedveld, 1992). Our recent research on a rabbit OA model (Bao et al., 2009) highlighted the dynamic variation in patterns of this enzyme system during different stages of OA. Like the expression patterns of the cysteine proteinases/cystatin C system described in previous section, uPA and PAI-1 transcript levels increased significantly in the early stage of OA and were highest in the middle stage. Their expression levels were both downregulated rapidly to their lowest levels in the terminal stage of OA. These observations could explain why the weight-bearing zones undergoing active cartilage destruction, which contained higher concentrations of these enzymes, were severely eroded and even depleted, while the remaining articular cartilage in nonweight-bearing areas showed very low levels of these enzymes.

5. DHEA ATTENUATES CARTILAGE DEGRADATION BY REGULATING THE WNT/β-CATENIN PATHWAY

Wnts are glycoproteins that are further modified through the addition of a lipid moiety, which influences the interactions of these proteins with receptors and affects their mobility. Wnt proteins activate a number of different signaling cascades; the best understood pathway is the so-called canonical Wnt signaling pathway, which results in the translocation of β-catenin to the nucleus, leading to the transcription of target genes (Yu & Virshup, 2014). Wnts are essential for articular cartilage and bone homeostasis, but excessive Wnt signaling is associated with a loss of the stable articular chondrocyte phenotype and can lead to cartilage damage (Lories, Corr, & Lane, 2013). Evidence has revealed that canonical Wnt signaling is responsible for the inhibition of collagen II expression, activation of MMP-13 expression, and regulation of Wnt7a and c-Jun expression during chondrocyte dedifferentiation in vitro (Sassi et al., 2014), supporting this signaling pathway's participation in cartilage ECM degradation in OA. On the other hand, inhibition of Wnt/β-catenin signals by its specific antagonists, dickkopf-related protein-1 (DKK1), and secreted frizzled-related protein 1 (sFRP1) can prevent the adverse effects on articular cartilage homeostasis by decreasing hypertrophic differentiation (Leijten et al., 2012; van den Bosch et al., 2014). In alignment with the above observations, our recent study confirmed the critical role of Wnt/β-catenin signaling in the pathogenesis of OA. In vivo, articular cartilage loss was associated with a trend toward increased β-catenin levels in damaged cartilage in a rabbit model of OA, and DHEA treatment decreased the expression of β-catenin and Wnt1 in cartilage and slowed the structural damage to cartilage compared with damage progression in vehicle-treated cartilage. In vitro, β-catenin-transfected chondrocytes strongly stimulated the expression of matrix-degrading enzymes, including MMP-3 and MMP-13. The effect of DHEA on Wnt/β-catenin signaling was further investigated using DKK1, the specific inhibitor of this signaling pathway. The protective effect of DHEA was significantly decreased when Wnt/β-catenin signaling was activated, while inactivating Wnt/β-catenin signaling enhanced the effects of DHEA, suggesting that the mechanism underlying DHEA-mediated protection against OA probably occurs via regulation of the Wnt/β-catenin pathway (Li et al., 2014). However, the direct relationship between DHEA and the Wnt/β-catenin pathway is still unclear. For instance, does DHEA complete with Wnts in binding to the same membrane receptor? Does DHEA negatively influence the phosphorylation of β-catenin and inhibit its nuclear translocation, consequently decreasing the downstream gene expression of catabolic enzymes? Further research focused on such issues remains to be conducted.

6. THE PATTERN OF THE ARTICULAR CARTILAGE-MODIFYING EFFECTS OF DHEA

Although DHEA shows a disease-modifying effect on cartilage, including modulating the balance of multiple enzyme systems and thus reducing structural damage to OA, there is still limited evidence about the details of the exact pattern of this structure-protecting effect. To help clarify how this steroid protects cartilage, we recently designed a study in which cartilage damage in different stages of OA was establish in the rabbit model. Moreover, in each disease stage, cartilage at four regions of the whole knee joint, namely, the medial femoral condyle (MFC), medial tibial plateau (MTP), and lateral knee compartment (lateral femoral condyle (LFC) and lateral tibial plateau (LTP)), were histomorphometrically analyzed. Our data showed that DHEA intervention effectively inhibited the progression of existing cartilage degeneration in both knee compartments in moderate OA and in the lateral knee compartment in advanced OA. Safranin O staining showed that the loss of proteoglycan could be prevented by DHEA administration. However, no effects were detected in advanced OA at the MFC and MTP of the joint. This finding indicates that DHEA administration does not greatly affect the medial compartment, which endures the most stress force when OA progresses to the advanced stage. This implies that the disease-modifying efficacy of DHEA against OA is time-dependent and site-specific (Huang, Bao, Jennings, & Wu, 2015). Although we did not investigate the detailed mechanisms of this pattern of structural modification in this study, we speculated on the mechanism based on previous studies. As mentioned earlier, the activation of the Wnt/β-catenin signaling pathway contributes to OA development, and the inhibition of this signaling might be a mechanism by which DHEA protects against OA (Li et al., 2014). Moreover, there is also evidence showing that Wnt/β-catenin signaling activation is enhanced in advanced OA with severe joint damage compared with signaling in the early and middle stages of the disease (Dell'Accio, De Bari, Eltawil, Vanhummelen, & Pitzalis, 2008; Lories et al., 2007). Thus, we can explain why DHEA administration did not greatly affect the medial compartment in our most recent research as follows: in this region, the cartilage exhibited the most severe damage, and the activation of Wnt/β-catenin signaling was upregulated.

7. PERSPECTIVE

OA is a disabling degenerative joint disease that affects more adults than any other rheumatic disease, and it is the main cause of pain and disability among the elderly. Emerging evidence suggests that OA is a disease of the whole joint, affecting many of its musculoskeletal components, including the articular cartilage, synovium, tendons, muscles, ligaments,

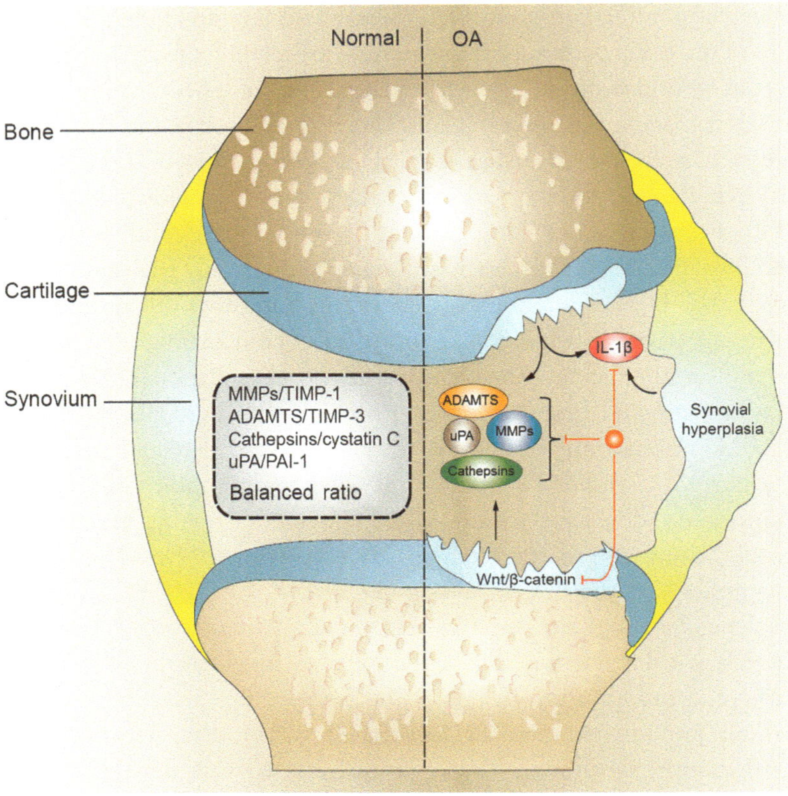

Fig. 1 Schematic representation of the mechanisms of action of DHEA against OA. In a normal joint, anabolic and catabolic activities (MMPs/TIMP-1, ADAMTS/TIMP-3, cathepsins/cystatin C, and uPA/PAI-1) are maintained in a proper balance by the organism, which is crucial for the maintenance of cartilage tissue integrity and for the repair of molecular damage sustained during daily usage. Under conditions present in OA, catabolic activities exceed anabolic activities, resulting in the progressive loss of cartilage ECM. This involves the activation of Wnt/β-catenin pathways, excess proinflammatory cytokine secretion by both the cartilage and synovium, and the overproduction of ECM-degrading enzymes such as MMPs and ADAMTS along with compromised expression of their endogenous inhibitors.

and subchondral bone. Effective development of disease-modifying osteoarthritis drugs (DMOADs) will need to consider the contribution of each of these joint components to the disease and the potential for restoring the original structure and function of damaged cartilage. DHEA is an ideal candidate among the DMOADs still under investigation as it has favorable effects on articular cartilage and the synovium through various molecular mechanisms, as described earlier (Fig. 1). Although the results described in this review demonstrate evidence of the antiosteoarthritic effect of DHEA in different animal species, there are significant limitations to applying these results to the human situation. The experimental models showed chondral changes after surgically induced instability, and pathologies in these models may develop rapidly compared with OA progression in humans; human OA progression is slow and may occur over a period of 15–30 years. Therefore, the effects of DHEA on the changes observed in our animal models may not be totally generalizable to the slow and progressive damage that occurs in degenerative arthritis in humans. Hence, extensive randomized, controlled trials are warranted to evaluate further the efficacy of DHEA in treating OA patients.

REFERENCES

Alves, V. B., Basso, P. J., Nardini, V., Silva, A., Chica, J. E., & Cardoso, C. R. (2016). Dehydroepiandrosterone (DHEA) restrains intestinal inflammation by rendering leukocytes hyporesponsive and balancing colitogenic inflammatory responses. *Immunobiology, 221*, 934–943.

Auci, D., Kaler, L., Subramanian, S., Huang, Y., Frincke, J., Reading, C., et al. (2007). A new orally bioavailable synthetic androstene inhibits collagen-induced arthritis in the mouse: Androstene hormones as regulators of regulatory T cells. *Annals of the New York Academy of Sciences, 1110*, 630–640.

Backlund, A., Holmdahl, M., Mattsson, R., Hakansson, K., Lindstrom, V., Nandakumar, K. S., et al. (2011). Cystatin C influences the autoimmune but not inflammatory response to cartilage type II collagen leading to chronic arthritis development. *Arthritis Research & Therapy, 13*, R54.

Baici, A., Horler, D., Lang, A., Merlin, C., & Kissling, R. (1995). Cathepsin B in osteoarthritis: Zonal variation of enzyme activity in human femoral head cartilage. *Annals of Rheumatic Diseases, 54*, 281–288.

Baici, A., Lang, A., Horler, D., Kissling, R., & Merlin, C. (1995). Cathepsin B in osteoarthritis: Cytochemical and histochemical analysis of human femoral head cartilage. *Annals of Rheumatic Diseases, 54*, 289–297.

Bao, J. P., Chen, W. P., Feng, J., Zhao, J., Shi, Z. L., Huang, K., et al. (2009). Variation patterns of two degradation enzyme systems in articular cartilage in different stages of osteoarthritis: Regulation by dehydroepiandrosterone. *Clinica Chimica Acta, 408*, 1–7.

Bayram, B., Sayin, E., Erkasap, N., Onlu, H., Ozkurt, M., Sahin, F., et al. (2012). Lack of association between plasminogen activator inhibitor type-1 (PAI-1) gene 4G/5G polymorphism and osteoarthritis. *Rheumatology International, 32*, 259–262.

Ben-Aderet, L., Merquiol, E., Fahham, D., Kumar, A., Reich, E., Ben-Nun, Y., et al. (2015). Detecting cathepsin activity in human osteoarthritis via activity-based probes. *Arthritis Research & Therapy, 17*(69).

Brahimaj, A., Muka, T., Kavousi, M., Laven, J. S., Dehghan, A., & Franco, O. H. (2017). Serum dehydroepiandrosterone levels are associated with lower risk of type 2 diabetes: The Rotterdam study. *Diabetologia, 60*, 98–106.

Brommer, E. J., Dooijewaard, G., Dijkmans, B. A., & Breedveld, F. C. (1992). Plasminogen activators in synovial fluid and plasma from patients with arthritis. *Annals of Rheumatic Diseases, 51*, 965–968.

Burrage, P. S., & Brinckerhoff, C. E. (2007). Molecular targets in osteoarthritis: Metalloproteinases and their inhibitors. *Current Drug Targets, 8*, 293–303.

Burrage, P. S., Mix, K. S., & Brinckerhoff, C. E. (2006). Matrix metalloproteinases: Role in arthritis. *Frontiers in Bioscience-Landmark*, *11*, 529–543.

Chen, J., Xu, L., & Huang, C. (2014). DHEA inhibits vascular remodeling following arterial injury: A possible role in suppression of inflammation and oxidative stress derived from vascular smooth muscle cells. *Molecular and Cellular Biochemistry*, *388*, 75–84.

Cheng, M. L., Chi, L. M., Wu, P. R., & Ho, H. Y. (2016). Dehydroepiandrosterone-induced changes in mitochondrial proteins contribute to phenotypic alterations in hepatoma cells. *Biochemical Pharmacology*, *117*, 20–34.

Chu, S. C., Yang, S. F., Tzang, B. S., Hsieh, Y. S., Lue, K. H., & Lu, K. H. (2010). Cathepsin B and cystatin C play an inflammatory role in gouty arthritis of the knee. *Clinica Chimica Acta*, *411*, 1788–1792.

Connor, J. R., LePage, C., Swift, B. A., Yamashita, D., Bendele, A. M., Maul, D., et al. (2009). Protective effects of a cathepsin K inhibitor, SB-553484, in the canine partial medial meniscectomy model of osteoarthritis. *Osteoarthritis and Cartilage*, *17*, 1236–1243.

Dancevic, C. M., & McCulloch, D. R. (2014). Current and emerging therapeutic strategies for preventing inflammation and aggrecanase-mediated cartilage destruction in arthritis. *Arthritis Research & Therapy*, *16*, 429.

Daniell, H. W. (2017). Potential prevention by oral DHEA of superficial tears in elderly atrophic skin. *Journal of Steroid Biochemistry and Molecular Biology*, *171*, 155–156.

Dell'Accio, F., De Bari, C., Eltawil, N. M., Vanhummelen, P., & Pitzalis, C. (2008). Identification of the molecular response of articular cartilage to injury, by microarray screening: Wnt-16 expression and signaling after injury and in osteoarthritis. *Arthritis and Rheumatism*, *58*, 1410–1421.

Duffy, M. J., McGowan, P. M., Harbeck, N., Thomssen, C., & Schmitt, M. (2014). uPA and PAI-1 as biomarkers in breast cancer: Validated for clinical use in level-of-evidence-1 studies. *Breast Cancer Research*, *16*, 428.

Gupta, D., & Morley, J. E. (2014). Hypothalamic-pituitary-adrenal (HPA) axis and aging. *Comprehensive Physiology*, *4*, 1495–1510.

Haller, J. M., Swearingen, C. A., Partridge, D., McFadden, M., Thirunavukkarasu, K., & Higgins, T. F. (2015). Intraarticular matrix metalloproteinases and aggrecan degradation are elevated after articular fracture. *Clinical Orthopaedics and Related Research*, *473*, 3280–3288.

Han, W., Fan, S., Bai, X., & Ding, C. (2017). Strontium ranelate, a promising disease modifying osteoarthritis drug. *Expert Opinion on Investigational Drugs*, *26*, 375–380.

Hayami, T., Zhuo, Y., Wesolowski, G. A., Pickarski, M., & Duong, L. T. (2012). Inhibition of cathepsin K reduces cartilage degeneration in the anterior cruciate ligament transection rabbit and murine models of osteoarthritis. *Bone*, *50*, 1250–1259.

Hirokawa, K., Ohira, T., Nagayoshi, M., Kajiura, M., Imano, H., Kitamura, A., et al. (2016). Dehydroepiandrosterone-sulfate is associated with cardiovascular reactivity to stress in women. *Psychoneuroendocrinology*, *69*, 116–122.

Howes, J. M., Bihan, D., Slatter, D. A., Hamaia, S. W., Packman, L. C., Knauper, V., et al. (2014). The recognition of collagen and triple-helical toolkit peptides by MMP-13: Sequence specificity for binding and cleavage. *Journal of Biological Chemistry*, *289*, 24091–24101.

Huang, K., Bao, J. P., Jennings, G. J., & Wu, L. D. (2015). The disease-modifying effect of dehydroepiandrosterone in different stages of experimentally induced osteoarthritis: A histomorphometric study. *BMC Musculoskeletal Disorders*, *16*, 178.

Huang, K., & Wu, L. D. (2010). Suppression of aggrecanase: A novel protective mechanism of dehydroepiandrosterone in osteoarthritis? *Molecular Biology Reports*, *37*, 1241–1245.

Huang, K., Zhang, C., Zhang, X. W., Bao, J. P., & Wu, L. D. (2011). Effect of dehydroepiandrosterone on aggrecanase expression in articular cartilage in a rabbit model of osteoarthritis. *Molecular Biology Reports*, *38*, 3569–3572.

Jo, H., Ahn, H. J., Kim, E. M., Kim, H. J., Seong, S. C., Lee, I., et al. (2004). Effects of dehydroepiandrosterone on articular cartilage during the development of osteoarthritis. *Arthritis and Rheumatism*, *50*, 2531–2538.
Jo, H., Park, J. S., Kim, E. M., Jung, M. Y., Lee, S. H., Seong, S. C., et al. (2003). The in vitro effects of dehydroepiandrosterone on human osteoarthritic chondrocytes. *Osteoarthritis and Cartilage*, *11*, 585–594.
Kato, T., Miyaki, S., Ishitobi, H., Nakamura, Y., Nakasa, T., Lotz, M. K., et al. (2014). Exosomes from IL-1beta stimulated synovial fibroblasts induce osteoarthritic changes in articular chondrocytes. *Arthritis Research & Therapy*, *16*, R163.
Kim, K. S., Lee, Y. A., Choi, H. M., Yoo, M. C., & Yang, H. I. (2012). Implication of MMP-9 and urokinase plasminogen activator (uPA) in the activation of pro-matrix metalloproteinase (MMP)-13. *Rheumatology International*, *32*, 3069–3075.
Knudsen, B. S., Harpel, P. C., & Nachman, R. L. (1987). Plasminogen activator inhibitor is associated with the extracellular matrix of cultured bovine smooth muscle cells. *Journal of Clinical Investigation*, *80*, 1082–1089.
Kozawa, E., Cheng, X. W., Urakawa, H., Arai, E., Yamada, Y., Kitamura, S., et al. (2016). Increased expression and activation of cathepsin K in human osteoarthritic cartilage and synovial tissues. *Journal of Orthopaedic Research*, *34*, 127–134.
Kvasnickova, H., Hampl, R., & Vondra, K. (2015). DHEA, DHEAS and prolactin correlate with glucose control parameters in women of fertile age with type-1 diabetes mellitus. *Physiological Research*, *64*(Suppl. 2), S255–8.
Kwaan, H. C., Mazar, A. P., & McMahon, B. J. (2013). The apparent uPA/PAI-1 paradox in cancer: More than meets the eye. *Seminars Thrombosis. Hemostasis*, *39*, 382–391.
Lafarge, J. C., Naour, N., Clement, K., & Guerre-Millo, M. (2010). Cathepsins and cystatin C in atherosclerosis and obesity. *Biochimie*, *92*, 1580–1586.
Larkin, J., Lohr, T. A., Elefante, L., Shearin, J., Matico, R., Su, J. L., et al. (2015). Translational development of an ADAMTS-5 antibody for osteoarthritis disease modification. *Osteoarthritis and Cartilage*, *23*, 1254–1266.
Lee, M. J., Kim, E. H., Lee, S. A., Kang, Y. M., Jung, C. H., Yoon, H. K., et al. (2015). Dehydroepiandrosterone prevents linoleic acid-induced endothelial cell senescence by increasing autophagy. *Metabolism*, *64*, 1134–1145.
Lee, E. R., Lamplugh, L., Kluczyk, B., Leblond, C. P., & Mort, J. S. (2009). Neoepitopes reveal the features of type II collagen cleavage and the identity of a collagenase involved in the transformation of the epiphyses anlagen in development. *Developmental Dynamics*, *238*, 1547–1563.
Leijten, J. C., Emons, J., Sticht, C., van Gool, S., Decker, E., Uitterlinden, A., et al. (2012). Gremlin 1, frizzled-related protein, and Dkk-1 are key regulators of human articular cartilage homeostasis. *Arthritis and Rheumatism*, *64*, 3302–3312.
Li, J., Anemaet, W., Diaz, M. A., Buchanan, S., Tortorella, M., Malfait, A. M., et al. (2011). Knockout of ADAMTS5 does not eliminate cartilage aggrecanase activity but abrogates joint fibrosis and promotes cartilage aggrecan deposition in murine osteoarthritis models. *Journal of Orthopaedic Research*, *29*, 516–522.
Li, Y., Li, K., Han, X., Mao, C., Zhang, K., Zhao, T., et al. (2016). The imbalance between TIMP3 and matrix-degrading enzymes plays an important role in intervertebral disc degeneration. *Biochemical and Biophysical Research Communications*, *469*, 507–514.
Li, W. J., Tang, L. P., Xiong, Y., Chen, W. P., Zhou, X. D., Ding, Q. H., et al. (2014). A possible mechanism in DHEA-mediated protection against osteoarthritis. *Steroids*, *89*, 20–26.
Lichte, P., Pfeifer, R., Werner, B. E., Ewers, P., Tohidnezhad, M., Pufe, T., et al. (2014). Dehydroepiandrosterone modulates the inflammatory response in a bilateral femoral shaft fracture model. *European Journal of Medical Research*, *19*, 27.
Lim, H., Min, D. S., Kang, Y., Kim, H. W., Son, K. H., & Kim, H. P. (2015). Inhibition of matrix metalloproteinase-13 expression in IL-1beta-treated articular chondrocytes by a

steroidal saponin, spicatoside A, and its cellular mechanisms of action. *Archives of Pharmacal Research, 38*, 1108–1116.
Little, C. B., Barai, A., Burkhardt, D., Smith, S. M., Fosang, A. J., Werb, Z., et al. (2009). Matrix metalloproteinase 13-deficient mice are resistant to osteoarthritic cartilage erosion but not chondrocyte hypertrophy or osteophyte development. *Arthritis and Rheumatism, 60*, 3723–3733.
Liu-Bryan, R. (2015). Inflammation and intracellular metabolism: New targets in OA. *Osteoarthritis and Cartilage, 23*, 1835–1842.
Lois, K., Kassi, E., Prokopiou, M., & Chrousos, G. P. (2000–2014). *Adrenal androgens and aging, Endotext [Internet]*. South Dartmouth, MA: MDText.com, Inc.
Lopez-Marure, R., Zapata-Gomez, E., Rocha-Zavaleta, L., Aguilar, M. C., Espinosa, C. M., Melendez, Z. J., et al. (2016). Dehydroepiandrosterone inhibits events related with the metastatic process in breast tumor cell lines. *Cancer Biology & Therapy, 17*, 915–924.
Lories, R. J., Corr, M., & Lane, N. E. (2013). To Wnt or not to Wnt: The bone and joint health dilemma. *Nature Reviews Rheumatology, 9*, 328–339.
Lories, R. J., Peeters, J., Bakker, A., Tylzanowski, P., Derese, I., Schrooten, J., et al. (2007). Articular cartilage and biomechanical properties of the long bones in Frzb-knockout mice. *Arthritis and Rheumatism, 56*, 4095–4103.
Maciewicz, R. A., Wotton, S. F., Etherington, D. J., & Duance, V. C. (1990). Susceptibility of the cartilage collagens types II, IX and XI to degradation by the cysteine proteinases, cathepsins B and L. *FEBS Letters, 269*, 189–193.
Maggio, M., De Vita, F., Fisichella, A., Colizzi, E., Provenzano, S., Lauretani, F., et al. (2015). DHEA and cognitive function in the elderly. *Journal of Steroid Biochemistry and Molecular Biology, 145*, 281–292.
Mantovani, V., Maccari, F., & Volpi, N. (2016). Chondroitin sulfate and glucosamine as disease modifying anti-osteoarthritis drugs (DMOADs). *Current Medicinal Chemistry, 23*, 1139–1151.
McDougall, J. J., Schuelert, N., & Bowyer, J. (2010). Cathepsin K inhibition reduces CTXII levels and joint pain in the guinea pig model of spontaneous osteoarthritis. *Osteoarthritis and Cartilage, 18*, 1355–1357.
Miller, R. E., Lu, Y., Tortorella, M. D., & Malfait, A. M. (2013). Genetically engineered mouse models reveal the importance of proteases as osteoarthritis drug targets. *Current Rheumatology Reports, 15*, 350.
Mochan, E., & Keler, T. (1984). Plasmin degradation of cartilage proteoglycan. *Biochimica et Biophysica Acta, 800*, 312–315.
Morko, J. P., Soderstrom, M., Saamanen, A. M., Salminen, H. J., & Vuorio, E. I. (2004). Up regulation of cathepsin K expression in articular chondrocytes in a transgenic mouse model for osteoarthritis. *Annals of Rheumatic Diseases, 63*, 649–655.
Mort, J. S., Beaudry, F., Theroux, K., Emmott, A. A., Richard, H., Fisher, W. D., et al. (2016). Early cathepsin K degradation of type II collagen in vitro and in vivo in articular cartilage. *Osteoarthritis and Cartilage, 24*, 1461–1469.
Mravcova, M., Chovanova, L., Paulikova, L., Vlcek, M., Rovensky, J., Killinger, Z., et al. (2015). Genetics of neuroendocrine factors in rheumatoid arthritis. *Hormone and Metabolic Research, 47*, 411–417.
Offner, H., Firestein, G. S., Boyle, D. L., Pieters, R., Frincke, J. M., Garsd, A., et al. (2009). An orally bioavailable synthetic analog of an active dehydroepiandrosterone metabolite reduces established disease in rodent models of arthritis. *Journal of Pharmacology and Experimental Therapeutics, 329*, 1100–1109.
Okamoto, K., Kiga, N., Shinohara, Y., Tojyo, I., & Fujita, S. (2015). Effect of interleukin-1beta and dehydroepiandrosterone on the expression of lumican and fibromodulin in

fibroblast-like synovial cells of the human temporomandibular joint. *European Journal of Histochemistry*, *59*(2440).

Park, S. G., Hwang, S., Kim, J. S., Park, K. C., Kwon, Y., & Kim, K. C. (2017). The association between dehydroepiandrosterone sulfate (DHEA-S) and bone mineral density in Korean men and women. *Journal of Bone. Metabolism*, *24*, 31–36.

Perić, B., Kruljac, I., Šundalić, S., Pećina, H. I., Jović, A., Štefanović, M., et al. (2016). Obesity and hypercholesterolemia in patients with prolactinomas: Could DHEA-S and growth hormone be the missing link? *Endocrine Research*, *41*, 200–206.

Prough, R. A., Clark, B. J., & Klinge, C. M. (2016). Novel mechanisms for DHEA action. *Journal of Molecular Endocrinology*, *56*, R139–55.

Qu, H., Li, J., Wu, L. D., & Chen, W. P. (2016). Trichostatin A increases the TIMP-1/MMP ratio to protect against osteoarthritis in an animal model of the disease. *Molecular Medicine Reports*, *14*, 2423–2430.

Ries, C. (2014). Cytokine functions of TIMP-1. *Cellular and Molecular Life Sciences*, *71*, 659–672.

Roos, E. M., & Arden, N. K. (2016). Strategies for the prevention of knee osteoarthritis. *Nature Reviews Rheumatology*, *12*, 92–101.

Rutkowski, K., Sowa, P., Rutkowska-Talipska, J., Kuryliszyn-Moskal, A., & Rutkowski, R. (2014). Dehydroepiandrosterone (DHEA): Hypes and hopes. *Drugs*, *74*, 1195–1207.

Sassi, N., Laadhar, L., Allouche, M., Zandieh-Doulabi, B., Hamdoun, M., Klein-Nulend, J., et al. (2014). The roles of canonical and non-canonical Wnt signaling in human de-differentiated articular chondrocytes. *Biotechnic & Histochemistry*, *89*, 53–65.

Savineau, J. P., Marthan, R., & Dumas, D. L. R. E. (2013). Role of DHEA in cardiovascular diseases. *Biochemical Pharmacology*, *85*, 718–726.

Sheu, S. Y., Ho, S. R., Sun, J. S., Chen, C. Y., & Ke, C. J. (2015). Arthropod steroid hormone (20-Hydroxyecdysone) suppresses IL-1beta-induced catabolic gene expression in cartilage. *BMC Complementary and Alternative Medicine*, *15*, 1.

Song, R. H., Tortorella, M. D., Malfait, A. M., Alston, J. T., Yang, Z., Arner, E. C., et al. (2007). Aggrecan degradation in human articular cartilage explants is mediated by both ADAMTS-4 and ADAMTS-5. *Arthritis and Rheumatism*, *56*, 575–585.

Sun, Y., Haines, N., Roberts, A., Ruffolo, M., Mauerhan, D. R., Mihalko, K. L., et al. (2015). Disease-modifying effects of phosphocitrate and phosphocitrate-beta-ethyl ester on partial meniscectomy-induced osteoarthritis. *BMC Musculoskeletal Disorders*, *16*(270).

Thysen, S., Luyten, F. P., & Lories, R. J. (2015). Targets, models and challenges in osteoarthritis research. *Disease Models & Mechanisms*, *8*, 17–30.

Troeberg, L., & Nagase, H. (2012). Proteases involved in cartilage matrix degradation in osteoarthritis. *Biochimica et Biophysica Acta*, *1824*, 133–145.

van den Bosch, M. H., Blom, A. B., van Lent, P. L., van Beuningen, H. M., Blaney, D. E., van der Kraan, P. M., et al. (2014). Canonical Wnt signaling skews TGF-beta signaling in chondrocytes towards signaling via ALK1 and Smad 1/5/8. *Cellular Signalling*, *26*, 951–958.

Verma, P., Dalal, K., & Chopra, M. (2016). Pharmacophore development and screening for discovery of potential inhibitors of ADAMTS-4 for osteoarthritis therapy. *Journal of Molecular Modeling*, *22*, 178.

Wang, B. L., Tu, Y. Y., Fu, J. F., Zhong, Y. X., Fu, G. Q., Tian, X. X., et al. (2011). Unbalanced MMP/TIMP-1 expression during the development of experimental pulmonary fibrosis with acute paraquat poisoning. *Molecular Medicine Reports*, *4*, 243–248.

Watanabe, T., Ashida, K., Goto, K., Nawata, H., Takayanagi, R., Yanase, T., et al. (2015). Dehydroepiandrosterone-enhanced dual specificity protein phosphatase (DDSP) prevents diet-induced and genetic obesity. *Biochemical and Biophysical Research Communications*, *468*, 196–201.

Wiley, M. R., Durham, T. B., Adams, L. A., Chambers, M. G., Lin, C., Liu, C., et al. (2016). Use of osmotic pumps to establish the pharmacokinetic-pharmacodynamic relationship and define desirable human performance characteristics for aggrecanase inhibitors. *Journal of Medicinal Chemistry*, *59*, 5810–5822.

Withrow, J., Murphy, C., Liu, Y., Hunter, M., Fulzele, S., & Hamrick, M. W. (2016). Extracellular vesicles in the pathogenesis of rheumatoid arthritis and osteoarthritis. *Arthritis Research & Therapy*, *18*, 286.

Wu, L. D., Yu, H. C., Xiong, Y., & Feng, J. (2006). Effect of dehydroepiandrosterone on cartilage and synovium of knee joints with osteoarthritis in rabbits. *Rheumatology International*, *27*, 79–85.

Yang, C. Y., Chanalaris, A., & Troeberg, L. (2017). ADAMTS and ADAM metalloproteinases in osteoarthritis—Looking beyond the 'usual suspects'. *Osteoarthritis and Cartilage*, *25*, 1000–1009.

Yu, J., & Virshup, D. M. (2014). Updating the Wnt pathways. *Bioscience Reports*, *34*, 593.

Zhan, J., Wu, L. Q., Gu, Z. Y., Zhang, Y. K., & Hu, J. A. (2006). Effect of disc displacement on mRNA expression of urokinase plasminogen activator and its inhibitor-1 in synovial tissues. *Hua Xi Kou Qiang Yi Xue Za Zhi*, *24*, 63–66.

CHAPTER FIFTEEN

Dehydroepiandrosterone and Addiction

Gal Yadid*^,†,1, Hadas Ahdoot-Levi*, Tzofnat Bareli*, Rachel Maayan‡, Abraham Weizman‡,§

*The Mina and Everard Goodman Faculty of Life Sciences, Bar-Ilan University, Ramat Gan, Israel
†The Leslie and Susan Gonda (Goldschmied) Multidisciplinary Brain Research Center, Bar-Ilan University, Ramat Gan, Israel
‡Laboratory of Biological Psychiatry, Felsenstein Medical Research Center, Research Unit, Petah Tikva, Israel
§Geha Mental Health Center, Tel-Aviv University, Tel Aviv, Israel
[1]Corresponding author: e-mail address: yadidg@gmail.com

Contents

Abstract

Drug addiction has a great negative influence on society, both social and economic burden. It was widely thought that addicts could choose to stop using drugs if only they had some self-control and principles. Nowadays, science has changed this view, defining drug addiction as a complex brain disease that affects behavior in many ways, both biological and psychological. Currently there is no ground-breaking reliable treatment for drug addiction. For more than a decade we are researching an alternative approach for intervention with drug craving and relapse to its usage, using DHEA, a well-being

Vitamins and Hormones, Volume 108
ISSN 0083-6729
https://doi.org/10.1016/bs.vh.2018.04.001

and antiaging food supplement. In this chapter we navigate through the significant therapeutic effect of DHEA on the brain circuits that control addiction and on behavioral performance both in animal models and addicts. We suggest that an integrative program of add-on DHEA treatment may further enable to dynamically evaluate the progress of rehabilitation of an individual patient, in a comprehensive assessment. Such a program may boost and support the detoxification and rehabilitation process, and help patients regain a normal life in a shorter amount of time.

1. INTRODUCTION

Drug addiction has a great influence on society. In the United States alone, the annual societal cost of substance use disorders (SUDs) amounts to about $740 billion in medical care spending and productivity losses, and SUDs are listed among the top 10 nongenetic causes of death globally (McCollister et al., 2017). Drug addiction is also a growing social and psychological problem (Buchanan, 2006; Substance Abuse and Mental Health Services Administration (US) & Office of the Surgeon General (US), 2016). For many years, society considered drug addicts to be people with weak willpower. It was widely thought that addicts could choose to stop using drugs if only they had some self-control and principles. Nowadays, science has changed this view, defining drug addiction as a complex brain disease that affects behavior in many ways, both biological and environmental.

1.1 The Complexity of Drug Addiction

Drug addiction is a complex disease, involving mind and body as in other mental disorders. It is characterized by compulsive drug seeking and intense drug craving, against which the person cannot resist, even in the face of a dire outcome. Drug addiction is considered to be a "relapsing" disease, since an addict may return to drugs even after prolonged abstinence. Drug relapse indicates that it is difficult to treat drug addicts because of the many long-lasting and permanent changes that occur in the brain while using drugs. It seems that addiction is initially a dopamine (DA)-dependent disorder, in which the positive reward provided by the drug is mediated through activation of the mesolimbic DA system (Di Chiara, 1999). This system comprises DA neurons with cell bodies in the ventral tegmental area (VTA) of the midbrain and projection areas of these neurons in the limbic forebrain and in particular, the nucleus accumbens (NAc). This VTA–NAc circuit is a key detector of the reward stimulus; drug-induced changes in these regions may modulate sensitivity to the reward of drug exposure

(Self & Choi, 2004). In addition, other relevant brain regions have been studied, including the hippocampus, which is crucial for the memory of the context of drug exposure and withdrawal, and the hypothalamus, which is important in mediating effects of drugs on the body's physiological state. In addition, current studies indicate that the "cognitive arm" is weakened; this is crucial for drug relapse. The frontal regions of the cerebral cortex, which provide executive control over drug use, are often severely deregulated in addicts. Neurons in the striatum play a role in the expectation and detection of reward. These regions do not function separately, but are part of a complex and highly integrated circuit network that is profoundly altered as a result of chronic exposure to drugs (Chao & Nestler, 2004; Martin-Soelch et al., 2001; Nestler & Malenka, 2004). In recent years, researchers have looked for new mechanisms and factors that interfere with addiction. For example, microRNAs are epigenetic regulators of gene expressions that may play a role in reprogramming of the brain, which in turn leads to the motivational effects of addictive drugs (Kenny, 2014; Smith & Kenny, 2017). In addition, the β-endorphin and enkephalin systems are associated with cocaine-induced behavioral sensitization (Dikshtein et al., 2013; Mongi-Bragato et al., 2016). Even gut bacteria (microbiome) may be involved in drug addiction (Kiraly et al., 2016). Based on previous studies, there is strong evidence that bacteria may respond directly to stress-related host signals, vice versa (Lyte, Vulchanova, & Brown, 2011). Bercik showed that strain-specific anxiety-like behaviors could be transposed simply by transplanting the microbiome between strains of mice (Bercik et al., 2011). In addition, Burokas demonstrated that chronic prebiotic treatment exhibited both antidepressant and anxiolytic effects in stressed mice (Burokas et al., 2016). Moreover, it has been known for many years that stress and depression may be risk factors for the development of addiction. Both are often comorbid with addiction (Polter & Kauer, 2014). A piece of the puzzle that is currently garnering much attention and is shifting the research focus is reward-related learning, since memorizing substance-associated cues is the most powerful cause of relapse (Massart et al., 2015) (Fig. 1).

1.2 Addiction and Stress

All of the above presents a complex picture of "addiction disease" that is associated by response to stress. Consuming a substance of abuse is often a stressful event. Nonetheless, additional external stress can cause recurrence of drug usage. In other words, stress can be a meaningful moderator of the

Fig. 1 DHEA integration with the Addiction Puzzle.

addictive process (see "Stress psychobiology in the context of addiction medicine: from drugs of abuse to behavioral addictions," 2016). Koob et al. describe addiction as a three-stage cycle binge/intoxication, withdrawal/negative effect, and preoccupation/anticipation that worsens over time and involves allostatic changes in hedonic function via changes in the brain reward and stress systems (Kwako & Koob, 2017). They claim that the negative emotional state that drives negative reinforcement derives from dysregulation of key neurochemical elements involved in the brain stress systems within the frontal cortex, NAc, and extended amygdala (Koob et al., 2014). These brain elements include a variety of neurochemical factors: corticotropin-releasing factor, opioids, norepinephrine, vasopressin, hypocretin, and substance P. Thus, brain stress-response systems are hypothesized to be activated by acute excessive drug intake, sensitized during repeated withdrawal, and to persist into protracted abstinence, contributing to the development and persistence of addiction.

Drug addiction is treatable but not fully curable (McLellan, Lewis, O'Brien, & Kleber, 2000). Scientists agree that the most effective treatment often combines medications with behavioral therapies. Treatment can manage withdrawal symptoms, prevent relapse, and treat cooccurring conditions. However, it is likely that reverting to drug use will be unavoidable. Hence, preclinical and clinical researchers should continue to develop new safe and effective treatments for SUD. In this chapter, we advocate DHEA as a conceivable novel treatment for drug abuse.

1.3 Addiction and DHEA

DHEA is a steroid produced in the gonads and in the adrenal gland. In addition it is produced in the brain and acts as a neurotransmitter, hence, it can be defined as a "neurosteroid" (Shah, Chin, Schmidt, & Soma, 2011). DHEA is currently used in Western countries as a food supplement. It has antiaging and antidepressant properties, and contributes to increasing motivation and a general well-being (Huerta-García, Montiél-Dávalos, Alfaro-Moreno, Gutiérrez-Iglesias, & López-Marure, 2013).

Studies indicate that DHEA administration improves memory and cognitive processing (Huerta-García et al., 2013), induces neurogenesis and neural survival, and attenuates levels of the stress hormone cortisol (Ulmann et al., 2009; Yadid, Sudai, Maayan, Gispan, & Weizman, 2010). In healthy men and women, DHEA has been shown to induce relaxation and a higher capability of handling stressful situations (Ulmann et al., 2009).

In a research monograph written by Wilkins et al. at the end of the previous millennium (1996), the authors suggested that plasma levels of DHEA sulfate (DHEAS) may be used to discriminate between treatment outcomes in groups of cocaine addicts (Wilkins et al., 2005). They found that those who had high basal levels of DHEAS remained abstinent following treatment for cocaine dependence/detoxification. Later, Buydens-Branchey et al. compared levels of cortisol and DHEA in cocaine addicts during abstinence (Buydens-Branchey, Branchey, Hudson, & Dorota Majewska, 2002). They demonstrated that levels of cortisol were highest on day 6 of abstinence and then subsequently decreased. DHEAS levels were low on day 6 and highest on day 18 of abstinence. Analyses revealed a significant effect of frequency of use of the drug. More sustained cocaine use was associated with higher cortisol levels and less pronounced cortisol decline after discontinuation of cocaine use, but drug intake variables had no influence on DHEAS levels. During cocaine abstinence, cortisol levels declined more noticeably, whereas DHEAS/cortisol ratios rose more dramatically in aggressive vs nonaggressive addicts. Other studies (Wilkins et al., 2005) identified each of the patient outcome groups by levels of circulating DHEAS and distressed mood at treatment entry. Results showed that cocaine addicts with high circulatory DHEAS levels relapsed less to cocaine abuse after 3 weeks of nonpharmacological treatment, during a 6-month follow-up. The authors suggested that in abstaining patients, distressed mood during withdrawal may have been mitigated through antidepressant-like actions of enhanced endogenous DHEAS activity (Brzoza et al., 2008; Genud et al., 2009; Wolkowitz & Reus, 2003), thus contributing to continuing abstinence and treatment retention. Patients with high levels of distressed mood at treatment entry and low DHEAS levels may express higher susceptibility to leaving the rehabilitation or may more easily relapse. Hence, correlation between endogenous DHEA levels and treatment outcome of addicts suggests the possibility that patients may benefit from adjunctive pharmacotherapy with DHEA.

1.4 A New Horizon: Screening for Potential Treatments

1.4.1 An Immediate Need for Antiaddictive Treatment

Addiction to drugs is still unanswered by the pharmaceutical industry. Treatment of the urge to use drugs and relapse is considered extremely difficult, especially with regard to cocaine, which is a highly dangerous and aggressive drug of abuse. For over 30 years, no effective remedy has been found nor has any innovative treatment been offered to cocaine addicts. So far, the treatment offered is mainly psychological, provided in rehab villages. The only pharmacological treatment currently offered around the world in order to try and support addicts is the controlled, clinical administration of substitute drugs. In other words, the common therapeutic approach is detoxification followed by offering an alternative substance that replaces the addictive substance, such as Subutex and Adolan (Tomkins & Sellers, 2001). However, the addict may develop dependence to this replacement. A safer but longer rehabilitation program uses psychological and behavioral treatment in an enriched environment, such as that provided in rehab centers (McLellan & Weisner, 1996).

1.4.2 Use of Animal Models for Testing New Medications

Two main animal models are used for studying addiction. The first model is conditioned place preference (CPP), in which substance injections are paired with sensorially distinguished compartments of an apparatus, so that the animal can associate the drug-induced changes it experiences with environmental cues provided by the apparatus (Aguilar, Rodríguez-Arias, & Miñarro, 2009; Mucha, Van Der Kooy, O'Shaughnessy, & Bucenieks,

1982). Another animal model is self-administration, in which animals are trained to press a lever, thereby self-administering substances of abuse. Animals thus become substance-dependent, developing behavioral, and neurological changes that simulate human addiction (for review, see Haney, 2009; Panlilio & Goldberg, 2007). Clinical and laboratory observations suggest that addiction represents the pathological usurpation of neural processes that normally serve reward-related learning. This process results in brain plasticity that is hypothesized to be driven by the emotional stress-inducing capacity of the drug.

2. DHEA TREATMENT FOR SUBSTANCES OF ABUSE

For over a decade, treatment of drug addiction with DHEA has been examined, with ground-breaking insights. This ongoing research, although reaching the clinic, is based on work with animal models. The effect of DHEA has been examined in terms of behavior in the various stages of addiction: acquisition (initiation), maintenance, extinction, and reinstatement (relapse). The added value of animal studies is the finding of biological and structural effects of DHEA treatment.

2.1 Effect of DHEA Treatment on Acquisition of Substance Dependence

In a study investigating the influence of high levels of brain DHEA on the acquisition of cocaine intake in rats, the authors showed that DHEA pretreatment and with the concomitant use cocaine, attenuated cocaine-seeking behavior, and elevated the levels of dopamine and serotonin in several brain regions relevant to cocaine addiction (Maayan et al., 2006).

Chronic cocaine self-administration induced elevations in brain DHEA, its sulfate ester, DHEAS, and pregnenolone. The increased brain DHEA following cocaine self-administration may serve as a compensatory protective mechanism geared to attenuating the craving for cocaine. Such anticraving activity is further enhanced by DHEA treatment before and during cocaine self-administration, accompanied by normalization of DHEA levels in the brain (Maayan et al., 2006).

DHEA may be involved in one or more of these phases by affecting different neural mechanisms such as the hypothalamic–pituitary–adrenal axis, the activity of glutamate receptors (Glu-R; both NMDA and non-NMDA as AMPA/kainate), nicotinic receptor, sigma-1 receptor (σ1R), in addition to the enhancement of DA release.

Cocaine simultaneously affects mood and the neuroendocrine hormones such as adrenocorticotropic hormone (ACTH) and corticosterone (CORT) (Levy et al., 1991; Mendelson et al., 2002; Moldow & Fischman, 1987; Saphier, Welch, Farrar, & Goeders, 1993). Goeders (1997) claimed that CORT is necessary during acquisition, and that substance self-administration does not occur unless this stress-related hormone is increased above the threshold critical for reward. Drugs blocking central corticotropin-releasing hormone (CRH) receptors or decreasing ACTH and CORT can attenuate self-administration (Goeders, 1997, 2002a, 2002b). In addition, other clinical and preclinical studies link stress to drug dependence since stress is accompanied by the release of neuroactive hormones which activate brain circuitries similar to psychostimulants such as cocaine (Majewska, 2002). Studies using several strains of rats showed that stress and elevated glucocorticoid increased acquisition and maintenance of psychostimulant use (Goeders, 1997; Piazza & Le Moal, 1998; Piazza et al., 1991). It was also reported that extrahypothalamic brain CRH systems may be involved in behavioral and physiological manifestations of drug-seeking behavior induced by environmental stressors (Sarnyai, Shaham, & Heinrichs, 2001). DHEA treatment caused a decrease in both CORT and ACTH levels at similar doses (Maayan et al., 2006, 2005). Such a decrease may explain, at least in part, the attenuation in cocaine-seeking behavior.

However, new findings indicate that the CORT level is directly related to DHEA attenuation of drug-seeking behavior, and that other mechanisms may exist. In addition, recent work (Guerin, Schmoutz, & Goeders, 2014) showed that adrenalectomy did not significantly affect the maintenance phase of cocaine self-administration. Since an earlier study found that adrenalectomy blunted the acquisition of cocaine self-administration (Goeders & Guerin, 1996), CORT levels may exert greater influence on drug self-administration during the acquisition phase than during the maintenance phase.

An additional mechanism to explain DHEA effect on substance-seeking behavior is the interrelationship of σ1R, DHEA, and reward. σ1R is localized in the endoplasmic reticulum and is used as a chaperone for diverse proteins including IP3R, BiP, K+ channel subunit, and RAC1 (Aydar, Palmer, Klyachko, & Jackson, 2002; Hayashi & Su, 2001). Direct interaction between σ1R and other proteins is important for σ1R cell function (Hayashi & Su, 2007). Studies found that cocaine binds to σ1R receptors with high affinity (Chen, Hajipour, Sievert, Arbabian, & Ruoho, 2007); hence, the σ1R receptor may be essential to substance addiction treatment (Miyatake, Furukawa, Matsushita, Higuchi, & Suwaki, 2004; Stefanski et al., 2004; Ujike, Kanzaki, Okumura, Akiyama, & Otsuki, 1992;

Ujike, Kuroda, & Otsuki, 1996). σ1R selective antagonists, when given simultaneously with cocaine, decrease the effect of locomotion caused by cocaine and can prevent cocaine addiction (Ujike et al., 1996). On the other hand, treatment with dehydroepiandrosterone (DHEA), a σ1R receptor agonist, albeit chronically, attenuates cocaine self-administration and relapse to cocaine-seeking behavior in rats (Doron, Fridman, Gispan-Herman, et al., 2006). Additionally, σ1R agonists protect from nerve damage (Nguyen et al., 2015). It remains to be determined whether agonists or antagonists are the better choice for treatment. Exposure to cocaine causes σ1R gene expression changes in the nucleus accumbens (NAc), a region associated with reward and addictive disorders (Liu & Matsumoto, 2008). The above evidence suggests the σ1R receptor as a target for treating cocaine addiction (Matsumoto, Liu, Lerner, Howard, & Brackett, 2003; Maurice et al., 2002).

Although it was originally thought that the sigma receptor was a subtype of the opioid receptor, it is now clear that both σ1R and σ2R are unique brain proteins. σ1R regulates glutamate NMDAR function and the release of neurotransmitters such as DA, and thus may be involved in neuropsychiatric disorders and reward (Hayashi & Su, 2003; Su & Hayashi, 2003). σ1R is critically involved in the rewarding effect of cocaine as measured in the CPP procedure in mice (Romieu, Martin-Fardon, Bowen, & Maurice, 2003), and σ1R agonist pretreatment facilitated cocaine self-administration (Hiranita, Soto, Tanda, & Katz, 2010). Sutton et al. found that rats repeatedly exposed to cocaine develop glutamate deficits in the NAc (Sutton et al., 2003). DHEAS, which promotes presynaptic glutamate release in the prelimbic cortex via activation of σ1R (Dong et al., 2007), may alleviate drug-seeking behavior.

Short- and long-term memory should be considered as another process involved in drug-seeking behavior beyond reward and stress. An alternative neuronal system affected by DHEA is the NMDA- and AMPA-glutaminergic system—which is involved in learning and memory.

DHEA is a positive modulator of the NMDA-R (Dubrovsky, 2005; Mellon & Griffin, 2002a, 2002b; Rupprecht & Holsboer, 1999) and may attenuate cocaine-induced effects by enhancing glutamate activity. A decrease in brain extracellular glutamate was obtained during the extinction from drug usage and was suggested to contribute to the susceptibility to relapse (Baker et al., 2003). DHEA may enable acquisition of new memories dissociating drug from the cue. Glutamatergic activity is connected to decision-making and working memory processes (Finn, 2002). Hence, DHEA may produce an inhibitory cognitive control in present of incentives and attenuate relapse to drug usage.

In addition, AMPA-R antagonists attenuate behavioral sensitization and self-administration in mice (Jackson, Mead, Rocha, & Stephens, 1998; Reeves, Thiruchelvam, & Cory-Slechta, 2004), and PREG sulfate acts as an AMPA-R negative modulator (Dubrovsky, 2005). Therefore, one may assume that DHEAS (partially converted from DHEA) acts similarly. This is supported by Kimonides et al. (1998), who found that DHEAS attenuates AMPA and kainate neurotoxicity in hippocampal neurons (Kimonides, Khatibi, Svendsen, Sofroniew, & Herbert, 1998). It is possible that as an antagonist of the AMPA-R, DHEAS may also attenuate cocaine self-administration and contribute to the glutamate homeostasis suggested by Kalivas (2009).

The rewarding effect of cocaine which promotes its reinforcement is mediated by the blockade of the presynaptic DA transporter, thus increasing transient DA activity in the mesolimbic or mesocortical DA reward system. DHEA caused an increase in the content of DA and 5-HT and their metabolites in the VTA and NAc following chronic cocaine intake. It is possible that such an increase in DA homeostasis may attenuate the need for cocaine intake and cocaine seeking-behavior (Maayan et al., 2006).

In one study (Maayan et al., 2006), rats initiated cocaine self-administration already with high basal levels of both peripheral and brain DHEA, which was achieved by pretreatment followed by continuous concomitant treatment with DHEA. This prevented drug consumption and drug seeking (see Fig. 2). Thus, it can be assumed that timing is a crucial factor in the anticraving effect of DHEA. Subjects with high basal DHEA may be less prone to developed drug addiction. This observation is supported by Reddy and Kulkarni (1997a, 1997b) who showed that DHEA treatment prevented the development of tolerance to benzodiazepines (BZ) in mice. In addition, DHEAS treatment prevents the development of morphine tolerance and attenuates abstinent behavior in mice (Reddy & Kulkarni, 1997a, 1997b; Ren, Noda, Mamiya, Nagai, & Nabeshima, 2004). Assuming that overlapping mechanisms are involved in brain response to different substances of abuse, these studies may support each other, since DHEA is converted partially to DHEAS both peripherally and in the brain by sulfatransferase (Maayan et al., 2006, 2005). The authors emphasize that control and DHEA-treated rats displayed similar levels of lever-pressing for water reinforcement during the maintenance phase of water self-administration, verifying that DHEA treatment specifically disrupted lever-pressing for cocaine reinforcement and not by impairment of physical performance.

Rats showed no interest in pressing the active lever even when the dose of infused cocaine was nullified; supporting the assumption that DHEA

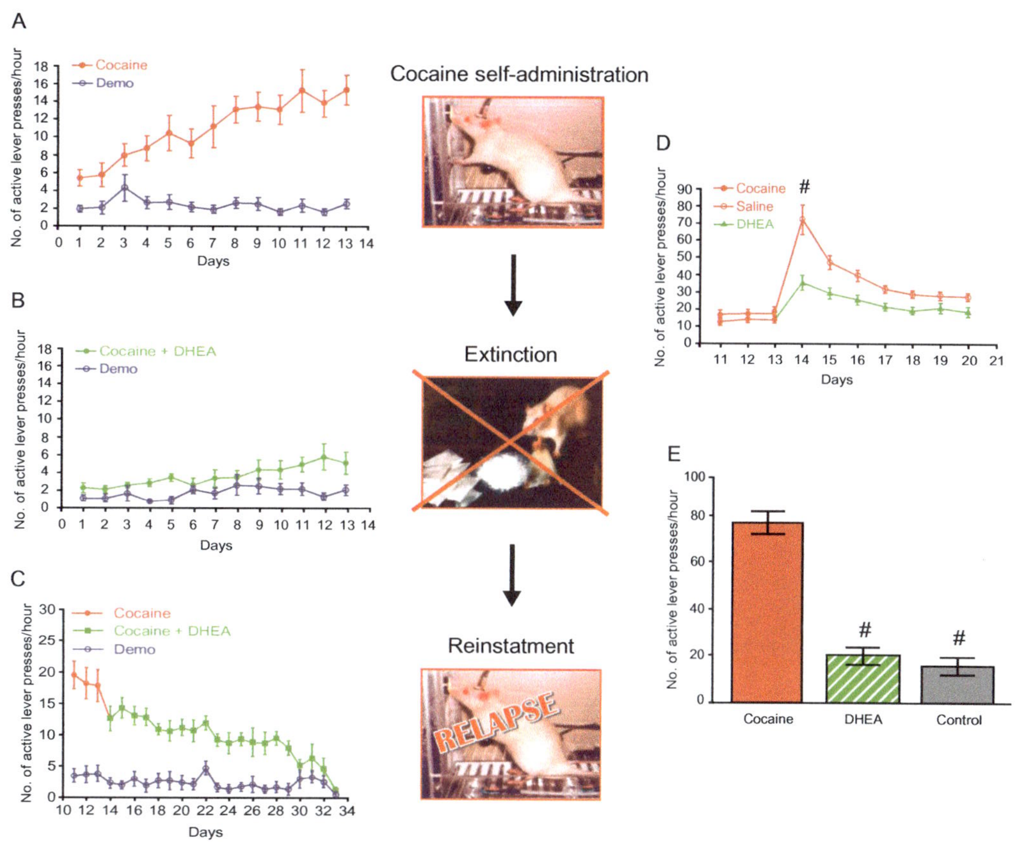

Fig. 2 See legend on next page.

probably has an anticraving effect and not a potentiating one. One cannot exclude the possibility that DHEA on its own may be a substitute to cocaine though no such effect has been found in the CPP model (Romieu et al., 2003). Consequently, this attenuation in cocaine-seeking behavior by DHEA may be explained either by it having an anticraving effect (the rats show no interest in cocaine) or by it potentiating the effect of cocaine (need for lower doses of cocaine in order to get the same reinforcement). However, a descending phase of the dose response proved that DHEA treatment did not overpotentiate activation of lever presses when cocaine infusion was decreased or withheld (Maayan et al., 2006).

Fig. 2 Effect of DHEA on cocaine-seeking behavior and relapse-preclinical experiments. Rats were operated and implanted with i.v. catheters. After 10 days of rehabilitation they were trained to self-administer cocaine (1 mg/kg) in a FR-1 paradigm. (A) An animal model for drug addiction: A typical progress of drug self-administration progress is depicted, from acquisition to maintenance, which is fully controlled by the animal. (B) Effect of DHEA on cocaine acquisition. DHEA (2 mg/kg, B) was administrated prior consecutively 5 days before training and during training, 90 min before entering the rat into the self-administration chamber. Presses on the active-, nonactive levers (demo) were recorded. A marked effect of DHEA pretreatment on drug-seeking behavior was noticed. (C) Effect of DHEA on cocaine maintenance. After rats reached stable maintenance (the figure indicate the three last days of maintenance before treatment; *red symbols*) and during DHEA (2 mg/kg) administration consecutively during maintenance (access to the drug intake was available), 90 min before entering the rat into the self-administration chamber. Presses on the active-, nonactive (demo) levers were recorded. A longitudinal decrease in cocaine self-administration is depicted. (D) Effect of DHEA on cocaine during drug extinction (washout). After rats maintained 10 days of stable drug self-administration (the figure indicate the three last days of maintenance before treatment; *red symbols*), cocaine was extinct and DHEA (2 mg/kg) was injected, 90 min before entering the rat into the self-administration chamber. Presses on the active-, nonactive (demo) levers were recorded. An immediate significant ($^{\#}P < 0.01$) decrease in cocaine craving is depicted. (E) Effect of DHEA on cocaine priming-induced reinstatement. Rats were treated with DHEA (2 mg/kg) or saline consecutively during maintenance 90 min before entering the into the self-administration chamber, when the drug was available. After reaching the abstinence criterion (pressing <10% of maintenance), rats were reinstated with 10 mg cocaine i.v. and entered the chamber without accesses to cocaine. Craving was evaluated by measuring their presses on the active lever. As shown, relapse was abolished ($^{\#}P < 0.01$) by DHEA treatment. *Panel (B) modified from Maayan, R., Lotan, S., Doron, R., Shabat-Simon, M., Gispan-Herman, I., Weizman, A., et al. (2006). Dehydroepiandrosterone (DHEA) attenuates cocaine-seeking behavior in the self-administration model in rats.* European Neuropsychopharmacology, 16*(5), 329–339. https://doi.org/10.1016/j.euroneuro.2005.10.002. Panel (C) modified from Doron, R., Fridman, L., Gispan-Herman, I., Maayan, R., Weizman, A., & Yadid, G. (2006). DHEA, a neurosteroid, decreases cocaine self-administration and reinstatement of cocaine-seeking behavior in rats.* Neuropsychopharmacology, 31*(10), 2231–6. https://doi.org/10.1038/sj.npp.1301013.*

2.2 Effect of DHEA Treatment on Maintenance of Drug-Seeking Behaviors

It has been successfully shown that chronic exposure to exogenous DHEA (2 mg/kg) attenuated cocaine self-administration and decreased the cocaine-seeking behavior of the rats to <20% of their maintenance levels (Doron, Fridman, Gispan-Herman, et al., 2006; see Fig. 2).

No surge in active lever response accompanied initiation of DHEA treatment, which would have been expected if DHEA blocked the reward effect of cocaine. This result is interesting, since most difficulties of addicts to stay on an antidrug program occur during the first weeks, when expectation for reward is highest (Self & Choi, 2004). Indeed, when higher doses of DHEA (10–20 mg/kg) were coadministered with cocaine for a short period of time (4 days), a markedly increased cocaine-induced CPP was reported (Romieu et al., 2003). Extinction of responding maintained by appetitive reward has been suggested to induce multiple stress events including activation of the hypothalamus–pituitary–adrenal axis, glucocorticoid secretion, and central β-endorphin release (Shaham, Shalev, Lu, De Wit, & Stewart, 2003; Vescovi, Coiro, Volpi, & Passeri, 1992). Hence, it was previously suggested that alleviation of distress associated with cocaine withdrawal may facilitate achieving abstinence (Self & Choi, 2004; Wilkins et al., 2005). Furthermore, during withdrawal, cocaine addicts have high plasma cortisol levels, which peak 6 days after initiation of the withdrawal and then gradually decrease (Buydens-Branchey et al., 2002). As the levels of plasma cortisol decrease, the levels of DHEAS gradually increase.

Only patients in which a spontaneous increase in DHEAS levels was observed were identified as being successful at abstaining from cocaine usage over time (Wilkins et al., 2005). Although a connection between DHEA/DHEAS, low distressed mood levels, and changes in CNS salience during withdrawal was suggested (Wilkins et al., 2005), it seems that appropriate DHEA dose is critical for relapse prevention (Shoptaw et al., 2004). The effect of exogenously applied DHEA (2 mg/kg) on cocaine-seeking behavior may be due to its conversion to DHEAS in the serum and brain (Maayan et al., 2006, 2005), in addition to increases of other bioactive neurosteroids (Dubrovsky, 2005). These neurosteroids may interact with various neurosystems involved in mood and drug-seeking behaviors, such as glutamatergic, GABAergic, and dopaminergic systems. Since DHEA can function as an antidepressant in both animals and humans (Maayan et al., 2006, 2005; Wolkowitz et al., 1997), it may lower the depression/distress involved with cocaine withdrawal (Self & Choi, 2004), similar to the β-endorphin-induced lowering of frustration during extinction (Roth-Deri et al., 2004, 2003).

An additional possible mechanism for the DHEA-induced attenuation of the maintenance of cocaine-seeking behavior suggests the involvement of σ1R. Based on data indicating that DHEA acts as an agonist of these receptors (Maurice et al., 1999), it is possible that long administration of cocaine downregulates σ1R and DHEA acts in a compensatory neuroadaptation mechanism that leads to attenuation of cocaine-seeking behavior (Ben-Ami, Kinor, Perelman, & Yadid, 2006).

2.3 Effect of DHEA on Extinction of Drug-Seeking Behaviors

In order to learn more about the effect of DHEA on the extinction phase after cocaine self-administration, an experiment was conducted in which two groups of rats were trained to self-administer cocaine until stable maintenance levels were attained. On the following days, rats were exposed to cocaine extinction in order to assess their craving to the drug. The conditions during the extinction phase were the same as during training with the exception that the cocaine syringes were removed, and that rats were injected either with DHEA (2 mg/kg i.p.) or with saline as a vehicle (same volume) 90 min prior to placement in the operant chambers.

Results showed (Fig. 2) that saline-treated rats pressed on the active lever significantly more on the first day of extinction. Moreover, the rate of return to baseline (maintenance) during the extinction period was different. DHEA-treated rats returned to baseline from day 2 forward, whereas saline-treated rats never returned to baseline behavior.

The most significant and important effect of DHEA treatment during drug washout was in the long term. After a total of 34 days since last exposure to cocaine (27 days from the end of treatment), rats received a priming injection of cocaine before being placed in the self-administration chambers. Active lever responses, reinforcements, and inactive lever responses were recorded. In the relapse test, number of active lever responses was significantly lower in the DHEA-treated group compared to saline-treated group (Fig. 2).

Since the influence of DHEA treatment is much longer than just the short term immediately after treatment (Doron, Fridman, Gispan-Herman, et al., 2006), the authors tested its influence on the dentate gyrus in the hippocampus.

A main factor in brain plasticity is neurogenesis. Neurogenesis is the process by which new neurons are created from progenitor cells. Adult neurogenesis is restricted mainly to two areas of the central nervous system: the subventricular zone and the subgranular zone of the dentate gyrus in the hippocampus.

High-dose cocaine intake (1.5 mg/kg) significantly reduces proliferation and neurogenesis levels (Sudai et al., 2011), in the hippocampus of adult rat. Significantly more newly formed cells (BrdU+) were found 24h after the end of treatment in the dentate gyrus of cocaine self-administrated rats treated with DHEA than in cocaine self-administrated rats treated with saline (data not published).

A month later, the number of new cells that eventually survived and became adult neurons was assessed and proved that DHEA repaired the decrease in neurogenesis as a result of cocaine consumption.

Not all of the new cells observed during proliferation eventually turn into neurons. Significant numbers of the new progenitor cells in the dentate gyrus turn into glia cells. The possible involvement of astrogliosis in addiction is intriguing and has been suggested in some other neurodegenerative diseases (Colangelo, Alberghina, & Papa, 2014).

Regarding stress, circulating CORT secretion is increased during cocaine withdrawal (Erb, Shaham, & Stewart, 1998; Goeders & Clampitt, 2002; Mantsch et al., 2008; Shalev, Marinelli, Baumann, Piazza, & Shaham, 2003). DHEA given only throughout extinction, although having no effect on CORT levels, significantly attenuated cocaine-seeking behavior.

2.4 Effect of DHEA Treatment on Reinstatement to Drug-Seeking Behaviors

There is very little literature on neuroactive steroids and drug reinstatement. Nie and Janak (2003) trained rats to press levers for 0.1 mL of 10% ethanol or 5% sucrose solutions. Response was then extinguished, and subjects were tested for reinstatement of lever-press response (Nie & Janak, 2003). Allopregnanolone promoted response for ethanol, but not sucrose, following a period of abstinence, suggesting that GABA-A receptor modulation may contribute to processes involved in reinstatement of ethanol-seeking behavior. Another study reported that progesterone, and to a greater extent allopregnanolone, decreased cocaine-primed reinstatement in females, while having no effect on cocaine-primed reinstatement in males (Anker, Holtz, Zlebnik, & Carroll, 2009).

The effect of DHEA on cocaine-induced reinstatement of drug seeking in rats exposed to withdrawal conditions was also examined. Rats receiving DHEA (2 mg/kg) daily showed a minimal response to acute priming with cocaine. This may suggest that DHEA can protect against relapse to cocaine usage following reexposure to the drug (Doron, Fridman, Gispan-Herman, et al., 2006).

Substance dependence is accompanied by usurpation of natural memory mechanisms resulting in long-lasting memories of the drug experience (Geier & Luna, 2009; Kauer & Malenka, 2007; Koob, 2009; Peters, Kalivas, & Quirk, 2009; Wang, 2008). Elevating endogenous glutamate or glutamate receptors reduces the propensity for relapse (Baker et al., 2003; Sutton et al., 2003). Therefore, the consequence of increasing glutamate activity may be eradication of memories associated with substance reward-related learning and gaining of new memories linked to drug extinction. If glutamate activity is increased by DHEA in the addicted brain, it may help drug extinction and attenuate relapse and may be in accordance with the glutamate homeostasis suggested by Kalivas (2009).

In addition, Sutton et al. (2003) found that increasing the GluR1 subunit of AMPA receptors in the NAc may promote extinction of cocaine seeking (Sutton et al., 2003). AMPA-R antagonists attenuate behavioral sensitization and self-administration in mice (Jackson et al., 1998; Reeves et al., 2004). The neurosteroid sulfate, PREG-S, acts as an AMPA-R negative modulator (Dubrovsky, 2005). Therefore, one may assume that DHEAS acts similarly. For reinstatement, it was found that CORT plays a role in stress-induced, cue-induced, environmental cue-induced, and food deprivation-induced reinstatement, whereas it has only a partial role in cocaine-induced reinstatement (Erb et al., 1998; Goeders & Clampitt, 2002; Mantsch et al., 2008; Shalev et al., 2003).

2.5 DHEA Affects Addiction Beyond Attenuating Stress

It was demonstrated above that DHEA treatment attenuates cocaine acquisition, extinction, and reinstatement (Doron, Fridman, Gispan-Herman, et al., 2006; Maayan et al., 2006; Yadid, Redlus, Barnea, & Doron, 2012). But the question is whether this effect is caused indirectly, through decreasing CORT levels (Maayan et al., 2005), or by DHEA itself.

Research has shown that rats treated with DHEA or simultaneous DHEA and CORT showed a significantly lower response to cocaine compared to control rats throughout all phases of the study: acquisition, maintenance, extinction, and reinstatement. However, the plasma level of CORT in DHEA-treated rats was low, whereas its level in DHEA- and CORT-treated rats was maintained at control levels. It was previously shown that a persistent decrease in brain DHEA levels may be associated with the decreased ability to cope with mood fluctuations and stress that accompany heightened drug craving (Yadid et al., 2012; Yadid, Sudai, Maayan, Gispan, & Weizman, 2010). DHEA may lower the distress associated with cocaine withdrawal (Self & Choi, 2004), similar to the β-endorphin-induced decrease in frustration during extinction (Roth-Deri et al., 2004, 2003). Thus, exogenous, long-term administration of DHEA may also have a beneficial effect on stress and anxiety, which increase particularly during heightened drug craving at extinction and reinstatement (Yadid et al., 2012, 2010). This may contribute to the attenuating effect of DHEA on cocaine-seeking behavior.

DHEA treatment attenuated cocaine self-administration and cocaine seeking even when CORT levels were constantly maintained in the normal range. Therefore, the effect of DHEA on cocaine intake and seeking is likely independent of fluctuations in CORT levels. Moreover, DHEA attenuated self-administration when given as pretreatment, at extinction or at reinstatement, emphasizing its potential as an effective anticraving agent.

3. DHEA IN CLINICAL TRIALS

3.1 DHEA as a Treatment for Mental Disorders

There are many manuscripts on psychiatric disorders, such as depression and anxiety that are comorbid with addiction. DHEA and its sulfated metabolite generated curiosity among researchers of psychiatry disorders as early as the 1990s. Wolkowitz and colleagues reported that DHEA may have antidepressant and promemory effects after they delivered it to six elderly patients with major depression (Wolkowitz et al., 1997). They also initiated a double-blind placebo study to test the efficacy of DHEA (Wolkowitz et al., 1999). They found that half of the DHEA-treated patients showed a 50% decrease in depressive symptoms. Most of them were resistant to treatment with traditional antidepressant medication alone. In 2006, a randomized, double-blind controlled study evaluated the effect of DHEA (200 mg/day) on schizophrenia symptoms (Ritsner, Gibel, Ratner, Tsinovoy, & Strous, 2006). The results showed no significant improvement in clinical symptoms. There was, however, an improvement in positive and negative symptom scale ratings compared with baseline and in cognitive functions of visual sustained attention and visual and movement skills compared with placebo conditions. In another study, 100 mg/day DHEA were sufficient to improve extrapyramidal symptoms in patients with schizophrenia (Nachshoni et al., 2005). DHEA was also tested in Alzheimer disease (Wolkowitz et al., 2003). Although it did not significantly improve cognitive performance, the authors suggested that the study was underpowered and DHEA should be tested in larger-scale and longer-term studies. The common denominator of most of these cases is that the levels of DHEA and DHEAS decrease over time in these diseases. Additionally, in human drug addicts, levels of these hormones were found to decrease during abstinence. These findings suggest that increased circulating DHEAS levels may be of benefit in these diseases, as well as in drug addiction.

3.2 Using DHEA as a Treatment for Detoxification of Polydrug Users

There are many clinical studies that assess the involvement of DHEA and its sulfate form (DHEAS) in SUD (see Section 1.3). However, only a few of them examine the effect of DHEA as a treatment for SUD. In a double-blind controlled study Maayan et al. (2008) reported that providing DHEA to heroin addicts as an add-on compound to their routine medication protocol was mostly effective (60% of treated subjects) in patients who had not previously used either cocaine or BZ and who had experienced only a few withdrawal programs (Maayan et al., 2008; see Fig. 3). In another placebo-controlled pilot, Shoptaw et al. (2004) suggested that DHEA treatment resulted in worse treatment outcome in cocaine addicts (Shoptaw et al., 2004). Therefore, more clinical trials should be carried out to test the efficacy of DHEA as a treatment for addiction detoxification.

3.3 Effect of DHEA Add-on Therapy on Rehabilitation of Polydrug Users

In view of these previous clinical and preclinical findings, whether DHEA will prove to have efficacy as an adjunctive treatment in drug users must be further examined. Consequently, Yadid's laboratory proposed a double-blind, placebo-controlled study to test the long-term effect of DHEA on addiction indices that succeeded to improve decision-making and lower relapse rates significantly (Ohana et al., 2016; see also Fig. 3). As mentioned in Section 1.3, levels of DHEA and DHEAS were found to decrease during abstinence in human drug addicts (Buydens-Branchey et al., 2002; Wilkins et al., 2005), and this decrease was found to predict later drug reuse

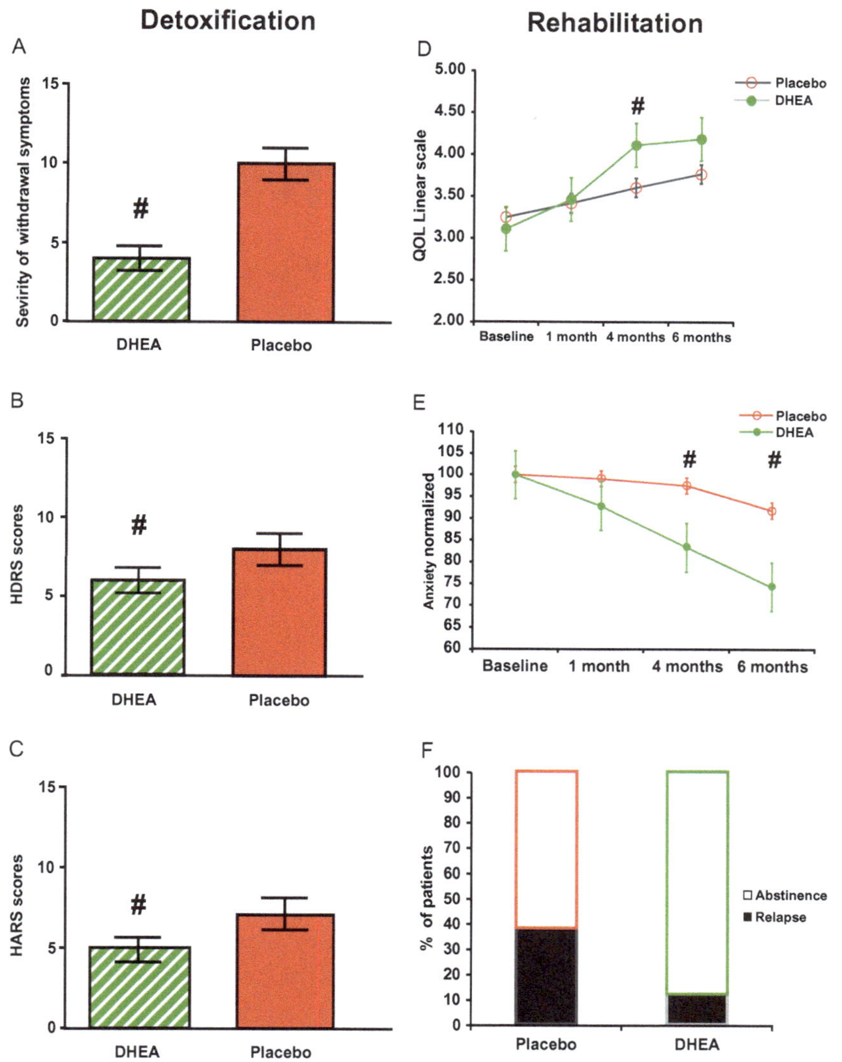

Fig. 3 Effect of DHEA on cocaine-seeking behavior and relapse-clinical trials. Treatment in a Detoxification center: DHEA effect as a complementary treatment was evaluated in opiate addicts undergoing detoxification. DHEA (100 mg/day) or placebo was added as a to the routine medication protocol in a randomized, double-blind controlled study. Severity of withdrawal symptoms (A), depression (B), and anxiety (C) scores were measured as a follow-up for 12 months. #$P < 0.05$ indicate statistically significant. *HARS*, Hamilton anxiety scores; *HDRS*, Hamilton depression scores. Treatment in a rehabilitation center: A double-blind, placebo-controlled study examined the effect of DHEA in adult polydrug users taking part in a detoxification program enriched with intensive

(Continued)

(Wilkins et al., 2005). This has led to the suggestion that increased circulating DHEAS levels may enhance brain resiliency during withdrawal by lowering the distressed mood levels of addicts (Doron, Fridman, Gispan-Herman, et al., 2006; Doron, Fridman, & Yadid, 2006; Wilkins et al., 2005). In research conducted by Ohana et al., participants (79% males, mean age 28) consumed DHEA (100 mg/day) or placebo daily for at least 30 days during their routine treatment. Of the 121 initial volunteers, 64 participated (34 placebo, 30 DHEA) for at least 1 month. The initial evaluation meeting with the participants consisted of two sessions in which they provided demographic details, performed psychological tests, and gave blood samples. In addition, further assessments involving psychological testing and blood samples were collected after 1, 4.5, and 6 months (study termination). They found that DHEA treatment resulted in an increase in DHEAS 1 month following treatment, and the level of DHEAS predicted relapse in the follow-up assessment. Additionally, based on the PANAS scale, DHEA seemed to decrease negative effect during treatment. Importantly, in a 16-month follow-up, the reuse rates in the DHEA condition were about a third compared with placebo. It seems that these results highlight the potential relevance of findings in animal studies of DHEA to recovery following addiction of human addicts. The current findings offer early confirmation of the potential long-term effect of DHEA on drug reuse.

4. EPILOG

The essence of this chapter is to suggest the use of dehydroepiandrosterone (DHEA), a natural neurosteroid shown to prevent relapse to drug use. *DHEA add-on treatment* is an innovative and highly effective intervention, which increases the percentage of rehabilitated addicts, more rapidly and for prolonged periods of time. Hence, we propose that *DHEA add-on treatment*

Fig. 3—Cont'd psychosocial interventions and aftercare. During treatment, participants consumed DHEA (100 mg/day) or placebo daily for at least 30 days. While in treatment, DHEA significantly improved quality of life (QOL; D), anxiety (E), and release rates in a 16-month follow-up about a third compared to placebo (F). $^{\#}P<0.01$ indicate significance between treatment groups. *Modified from Maayan, R., Touati-Werner, D., Shamir, D., Yadid, G., Friedman, A., Eisner, D., et al. (2008). The effect of DHEA complementary treatment on heroin addicts participating in a rehabilitation program: A preliminary study.* European Neuropsychopharmacology, 18*(6), 406–413. https://doi.org/10.1016/j.euroneuro.2007.12.003.*

may be seamlessly incorporated into existing rehabilitation programs, complementing, and enriching the various therapies in each center.

Drug abuse and addiction have negative consequences for individuals as well as for society as a whole. Addiction is a complex disease with high relapse rates and no reliable treatment. DHEA, together with a unique software and novel algorithms that can integrate many output parameters from the patient dynamically, may enable to evaluate the progress of rehabilitation of an individual patient, in a comprehensive assessment. Such a program may boost and support the detoxification and rehabilitation process, and help patients regain a normal life in a shorter amount of time.

ACKNOWLEDGMENTS

This study was supported by a fellowship from the Israel Anti-Drug Authority and the Wolf Foundation, Israel. The research described here was conducted as part of the Ph.D. dissertation of H.A.-L. and T.B.

The authors wish to thank Cheryl Balshayi for critically reading the manuscript and Ori Yadid for his help with graphical illustrations.

REFERENCES

Aguilar, M. A., Rodríguez-Arias, M., & Miñarro, J. (2009). Neurobiological mechanisms of the reinstatement of drug-conditioned place preference. *Brain Research Reviews, 59*, 253–277. https://doi.org/10.1016/j.brainresrev.2008.08.002.

Anker, J. J., Holtz, N. A., Zlebnik, N., & Carroll, M. E. (2009). Effects of allopregnanolone on the reinstatement of cocaine-seeking behavior in male and female rats. *Psychopharmacology, 203*(1), 63–72. https://doi.org/10.1007/s00213-008-1371-9.

Aydar, E., Palmer, C. P., Klyachko, V. A., & Jackson, M. B. (2002). The sigma receptor as a ligand-regulated auxiliary potassium channel subunit. *Neuron, 34*(3), 399–410. https://doi.org/10.1016/S0896-6273(02)00677-3.

Baker, D. A., McFarland, K., Lake, R. W., Shen, H., Tang, X.-C., Toda, S., et al. (2003). Neuroadaptations in cystine-glutamate exchange underlie cocaine relapse. *Nature Neuroscience, 6*(7), 743–749. https://doi.org/10.1038/nn1069.

Ben-Ami, O., Kinor, N., Perelman, A., & Yadid, G. (2006). Dopamine-1 receptor agonist, but not cocaine, modulates sigma(1) gene expression in SVG cells. *Journal of Molecular Neuroscience, 29*(2), 169–176. https://doi.org/10.1385/JMN:29:2:169.

Bercik, P., Denou, E., Collins, J., Jackson, W., Lu, J., Jury, J., et al. (2011). The intestinal microbiota affect central levels of brain-derived neurotropic factor and behavior in mice. *Gastroenterology, 141*(2), 599–609. https://doi.org/10.1053/j.gastro.2011.04.052.

Brzoza, Z., Kasperska-Zajac, A., Badura-Brzoza, K., Matysiakiewicz, J., Hese, R. T., & Rogala, B. (2008). Decline in dehydroepiandrosterone sulfate observed in chronic urticaria is associated with psychological distress. *Psychosomatic Medicine, 70*(6), 723–728. https://doi.org/10.1097/PSY.0b013e31817bcc8d.

Buchanan, J. (2006). Understanding problematic drug use: A medical matter or a social issue? *British Journal of Community Justice, 4*(2), 387–397. Retrieved from http://epubs.glyndwr.ac.uk/siru.

Burokas, A., Arboleya, S., Moloney, R. D., Peterson, V. L., Murphy, K., Clarke, G., et al. (2016). Targeting the microbiota-gut-brain axis: Prebiotics have anxiolytic and antidepressant-like effects and reverse the impact of chronic stress in mice. *Biological Psychiatry*, *82*, 456–457. https://doi.org/10.1016/j.biopsych.2016.12.031.

Buydens-Branchey, L., Branchey, M., Hudson, J., & Dorota Majewska, M. (2002). Perturbations of plasma cortisol and DHEA-S following discontinuation of cocaine use in cocaine addicts. *Psychoneuroendocrinology*, *27*(1–2), 83–97. https://doi.org/10.1016/S0306-4530(01)00037-3.

Chao, J., & Nestler, E. J. (2004). Molecular neurobiology of drug addiction. *Annual Review of Medicine*, *55*, 113–132. https://doi.org/10.1146/annurev.med.55.091902.103730.

Chen, Y., Hajipour, A. R., Sievert, M. K., Arbabian, M., & Ruoho, A. E. (2007). Characterization of the cocaine binding site on the sigma-1 receptor. *Biochemistry*, *46*(11), 3532–3542. https://doi.org/10.1021/bi061727o.

Colangelo, A. M., Alberghina, L., & Papa, M. (2014). Astrogliosis as a therapeutic target for neurodegenerative diseases. *Neuroscience Letters*, *565*, 59–64. https://doi.org/10.1016/j.neulet.2014.01.014.

Di Chiara, G. (1999). Drug addiction as dopamine-dependent associative learning disorder. *European Journal of Pharmacology*, *375*(1–3), 13–30. Retrieved from http://www.ncbi.nlm.nih.gov/pubmed/10443561.

Dikshtein, Y., Barnea, R., Kronfeld, N., Lax, E., Roth-Deri, I., Friedman, A., et al. (2013). β-Endorphin via the delta opioid receptor is a major factor in the incubation of cocaine craving. *Neuropsychopharmacology*, *38*(12), 2508–2514. https://doi.org/10.1038/npp.2013.155.

Dong, L. Y., Cheng, Z. X., Fu, Y. M., Wang, Z. M., Zhu, Y. H., Sun, J. L., et al. (2007). Neurosteroid dehydroepiandrosterone sulfate enhances spontaneous glutamate release in rat prelimbic cortex through activation of dopamine D1 and sigma-1 receptor. *Neuropharmacology*, *52*(3), 966–974. https://doi.org/10.1016/j.neuropharm.2006.10.015.

Doron, R., Fridman, L., Gispan-Herman, I., Maayan, R., Weizman, A., & Yadid, G. (2006). DHEA, a neurosteroid, decreases cocaine self-administration and reinstatement of cocaine-seeking behavior in rats. *Neuropsychopharmacology*, *31*(10), 2231–2236. https://doi.org/10.1038/sj.npp.1301013.

Doron, R., Fridman, L., & Yadid, G. (2006). Dopamine-2 receptors in the arcuate nucleus modulate cocaine-seeking behavior. *Neuroreport*, *17*(15), 1633–1636. https://doi.org/10.1097/01.wnr.0000234755.88560.c7.

Dubrovsky, B. O. (2005). Steroids, neuroactive steroids and neurosteroids in psychopathology. *Progress in Neuro-Psychopharmacology & Biological Psychiatry*, *29*, 169–192. https://doi.org/10.1016/j.pnpbp.2004.11.001.

Erb, S., Shaham, Y., & Stewart, J. (1998). The role of corticotropin-releasing factor and corticosterone in stress- and cocaine-induced relapse to cocaine seeking in rats. *The Journal of Neuroscience: The Official Journal of the Society for Neuroscience*, *18*(14), 5529–5536. Retrieved from http://www.ncbi.nlm.nih.gov/pubmed/9651233.

Finn, P. R. (2002). Motivation, working memory, and decision making: A cognitive-motivational theory of personality vulnerability to alcoholism. *Behavioral and Cognitive Neuroscience Reviews*, *1*(3), 183–205. https://doi.org/10.1177/1534582302001003001.

Geier, C., & Luna, B. (2009). The maturation of incentive processing and cognitive control. *Pharmacology, Biochemistry, and Behavior*, *93*, 212–221. https://doi.org/10.1016/j.pbb.2009.01.021.

Genud, R., Merenlender, A., Gispan-Herman, I., Maayan, R., Weizman, A., & Yadid, G. (2009). DHEA lessens depressive-like behavior via GABA-ergic modulation of the mesolimbic system. *Neuropsychopharmacology*, *34*(3), 577–584. https://doi.org/10.1038/npp.2008.46.

Goeders, N. E. (1997). A neuroendocrine role in cocaine reinforcement. *Psychoneuroendocrinology*, *22*(4), 237–259. https://doi.org/10.1016/S0306-4530(97)00027-9.
Goeders, N. E. (2002a). Stress and cocaine addiction. *The Journal of Pharmacology and Experimental Therapeutics*, *301*(3), 785–789. https://doi.org/10.1124/jpet.301.3.785.
Goeders, N. E. (2002b). The HPA axis and cocaine reinforcement. *Psychoneuroendocrinology*, *27*, 13–33. https://doi.org/10.1016/S0306-4530(01)00034-8.
Goeders, N. E., & Clampitt, D. M. (2002). Potential role for the hypothalamo-pituitary-adrenal axis in the conditioned reinforcer-induced reinstatement of extinguished cocaine seeking in rats. *Psychopharmacology*, *161*(3), 222–232. https://doi.org/10.1007/s00213-002-1007-4.
Goeders, N. E., & Guerin, G. F. (1996). Effects of surgical and pharmacological adrenalectomy on the initiation and maintenance of intravenous cocaine self-administration in rats. *Brain Research*, *722*(1–2), 145–152. https://doi.org/10.1016/0006-8993(96)00206-5.
Guerin, G. F., Schmoutz, C. D., & Goeders, N. E. (2014). The extra-adrenal effects of metyrapone and oxazepam on ongoing cocaine self-administration. *Brain Research*, *1575*, 45–54. https://doi.org/10.1016/j.brainres.2014.05.039.
Haney, M. (2009). Self-administration of cocaine, cannabis and heroin in the human laboratory: Benefits and pitfalls. *Addiction Biology*, *14*, 9–21. https://doi.org/10.1111/j.1369-1600.2008.00121.x.
Hayashi, T., & Su, T. P. (2001). Regulating ankyrin dynamics: Roles of sigma-1 receptors. *Proceedings of the National Academy of Sciences of the United States of America*, *98*(2), 491–496. https://doi.org/10.1073/pnas.98.2.491.
Hayashi, T., & Su, T. P. (2003). Intracellular dynamics of sigma-1 receptors (sigma(1) binding sites) in NG108-15 cells. *The Journal of Pharmacology and Experimental Therapeutics*, *306*(2), 726–733. https://doi.org/10.1124/jpet.103.051292.
Hayashi, T., & Su, T. P. (2007). Sigma-1 receptor chaperones at the ER-mitochondrion interface regulate Ca2+ signaling and cell survival. *Cell*, *131*(3), 596–610. https://doi.org/10.1016/j.cell.2007.08.036.
Hiranita, T., Soto, P. L., Tanda, G., & Katz, J. L. (2010). Reinforcing effects of sigma-receptor agonists in rats trained to self-administer cocaine. *The Journal of Pharmacology and Experimental Therapeutics*, *332*(2), 515–524. https://doi.org/10.1124/jpet.109.159236.
Huerta-García, E., Montiél-Dávalos, A., Alfaro-Moreno, E., Gutiérrez-Iglesias, G., & López-Marure, R. (2013). Dehydroepiandrosterone protects endothelial cells against inflammatory events induced by urban particulate matter and titanium dioxide nanoparticles. *BioMed Research International*, *2013*, 382058. https://doi.org/10.1155/2013/382058.
Jackson, A., Mead, A. N., Rocha, B. A., & Stephens, D. N. (1998). AMPA receptors and motivation for drug: Effect of the selective antagonist NBQX on behavioural sensitization and on self-administration in mice. *Behavioural Pharmacology*, *9*(5–6), 457–467.
Kalivas, P. W. (2009). The glutamate homeostasis hypothesis of addiction. *Nature Reviews Neuroscience*, *10*(8), 561–572. https://doi.org/10.1038/nrn2515.
Kauer, J. A., & Malenka, R. C. (2007). Synaptic plasticity and addiction. *Nature Reviews Neuroscience*, *8*(11), 844–858. https://doi.org/10.1038/nrn2234.
Kenny, P. J. (2014). Epigenetics, microRNA, and addiction. *Dialogues in Clinical Neuroscience*, *16*(3), 335–344. Retrieved from http://www.ncbi.nlm.nih.gov/pubmed/25364284.
Kimonides, V. G., Khatibi, N. H., Svendsen, C. N., Sofroniew, M. V., & Herbert, J. (1998). Dehydroepiandrosterone (DHEA) and DHEA-sulfate (DHEAS) protect hippocampal neurons against excitatory amino acid-induced neurotoxicity. *Proceedings of the National Academy of Sciences of the United States of America*, *95*(4), 1852–1857. https://doi.org/10.1073/pnas.95.4.1852.

Kiraly, D. D., Walker, D. M., Calipari, E. S., Labonte, B., Issler, O., Pena, C. J., et al. (2016). Alterations of the host microbiome affect behavioral responses to cocaine. *Scientific Reports*, *6*(1), 35455. https://doi.org/10.1038/srep35455.

Koob, G. F. (2009). Dynamics of neuronal circuits in addiction: Reward, antireward, and emotional memory. *Pharmacopsychiatry*, *42*(S 01), S32–S41. https://doi.org/10.1055/s-0029-1216356.

Koob, G. F., Buck, C. L., Cohen, A., Edwards, S., Park, P. E., Schlosburg, J. E., et al. (2014). Addiction as a stress surfeit disorder. *Neuropharmacology*, *76*, 370–382. https://doi.org/10.1016/j.neuropharm.2013.05.024.

Kwako, L. E., & Koob, G. F. (2017). Neuroclinical framework for the role of stress in addiction. *Chronic Stress*, *1*, 247054701769814. https://doi.org/10.1177/2470547017698140.

Levy, A. D., Li, Q. A., Kerr, J. E., Rittenhouse, P. A., Milonas, G., Cabrera, T. M., et al. (1991). Cocaine-induced elevation of plasma adrenocorticotropin hormone and corticosterone is mediated by serotonergic neurons. *The Journal of Pharmacology and Experimental Therapeutics*, *259*(2), 495–500.

Liu, Y., & Matsumoto, R. R. (2008). Alterations in fos-related antigen 2 and sigma1 receptor gene and protein expression are associated with the development of cocaine-induced behavioral sensitization: Time course and regional distribution studies. *The Journal of Pharmacology and Experimental Therapeutics*, *327*(1), 187–195. https://doi.org/10.1124/jpet.108.141051.

Lyte, M., Vulchanova, L., & Brown, D. R. (2011). Stress at the intestinal surface: Catecholamines and mucosa-bacteria interactions. *Cell and Tissue Research*, *343*, 23–32. https://doi.org/10.1007/s00441-010-1050-0.

Maayan, R., Lotan, S., Doron, R., Shabat-Simon, M., Gispan-Herman, I., Weizman, A., et al. (2006). Dehydroepiandrosterone (DHEA) attenuates cocaine-seeking behavior in the self-administration model in rats. *European Neuropsychopharmacology*, *16*(5), 329–339. https://doi.org/10.1016/j.euroneuro.2005.10.002.

Maayan, R., Morad, O., Dorfman, P., Overstreet, D. H., Weizman, A., & Yadid, G. (2005). The involvement of dehydroepiandrosterone (DHEA) and its sulfate ester (DHEAS) in blocking the therapeutic effect of electroconvulsive shocks in an animal model of depression. *European Neuropsychopharmacology: The Journal of the European College of Neuropsychopharmacology*, *15*(3), 253–262. https://doi.org/10.1016/j.euroneuro.2004.10.005.

Maayan, R., Touati-Werner, D., Shamir, D., Yadid, G., Friedman, A., Eisner, D., et al. (2008). The effect of DHEA complementary treatment on heroin addicts participating in a rehabilitation program: A preliminary study. *European Neuropsychopharmacology*, *18*(6), 406–413. https://doi.org/10.1016/j.euroneuro.2007.12.003.

Majewska, M. D. (2002). HPA axis and stimulant dependence: An enigmatic relationship. *Psychoneuroendocrinology*, *27*(1–2), 5–12. https://doi.org/10.1016/S0306-4530(01)00033-6.

Mantsch, J. R., Baker, D. A., Serge, J. P., Hoks, M. A., Francis, D. M., & Katz, E. S. (2008). Surgical adrenalectomy with diurnal corticosterone replacement slows escalation and prevents the augmentation of cocaine-induced reinstatement in rats self-administering cocaine under long-access conditions. *Neuropsychopharmacology*, *33*(4), 814–826. https://doi.org/10.1038/sj.npp.1301464.

Martin-Soelch, C., Chevalley, A. F., Künig, G., Missimer, J., Magyar, S., Mino, A., et al. (2001). Changes in reward-induced brain activation in opiate addicts. *European Journal of Neuroscience*, *14*(8), 1360–1368. https://doi.org/10.1046/j.0953-816X.2001.01753.x.

Massart, R., Barnea, R., Dikshtein, Y., Suderman, M., Meir, O., Hallett, M., et al. (2015). Role of DNA methylation in the nucleus accumbens in incubation of cocaine craving. *The Journal of Neuroscience: The Official Journal of the Society for Neuroscience*, *35*(21), 8042–8058. https://doi.org/10.1523/JNEUROSCI.3053-14.2015.

Matsumoto, R. R., Liu, Y., Lerner, M., Howard, E. W., & Brackett, D. J. (2003). Sigma receptors: Potential medications development target for anti-cocaine agents. *European Journal of Pharmacology*, *469*(1–3), 1–12. Retrieved from http://www.ncbi.nlm.nih.gov/pubmed/12782179.

Maurice, T., Martin-Fardon, R., Romieu, P., & Matsumoto, R. R. (2002). Sigma1 (σ1) receptor antagonists represent a new strategy against cocaine addiction and toxicity. *Neuroscience and Biobehavioral Reviews*, *26*, 499–527. https://doi.org/10.1016/S0149-7634(02)00017-9.

Maurice, T., Phan, V. L., Urani, A., Kamei, H., Noda, Y., & Nabeshima, T. (1999). Neuroactive neurosteroids as endogenous effectors for the sigma1 (sigma1) receptor: Pharmacological evidence and therapeutic opportunities. *Japanese Journal of Pharmacology*, *81*(2), 125–155. https://doi.org/10.1254/jjp.81.125.

McCollister, K., Yang, X., Sayed, B., French, M. T., Leff, J. A., & Schackman, B. R. (2017). Monetary conversion factors for economic evaluations of substance use disorders. *Journal of Substance Abuse Treatment*, *81*, 25–34. https://doi.org/10.1016/j.jsat.2017.07.008.

McLellan, A. T., Lewis, D. C., O'Brien, C. P., & Kleber, H. D. (2000). Drug dependence, a chronic medical illness. *JAMA*, *284*(13), 1689. https://doi.org/10.1001/jama.284.13.1689.

McLellan, A. T., & Weisner, C. (1996). *Achieving the public health and safety potential of substance abuse treatments. In Drug policy and human nature* (pp. 127–154). Boston, MA: Springer. *https://doi.org/10.1007/978-1-4899-3591-5_6.*

Mellon, S. H., & Griffin, L. D. (2002a). Neurosteroids: Biochemistry and clinical significance. *Trends in Endocrinology and Metabolism: TEM*, *13*(1), 35–43. Retrieved from http://www.ncbi.nlm.nih.gov/pubmed/11750861.

Mellon, S. H., & Griffin, L. D. (2002b). Synthesis, regulation, and function of neurosteroids. *Endocrine Research*, *28*(4), 463. Retrieved from http://www.ncbi.nlm.nih.gov/pubmed/12530649.

Mendelson, J. H., Mello, N. K., Sholar, M. B., Siegel, A. J., Mutschler, N., & Halpern, J. (2002). Temporal concordance of cocaine effects on mood states and neuroendocrine hormones. *Psychoneuroendocrinology*, *27*(1–2), 71–82. https://doi.org/10.1016/S0306-4530(01)00036-1.

Miyatake, R., Furukawa, A., Matsushita, S., Higuchi, S., & Suwaki, H. (2004). Functional polymorphisms in the sigma1 receptor gene associated with alcoholism. *Biological Psychiatry*, *55*(1), 85–90. https://doi.org/10.1016/j.biopsych.2003.07.008.

Moldow, R. L., & Fischman, A. J. (1987). Cocaine induced secretion of ACTH, beta-endorphin, and corticosterone. *Peptides*, *8*(5), 819–822. https://doi.org/10.1016/0196-9781(87)90065-9.

Mongi-Bragato, B., Zamponi, E., García-Keller, C., Assis, M. A., Virgolini, M. B., Mascó, D. H., et al. (2016). Enkephalin is essential for the molecular and behavioral expression of cocaine sensitization. *Addiction Biology*, *21*(2), 326–338. https://doi.org/10.1111/adb.12200.

Mucha, R. F., Van Der Kooy, D., O'Shaughnessy, M., & Bucenieks, P. (1982). Drug reinforcement studied by the use of place conditioning in rat. *Brain Research*, *243*(1), 91–105. https://doi.org/10.1016/0006-8993(82)91123-4.

Nachshoni, T., Ebert, T., Abramovitch, Y., Assael-Amir, M., Kotler, M., Maayan, R., et al. (2005). Improvement of extrapyramidal symptoms following dehydroepiandrosterone (DHEA) administration in antipsychotic treated schizophrenia patients: A randomized, double-blind placebo controlled trial. *Schizophrenia Research*, *79*(2–3), 251–256. https://doi.org/10.1016/j.schres.2005.07.029.

Nestler, E. J., & Malenka, R. C. (2004). The addicted brain. *Scientific American*, *290*(3), 78–85. https://doi.org/10.1038/scientificamerican0304-78.

Nguyen, L., Lucke-Wold, B. P., Mookerjee, S. A., Cavendish, J. Z., Robson, M. J., Scandinaro, A. L., et al. (2015). Role of sigma-1 receptors in neurodegenerative diseases. *Journal of Pharmacological Sciences*, *127*(1), 17–29. https://doi.org/10.1016/j.jphs.2014.12.005.

Nie, H., & Janak, P. H. (2003). Comparison of reinstatement of ethanol- and sucrose-seeking by conditioned stimuli and priming injections of allopregnanolone after extinction in rats. *Psychopharmacology*, *168*(1–2), 222–228. https://doi.org/10.1007/s00213-003-1468-0.

Ohana, D., Maayan, R., Delayahu, Y., Roska, P., Ponizovsky, A. M., Weizman, A., et al. (2016). Effect of dehydroepiandrosterone add-on therapy on mood, decision making and subsequent relapse of polydrug users. *Addiction Biology*, *21*(4), 885–894. https://doi.org/10.1111/adb.12241.

Panlilio, L. V., & Goldberg, S. R. (2007). Self-administration of drugs in animals and humans as a model and an investigative tool. *Addiction*, *102*, 1863–1870. https://doi.org/10.1111/j.1360-0443.2007.02011.x.

Peters, J., Kalivas, P. W., & Quirk, G. J. (2009). Extinction circuits for fear and addiction overlap in prefrontal cortex. *Learning & Memory*, *16*(5), 279–288. https://doi.org/10.1101/lm.1041309.

Piazza, P. V., & Le Moal, M. (1998). The role of stress in drug self-administration. *Trends in Pharmacological Sciences*, *19*, 67–74. https://doi.org/10.1016/S0165-6147(97)01115-2.

Piazza, P. V., Maccari, S., Deminiere, J. M., Le Moal, M., Mormede, P., & Simon, H. (1991). Corticosterone levels determine individual vulnerability to amphetamine self-administration. *Proceedings of the National Academy of Sciences of the United States of America*, *88*(6), 2088–2092. https://doi.org/10.1073/pnas.88.6.2088.

Polter, A. M., & Kauer, J. A. (2014). Stress and VTA synapses: Implications for addiction and depression. *European Journal of Neuroscience*, *39*(7), 1179–1188. https://doi.org/10.1111/ejn.12490.

Reddy, D. S., & Kulkarni, S. K. (1997a). Chronic neurosteroid treatment prevents the development of morphine tolerance and attenuates abstinence behavior in mice. *European Journal of Pharmacology*, *337*(1), 19–25. https://doi.org/10.1016/S0014-2999(97)01294-6.

Reddy, D. S., & Kulkarni, S. K. (1997b). Reversal of benzodiazepine inverse agonist FG 7142-induced anxiety syndrome by neurosteroids in mice. *Methods and Findings in Experimental and Clinical Pharmacology*, *19*(10), 665–681.

Reeves, R., Thiruchelvam, M., & Cory-Slechta, D. A. (2004). Expression of behavioral sensitization to the cocaine-like fungicide triadimefon is blocked by pretreatment with AMPA, NMDA and DA D1 receptor antagonists. *Brain Research*, *1008*(2), 155–167. https://doi.org/10.1016/j.brainres.2004.01.079.

Ren, X., Noda, Y., Mamiya, T., Nagai, T., & Nabeshima, T. (2004). A neuroactive steroid, dehydroepiandrosterone sulfate, prevents the development of morphine dependence and tolerance via c-fos expression linked to the extracellular signal-regulated protein kinase. *Behavioural Brain Research*, *152*(2), 243–250. https://doi.org/10.1016/j.bbr.2003.10.013.

Ritsner, M. S., Gibel, A., Ratner, Y., Tsinovoy, G., & Strous, R. D. (2006). Improvement of sustained attention and visual and movement skills, but not clinical symptoms, after dehydroepiandrosterone augmentation in schizophrenia: A randomized, double-blind, placebo-controlled, crossover trial. *Journal of Clinical Psychopharmacology*, *26*, 495–499. https://doi.org/10.1097/01.jcp.0000237942.50270.35.

Romieu, P., Martin-Fardon, R., Bowen, W. D., & Maurice, T. (2003). Sigma 1 receptor-related neuroactive steroids modulate cocaine-induced reward. *The Journal of Neuroscience*, *23*(9), 3572–3576. https://doi.org/23/9/3572. [pii].

Roth-Deri, I., Schindler, C. J., & Yadid, G. (2004). A critical role for beta-endorphin in cocaine-seeking behavior. *Neuroreport*, *15*(3), 519–521. https://insights.ovid.com/pubmed?pmid=15094515.

Roth-Deri, I., Zangen, A., Aleli, M., Goelman, R. G., Pelled, G., Nakash, R., et al. (2003). Effect of experimenter-delivered and self-administered cocaine on extracellular beta-endorphin levels in the nucleus accumbens. *Journal of Neurochemistry*, *84*(5), 930–938. https://doi.org/10.1046/j.1471-4159.2003.01584.x.

Rupprecht, R., & Holsboer, F. (1999). Neuroactive steroids: Mechanisms of action and neuropsychopharmacological perspectives. *Trends in Neurosciences*, *22*, 410–416. https://doi.org/10.1016/S0166-2236(99)01399-5.

Saphier, D., Welch, J. E., Farrar, G. E., & Goeders, N. E. (1993). Effects of intracerebroventricular and intrahypothalamic cocaine administration on adrenocortical secretion. *Neuroendocrinology*, *57*(1), 54–62.

Sarnyai, Z., Shaham, Y., & Heinrichs, S. C. (2001). The role of corticotropin-releasing factor in drug addiction. *Pharmacological Reviews*, *53*(2), 209–243. https://doi.org/10.2165/11587790-000000000-00000.

Self, D. W., & Choi, K. H. (2004). Extinction-induced neuroplasticity attenuates stress-induced cocaine seeking: A state-dependent learning hypothesis. *Stress*, 7, 145–155. https://doi.org/10.1080/10253890400012677.

Shah, A. H., Chin, E. H., Schmidt, K. L., & Soma, K. K. (2011). DHEA and estradiol levels in brain, gonads, adrenal glands, and plasma of developing male and female European starlings. *Journal of Comparative Physiology. A, Neuroethology, Sensory, Neural, and Behavioral Physiology*, *197*(10), 949–958. https://doi.org/10.1007/s00359-011-0655-4.

Shaham, Y., Shalev, U., Lu, L., De Wit, H., & Stewart, J. (2003). The reinstatement model of drug relapse: History, methodology and major findings. *Psychopharmacology*, *168*, 3–20. https://doi.org/10.1007/s00213-002-1224-x.

Shalev, U., Marinelli, M., Baumann, M. H., Piazza, P. V., & Shaham, Y. (2003). The role of corticosterone in food deprivation-induced reinstatement of cocaine seeking in the rat. *Psychopharmacology*, *168*(1–2), 170–176. https://doi.org/10.1007/s00213-002-1200-5.

Shoptaw, S., Majewska, M. D., Wilkins, J., Twitchell, G., Yang, X., & Ling, W. (2004). Participants receiving dehydroepiandrosterone during treatment for cocaine dependence show high rates of cocaine use in a placebo-controlled pilot study. *Experimental and Clinical Psychopharmacology*, *12*(2), 126–135. https://doi.org/10.1037/1064-1297.12.2.126.

Smith, A. C. W., & Kenny, P. J. (2017). MicroRNAs regulate synaptic plasticity underlying drug addiction. *Genes, Brain, and Behavior*, *17*, 2. https://doi.org/10.1111/gbb.12424.

Stefanski, R., Justinova, Z., Hayashi, T., Takebayashi, M., Goldberg, S. R., & Su, T. P. (2004). Sigma1 receptor upregulation after chronic methamphetamine self-administration in rats: A study with yoked controls. *Psychopharmacology*, *175*(1), 68–75. https://doi.org/10.1007/s00213-004-1779-9.

Stress psychobiology in the context of addiction medicine: from drugs of abuse to behavioral addictions. (2016). *223*, 43–62. https://doi.org/10.1016/BS.PBR.2015.08.001.

Su, T.-P., & Hayashi, T. (2003). Understanding the molecular mechanism of sigma-1 receptors: Towards a hypothesis that sigma-1 receptors are intracellular amplifiers for signal transduction. *Current Medicinal Chemistry*, *10*(20), 2073–2080. https://doi.org/10.2174/0929867033456783.

Substance Abuse and Mental Health Services Administration (US) & Office of the Surgeon General (US). (2016). *Facing addiction in America. Facing addiction in America: The surgeon general's report on alcohol, drugs, and health.* US Department of Health and Human Services. Retrieved from http://www.ncbi.nlm.nih.gov/pubmed/28252892.

Sudai, E., Croitoru, O., Shaldubina, A., Abraham, L., Gispan, I., Flaumenhaft, Y., et al. (2011). High cocaine dosage decreases neurogenesis in the hippocampus and impairs working memory. *Addiction Biology*, *16*(2), 251–260. https://doi.org/10.1111/j.1369-1600.2010.00241.x.

Sutton, M. A., Schmidt, E. F., Choi, K. H., Schad, C. A., Whisler, K., Simmons, D., et al. (2003). Extinction-induced upregulation in AMPA receptors reduces cocaine-seeking behaviour. *Nature*, *421*(6918), 70–75. https://doi.org/10.1038/nature01249.

Tomkins, D. M., & Sellers, E. M. (2001). Addiction and the brain: The role of neurotransmitters in the cause and treatment of drug dependence. *CMAJ: Canadian Medical Association Journal = Journal de l'Association Medicale Canadienne, 164*(6), 817–821. Retrieved from http://www.ncbi.nlm.nih.gov/pubmed/11276551.

Ujike, H., Kanzaki, A., Okumura, K., Akiyama, K., & Otsuki, S. (1992). Sigma antagonist BMY 14802 prevents methamphetamine-induced sensitization. *Life Sciences, 50*(16), 129–134. https://doi.org/10.1016/0024-3205(92)90466-3.

Ujike, H., Kuroda, S., & Otsuki, S. (1996). σ Receptor antagonists block the development of sensitization to cocaine. *European Journal of Pharmacology, 296*(2), 123–128. https://doi.org/10.1016/0014-2999(95)00693-1.

Ulmann, L., Rodeau, J. L., Danoux, L., Contet-Audonneau, J. L., Pauly, G., & Schlichter, R. (2009). Dehydroepiandrosterone and neurotrophins favor axonal growth in a sensory neuron–keratinocyte coculture model. *Neuroscience, 159*(2), 514–525. https://doi.org/10.1016/J.NEUROSCIENCE.2009.01.018.

Vescovi, P. P., Coiro, V., Volpi, R., & Passeri, M. (1992). Diurnal variations in plasma ACTH, cortisol and beta-endorphin levels in cocaine addicts. *Hormone Research, 37*(6), 221–224. https://doi.org/10.1159/000182316.

Wang, Y. T. (2008). Probing the role of AMPAR endocytosis and long-term depression in behavioural sensitization: Relevance to treatment of brain disorders, including drug addiction. *British Journal of Pharmacology, 153*(Suppl(July 2016)), S389–S395. https://doi.org/10.1038/sj.bjp.0707616.

Wilkins, J. N., Majewska, M. D., Van Gorp, W., Li, S. H., Hinken, C., Plotkin, D., et al. (2005). DHEAS and POMS measures identify cocaine dependence treatment outcome. *Psychoneuroendocrinology, 30*(1), 18–28. https://doi.org/10.1016/j.psyneuen.2004.04.006.

Wolkowitz, O. M., Kramer, J. H., Reus, V. I., Costa, M. M., Yaffe, K., Walton, P., et al. (2003). DHEA treatment of Alzheimer's disease: A randomized, double-blind, placebo-controlled study. *Neurology, 60*(7), 1071–1076. https://doi.org/10.1212/01.WNL.0000052994.54660.58.

Wolkowitz, O. M., & Reus, V. I. (2003). *Neurotransmitters, neurosteroids and neurotrophins: New models of the pathophysiology and treatment of depression*. PubMed—NCBI. Retrieved November 20, 2017, from https://www.ncbi.nlm.nih.gov/m/pubmed/12872201/.

Wolkowitz, O. M., Reus, V. I., Keebler, A., Nelson, N., Friedland, M., Brizendine, L., et al. (1999). Double-blind treatment of major depression with dehydroepiandrosterone. *American Journal of Psychiatry, 156*(4), 646–649. https://doi.org/10200751.

Wolkowitz, O. M., Reus, V. I., Roberts, E., Manfredi, F., Chan, T., Raum, W. J., et al. (1997). Dehydroepiandrosterone (DHEA) treatment of depression. *Biological Psychiatry, 41*(3), 311–318. https://doi.org/S0006322396000431.

Yadid, G., Redlus, L., Barnea, R., & Doron, R. (2012). Modulation of mood states as a major factor in relapse to substance use. *Frontiers in Molecular Neuroscience, 5*, 81. https://doi.org/10.3389/fnmol.2012.00081.

Yadid, G., Sudai, E., Maayan, R., Gispan, I., & Weizman, A. (2010). The role of dehydroepiandrosterone (DHEA) in drug-seeking behavior. *Neuroscience & Biobehavioral Reviews, 35*(2), 303–314. https://doi.org/10.1016/j.neubiorev.2010.03.003.

CHAPTER SIXTEEN

Dehydroepiandrosterone and Dehydroepiandrosterone-Sulfate and Emotional Processing

Sónia do Vale[*,†,1], **Carles Escera**[‡,§,¶]
[*]Hospital de Santa Maria, Centro Hospitalar Lisboa Norte, EPE, Lisboa, Portugal
[†]Faculdade de Medicina, Universidade de Lisboa, Lisboa, Portugal
[‡]Institute of Neurosciences, University of Barcelona, Barcelona, Catalonia, Spain
[§]Brainlab—Cognitive Neuroscience Research Group, University of Barcelona, Barcelona, Catalonia, Spain
[¶]Institut de Recerca Sant Joan de Déu, Barcelona, Catalonia, Spain
[1]Corresponding author: e-mail address: sonia.vale@chln.min-saude.pt

Contents

Abstract

Steroid hormones are important regulators of brain development, physiological function, and behavior. Among them, dehydroepiandrosterone (DHEA) and dehydroepiandrosterone-sulfate (DHEAS) also do modulate emotional processing and may have mood enhancement effects. This chapter reviews the studies that bear relation to DHEA and DHEAS [DHEA(S)] and brain emotional processing and behavior. A brief introduction to the mechanisms of action and variations of DHEA(S) levels throughout life has also been forward in this chapter. Higher DHEA(S) levels may reduce activity in brain regions involved in the generation of negative emotions and modulate activity in regions involved in regulatory processes. At the electrophysiological level, higher DHEA-to-cortisol and DHEAS-to-DHEA ratios were related to shorter P300 latencies and shorter P300 amplitudes during the processing of negative stimuli, suggesting less interference of negative stimuli with the task and less processing of the negative information, which in turn may suggest a protective mechanism against negative information overload. Present knowledge indicates that

Vitamins and Hormones, Volume 108
ISSN 0083-6729
https://doi.org/10.1016/bs.vh.2018.01.022

DHEA(S) may play a role in cortical development and plasticity, protecting against negative affect and depression, and at the same time enhancing attention and overall working memory, possibly at the cost of a reduction in emotional processing, emotional memory, and social understanding.

1. INTRODUCTION

In humans, dehydroepiandrosterone-sulfate (DHEAS) is the most abundant hormone in the peripheral circulation (Baulieu & Robel, 1998; Berr, Lafont, Debuire, Dartigues, & Baulieu, 1996) and levels of dehydroepiandrosterone (DHEA) and DHEAS are even higher in brain than in peripheral circulation (Lacroix et al., 1987). Both forms of the hormone are synthesized mostly in the adrenals, but they can also be synthesized in the central nervous system (Baulieu & Robel, 1998). However, DHEA and DHEAS [here, jointly referred to as DHEA(S)] regulation and physiological effects are still a matter of debate. DHEA(S) concentrations rise with adrenarche, coinciding with the development of brain regions involved in cognitive control and emotional processing (Nguyen et al., 2013). Levels dramatically decrease with aging, reaching a nadir by the time cognitive impairment and depression become more prevalent. DHEA(S) in general present a neurostimulatory and antiglucocorticoid effect, among other biological actions (Maninger, Wolkowitz, Reus, Epel, & Mellon, 2009), and they also seem to play a role in brain organization and plasticity (Compagnone & Mellon, 1998; Nguyen et al., 2013).

At behavioral level, higher DHEA(S) or DHEA(S)-to-cortisol ratios were related to improved attention, cognition, and mood (Barrett-Connor & Edelstein, 1994; Barrett-Connor, von Mühlen, Laughlin, & Kripke, 1999). The previous findings raised the hypothesis that restoring DHEA to youthful levels might protect the brain from cognitive decline and improve cognition and mood. In this regard, DHEA(S) were evaluated in the treatment of neuropsychiatric disorders, with published reports appearing as early as 1952 (Strauss, Sands, Robinson, Tindall, & Stevenson, 1952). Until the present date, however, they failed to show any consistent beneficial effect of DHEA administration on cognition but suggested beneficial effects on mood. The evidence on the neurobiological actions of DHEA(S) is growing. Some studies have addressed the effects of DHEA(S) on brain processing at the electrophysiological and functional neuroimaging levels. This chapter addresses a short revision of DHEA(S) changes throughout life and their mechanisms of action, followed by DHEA(S) relations to emotional processing at the neuroimaging, electrophysiological, and behavioral levels.

2. DHEA(S) LEVELS ALONG THE LIFE SPAN

After birth, DHEA(S) levels decline during the first year of life and are maintained at low levels until the age of 5–8 years (Sulcová, Hill, Hampl, & Stárka, 1997). By the time adrenarche starts, increasing amounts of DHEA(S) are secreted from the *zona reticularis* of the adrenal cortex, increasing their serum concentrations. Those levels continue to rise through early adulthood. Of note, there is a 20-fold increase in DHEAS levels between the onset of adrenarche and midpuberty. DHEA is lipophilic and may readily cross the blood–brain barrier, while DHEAS is hydrophilic and its transport to the central nervous system is not completely understood.

Whatever the mechanisms involved, during adrenarche, puberty, and early adulthood, there is a parallel increase in DHEA(S) levels in blood and the cerebrospinal fluid (CSF), and then those levels progressively decrease with aging, while cortisol levels increase from childhood to young adults and the elderly (Guazzo, Kirkpatrick, Goodyear, Shiers, & Herbert, 1996; Kancheva et al., 2011). Cortisol-to-DHEA(S) ratios in the CSF, decrease during adolescence and young adulthood, but increase in the elderly (Guazzo et al., 1996). DHEAS-to-DHEA ratios also increase during adolescence and young adulthood, both in the peripheral circulation and in the CSF (Guazzo et al., 1996). Although its functional relevance is still under debate, the changes in DHEA(S) levels through life are unique and mismatch those of other steroid hormones. Several epidemiologic associations have been found, raising interesting hypothesis about DHEA teleological meaning, but in most cases, causal relationships are yet to be established. Besides humans, adrenarche was shown to occur only in great apes and, to some extent, also in Old World monkeys, suggesting that it evolved to promote brain development (Campbell, 2006).

During adulthood and aging, males maintain higher DHEAS levels in the peripheral circulation than females, whereas unconjugated DHEA levels are not higher in males than in females (Sulcová et al., 1997). By the age of 65–70 years, DHEA(S) levels in the peripheral circulation are about 20% of those in the mid-twenties and there is an attenuated age-related decline in CSF levels (Guazzo et al., 1996; Tannenbaum, Barrett-Connor, Laughlin, & Platt, 2004). Those lower DHEA(S) levels parallel the age of cognitive decline and more frequent depression. Besides, low DHEA(S) levels were also related to higher morbidity and mortality ratios even when corrected for age (Berr et al., 1996). Some authors propose that DHEA(S) are inactive precursors and that their actions are mainly exerted after conversion to active androgens and estrogens in target tissues (intracrinology) (Labrie, Bélanger, Cusan, & Candas, 1997). However, numerous studies also suggest specific central nervous system effects, independent of sex steroids (Maninger et al., 2009).

3. DHEA(S) MECHANISMS OF ACTION, CELLULAR EFFECTS, AND REACTIVITY

Although DHEA(S) are steroid hormones, no high-affinity nuclear receptor has been found for them. However, these molecules have affinity for several membrane receptors, namely, ion-gated channel neurotransmitter receptors, which may contribute to their central nervous system effects. DHEA(S) have a general neurostimulatory effect: presynaptic actions include glutamate, acetylcholine, and norepinephrine release and postsynaptic actions include sigma-1 receptor agonism, *N*-methyl-D-aspartate (NMDA) receptor agonism, γ-aminobutyric acid type A (GABA-A) receptor antagonism, and inhibition of voltage-gated calcium currents (Dong & Zheng, 2011; Maninger et al., 2009). Although they may influence the release of several neurotransmitters at the synaptic terminals, they act primarily through postsynaptic receptors (Baulieu & Robel, 1998; Dong & Zheng, 2011).

DHEA(S) act as noncompetitive antagonists at the GABA-A receptor (Imamura & Prasad, 1998) and as positive allosteric modulators of the NMDA receptor (Compagnone & Mellon, 1998). They may either act directly at the NMDA receptor level or bind to sigma-1 receptors with posterior NMDA receptor activation and sigma-1 signaling may involve several intracellular pathways (Dong & Zheng, 2011). DHEA(S) GABAergic receptor antagonism and sigma-1 receptor agonism might mediate some antidepressive effects of these molecules. In fact, the GABAergic system can mediate depression (Genud et al., 2009) and sigma-1 receptor agonists do cause antidepressant-like effects on rats (Urani, Roman, Phan, Su, & Maurice, 2001). On the other hand, working memory rely on glutamatergic transmission (Yuen et al., 2011), consistent with the hypothesis that DHEA(S) may simultaneously enhance working memory. Sulfated steroids, however, possibly act as endogenous neuromodulators (Gibbs, Russek, & Farb, 2006) and some distinct effects of DHEA and DHEAS have been described. Namely, DHEAS has more potent NMDA agonism and especially GABAminergic antagonism action than DHEA (Imamura & Prasad, 1998). Also, DHEA and DHEAS concentrations vary across different regions of the central nervous system (Dong & Zheng, 2011; Maninger et al., 2009).

Overall, effects of DHEA(S) are complex, depending on factors such as the brain region involved, synaptic functional state, hormone concentration, and acute vs chronic stimulation (Zheng, 2009). For example, DHEAS reduces inhibitory postsynaptic currents at the ventral midbrain by GABA-A receptor antagonism, inhibits persistent sodium currents in the prefrontal cortex (PFC) by sigma-1 receptor agonism, enhances neuronal excitability and facilitates long-term potentiation induction in the hippocampus and cortex through NMDA signaling, and also inhibits voltage-gated calcium currents in the hippocampus CA1 region (Dong & Zheng, 2011). In the pyramidal cells of the prelimbic cortex, low DHEAS concentrations may inhibit presynaptic serotonin-evoked glutamate release, whereas higher DHEAS concentrations can promote spontaneous glutamate release (Dong et al., 2007).

Biological actions of DHEA(S) involve neuroprotection, neurogenesis, neurite growth, and modulation of apoptosis (Maninger et al., 2009). DHEA(S) administration modulated the levels of brain-derived neurotrophic factor, suggesting a role in neurogenesis (Naert, Maurice, Tapia-Arancibia, & Givalois, 2007). Sigma-1 receptor stimulation by DHEA increased neurogenesis in the hippocampal dentate gyrus of mice and ameliorated depressive-like behaviors (Moriguchi et al., 2013). Using embryonic neocortical neurons, DHEA preferentially increased axonal length and DHEAS increased dendritic growth, which could lead to the formation of specific neuronal synapses in the rat neocortex (Compagnone & Mellon, 1998).

Several lines of evidence point toward antiglucocorticoid effects of DHEA(S). However, the mechanism of this antiglucocorticoid effect is not clear. Possible mechanisms are the inhibition of 11β-hydroxysteroid dehydrogenase (11β-HSD) type 1 and upregulation of 11β-HSD type 2, therefore reducing the conversion of cortisone to cortisol (which is the active glucocorticoid) and enhancing the conversion of cortisol to cortisone, respectively (Balazs, Schweizer, Frey, Rohner-Jeanrenaud, & Odermatt, 2008). Of note, these effects are not mediated by DHEA metabolism to estradiol or to testosterone (Balazs et al., 2008). On the other hand, DHEA can be converted into 7α-hydroxy-DHEA or 7β-hydroxy-DHEA in the brain, which are competitive inhibitors of 11β-HSD type 1 (Hennebert, Chalbot, Alran, & Morfin, 2007). Contrary to what might be expected, there is no evidence that DHEA or its active metabolites interfere with glucocorticoid hormone binding with its receptor. Nevertheless, prolonged treatment with DHEA may reduce the glucocorticoid receptor-dependent transcriptional activity (Saponaro, Guarnieri, Pescarmona, & Silvagno, 2007), and it was suggested that DHEA could perhaps interfere with the glucocorticoid receptor binding to the glucocorticoid responsive element sequence, counteracting cortisol effects (Buoso et al., 2011).

DHEA is synthesized in response to corticotrophin (ACTH), even if another unknown and more significant stimulating factor may exist. Accordingly, several studies have shown that DHEA(S) are synthesized in response to acute psychological stressful stimuli. Nevertheless, while acute stress is related to an increase in DHEA(S) levels (Lennartsson, Kushnir, Bergquist, & Jonsdottir, 2012), chronic stress decreases baseline DHEA(S) levels (Jeckel et al., 2010), and DHEA(S) response to a superimposed psychological stress (Lennartsson, Theorell, Kushnir, Bergquist, & Jonsdottir, 2013). DHEA(S) response to emotional stimuli is mostly unexplored. A few studies reported increases in CRH or DHEA(S) levels with either positive, neutral, or negative emotional stimuli (do Vale et al., 2015; Hamilton & Meston, 2011; Martins et al., 2010). Therefore, a DHEA(S) response to emotional stimuli is suggested, even if it is unspecific or still not defined.

Although CRH stimulates corticotrophin secretion and in turn corticotrophin is a stimulus for cortisol and DHEA secretion (do Vale, Martins, Fagundes, & Carmo, 2011; Nieschlag et al., 1973), thus explaining the direct relation between cortisol and DHEA levels (both baseline and responses to stimuli), results suggest that DHEA is not just a stress hormone and that some differential regulation of DHEA and cortisol exists (do Vale et al., 2015). Multiple results even suggest that DHEA may be an antistress hormone, and Boudarene, Legros, and Timsit-Berthier (2002) observed that high-anxiety individuals have a preferential cortisol response, whereas low-anxiety trait individuals show a preferential DHEA(S) response to stress. These authors hypothesize that the antagonism in DHEAS and cortisol might be related to a competition in their synthesis and release by the adrenal gland (Boudarene et al., 2002). We also observed that patients with higher baseline DHEAS levels had lower peak/baseline cortisol response to CRH infusion (do Vale et al., 2011).

4. DHEA(S) AND EMOTIONAL PROCESSING: NEUROIMAGING CORRELATES

The limbic system plays a central role in emotional processing, as initially suggested by the works of Broca (1878), Papez (1937), and MacLean (1952). Currently, both limbic (amygdalae, thalamus, hypothalamus, hippocampus, fornix, mammillary bodies, olfactory bulbs, and cingulate gyrus) and nonlimbic structures (basal ganglia, PFC, ventral striatum, insular cortex, and cerebellum) are thought to be involved in emotion processing (Lindquist, 2012). The amygdalae are involved in detecting, evaluating, and learning emotionally important aspects of the surroundings, particularly those involving fear and other negative emotions; the hippocampus is mainly involved in memory encoding, processing, and consolidation, being also implicated in memory retrieval that is used to evaluate affective stimuli (Brosch, Scherer, Grandjean, & Sander, 2013); the PFC modulates emotions and behaviors by anticipating the consequences of our actions, maintains emotions over time, and organizes behavior toward specific goals (Etkin, Egner, & Kalisch, 2011); and the insula integrates somatosensory and emotional information, being critical for the bodily experience of emotion (Craig, 2009). Evidence suggests DHEA(S) effects on emotional processing. Several neuroimaging studies have been performed in the last decade, providing important information on the neurobiology and neurochemistry of emotions. A small number of those studies explored the neural network correlates of DHEA(S) in what concerns human emotion processing (Table 1).

The National Institutes of Health Study of Normal Brain Development, longitudinally explored the relations of DHEA to corticolimbic development and cognition in children and adolescents with 4–24 years of age (Evans, 2006). Magnetic resonance imaging (MRI) scans were used in those studies and revealed that DHEA was related to a prepubertal increase in cortical thickness of regions involved in cognitive control in both genders (Nguyen et al., 2013). Those relations between DHEA and cortical thickness relate to behavioral changes supported by these same regions (Nguyen et al., 2013). Between ages 4 and 8 years old, DHEA was related to cortical thickness of the left dorsolateral PFC, eventually related to the improved language and behavioral control (Nguyen et al., 2013; Szaflarski et al., 2006). Between 7 and 12 years old, DHEA levels were related to the cortical thickness at the right temporoparietal junction,

Table 1 DHEA and DHEAS [DHEA(S)] and Emotional Processing: Neuroimaging, Electrophysiological, and Behavioral Correlates
Neuroimaging correlates of higher DHEA levels

Findings	Behavioral Implications	References
Premature adrenarche		
Decreased cingulate cortex activity to emotional faces	Decreased ability to process affective stimuli	Whittle et al. (2015)
Decreased insular, striatal, and prefrontal cortex activation to fearful facial expressions	Difficulties in reward-related processing and the regulation of fearful stimuli responses	
Decreased insular but increased prefrontal activation and increased externalizing symptoms to happy facial expressions	Deficient processing of bodily states associated with positive affect	
Middle childhood and adolescence		
Increase in cortical thickness of the left dorsolateral prefrontal cortex, right temporoparietal junction, and right entorhinal cortex	Improvement of behavioral control, mentalizing abilities, and development of emotional processing and emotional memory	Nguyen et al. (2013)
Negative amygdala–occipital covariance and improved visual abilities; negative amygdala–parietal covariance and improved visuomotor processing	Reduced detection of emotional stimuli. Improvement of attention through the conscious inhibition of emotions	Nguyen et al. (2016)
Negative amygdala–prefrontal covariance and attention impairment	Reduced detection of emotional stimuli. Decreased attentional bias to emotional stimuli	
Negative insular–hippocampal and occipital–hippocampal covariance, lower spatial recognition memory, but higher overall working memory	Impairment of hippocampal-to-cortical connections, resulting in impaired encoding and processing of spatial and social cues, but improvement of cortical functions related to attention and working memory, by disconnecting insular and hippocampal areas	Nguyen (2017)

Adults		
During emotion processing: decreased amygdala and hippocampus activity, increased amygdala–hippocampus connectivity, enhanced cingulated cortex activity, reduced emotional memory	Reduced generation of negative emotions, reduced emotional memory, and enhancement of regulatory processes	Sripada et al. (2013)
Resting state: decreased amygdala–periamygdala and amygdala–insula connectivity and decreased negative affect	Enhanced shift in the balance between salience and default network	Sripada, Welsh, Marx, and Liberzon (2014)
Electrophysiological correlates of DHEA(S)		
Findings	**Suggested implications**	**References**
Higher DHEA-to-cortisol and DHEAS-to-DHEA ratios were related to shorter visual P300 latencies under negative emotional context	Higher proportion of DHEA respective to cortisol and a higher proportion of sulfated to nonsulfated form of DHEA may protect from interference caused by the processing of the negative emotional stimuli on task performance	do Vale et al. (2015)
Higher DHEAS-to-DHEA ratios were related to reduced visual P300 amplitude elicited by a negative emotional context	Higher proportion of DHEAS in relation to DHEA may reduce the processing of the task-irrelevant negative content of the stimuli	
Behavioral correlates		
Findings	**Suggested implications**	**References**
Higher DHEAS levels in children and adolescents with disruptive behavior and callous-unemotional primary psychopathy	Higher DHEAS levels may impair emotional and social understanding	van Goozen, Matthys, Cohen-Kettenis, Thijssen, and van Engeland (1998), Whittle et al. (2015), and Kimonis, Goulter, Hawes, Wilbur, and Groer (2017)

Continued

Table 1 DHEA and DHEAS [DHEA(S)] and Emotional Processing: Neuroimaging, Electrophysiological, and Behavioral Correlates—cont'd

Behavioral correlates

Findings	**Suggested implications**	**References**
Higher DHEAS levels in adults with greater impulsivity, sensation-seeking, and monotony-avoidant traits	Higher DHEAS may be related to more disinhibited behavior in the social sphere, interest in activities involving some danger and an increased need for change	af Klinteberg et al. (1992)
Higher DHEAS concentrations and DHEA(S)-to-cortisol ratios were related to lower depression scores	Higher DHEAS levels and higher DHEA(S) in relation to cortisol levels, may protect from depression	Goodyer, Herbert, and Altham (1998), Barrett-Connor et al. (1999), Young, Gallagher, and Porter (2002), Hu et al. (2015), and Mocking et al. (2015)
Small trials of DHEA therapy reported positive effects on mood	DHEA(S) may be a potential antidepressant drug	Wolkowitz et al. (1999), Schmidt et al. (2005), and Strous et al. (2007)

consistent with the improvement of mentalizing abilities (Nguyen et al., 2013; Saxe, Whitfield-Gabrieli, Scholz, & Pelphrey, 2009). Between 4 and 13 years old, DHEA was related to changes in cortical thickness of the right entorhinal cortex, which could eventually be related to the development of emotional processing and emotional memory (Grieve, Korgaonkar, Clark, & Williams, 2011; Nguyen et al., 2013).

In fact, during middle childhood and adolescence DHEA is involved in the regulation of neurogenesis and apoptosis in structures involved in emotional processing, namely, the amygdalae and hippocampus (Nguyen, 2017; Nguyen et al., 2013). DHEA covariates with amygdala–occipital, amygdala–parietal, and amygdala–prefrontal networks, independently of estradiol or testosterone levels (Nguyen et al., 2016; see Fig. 1). Higher DHEA levels were related to a negative amygdala–occipital covariance and this covariance was related with improved visual and decoding abilities. Higher DHEA levels were related to a negative amygdala–parietal covariance and higher visuomotor processing. Finally, higher DHEA levels were related to a negative amygdala–prefrontal covariance and attention impairment. Overall, these results suggest that DHEA may improve attention through the conscious inhibition of emotions (Etkin et al., 2011; Nguyen, 2017; Nguyen et al., 2016). DHEA-related reduction in amygdala–prefrontal networks might eventually contribute to improvements in visual attention by decreasing the attentional bias related to affect modulation, that is to say, the salience of emotional stimuli (Nguyen et al., 2016).

Insular–hippocampal and occipital–hippocampal structural covariance also varied as a function of DHEA levels (Nguyen, 2017). Higher DHEA levels were related to negative insular–hippocampal and occipital–hippocampal structural covariance and this was associated with lower spatial recognition memory, while the relation to negative insular–hippocampal covariance was also associated with higher overall working memory, supporting the hypothesis that DHEA may optimize cortical functions related to attention and working memory, perhaps by disconnecting insular and hippocampal areas (Nguyen, 2017). The authors also hypothesize that DHEA may impair the development of bottom-up, hippocampal-to-cortical connections, resulting in impaired encoding of spatial cues (Nguyen, 2017). Hippocampus–PFC connectivity is important for emotion regulation and memory, and stress modulates that connectivity (Godsil, Kiss, Spedding, & Jay, 2013). Depression and cognitive impairment show several abnormalities and reduced plasticity in the hippocampus–PFC pathway (Godsil et al., 2013). Notably, late-life major depression, that in clinical studies was related to lower

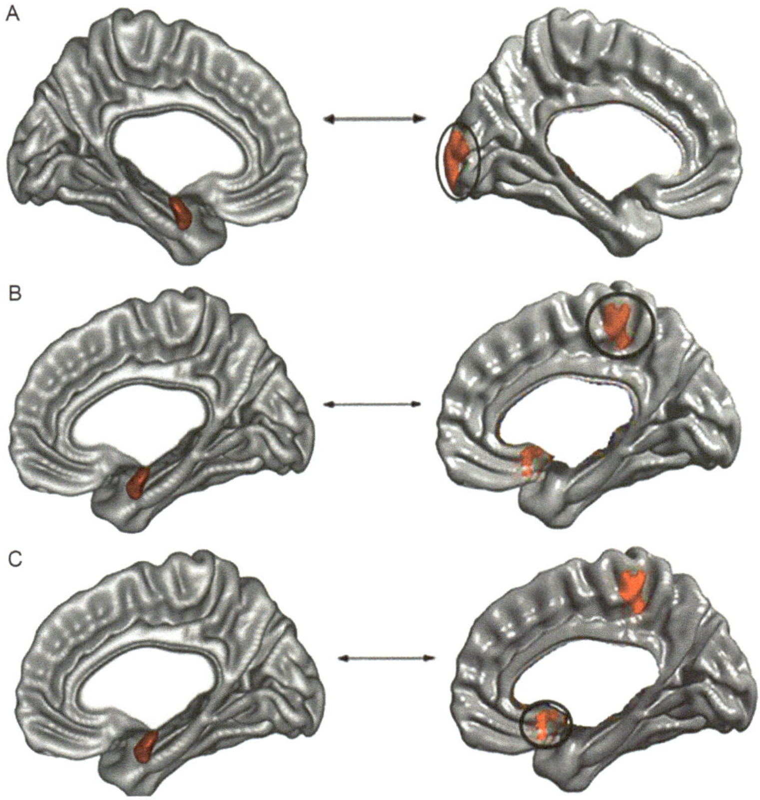

Fig. 1 DHEA-related corticoamygdalar structural covariance. The National Institutes of Health Study of Normal Brain Development, longitudinally evaluated typically developing children. A sample of 216 participants (6–22 years old) was used to explore the relationship between DHEA levels, corticoamygdalar covariance, and visual attention. The brain figures show the regions where the structural covariance between cortical thickness and amygdalar volume significantly varied according to DHEA levels: left occipital pole (A), right somatosensory cortex (B), and right anterior cingulate cortex (C). Higher DHEA levels were associated with a negative covariance between the volume of the amygdala and the thickness of these cortical regions. *Reprinted from Nguyen, T. V., Gower, P., Albaugh, M. D., Botteron, K. N., Hudziak, J. J., Fonov, V. S., et al. (2016). The developmental relationship between DHEA and visual attention is mediated by structural plasticity of cortico-amygdalar networks.* Psychoneuroendocrinology, 70, *122–133, Copyright (2016) with permission from Elsevier.*

DHEA(S) levels and DHEA-to-cortisol ratios, showed increased hippocampal–frontal connectivity in the resting state and abnormal activation with negative stimuli (Goveas et al., 2011).

Taken together, evidence suggests that during adrenarche/pubertal development, higher DHEA levels might reduce the influence of amygdala and hippocampal afferents on cortical function. That would inhibit primarily amygdala-based function such as the detection of emotional stimuli, as well as primarily hippocampal-based function such as encoding and processing of spatial and social cues (Nguyen, 2017). This DHEA-related inhibition could improve cognition, namely, attention and working memory, at the cost of an impairment in emotional processing and social understanding. There are, however, other studies that failed to find significant relations between DHEA and the social emotion network in puberty (Goddings, Heyes, Bird, Viner, & Blakemore, 2012; Klapwijk et al., 2013).

The previous relations would agree with the findings at clinical level, in which higher rates of disruptive behaviors were observed in relation with increased DHEA levels during adrenarche (Nguyen, 2017; Pajer et al., 2006). Whittle et al. (2015) studied the effects of early adrenarche and found a relation between higher DHEA levels in girls and reduced cingulate cortex activity by fearful, angry, and happy faces, suggestive of decreased ability to process affective stimuli; reduced insular, striatal, and PFC activation to fearful facial expressions, suggestive of difficulties in reward-related processing and the regulation of fearful stimuli responses; and reduced insular but increased prefrontal activation and increased externalizing symptoms to happy facial expressions. The relation between higher DHEA levels and increased externalizing symptoms was partially mediated by reduced posterior insular activation to happy faces (Whittle et al., 2015). Again, suggestive that the time of adrenarche may be important for emotional processing, is the fact that girls and boys with premature adrenarche have higher prevalence of oppositional defiant disorder, and aggressive or disruptive antisocial behaviors (van Goozen et al., 1998). Of note, those prepubertal boys had no differences in testosterone levels when compared to controls (van Goozen et al., 1998). Also, it is interesting to note that some opposite effects of DHEA and testosterone were described on brain development during adrenarche and puberty. In that period, DHEA increases and testosterone decreases cortical thickness of regions that integrate multimodal information (Nguyen et al., 2016, 2013), eventually suggesting the importance of adrenarche and puberty temporal coordination.

In adults, Sripada et al. (2013) performed a functional MRI study to investigate the effect of DHEA administration on emotion regulation neurocircuits. They administered 400 mg DHEA ($n = 14$) or placebo ($n = 15$) to healthy men (18–34 years old) and used functional MRI during the performance of the shifted-attention emotion appraisal task. They found that DHEA administration reduced the activity in the amygdala and hippocampus, increased the connectivity between these two regions, and enhanced the activity in the rostral anterior cingulated cortex during emotion processing. Moreover, DHEA administration reduced the negative affect at the trend level and significantly reduced the memory accuracy for emotional stimuli. Hence, in agreement with previous results in younger ages, those results in young adult men suggest that DHEA may downregulate the negative emotions induced by aversive stimuli by reducing the activity in regions associated with the generation of negative emotions (amygdala and hippocampus). These results, however, also suggest enhanced activity in regions linked to regulatory processes (cingulate cortex), which disagrees with the findings in girls with premature adrenarche (Whittle et al., 2015). Whether the differences in age (with greater maturity of affective prefrontal control in adults), gender, or considering acute increases vs baseline levels of DHEA could contribute to the apparently discordant result is not known. Sripada et al. (2013) did not exclude the possibility that the effects of DHEA administration were mediated by other DHEA derivatives, namely, DHEAS and androsterone. Of note, DHEA administration produced an increase in both DHEA and DHEAS levels in serum and saliva, without a simultaneous change in cortisol levels.

Sripada et al. (2014) also studied the effects of DHEA in resting networks and found that DHEA administration to healthy men reduced the connectivity between amygdala and periamygdala and between amygdala and insula. In this study, reductions of amygdala to precuneus connectivity were associated with less self-reported negative affect. Therefore, DHEA effects on emotion processing neurocircuits differed from its effects on resting networks, and the authors purposed that DHEA may shift the balance between salience network and default network, which could have implications in the neurocircuitry of anxiety (Sripada et al., 2014).

Overall, results suggest that during middle childhood and adolescence, DHEA may decrease the influence of amygdalar afferents on cortical functions, which might contribute to improved attention and working memory (Nguyen, 2017). Those effects seem independent from testosterone and occur in both genders. Nevertheless, eventually in relation to the reduced emotional processing and consequent reduction in social and emotional understanding, subjects with higher DHEA(S) levels also present higher rates of disruptive behaviors (van Goozen et al., 1998; Whittle et al., 2015). Accordingly, DHEA administration to adults also reduced the amygdala and hippocampus activity, thus reducing the negative affect and the memory for emotional stimuli (Sripada et al., 2013). These results agree with the clinical findings suggesting antidepressant effects of DHEA(S). However, the limited number of studies addressing the effects of DHEA(S) on the emotional network, prevent definitive conclusions.

5. DHEA(S) AND EMOTIONAL PROCESSING: ELECTROPHYSIOLOGICAL CORRELATES

By 1884, William James defined emotions as subjective experiences. He proposed that stimuli trigger activity in the autonomic nervous system, and that activity would raise an emotional experience in the brain. In his theory, the physiological response was the emotion (James, 1884). Carl Lange proposed a similar theory arguing that "we feel sad because we cry, angry because we strike, afraid because we tremble, and neither we cry, strike, nor tremble because we are sorry, angry, or fearful" (Cannon, 1987). Currently, emotions are believed to involve several components, such as subjective experiences, cognitive processes, expressive and instrumental behaviors, and psychophysiological changes (Weisfeld & Goetz, 2013). The processing of complex emotional information in one sensory modality can affect emotion processing of another modality during early stages of neural processing, as well as self-reported emotions (Gerdes, Wieser, & Alpers, 2014). Hormones and neurotransmitters are known to be involved in emotions' modulation, namely, cortisol, noradrenaline, dopamine, serotonin, oxytocin, and GABA (Cadle & Zoladz, 2015). However, in humans, DHEA(S) relations to emotional processing at the electrophysiological level are mostly unexplored.

One of the most studied electrophysiological correlates of human brain functions is the so-called P300. The P300 (P3) is a family of event-related potential (ERP) components that reflect the cognitive updating of information, activation of working memory, categorization of comparisons to models stored in memory, and contextual updating of information (Verleger, Grauhan, & Śmigasiewicz, 2016). Increased attention allocation elicits increases in P300 amplitudes, whereas irrelevant stimuli elicit deflections of reduced amplitude. The P300 latencies depend on the process of stimulus classification. Emotional stimuli usually elicit enhanced P300 amplitudes (Schupp et al., 2004) and delayed P300 latencies for negative emotional targets when compared to neutral stimuli (Fichtenholtz, Hopfinger, Graham, Detwiler, & LaBar, 2007).

We studied the relation between DHEA(S) and brain processing under negative and neutral context, in healthy young women. This study revealed relations between DHEA(S) and brain processing under an emotionally negative context induced by images of fearful faces, suggesting that these neurosteroids may modulate the processing of emotionally negative information (do Vale et al., 2015; see Table 1 and Fig. 2). A higher DHEA-to-cortisol ratio was related to reduced visual P300 latencies in the negative emotional context, suggesting that DHEA may protect from interference of the implicit negative content of the stimuli on task performance (do Vale et al., 2015). Under negative emotional context, higher DHEAS-to-DHEA ratios were also related to shorter visual P300 latencies, thus suggesting that a higher proportion of DHEA respective to cortisol and a higher proportion of sulfated to nonsulfated form of DHEA may provide protection from any interference caused by the processing of the implicit negative emotional content of the stimuli (do Vale et al., 2015).

A prolonged P300 latency was previously associated with an attentional bias to negative stimuli in high trait anxiety subjects (Wang et al., 2013) and hydrocortisone administration increased the processing of angry faces in highly anxious individuals as indicated by increased amplitudes of early (P150) and late (P3) ERPs (van Peer et al., 2007). On the contrary, we examined subjects with low trait anxiety and found that higher DHEA-to-cortisol ratios were related to shorter P300 latencies in the negative condition (do Vale et al., 2015). Higher DHEA-to-cortisol may be related to less interference with the task caused by the implicit stressful content of the stimuli, in agreement with the hypothesis of anticortisol effects of DHEA and suggesting a protective mechanism against the deleterious consequences of stress (do Vale et al., 2015). In a different setting, a study by Alhaj, Massey, and McAllister-Williams (2006) also suggests anticortisol effects of DHEA. Those authors administered DHEA to young men and used low-resolution brain electromagnetic tomography (LORETA) imaging. In this study, DHEA treatment reduced cortisol levels, improved episodic memory recollection and mood, and modified the ERPs associated with retrieval (Alhaj et al., 2006).

A relation between DHEA(S) levels and visual P300 amplitude was also found. Participants with higher DHEAS-to-DHEA ratios had lower visual P300 amplitude increments attributed to the negative emotional context (as compared to the neutral one), suggesting that a higher proportion of DHEAS in relation to DHEA might eventually reduce the processing of the task-irrelevant negative content of the stimuli (do Vale et al., 2015). Notably, enhanced negative affect is a characteristic of depressive disorders and negative mood can be experimentally induced using various procedures,

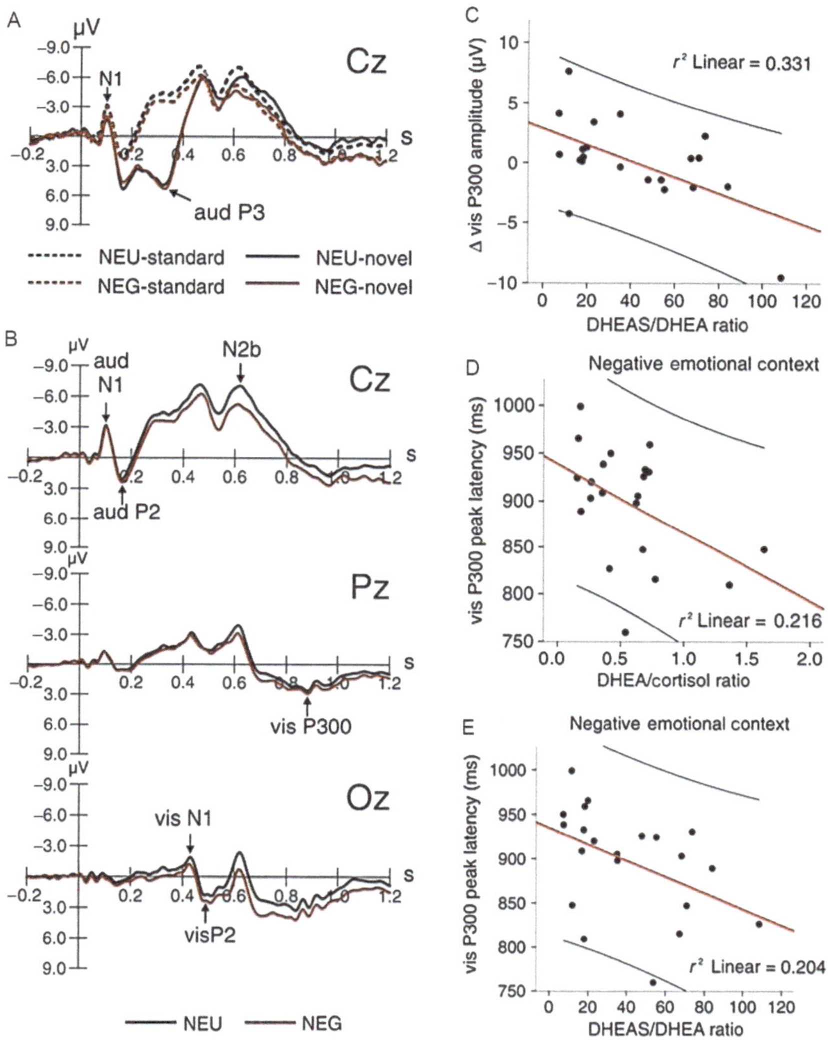

Fig. 2 DHEA(S) relations to emotional processing at the electrophysiological level. Twenty-one healthy young women (18–26 years old) performed an auditory–visual distraction task, with emotional context manipulation. Baseline DHEA (pg/mL), DHEAS (pg/min), and cortisol (pg/mL) were measured in the saliva. (A) Grand average waveforms at Cz for each emotional context (neutral and negative) and type of sound (standard and novel). (B) Grand average event-related potentials (ERPs) for neutral and negative context (only trials without auditory distraction). (C) Higher DHEAS-to-DHEA ratios were related to reduced interference with the task due to the processing of the implicit negative content of the stimuli as shown by reduced visual P300 amplitudes. (D) and (E) Higher DHEA-to-cortisol and DHEAS-to-DHEA ratios were related to shorter visual P300 peak latencies under negative emotional context. *NEG*, negative emotional context; *NEU*, neutral emotional context. *Δ vis P300 amplitude*, mean visual P300 amplitude in the negative minus neutral context. *Reprinted from do Vale, S., Selinger, L., Martins, J. M., Bicho, M., do Carmo, I., & Escera, C. (2015). Dehydroepiandrosterone (DHEA) and dehydroepiandrosterone-sulfate (DHEAS) and emotional processing—A behavioral and electrophysiological approach.* Hormones and Behavior, 73, *94–103, Copyright (2015) with permission from Elsevier.*

namely, by the simple presentation of images of faces with emotional expressions (Schneider, Gur, Gur, & Muenz, 1994). The mechanisms underlying mood persistence in major depressive disorder remain poorly understood. However, cognitive theories hypothesize that depressed patients have cognitive biases for emotional information, which help perpetuate depressive symptoms (Gotlib & Joormann, 2010). In this context, higher DHEAS-to-DHEA levels might be involved in protective mechanisms against a negative attention bias (do Vale et al., 2015) and undermine the relation between higher DHEAS levels and lower frequencies of depression, lower depression ratings, and better wellbeing scores found at the clinical level (Barrett-Connor et al., 1999). In line with this hypothesis, subjects with higher DHEAS-to-DHEA ratio would process less, have less memory updating of the implicit negative content of the stimuli, and have a reduced attentional bias toward these stimuli, which in turn might eventually provide protection from depressive states (do Vale et al., 2015).

Emotional stimuli usually capture attention more effectively than non-emotional ones, thus having an impact on cognitive functions, even when they are task-irrelevant. It is hypothesized that protective mechanisms may exist, that filter out irrelevant sensory inputs protecting the higher brain functions from sensory overload (Braff & Geyer, 1990). We found a relation between higher DHEA-to-cortisol and DHEAS-to-DHEA ratios in women, and lower processing of the negative emotional stimuli and less disruption by the processing of negative emotional stimuli (do Vale et al., 2015). Taken together, DHEA(S) may eventually intervene in the protective mechanisms filtering out negative information overload (do Vale et al., 2015). It was hypothesized that DHEA(S) should also affect involuntary attention (distraction), but this study did not confirm such idea. DHEA(S) or DHEA(S)-to-cortisol ratios were not related to auditory distraction processing (as evaluated by novelty-P3) in low-anxiety women, under an emotionally negative context (do Vale et al., 2015). In fact, the influence of DHEA(S) and cortisol on attention may depend on their levels and the cognitive task involved (do Vale et al., 2014, 2016).

The findings discussed earlier suggest that the balance between DHEA, DHEAS, and cortisol may modulate the processing of negative emotions. These results agree with findings at the clinical level, in which higher DHEAS concentrations and DHEA-to-cortisol ratios but not DHEA levels alone were related to less frequent depression and less depressive mood (Barrett-Connor et al., 1999; Young et al., 2002).

6. DHEA(S) RELATIONS TO PERSONALITY, QUALITY OF LIFE, AND MOOD

Several relations of DHEA(S) to cognitive and neuropsychiatric parameters were described over several decades, namely, with personality, quality of life, anxiety, depression, and memory (Table 1). Unfortunately, in most cases these relations are neither systematic nor conclusive. Personality characteristics and the hypothalamus–pituitary–adrenal (HPA) axis phenotype develop in childhood and remain relatively stable over time (Bertagna, Coste, Raux-Demay, Letrait, & Strauch, 1994), even if changes may occur according to individual experiences (van Dijken et al., 1993; Walker & Dallman, 1993). In general, positive psychological traits are related to reduced HPA axis reactivity, in what concerns cortisol (Chida & Hamer, 2008). Yet, the relationship between DHEA(S) and personality is mostly unknown.

Lower DHEAS levels were described in individuals with higher type A behavior scores (Fava, Littman, & Halperin, 1987), while higher DHEAS levels might contribute to greater impulsivity, sensation-seeking, and monotony-avoidant traits. Therefore, DHEAS may be related to more disinhibited behavior in the social sphere, increased interest in activities involving some danger, and an increased need for change (af Klinteberg et al., 1992). Accordingly, higher DHEAS levels were found in children and adolescents with disruptive behavior (van Goozen et al., 1998; Whittle et al., 2015) and borderline personality disorder (Conti et al., 2014). Callous-unemotional primary psychopathy was related to higher DHEA levels, whereas callous-unemotional secondary psychopathy was related to lower DHEA or DHEA-to-cortisol levels (Kimonis et al., 2017). We also described a relation between DHEAS responses to the administration of CRH and both type A personality and behavior deviant scores (do Vale et al., 2011). However, several other studies found no relation between DHEA(S) levels and personality traits (Tsuchimine, Kaneda, Nakamura, & Norio, 2015).

Concerning quality of life, many studies found a direct relation between lower DHEA(S) levels and poor quality of life, namely, poor life satisfaction, psychosocial stress, and functional limitations (Berr et al., 1996; Scott, Salahuddin, Cooney, Svec, & Dinan, 1999). In male subjects, higher DHEAS levels were related to indulging in more leisure activities and greater enjoyment in these activities (Fava et al., 1992), while females had greater sexual gratification and higher vitality ratings (Persky et al., 1982). Higher DHEA(S) levels in older adults were related to reduced risk for the physical weakness and diseases that precede frailty, while larger decrements in DHEA(S) were related to loss of physical vitality (Rendina, Ryff, & Coe, 2017). Intravaginal DHEA treatment may improve sexual function in postmenopausal women (Scheffers, Armstrong, Cantineau, Farquhar, & Jordan, 2015). However, a relation between DHEA(S) and quality of life was not systematically found and there is no evidence to support the use of DHEA treatment to improve it (Scheffers et al., 2015; Wolf, Naumann, Hellhammer, & Kirschbaum, 1998).

Higher DHEA(S) or DHEA-to-cortisol levels were related to improved attention, lower perceived stress, and improved performance under stressful conditions (Morgan, Rasmusson, Piertzak, Coric, & Southwick, 2009; Wolf et al., 1997), suggesting the importance of the cortisol-to-DHEA ratio. Again, studies using DHEA administration to postmenopausal women or healthy elderly subjects do not support the use of DHEA to improve attention (Davis, Panjari, & Stanczyk, 2011). Of note, the use of supraphysiologic doses of DHEAS would raise the potential problem of enhanced brain excitability and seizures (Reddy & Kulkarni, 1998), which has not been reported with the lower doses used in studies performed in humans. Also, androgenic effects would be expected with the administrations of high DHEA(S) doses to women.

Higher DHEAS concentrations and DHEA(S)-to-cortisol ratios have been related to lower prevalence of depression, lower depression ratings, and lower depression recurrence (Barrett-Connor et al., 1999; Hu et al., 2015; Mocking et al., 2015; Young et al., 2002). Low serum DHEAS levels were also found in dysthymic patients (Markianos, Tripodianakis, Sarantidis, & Hatzimanolis, 2007) and schizophrenic patients with more negative symptoms (Goyal, Sagar, Ammini, Khurana, & Alias, 2004). Again, there are also other studies with contradictory results in the literature, showing either unaltered or increased levels of DHEA(S) in depressed patients (Heuser et al., 1998; Maninger et al., 2009). The measurement of DHEA(S) or cortisol individually may, however, have contributed to some apparently contradictory results, as the evidence for a relation between DHEA levels alone and depression is less consistent than the evidence that suggests a relation between higher DHEAS concentrations or higher DHEA(S)-to-cortisol ratios and lower depression ratings. In fact, several groups have found that DHEA-to-cortisol ratios, rather than the concentrations of either hormone alone, discriminated more accurately depressed from nondepressed individuals, with lower DHEA-to-cortisol ratios seen in depression (Young et al., 2002), in untreated depressed patients, and in patients who remained depressed after several months (Goodyer et al., 1998). Moreover, in a study assessing both DHEA and DHEAS, depressed patients had low DHEAS but normal DHEA concentrations (Scott et al., 1999). These results point toward the complexity of DHEA(S) effects and the importance of DHEA-to-cortisol and DHEAS-to-DHEA balance.

DHEA(S) antidepressant effects might be partially mediated via GABAergic receptors antagonism (Genud et al., 2009) and sigma-1 receptors agonism (Reddy, Kaur, & Kulkarni, 1998), with DHEAS expected to have a more potent excitatory action than DHEA. Evidence also suggests that DHEA(S) antiglucocorticoid effects may contribute to those antidepressant mechanisms (Kaminska, Harris, Gijsbers, & Dubrovsky, 2000), thus supporting the inverse relation found between DHEA-to-cortisol ratios and depression scores. Additionally, DHEA(S) effects in neurogenesis and neuroprotection, catecholamine, antiinflammatory, and antioxidant effects, could eventually contribute to, or partially mediate, the relationship between these hormones and depression.

Also supporting the hypothesis of DHEA antidepressant effects, is the fact that controlled trials of DHEA therapy have reported significant positive effects on mood (Schmidt et al., 2005; Wolkowitz et al., 1999). In schizophrenic patients, a decrease in negative symptoms was observed after DHEA administration, even if there were no changes in positive symptoms (Strous et al., 2007). Those mood improvements were related to increases in the circulating concentrations of DHEA and DHEAS and to increases in their ratios with cortisol, again suggesting antiglucocorticoid effects of DHEA. Overall, controlled trials of DHEA administration in depression have consistently reported significant antidepressant effects (Maninger et al., 2009). However, large-scale, long-term, and comparative studies (with established antidepressant drugs) are still missing. Such studies will be necessary before an eventual role for DHEA administration in depression management can be established.

7. SUMMARY AND FUTURE RESEARCH

This chapter revised DHEA(S) relations to emotional processing and behavior. These steroids modulate emotional processing, as supported by multiple studies at both basic and clinical levels. In general, evidence suggests that they protect from negative emotions and have an overall mood enhancement effect. However, several studies did not support these conclusions. DHEA(S) have neurostimulant and anticortisol effects, which contribute to their effects. During adrenarche, the increase in DHEA(S) levels is related to the development of cortical regions involved in cognitive control in both genders, suggesting a relation to the development of mentalizing and emotional processing (Nguyen et al., 2016, 2013). DHEA(S) may reduce amygdalar and hippocampal activation by emotional stimuli, therefore reducing the generation of negative emotions and the memory for emotional stimuli (Sripada et al., 2013, 2014). Besides, DHEA may reduce the influence of amygdalar afferents on cortical functions, which may eventually improve cognition, namely, attention and working memory, at the cost of an impairment in social understanding and reduced emotional processing (Nguyen et al., 2016, 2013).

DHEA(S) effects on emotional processing at the neuroimaging level and specially, at the electrophysiological level are, however, largely unexplored. These findings are based on a small number of studies, most of them being limited to specific age or gender groups. In women, higher DHEA-to-cortisol and DHEAS-to-DHEA ratios were related to shorter visual P300 latencies and reduced visual P300 amplitudes, suggesting less interference of the negative stimuli with the cognitive task and less processing of the negative information. The reduced processing of negative information might constitute a protective mechanism, eventually reducing the generation of negative emotions, and therefore underlie the relation between higher DHEAS or DHEA-to-cortisol ratios and lower depression ratings described at the clinical level (do Vale et al., 2015). In line with the mentioned findings, controlled trials using DHEA administration reported significant antidepressant effects, that were related to increases in DHEA(S) and DHEA(S)-to-cortisol ratios (Schmidt et al., 2005; Wolkowitz et al., 1999). Nevertheless, these are limited to a number of small studies. Larger, longer term, and comparative studies to other drugs are still missing, before a potential use of DHEA(S) as an antidepressive drug can be considered.

The study of DHEA(S) effects on human brain involves several challenges. We do not have a direct access to human brain DHEA(S) and these steroids are further metabolized locally; they can be interconverted into each other and into other steroids, being difficult to understand the effect of each steroid individually; their levels and biochemical/cellular effects vary with gender, developmental period, and brain structure; and animal findings cannot be directly extrapolated to humans. Therefore, in the study of DHEA(S) effects in emotional processing, we need to consider species, gender, life stage, brain structure, type of emotion, neurosteroid metabolome, and endocrine phenotype (reactivity) specificities. It will be important to establish the physiologic mechanism of action of DHEA(S) in each brain structure involved in emotion neurocircuits. Also, addressing specific physiological mechanisms underlying neuropsychiatric behaviors in which DHEA(S) may act could provide more consistent results about their effects and potential clinical utility. For instance, studying specific biochemical pathologies, altered brain activity or altered emotional processing subjacent to emotional behaviors (instead of generally considering depression or well-being) in relation to divergent DHEA(S) concentrations or effects in these circuits, at the neuroimaging, electrophysiological and clinical levels, could provide some important information.

There is no doubt that DHEA(S) influence emotional processing, even if their physiological effects are still being uncovered and therapeutic utility is not established. Only a limited number of studies explored the effects of DHEA(S) on emotional processing at the electrophysiological and neuroimaging levels in humans. Further studies will provide fundamental information regarding the effects of DHEA(S) on emotion processing and together with clinical research may eventually establish if there is a place for the use of DHEA(S) levels as clinical markers for emotional processing or if there is a place for the use of DHEA as a treatment.

ACKNOWLEDGMENTS

The authors acknowledge Dr. Nelson de Melo for the English language review. This work was partially supported by the Generalitat de Catalunya (SGR2017-974) and the ICREA Acadèmia Distinguished Professorship awarded to C.E.

REFERENCES

af Klinteberg, B., Hallman, J., Oreland, L., Wirsén, A., Levander, S. E., & Schalling, D. (1992). Exploring the connections between platelet monoamine oxidase activity and behavior. II. Impulsive personality without neuropsychological signs of disinhibition in air force pilot recruits. *Neuropsychobiology*, *26*, 136–145.

Alhaj, H. A., Massey, A. E., & McAllister-Williams, R. H. (2006). Effects of DHEA administration on episodic memory, cortisol and mood in healthy young men: A double-blind, placebo-controlled study. *Psychopharmacology*, *188*, 541–551.

Balazs, Z., Schweizer, R. A., Frey, F. J., Rohner-Jeanrenaud, F., & Odermatt, A. (2008). DHEA induces 11-HSD2 by acting on CCAAT/enhancer-binding proteins. *Journal of the American Society of Nephrology*, *19*, 92–101.

Barrett-Connor, E., & Edelstein, S. L. (1994). A prospective study of dehydroepiandrosterone sulfate and cognitive function in an older population: The Rancho Bernardo Study. *Journal of the American Geriatrics Society*, *42*, 420–423.

Barrett-Connor, E., von Mühlen, D., Laughlin, G. A., & Kripke, A. (1999). Endogenous levels of dehydroepiandrosterone sulfate, but not other sex hormones, are associated with depressed mood in older women: The Rancho Bernardo Study. *Journal of the American Geriatrics Society, 47*, 685–691.

Baulieu, E. E., & Robel, P. (1998). Dehydroepiandrosterone (DHEA) and dehydroepiandrosterone sulfate (DHEAS) as neuroactive neurosteroids. *Proceedings of the National Academy of Sciences, 95*, 4089–4091.

Berr, C., Lafont, S., Debuire, B., Dartigues, J. F., & Baulieu, E. E. (1996). Relationships of dehydroepiandrosterone sulfate in the elderly with functional, psychological, and mental status, and short-term mortality: A French community-based study. *Proceedings of the National Academy of Sciences of the United States of America, 93*, 13410–13415.

Bertagna, X., Coste, J., Raux-Demay, M. C., Letrait, M., & Strauch, G. (1994). The combined corticotropin-releasing hormone/lysine vasopressin test discloses a corticotroph phenotype. *The Journal of Clinical Endocrinology and Metabolism, 79*, 390–394.

Boudarene, M., Legros, J. J., & Timsit-Berthier, M. (2002). Study of the stress response: Role of anxiety, cortisol and DHEAs. *Encephale, 28*, 139–146.

Braff, D. L., & Geyer, M. A. (1990). Sensorimotor gating and schizophrenia. Human and animal model studies. *Archives of General Psychiatry, 47*, 181–188.

Broca, P. (1878). Anatomie comparée des circonvolutions cérébrales: Le grand lobe limbique. *Reviews in Anthropology, 1*, 385–498.

Brosch, T., Scherer, K., Grandjean, D., & Sander, D. (2013). The impact of emotion on perception, attention, memory, and decision-making. *Swiss Medical Weekly, 143*, w13786.

Buoso, E., Lanni, C., Molteni, E., Rousset, F., Corsini, E., & Racchi, M. (2011). Opposing effects of cortisol and dehydroepiandrosterone on the expression of the receptor for activated C kinase 1: Implications in immunosenescence. *Experimental Gerontology, 46*, 877–883.

Cadle, C. E., & Zoladz, P. R. (2015). Stress time-dependently influences the acquisition and retrieval of unrelated information by producing a memory of its own. *Frontiers in Psychology, 6*, 910.

Campbell, B. (2006). Adrenarche and the evolution of human life history. *American Journal of Human Biology, 18*, 569–589.

Cannon, W. B. (1987). The James-Lange theory of emotions: A critical examination and an alternative theory. *The American Journal of Psychology, 100*, 567–586.

Chida, Y., & Hamer, M. (2008). Chronic psychosocial factors and acute physiological responses to laboratory-induced stress in healthy populations: A quantitative review of 30 years of investigations. *Psychological Bulletin, 134*, 829–885.

Compagnone, N. A., & Mellon, S. H. (1998). Dehydroepiandrosterone: A potential signalling molecule for neocortical organization during development. *Proceedings of the National Academy of Sciences of the United States of America, 95*, 4678–4683.

Conti, E., Nacinovich, R., Bomba, M., Uccellini, O., Rossi, M. S., Casati, M., et al. (2014). Diazepam binding inhibitor and dehydroepiandrosterone sulphate plasma levels in borderline personality disorder adolescents. *Neuropsychobiology, 69*, 9–24.

Craig, A. D. (2009). Emotional moments across time: A possible neural basis for time perception in the anterior insula. *Philosophical Transactions of the Royal Society of London. Series B, Biological Sciences, 364*, 1933–1942.

Davis, S. R., Panjari, M., & Stanczyk, F. Z. (2011). Clinical review: DHEA replacement for postmenopausal women. *The Journal of Clinical Endocrinology and Metabolism, 96*, 1642–1653.

do Vale, S., Martins, J. M., Fagundes, M. J., & Carmo, I. (2011). Dehydroepiandrosterone-sulphate (DHEAS) is related to personality and stress response. *Neuro Endocrinology Letters, 32*, 442–448.

do Vale, S., Selinger, L., Martins, J. M., Bicho, M., do Carmo, I., & Escera, C. (2015). Dehydroepiandrosterone (DHEA) and dehydroepiandrosterone-sulfate (DHEAS) and emotional processing—A behavioral and electrophysiological approach. *Hormones and Behavior, 73*, 94–103.

do Vale, S., Selinger, L., Martins, J. M., Bicho, M., do Carmo, I., & Escera, C. (2016). Hormonal modulation of novelty processing in women: Enhanced under working memory load with high dehydroepiandrosterone-sulfate-to-dehydroepiandrosterone ratios. *Neuroscience Letters, 634*, 98–103.

do Vale, S., Selinger, L., Martins, J. M., Gomes, A. C., Bicho, M., do Carmo, I., et al. (2014). The relationship between dehydroepiandrosterone (DHEA), working memory and distraction—A behavioral and electrophysiological approach. *PLoS One, 9*, e104869.

Dong, L. Y., Cheng, Z. X., Fu, Y. M., Wang, Z. M., Zhu, Y. H., Sun, J. L., et al. (2007). Neurosteroid dehydroepiandrosterone sulfate enhances spontaneous glutamate release in rat prelimbic cortex through activation of dopamine D1 and sigma-1 receptor. *Neuropharmacology, 52*, 966–974.

Dong, Y., & Zheng, P. (2011). Dehydroepiandrosterone sulphate: Action and mechanism in the brain. *Journal of Neuroendocrinology, 24*, 215–224.

Etkin, A., Egner, T., & Kalisch, R. (2011). Emotional processing in anterior cingulate and medial prefrontal cortex. *Trends in Cognitive Sciences, 15*(2), 85–93.

Evans, A. C. (2006). The NIH MRI study of normal brain development. *NeuroImage, 30*, 184–202.

Fava, M., Littman, A., & Halperin, P. (1987). Neuroendocrine correlates of the type A behavior pattern: A review and new hypotheses. *International Journal of Psychiatry in Medicine, 17*, 289–307.

Fava, M., Littman, A., Lamon-Fava, S., Milani, R., Shera, D., MacLaughlin, R., et al. (1992). Psychological, behavioral and biochemical risk factors for coronary artery disease among American and Italian male corporate managers. *The American Journal of Cardiology, 70*, 1412–1416.

Fichtenholtz, H. M., Hopfinger, J. B., Graham, R., Detwiler, J. M., & LaBar, K. S. (2007). Happy and fearful emotion in cues and targets modulate event-related potential indices of gaze-directed attentional orienting. *Social Cognitive and Affective Neuroscience, 2*, 323–333.

Genud, R., Merenlender, A., Gispan-Herman, I., Maayan, R., Weizman, A., & Yadid, G. (2009). DHEA lessens depressive-like behavior via GABA-ergic modulation of the mesolimbic system. *Neuropsychopharmacology, 34*, 577–584.

Gerdes, A. B., Wieser, M. J., & Alpers, G. W. (2014). Emotional pictures and sounds: A review of multimodal interactions of emotion cues in multiple domains. *Frontiers in Psychology, 5*, 1351.

Gibbs, T. T., Russek, S. J., & Farb, D. H. (2006). Sulfated steroids as endogenous neuromodulators. *Pharmacology, Biochemistry, and Behavior, 84*, 555–567.

Goddings, A. L., Heyes, S. B., Bird, G., Viner, R. M., & Blakemore, S. J. (2012). The relationship between puberty and social emotion processing. *Developmental Science, 15*, 801–811.

Godsil, B. P., Kiss, J. P., Spedding, M., & Jay, T. M. (2013). The hippocampal-prefrontal pathway: The weak link in psychiatric disorders? *European Neuropsychopharmacology, 23*, 1165–1181.

Goodyer, I. M., Herbert, J., & Altham, P. M. (1998). Adrenal steroid secretion and major depression in 8- to 16-year-olds, III. Influence of cortisol/DHEA ratio at presentation on subsequent rates of disappointing life events and persistent major depression. *Psychological Medicine, 28*, 265–273.

Gotlib, I. H., & Joormann, J. (2010). Cognition and depression: Current status and future directions. *Annual Review of Clinical Psychology, 6*, 285–312.

Goveas, J., Xie, C., Wu, Z., Douglas Ward, B., Li, W., Franczak, M. B., et al. (2011). Neural correlates of the interactive relationship between memory deficits and depressive symptoms in nondemented elderly: Resting fMRI study. *Behavioural Brain Research*, *219*, 205–212.

Goyal, R. O., Sagar, R., Ammini, A. C., Khurana, M. L., & Alias, A. G. (2004). Negative correlation between negative symptoms of schizophrenia and testosterone levels. *Annals of the New York Academy of Sciences*, *1032*, 291–294.

Grieve, S. M., Korgaonkar, M. S., Clark, C. R., & Williams, L. M. (2011). Regional heterogeneity in limbic maturational changes: Evidence from integrating cortical thickness, volumetric and diffusion tensor imaging measures. *NeuroImage*, *55*, 868–879.

Guazzo, E. P., Kirkpatrick, P. J., Goodyear, I. M., Shiers, H. M., & Herbert, J. (1996). Cortisol, dehydroepiandrosterone (DHEA), and DHEA sulfate in the cerebrospinal fluid of man: Relation to blood levels and the effects of age. *The Journal of Clinical Endocrinology and Metabolism*, *81*, 3951–3960.

Hamilton, L. D., & Meston, C. M. (2011). The role of salivary cortisol and DHEA-S in response to sexual, humorous, and anxiety-inducing stimuli. *Hormones and Behavior*, *59*, 765–771.

Hennebert, O., Chalbot, S., Alran, S., & Morfin, R. (2007). Dehydroepiandrosterone 7alpha-hydroxylation in human tissues: Possible interference with type 1 11beta-hydroxysteroid dehydrogenase-mediated processes. *The Journal of Steroid Biochemistry and Molecular Biology*, *104*, 326–333.

Heuser, I., Deuschle, M., Luppa, P., Schweiger, U., Standhardt, H., & Weber, B. (1998). Increased diurnal plasma concentrations of dehydroepiandrosterone in depressed patients. *The Journal of Clinical Endocrinology and Metabolism*, *83*, 3130–3133.

Hu, Q., Zhang, S. Y., Liu, F., Zhang, Y. L., Zhu, D. M., & Zang, Y. Y. (2015). Clinical significance of decreased protein expression of dehydroepiandrosterone sulfate in the development of depression: A meta-analysis. *Journal of Affective Disorders*, *174*, 416–423.

Imamura, M., & Prasad, C. (1998). Modulation of GABA-gated chloride ion influx in the brain by dehydroepiandrosterone and its metabolites. *Biochemical and Biophysical Research Communications*, *243*, 771–775.

James, W. (1884). What is an emotion? *Mind*, *9*, 188–205.

Jeckel, C. M., Lopes, R. P., Berleze, M. C., Luz, C., Feix, L., Argimon, I. I., et al. (2010). Neuroendocrine and immunological correlates of chronic stress in 'strictly healthy' populations. *Neuroimmunomodulation*, *17*, 9–18.

Kaminska, M., Harris, J., Gijsbers, K., & Dubrovsky, B. (2000). Dehydroepiandrosterone sulfate (DHEAS) counteracts decremental effects of corticosterone on dentate gyrus LTP. Implications for depression. *Brain Research Bulletin*, *52*, 229–234.

Kancheva, R., Hill, M., Novák, Z., Chrastina, J., Kancheva, L., & Stárka, L. (2011). Neuroactive steroids in periphery and cerebrospinal fluid. *Neuroscience*, *191*, 22–27.

Kimonis, E. R., Goulter, N., Hawes, D. J., Wilbur, R. R., & Groer, M. W. (2017). Neuroendocrine factors distinguish juvenile psychopathy variants. *Developmental Psychobiology*, *59*, 161–173.

Klapwijk, E. T., Goddings, A. L., Heyes, S. B., Bird, G., Viner, R. M., & Blakemore, S. J. (2013). Increased functional connectivity with puberty in the mentalizing network involved in social emotion processing. *Hormones and Behavior*, *64*, 314–322.

Labrie, F., Bélanger, A., Cusan, L., & Candas, B. (1997). Physiological changes in dehydroepiandrosterone are not reflected by serum levels of active androgens and estrogens but of their metabolites: Intracrinology. *The Journal of Clinical Endocrinology and Metabolism*, *82*, 2403–2409.

Lacroix, C., Fiet, J., Benais, J. P., Gueux, B., Bonete, R., Villette, J. M., et al. (1987). Simultaneous radioimmunoassay of progesterone, androst-4-enedione, pregnenolone, dehydroepiandrosterone and 17-hydroxyprogesterone in specific regions of human brain. *Journal of Steroid Biochemistry*, *28*, 317–325.

Lennartsson, A. K., Kushnir, M. M., Bergquist, J., & Jonsdottir, I. H. (2012). DHEA and DHEA-S response to acute psychosocial stress in healthy men and women. *Biological Psychology*, *90*, 143–149.
Lennartsson, A. K., Theorell, T., Kushnir, M. M., Bergquist, J., & Jonsdottir, I. H. (2013). Perceived stress at work is associated with attenuated DHEA-S response during acute psychosocial stress. *Psychoneuroendocrinology*, *38*, 1650–1657.
Lindquist, K. A. (2012). The brain basis of emotion: A meta-analytic review. *The Behavioral and Brain Sciences*, *35*, 121–143.
Maclean, P. D. (1952). Some psychiatric implications of physiological studies on frontotemporal portion of limbic system (visceral brain). *Electroencephalography and Clinical Neurophysiology*, *4*(Suppl. 4), 407–418.
Maninger, N., Wolkowitz, O. M., Reus, V. I., Epel, E. S., & Mellon, S. H. (2009). Neurobiological and neuropsychiatric effects of dehydroepiandrosterone (DHEA) and DHEA sulfate (DHEAS). *Frontiers in Neuroendocrinology*, *30*, 65–91.
Markianos, M., Tripodianakis, J., Sarantidis, D., & Hatzimanolis, J. (2007). Plasma testosterone and dehydroepiandrosterone sulfate in male and female patients with dysthymic disorder. *Journal of Affective Disorders*, *101*, 255–258.
Martins, J. M., do Vale, S., Ferreira, F., Fagundes, M. J., do Carmo, I., Saldanha, C., et al. (2010). Plasma corticotropin releasing hormone during the feeling of induced emotions. *Neuroendocrinology Letters*, *31*, 250–255.
Mocking, R. J., Pellikaan, C. M., Lok, A., Assies, J., Ruhé, H. G., Koeter, M. W., et al. (2015). DHEAS and cortisol/DHEAS-ratio in recurrent depression: State, or trait predicting 10-year recurrence? *Psychoneuroendocrinology*, *59*, 91–101.
Morgan, C. A., 3rd, Rasmusson, A., Piertzak, R. H., Coric, V., & Southwick, S. M. (2009). Relationships among plasma dehydroepiandrosterone and dehydroepiandrosterone sulfate, cortisol, symptoms of dissociation, and objective performance in humans exposed to underwater navigation stress. *Biological Psychiatry*, *66*, 334–340.
Moriguchi, S., Shinoda, Y., Yamamoto, Y., Sasaki, Y., Miyajima, K., Tagashira, H., et al. (2013). Stimulation of the sigma-1 receptor by DHEA enhances synaptic efficacy and neurogenesis in the hippocampal dentate gyrus of olfactory bulbectomized mice. *PLoS One*, *8*, e60863.
Naert, G., Maurice, T., Tapia-Arancibia, L., & Givalois, L. (2007). Neuroactive steroids modulate HPA axis activity and cerebral brain-derived neurotrophic factor (BDNF) protein levels in adult male rats. *Psychoneuroendocrinology*, *32*, 1062–1078.
Nguyen, T. V. (2017). Developmental effects of androgens in the human brain. *Journal of Neuroendocrinology*. https://doi.org/10.1111/jne.12486. Epub ahead of print.
Nguyen, T. V., Gower, P., Albaugh, M. D., Botteron, K. N., Hudziak, J. J., Fonov, V. S., et al. (2016). The developmental relationship between DHEA and visual attention is mediated by structural plasticity of cortico-amygdalar networks. *Psychoneuroendocrinology*, *70*, 122–133.
Nguyen, T. V., McCracken, J. T., Ducharme, S., Cropp, B. F., Botteron, K. N., Evans, A. C., et al. (2013). Interactive effects of dehydroepiandrosterone and testosterone on cortical thickness during early brain development. *Journal of Neuroscience*, *33*, 10840–10848.
Nieschlag, E., Loriaux, D. L., Ruder, H. J., Zucker, I. R., Kirschner, M. A., & Lipsett, M. B. (1973). The secretion of dehydroepiandrosterone and dehydroepiandrosterone sulphate in man. *The Journal of Endocrinology*, *57*, 123–134.
Pajer, K., Tabbah, R., Gardner, W., Rubin, R. T., Czambel, R. K., & Wang, Y. (2006). Adrenal androgen and gonadal hormone levels in adolescent girls with conduct disorder. *Psychoneuroendocrinology*, *31*, 1245–1256.
Papez, J. W. (1937). A proposed mechanism of emotion. *The Journal of Neuropsychiatry and Clinical Neurosciences*, 7, 103–112.

Persky, H., Dreisbach, L., Miller, W. R., O'Brien, C. P., Khan, M. A., Lief, H. I., et al. (1982). The relation of plasma androgen levels to sexual behaviors and attitudes of women. *Psychosomatic Medicine*, *44*, 305–319.

Reddy, D. S., Kaur, G., & Kulkarni, S. K. (1998). Sigma (sigma1) receptor mediated anti-depressant-like effects of neurosteroids in the Porsolt forced swim test. *Neuroreport*, *9*, 3069–3073.

Reddy, D. S., & Kulkarni, S. K. (1998). Proconvulsant effects of neurosteroids pregnenolone sulfate and dehydroepiandrosterone sulfate in mice. *European Journal of Pharmacology*, *345*, 55–59.

Rendina, D. N., Ryff, C. D., & Coe, C. L. (2017). Precipitous dehydroepiandrosterone declines reflect decreased physical vitality and function. *The Journals of Gerontology. Series A, Biological Sciences and Medical Sciences*, *72*, 747–753.

Saponaro, S., Guarnieri, V., Pescarmona, G. P., & Silvagno, F. (2007). Long-term exposure to Dehydroepiandrosterone affects the transcriptional activity of the glucocorticoid receptor. *The Journal of Steroid Biochemistry and Molecular Biology*, *103*, 129–136.

Saxe, R. R., Whitfield-Gabrieli, S., Scholz, J., & Pelphrey, K. A. (2009). Brain regions for perceiving and reasoning about other people in school-aged children. *Child Development*, *80*, 1197–1209.

Scheffers, C. S., Armstrong, S., Cantineau, A. E., Farquhar, C., & Jordan, V. (2015). Dehydroepiandrosterone for women in the peri- or postmenopausal phase. *Cochrane Database of Systematic Reviews*, *1*, CD011066.

Schmidt, P. J., Daly, R. C., Bloch, M., Smith, M. J., Danaceau, M. A., St Clair, L. S., et al. (2005). Dehydroepiandrosterone monotherapy in midlife-onset major and minor depression. *Archives of General Psychiatry*, *62*, 154–162.

Schneider, F., Gur, R. C., Gur, R. E., & Muenz, L. R. (1994). Standardized mood induction with happy and sad facial expressions. *Psychiatry Research*, *51*, 19–31.

Schupp, H. T., Ohman, A., Junghofer, M., Weike, A. I., Stockburger, J., & Hamm, A. O. (2004). The facilitated processing of threatening faces: An ERP analysis. *Emotion (Washington, DC)*, *4*, 189–200.

Scott, L. V., Salahuddin, F., Cooney, J., Svec, F., & Dinan, T. G. (1999). Differences in adrenal steroid profile in chronic fatigue syndrome in depression and in health. *Journal of Affective Disorders*, *54*, 129–137.

Sripada, R. K., Marx, C. E., King, A. P., Rajaram, N., Garfinkel, S. N., Abelson, J. L., et al. (2013). DHEA enhances emotion regulation neurocircuits and modulates memory for emotional stimuli. *Neuropsychopharmacology*, *38*, 1798–1807.

Sripada, R. K., Welsh, R. C., Marx, C. E., & Liberzon, I. (2014). The neurosteroids allopregnanolone and dehydroepiandrosterone modulate resting-state amygdala connectivity. *Human Brain Mapping*, *35*, 3249–3261.

Strauss, E. B., Sands, D. E., Robinson, A. M., Tindall, W. J., & Stevenson, W. A. (1952). Use of dehydroisoandrosterone in psychiatric treatment; a preliminary survey. *British Medical Journal*, *2*, 64–66.

Strous, R. D., Stryjer, R., Maayan, R., Gal, G., Viglin, D., Katz, E., et al. (2007). Analysis of clinical symptomatology, extrapyramidal symptoms and neurocognitive dysfunction following dehydroepiandrosterone (DHEA) administration in olanzapine treated schizophrenia patients: A randomized, double-blind placebo controlled trial. *Psychoneuroendocrinology*, *32*, 96–105.

Sulcová, J., Hill, M., Hampl, R., & Stárka, L. (1997). Age and sex related differences in serum levels of unconjugated dehydroepiandrosterone and its sulphate in normal subjects. *The Journal of Endocrinology*, *154*, 57–62.

Szaflarski, J. P., Schmithorst, V. J., Altaye, M., Byars, A. W., Ret, J., Plante, E., et al. (2006). A longitudinal functional magnetic resonance imaging study of language development in children 5 to 11 years old. *Annals of Neurology*, *59*, 796–807.

Tannenbaum, C., Barrett-Connor, E., Laughlin, G. A., & Platt, R. W. (2004). A longitudinal study of dehydroepiandrosterone sulphate (DHEAS) change in older men and women: The Rancho Bernardo Study. *European Journal of Endocrinology, 151*, 717–725.

Tsuchimine, S., Kaneda, A., Nakamura, K., & Norio, Y. F. (2015). The relationships between androgens and novelty seeking in healthy Japanese men. *Psychiatry Research, 225*, 175–178.

Urani, A., Roman, F. J., Phan, V. L., Su, T. P., & Maurice, T. (2001). The antidepressant-like effect induced by sigma(1)-receptor agonists and neuroactive steroids in mice submitted to the forced swimming test. *The Journal of Pharmacology and Experimental Therapeutics, 298*, 1269–1279.

van Dijken, H. H., de Goeij, D. C., Sutanto, W., Mos, J., de Kloet, E. R., & Tilders, F. J. (1993). Short inescapable stress produces long-lasting changes in the brain-pituitary-adrenal axis of adult male rats. *Neuroendocrinology, 58*, 57–64.

van Goozen, S. H. M., Matthys, W., Cohen-Kettenis, P. T., Thijssen, J. H., & van Engeland, H. (1998). Adrenal androgens and aggression in conduct disorder prepubertal boys and normal controls. *Biological Psychiatry, 43*, 156–158.

van Peer, J. M., Roelofs, K., Rotteveel, M., van Dijk, J. G., Spinhoven, P., & Ridderinkhof, K. R. (2007). The effects of cortisol administration on approach-avoidance behavior: An event-related potential study. *Biological Psychology, 76*, 135–146.

Verleger, R., Grauhan, N., & Śmigasiewicz, K. (2016). Is P3 a strategic or a tactical component? Relationships of P3 sub-components to response times in oddball tasks with go, no-go and choice responses. *NeuroImage, 143*, 223–234.

Walker, C. D., & Dallman, M. F. (1993). Neonatal facilitation of stress-induced adrenocorticotropin secretion by prior stress: Evidence for increased central drive to the pituitary. *Endocrinology, 132*, 1101–1107.

Wang, S., Zhao, Y., Chen, S., Lin, G., Sun, P., & Wang, T. (2013). EEG biofeedback improves attentional bias in high trait anxiety individuals. *BMC Neuroscience, 14*, 115.

Weisfeld, G. E., & Goetz, S. M. (2013). Applying evolutionary thinking to the study of emotion. *Behavioral Sciences, 3*, 388–407.

Whittle, S., Simmons, J. G., Byrne, M. L., Strikwerda-Brown, C., Kerestes, R., Seal, M. L., et al. (2015). Associations between early adrenarche, affective brain function and mental health in children. *Social Cognitive and Affective Neuroscience, 10*, 1282–1290.

Wolf, O. T., Naumann, E., Hellhammer, D. H., & Kirschbaum, C. (1998). Effects of dehydroepiandrosterone replacement in elderly men on event-related potentials, memory, and well-being. *The Journals of Gerontology. Series A, Biological Sciences and Medical Sciences, 53*, M385–390.

Wolf, O. T., Neumann, O., Hellhammer, D. H., Geiben, A. C., Strasburger, C. J., Dressendörfer, R. A., et al. (1997). Effects of a two-week physiological dehydroepiandrosterone substitution on cognitive performance and well-being in healthy elderly women and men. *The Journal of Clinical Endocrinology and Metabolism, 82*, 2363–2367.

Wolkowitz, O. M., Reus, V. I., Keebler, A., Nelson, N., Friedland, M., Brizendine, L., et al. (1999). Double-blind treatment of major depression with dehydroepiandrosterone. *The American Journal of Psychiatry, 156*, 646–649.

Young, A. H., Gallagher, P., & Porter, R. J. (2002). Elevation of the cortisol-dehydroepiandrosterone ratio in drug free depressed patients. *The American Journal of Psychiatry, 159*, 1237–1239.

Yuen, E. Y., Liu, W., Karatsoreos, I. N., Ren, Y., Feng, J., McEwen, B. S., et al. (2011). Mechanisms for acute stress-induced enhancement of glutamatergic transmission and working memory. *Molecular Psychiatry, 16*, 156–170.

Zheng, P. (2009). Neuroactive steroid regulation of neurotransmitter release in the CNS: Action, mechanism and possible significance. *Progress in Neurobiology, 89*, 134–152.

Kolibri 360 GmbH
Ferdinand-Jühlke-Straße 7
99095 Erfurt, Deutschland
produktsicherheit@kolibri360.de